Praise for *Farmageddon*:

'This book is passionate and outraged, understandably so, not just because of the lunacy and cruelty of it all but because the authors believe it is unnecessary and even, in global terms, unprofitable' *Sunday Times*

'Committed, balanced and appalling . . . *Farmageddon* is a classic polemic that provides all the ammunition you'll ever need' *Observer*

'The great virtues of *Farmageddon* are its global reach and eyewitness accounts' *London Review of Books*

'*Farmageddon* is an excellent book: a fine overview of what's gone wrong, with case histories and possible solutions that give cause for hope' *Literary Review*

'Lymbery's book carries great emotional impact . . . *Farmageddon*'s central message is powerful: industrial farming is playing havoc with nature even while it fails at its main goal' *Times Literary Supplement*

'The title is shockingly appropriate. And yet, out of the mire, come the kind of realistic and compassionate solutions on which our prospects for a truly sustainable world depend' Jonathon Porritt

'This incredibly important book should be read by anyone who cares about people, the planet and particularly animals' Jilly Cooper

D0263239

A Note on the Authors

PHILIP LYMBERY is the CEO of leading international farm animal welfare organisation Compassion in World Farming and a prominent commentator on the effects of industrial farming. Under his leadership, Compassion's prestigious awards have included an *Observer* Ethical Award for Campaigner of the Year and BBC Radio 4 Food and Farming awards for Best Campaigner and Educator. He is a lifelong wildlife enthusiast and lives in rural Hampshire with his wife and stepson.

ISABEL OAKESHOTT is a political journalist and commentator. In 2012, she won Political Journalist of the Year at the UK Press Awards and she is the ghost writer for *Inside Out*, an explosive insider account of Gordon Brown's regime. She lives in the Cotswolds with her husband and three young children.

FARMAGEDDON

THE TRUE COST OF CHEAP MEAT

PHILIP LYMBERY WITH **ISABEL OAKESHOTT**

BLOOMSBURY
LONDON · NEW DELHI · NEW YORK · SYDNEY

To the memory of Peter and Anna Roberts

First published in Great Britain 2014
This paperback edition published 2015

Text copyright © 2014 Philip Lymbery and Isabel Oakeshott
Illustrations by Liane Payne

The moral right of the authors has been asserted

Bloomsbury Publishing plc
50 Bedford Square
London
WC1B 3DP

www.bloomsbury.com

Bloosmbury is a trademark of Bloomsbury Publishing Plc

Bloomsbury Publishing, London, New Delhi, New York and Sydney

A CIP catalogue record for this book is available from the British Library

ISBN 978 1 4088 4634 6

10 9 8 7 6 5 4 3 2 1

Typeset by Hewer Text UK Ltd, Edinburgh
Printed and bound in Great Britain by CPI Group (UK) Ltd, Croydon CR0 4YY

Contents

Preface to the Paperback Edition

Walking through the English countryside on a glorious autumn morning, it is hard to imagine that a battle is raging over the future of our food and the countryside. Dappled shades of green and brown line my path; glistening grass, gently kissed by the weak morning sun; heavy dew spits from my boots with every step. Winter thrushes, recently arrived from Scandinavia, feast on berries; cattle dot the hillside, grazing on the last of the summer's growth.

I live in the rural south of England where pasture, hedgerows and wildlife are very much part of the landscape. Yet, under the guise of 'sustainable intensification', battle lines have been drawn; a more industrial approach to farming, with little room for luxuries like animals out in the fields, is now seen as the way forward. After all, we need to feed a growing population – billions of extra mouths are expected on the planet within decades. This will mean, like it or not, they say, animals confined in mega-farms, disappearing from the landscape and replaced by crops grown in prairies with the aid of chemical pesticides and fertilisers.

Things have been moving in this direction for a while, but now the pace is quickening. The strain is already showing; farmland birds that were once common in Britain are at an all-time low; bees have declined below what is needed for the proper pollination of crops in Europe; and concern grows about the quality of food on supermarket shelves – where it comes from, how it is produced and what it's doing to our health.

What the intensive farming lobby doesn't acknowledge is that the system already produces enough to feed everybody – and plenty more besides. Industrial farming now makes up a third of global production and is responsible for the greatest damage and the greatest inefficiency. The biggest single area of food waste comes not from what we throw in the bin but from feeding crops that might feed human beings to industrially reared animals, losing much of its calorific value in the process.

And that really brings us to the crux of *Farmageddon*; far from being an uninterrupted series of warnings or horror stories, it is above all a story of hope. It shows that grazing animals on pasture, converting things people can't eat – grass and marginal lands – into things they can – meat, milk and eggs – is a far saner way of producing food.

Farmageddon was launched on a February night in a packed London bookshop, with Joanna Lumley declaring it 'food's *An Inconvenient Truth*'. Favourable reviews were received from much of the UK's broadsheet press along with much positive feedback from engaged audiences in North America, India, South Africa, and also in Brussels. In the UK I had the pleasure of meeting so many people concerned about the future of food. I found myself described in a myriad of ways, from the patron saint of chickens to one of industrial farming's fiercest critics. It was particularly pleasing to see the Presidency of the European Union grasping the nettle by holding a conference in Brussels called 'Averting Farmageddon: Sustainable Food for All'; an event opened by two government ministers and a senior UN official. Now we must keep up the momentum and make the message stick.

Along the way, I felt a shift in the discussion; a more questioning approach emerging toward the idea that intensification of farming is somehow efficient and necessary. As this book shows, neither is true.

In writing *Farmageddon*, I set out to explore the tentacles of the global food system. I was joined by the then Political Editor of the *Sunday Times*, Isabel Oakeshott, who came to the project with a combination of political savvy and the eyes of a new mother concerned about how best to feed her family. Together, we dived beneath the surface of the food industry to find out what's really going on. It was exhilarating, harrowing, sobering, eye-opening and often astounding, but above all it was life-affirming. It was a journey that until this book appeared few would ever have had the opportunity to undertake. I wish you *bon voyage* as you travel through its pages.

Philip Lymbery

October 2014

Preface

Mid-April in Pennsylvania, USA, and spring is in full swing. Birds are singing and daffodils celebrate in rampant profusion outside the front door of the white clapboard farmhouse. I gaze from the childhood bedroom window of the late Rachel Carson, the mother of the modern environmental movement, and look across the Allegheny valley where she grew up. I picture the young girl being inspired by the natural world around her: picking fruit from apple orchards, wandering nearby woods and hillsides, making countless discoveries as she went. Peering out into the morning light, I see two enormous chimney stacks belching smoke into the blue sky. Carson grew up in a world where industry and countryside existed side by side. But during her lifetime lines became blurred and industrial methods found their way into farming, with devastating consequences.

In 1962 Rachel Carson was the first to raise the alarm about the peril facing food and the countryside. Her book *Silent Spring* shone a spotlight on the effects of spraying the countryside with chemicals, part of agriculture's new industrialised approach.

I was on the last leg of a journey to see for myself the reality behind the marketing gloss of 'cheap' meat, to find out how the long tentacles of the global food system are wrapped around the food on our plate. I wanted to find out, half a century on, how things had changed, what notice we have taken, and what has happened to our food. It was a journey that had already taken me across continents, from the

California haze to the bright lights of Shanghai, from South America's Pacific coast and rainforests to the beaches of Brittany.

In the 1960s, Carson's clarion call was heard across the Atlantic by Peter Roberts, a dairy farmer from Hampshire, England. He was one of the first in Europe to talk about the invasion of intensive farming methods sweeping across from America. As he walked his fields and milked his cows, Roberts became uneasy at what was going on. He saw farm animals disappearing from the land into huge, windowless sheds, the farming press acting as cheerleader for the post-war agricultural revolution, his fellow farmers bombarded with messages ushering them along the industrial route. He felt something had to be done.

Angered by the institutionalised cruelty to animals on factory farms, Roberts approached the main animal charities of the day, urging them to get involved. He left disappointed: the charities were too busy focusing on cruelty to cats, dogs and horses. Despondent but undeterred, he shared his thoughts with a lawyer friend. 'Well Peter, at least you know where you stand,' the friend responded. 'You'll just have to take it up yourself.'

In 1967, Roberts founded the charity for which I now work: Compassion in World Farming. It was the autumn and the new organisation was run out of the family cottage; one man, his wife, Anna, and three small daughters against an industry driven by government policy, subsidised by taxpayers' money, guided by agricultural advisers and supported by a profusion of chemical, pharmaceutical and equipment companies. The odds against making any impact were huge.

The seeds of the problem were sown way back in the last century. During the 1940s, the world was at war, riven by what was perhaps the deadliest conflict in human history. The Second World War was to be a huge watershed moment, not only for global politics, but also heralding perhaps the greatest revolution in recent food and farming history. As bombs shook battlefields, the building blocks were being put in place for the industrialisation of the countryside. The means to make explosives out of thin air had been discovered three decades earlier by two German scientists who, in 1910, worked out how to convert atmospheric nitrogen into ammonia, a key ingredient in both artificial fertiliser and TNT.

During the Second World War, German scientists perfected the mass production of organophosphate nerve agents as chemical weapons, although they were never used. After the war, US companies adopted the technology for agricultural use. In the words of Carson, in 'developing agents of chemical warfare, some of the chemicals created in the laboratory were found to be lethal to insects . . . widely used to test chemicals as agents of death for man'. The scene was set for weapons of destruction to become the means for mass production in farming.

The Great Depression of the 1930s, a severe economic slump that lasted until the outbreak of war, led the US Congress to pass the first Farm Bill in 1933, a package of subsidy support for agriculture that remains to this day the federal government's main way of affecting how food is produced. It was introduced to help US farmers struggling with low crop prices due to flooded markets. It included a government commitment to buy up surplus grain, which took the brake off burgeoning production.

Some of the world's richest countries had experienced food shortages during the war years as supplies from overseas were hampered by enemy activity. It taught them a hard lesson in the benefits of self-sufficiency. When peace returned, many countries focused on boosting home-grown food supplies. In 1947 Britain passed the Agriculture Act, heralding government funding and encouragement for the new ways of mass production through the 'efficiencies' of intensification: getting more out of the same land using the latest chemicals, pharmaceuticals and machinery. In the US, the munitions plants of the American war machine were converted into artificial fertiliser factories. Pesticides derived from wartime nerve gas were used on the new enemy: agricultural insects. Plant-breeding techniques caused corn yields to take off, leading to cheap corn, and lots of it. So much so that corn became a cheap source of animal feed.

The industrialised nations had the means and the impetus for turning farming into a process of mass production, transforming food and the countryside with serious if unintended consequences. Quality was replaced by quantity as the main driver. Farmers were encouraged to meet minimum standards for the commodity market rather than trying to produce the best. Antibiotics were cleared for use in livestock, providing the means to dampen down disease arising from keeping too many animals in too small a space. The drugs came with the additional benefit of boosting growth rates which, along with hormones, helped fatten animals for slaughter faster.

Across the countryside, the old patchworks of mixed farms with their variety of crops and animals became a thing of the past, replaced by monocultures – farms specialising in the mass production of a single crop or animal. Farming in tune with nature was no longer necessary. The same crop could be grown on the same soil over and over. Artificial fertilisers provided a quick fix for flagging fields while unwelcome weeds, insects and other pests could be sprayed away with copious chemicals. Farm animals disappeared from the land into factory-like sheds; artificial fertilisers had usurped their role of replenishing tired soils through their manure in fields and orchards. There was talk of a new type of farming; of production-line methods applied to the rearing of animals; of animals living out their lives in darkness and immobility without sight of the sun. In her ground-breaking exposé of 1964, Ruth Harrison described a generation of men who saw in the animal they reared 'only its conversion factor into human food'.[1] Factory farming was born.

Successive governments saw to it that the new regime was widely adopted, blinkered to the hidden costs and investing significant resources in spreading the message. Everything became supercharged in the rush for production. Companies began specialising in fast-growing varieties of animal, like chickens that grow from tiny Easter chicks to grotesquely oversized adults in just six weeks – twice the speed of previous generations. An army of 'expert' advisers on the government payroll told farmers to get on board or face ruin. I remember Peter Roberts telling me about the day one of the farm advisers came knocking at his door. It was the early Sixties and they had a long conversation but the message was simple: if you want to boost your business, you'll have to move into intensive chicken rearing. He was told that meant specialising in chickens, lots of them, in large industrial sheds. He could buy the birds and their feed from a big company and when they were fully grown – which would not take long – he could sell them back to the same firm, which would have them slaughtered and find them a market. It would be sanitised, industrialised, integrated. All he'd have to do was sign a contract and grow the 'crop' of chickens.

Although he kept a few hundred chickens already, Roberts was uncomfortable. He felt it would mean relinquishing his power as a farmer to decide how things were done. It didn't feel right. That evening, he discussed it with his wife Anna. Her reaction was instant and instinctive: 'If you want to do this, Peter, I won't stop you, but I want you to know that I don't agree with it.' Unlike Roberts, many others succumbed to the sales patter.

Taxpayers' money was used to support farming's new direction, a legacy that lives on today. The much-criticised Common Agricultural Policy (CAP) of the European Union was set up in 1962 and now swallows up nearly half the EU's budget. Fifty billion euros a year are doled out in payments to complying farmers. Likewise, the Farm Bill in the US gives out around $30 billion[2] in the form of subsidies to farmers, with three-quarters going to just a tenth of farms – generally the wealthiest and biggest. Corn (maize) continues to be the most heavily subsidised crop, underpinning a cheap-meat culture based on the products of factory-farmed animals fed cereals and soya instead of grass and forage from the land.

Looking back, what wasn't so clear was the treadmill farmers were boarding: to produce more and more with less and less, so often for diminishing rewards. Inevitably, mass production led to a squeeze on the prices farmers earned for their work, and many farmers learned the hard way that the seductive new system was not all it was cracked up to be. Quite simply they went out of business.

Animal and crop rearing were once a happy partnership. Industrialisation divorced them. It saw the rise of 'barley barons' who would
grow cereals in great monocultures. Field sizes grew as hedges disappeared. Nature's protests at the death of diversity – insects and weeds
kept in check previously by natural means – were drowned out with
pesticides. The soil was forced to work harder and harder. Insects and
weeds were sprayed away, wildlife habitats diminished, and the growing fear of silent springs – the demise of birdsong in a desert of
industrial crops – was captured in Carson's whistle-blowing book.
Today there is scarcely a corner of the Earth that is not touched to
some extent by the spread of intensive agriculture.

In recent decades, things have changed, sometimes for the better. For
example, keeping calves in premature coffins – narrow veal crates – for
their entire lifetimes is banned throughout the EU; the toxic and hugely
damaging pesticide DDT has been banned for farm use worldwide.

But fifty years on from Carson and Roberts and their first cries of
alarm, the way food is produced again stands at a crossroads, captured
best by the proposal for a US-style mega-dairy in Lincolnshire,
England. The idea was to take 8,000 cows out of fields and to house
them permanently on concrete and sand. This was the new frontier in
the battle for the British countryside. It united local people, foodies,
celebrity chefs, environmental and civil society interests in opposition.
Eventually, the proposal was withdrawn. But the spectre of a new wave
of intensification in the countryside had been raised; was US-style
'mega-farming', with its massive scale and super-intensification, now
camped on Europe's lawn? How far had it spread already? And what
were the effects in the US itself?

I am privileged to be Chief Executive of Compassion in World
Farming, the charity Peter Roberts founded, and now the world's leading farm animal welfare organisation, with offices and representatives
across Europe, the USA, China and South Africa. In 2011, I was chal-
lenged by the charity's chairperson, Valerie James, to uncover why an
industry that started out with such good intentions – feeding nations
and the world – had gone so wrong, all too often appearing to put

profit before feeding people. How were people, animals and the planet being affected and what could be done about it? The idea for this book was born.

I set out to get under the skin of today's food system. I took on the role of investigative journalist, following leads and tip-offs; lifting the lid on intensive food production; always in my official capacity and sometimes using my Compassion in World Farming business card to dig myself out of awkward holes.

Over two years, I travelled with the *Sunday Times*'s political editor, Isabel Oakeshott, and a camera crew to explore the complex web of farming, fishing, industrial production and international trade that affects the food on our plate. I used my contacts across the world to pinpoint where to go and who to speak to. We drew up a list of countries and places to visit, based on their involvement in the globalised world of food. California was an obvious choice, not only for its cultural exports like Hollywood, but also for what some see as futuristic ways of farming. China is a rising power and the most populous country on the planet for people and pigs. Argentina is the world's greatest exporter of soya for animal feed. I wanted to see for myself how people, often in faraway lands, who provide the feed, the ingredients or the food on our plates are affected by the runaway industrialisation of the countryside. I was keen to hear firsthand from the people involved and those affected. This is their story as much as mine.

Philip Lymbery

Introduction
Old Macdonald

At the height of his powers, Chairman Mao launched a war on sparrows. On a mission to turbocharge China's productivity, the Communist leader decided that the birds were eating too much grain. One winter day in 1958, he mobilised the population of China to kill them off. The campaign was ruthlessly coordinated, as if the birds were any other enemy.

Instructions were issued, weapons assembled, and the media hammered home the importance of victory. At dawn on the specified day, young and old, in town and country, gathered to launch a simultaneous attack. Everyone had a role, from the old folk who stood under trees waving flags and banging pots and pans to terrify the birds, to the schoolgirls issued with rifles and trained how to shoot sparrows that took flight and the teenage boys who climbed trees and tore down nests, smashing eggs and killing baby birds.[1] Goaded into action by local party bureaucrats, spurred on by national anthems blasting out of Peking Radio, they threw themselves into the task.

Against such an onslaught, the birds didn't stand a chance. According to one newspaper report, by the end of day one, in Shanghai alone an 'estimated' 194,432 sparrows had been killed.[2] Across China, the sparrow population was decimated. Millions of birds lay dead.

Too late, the regime realised that the sparrows were not pests, pilfering the harvest, but vital to the food chain. When they disappeared, the bugs they once fed on thrived. The locust population spiralled out of control, the grasshoppers too. The insects devoured the crops and famine followed. So Chairman Mao called off the campaign and sparrows were once again left in peace. But it took decades for the species to recover. Meanwhile the balance of nature was so out of kilter that there was talk of having to import sparrows from the Soviet Union.

Imagine if the prime minister of Great Britain or the president of the United States tried something similar today. We would think they were out of their minds. Yet the effect of agricultural policy in Europe and the Americas in the past few decades has been almost exactly the same as Mao's purge. Tree sparrows – the same species that Mao targeted – have declined in Britain by 97 per cent over the last forty years, largely due to the intensification of agriculture. The figures for other well-loved birds like turtle doves and corn buntings are no less alarming. Modern farming has become so 'efficient' that the countryside is now too sterile to support native farmland birds. The situation is so critical that the British government is offering farmers payments to install bird feeders on their land to prevent certain species dying out through starvation.[3]

The collapse of native bird populations is just one of many disturbing consequences of an agricultural policy based on intensification. It's a process that has been under way for decades, and that some now want to deliberately accelerate in the name of 'sustainable intensification'. But where will it take us? The aim is to pound more flesh out of every farm animal, and extract ever higher yields from every acre of land, where money is poured into high-input intensive farming systems that rely on mass production to give a return on investment. The result has been the slow demise of the traditional mixed farm, on which animals and crops were rotated on grass and soil that largely replenished itself, and the ascendancy of farms that specialise in single crops sustained by fertilisers, or in livestock reared indoors.

Of course birds are not the only victims of this quiet revolution. The remorseless drive to get more for less is taking place at the expense of many other animals and insects; at great risk to public health; and often, at a heavy cost to people thousands of kilometres away.

This is not a 'poor animals' book – though chickens, pigs, cattle and fish have an appalling time on factory farms. Nor does it preach vegetarianism. It is not anti-meat, it is not anti-GM per se, and it is not anti-corporate. It dares to ask whether, in farming, big has to mean bad. It goes to the heart of the question of whether factory farms are the most 'efficient' way of providing meat and the only way to feed the world.

The insidious creep of industrial agriculture has taken place quietly, almost unnoticed except by communities immediately affected. Perhaps that's because so much of the business now goes on literally behind closed doors. Without fuss or fanfare, farm animals have slowly disappeared from fields, and moved into cramped, airless hangars and barns.

People may have a vague notion that things have changed, but they prefer to believe that farms are still wholesome places where chickens scratch around in the yard, a few pigs snooze and snort in muddy pens and contented cows chew the cud. It's a myth often peddled to children from an exceptionally young age. The fiction starts before they can walk or talk, with colourful picture books showing happy animals grazing by duck ponds in lush green fields. In these story books the ruddy-cheeked farmer and his wife are a picture of health, with a couple of bonny children and a mischievous-looking dog at their sides. At nursery school, the fake idyll is reinforced in nursery rhymes and story books. Then come the school trips and family days out to farms open to the public, where another altogether unreal image of a working farm is often portrayed. These visitor attractions offer tractor rides through meadows full of glorious spring flowers; the chance to pet newborn piglets and lambs; pony rides, donkey rides and even pig races – all in the most beautiful countryside. They are wonderful places of laughter and fun, but no more reflective of the average working farm than a schmaltzy Hollywood romance is of the average relationship.

In fact, only 8 per cent of farms in England today are 'mixed' – rearing more than one type of animal and also growing crops.[4] They face a desperate struggle to survive. They have all too often been replaced by farms that specialise in one thing only, whether it is producing cereals, eggs, chicken, milk, pork or beef. These places would make a dismal day out for anyone, and shock most schoolchildren. The Old Macdonald fallacy won't stay credible much longer.

Thankfully, Britain still has a fair proportion of farms where animals are allowed to do what nature intended: roam or graze on grass.[5] But if the policy of intensification continues unchallenged, soon the only farms rearing animals on grass in higher-welfare conditions will be tourist attractions, or rich men's playthings. Britain and Europe's farmers are still relative novices at the intensification game, but agricultural policy is encouraging them to adopt dubious and controversial practices already common in the USA and elsewhere. Without a change of tack, mega-piggeries, mega-dairies, 'battery'-reared beef and genetically engineered crops – and animals – will soon be the norm.

To anyone who travels to places where such systems are well established, the repercussions are plain to see. For the countryside, it often means a landscape so barren and depleted that little except the animal or crop at the centre of the production operation is allowed to thrive. For farm animals, intensification often means terrible suffering and results in poorer-quality produce. Some 70 billion farm animals are produced worldwide every year, two-thirds of them now factory-farmed. They are kept permanently indoors and treated like production machines, pushed ever further beyond their natural limits, selectively bred to produce more milk or eggs, or to grow fat enough for slaughter at a younger and younger age. A typical factory-farmed dairy cow is forced to produce so much milk that she is often exhausted and useless by the tender age of five – at least a decade less than her natural lifespan.

Those unmoved by the suffering might find other reasons to look again at the waste and the woefully poor-quality, high-fat meat that result from these techniques. Since farm animals are no longer on the land and have no access to grass or forage, their feed must be transported to them, sometimes across several continents. Together they consume a third of the world's cereal harvest,[6] 90 per cent of its soya meal and up to 30 per cent of the global fish catch[7] – precious resources that could be fed direct to billions of hungry people.[8]

Meanwhile the barns they are reared in are often hotbeds of disease – small wonder when so many animals are crammed into such small spaces. It's a business that depends on vast quantities of antibiotics – half of all those used in the world.[9] One consequence has been the breeding of antibiotic-resistant 'superbugs' in humans and weird and deadly new viruses that have been linked to industrial farming.

Consumers become the scapegoats, the supposed beneficiaries of a benevolent industry producing 'what the consumer wants'. Yet consumers are forced to walk supermarket aisles blindfolded, often unable to tell what is grown more naturally from what is 'fresh' from the factory farm, thanks to an industry that resists better labelling. The way food is produced has a key bearing on its quality, not just from an ethical standpoint, but also in terms of its nutritional quality and how it tastes. Feeding animals grain, rather than letting them graze grass, often results in fatty meat. In short, consumers often don't know what they're buying from an industry that wants to keep it that way.

From time to time, a food scandal will blow the lid off a shadowy aspect of what's going on. The horsemeat scandal of 2013 confirmed consumer fears that they don't always know the full story behind the food they buy, when hot breaking news quickly degenerated into a furious blame-game. Horsemeat had been switched for beef, leaving the horse-loving nation of Britain stunned and distrustful. Keen to avoid taint from the torrent of revelations, the UK prime minister David Cameron blamed supermarkets, who blamed their suppliers, who pointed to distant traders in faraway lands. Consumers were left baffled and angry.

The alarm was first raised by the Irish Food Safety Authority, which revealed the finding of horsemeat in products labelled as beef. The supermarket giant Tesco, Britain's biggest, was one of the first to be involved when an 'Everyday Value' beefburger from the store turned out to contain 29 per cent horsemeat. The offending burger was manufactured in Ireland from meat thought to be of Polish origin. Other supermarkets were affected. Within days, 10 million burgers – enough calories to feed a million people for a day – had been removed from shelves by worried retailers.[10]

What was uncovered was a fraudulent labelling scam stretching the length of Europe.[11] Day after day, new revelations involved more big-name brands. Consumers reacted by shunning frozen burgers; UK sales fell by 43 per cent. Tesco placed full-page advertisements in national newspapers with the headline 'We apologise',[12] suffering its sharpest fall in market share for two decades.[13]

'Horsegate', as it became known, was all about trust. Consumer confidence had crashed and companies licked reputational wounds. Some admitted to having lost control of supply chains which, over the years, had grown longer and more complex, as food might pass through several hands before getting to the supermarket. Some blamed the incessant pressure for low prices during the global recession that started in 2007. 'We now need the supermarkets to stop scouring the world for the cheapest products they can find,' thundered the president of the National Farmers Union (NFU), Peter Kendall.[14]

Horsegate was the biggest scandal to hit British food since 'mad cow disease' or BSE, which two decades earlier caused a ten-year ban on British beef exports. BSE, caused by turning natural herbivores – cows – into carnivores, feeding them meat and bone meal, was a real own goal for industrial agriculture. It will not be the last.

Of course there are some winners from the system, like the companies peddling products that promise farmers ever-greater yields. The new technology can be effective in the short term, but sooner or later someone pays the price. In India, for example, around 200,000 farmers have killed themselves since 1997, typically after falling into debt.

They mortgage themselves up to the hilt to buy 'magic' genetically modified seed, then belatedly discover it is totally unsuitable for local conditions. The harvest fails. In the UK, a couple of dozen farmer suicides would trigger a national outcry; in India, the tragedy has unfolded almost in silence.

In the United States, while researching this book, I stood among thousands of acres of almond trees, all in perfectly regimented rows, breathing in air so heavy with chemical sprays that it smelled like washing-up liquid. There was not a blade of grass, nor butterfly, nor insect, to be seen. In the distance was one of the many mega-dairies in the state. Thousands of listless cows with udders the size of beachballs stood in the mud, waiting to be fed, milked, or injected with drugs. There was no shortage of land; no logical reason for them not to be on grass. The system wasn't even working for the farmers themselves. At a livestock market in a nearby town, a farmer wept as he told how a friend's mega-dairy had gone out of business and the despairing owner took his own life.

In Argentina I stood in a field of genetically modified soya as thousands of mosquitoes swarmed around my head. There was no stagnant water nor any of the conditions normally associated with such high numbers of insects. Something was wrong.

In Peru I saw a malnourished child, covered in sores associated with air pollution from the fish-processing industry, hearing from doctors that she could be healthy and well fed if only she were given the local anchovies destined for animal feed in Europe's factory farms.

In France we talked to the family of a worker who had succumbed to toxic fumes as he cleared luminous green algae from a once unspoilt beach. The gunge that now blights the coast of Brittany every summer is the highly visible face of pollution from the region's mega-pig farms.

In Britain, I helped campaign against the establishment of the country's first-ever mega-dairy of 8,000 cows. It was a battle we won – but for how long?

There is a widespread and deep-seated assumption that industrialising farming – treating the delicate art of rearing animals and working the land as if it were any other business, like making widgets or rubber tyres – is the only way to produce affordable meat. For too long, this basic premise has gone almost entirely unquestioned. Governments have rushed to create the conditions in which shoppers can buy a £2 chicken, thinking they're doing everyone a favour. Yet the reality behind how cheap meat is produced remains hidden.

This book looks into the unintended consequences of putting profit before feeding people. It asks how something with such good intentions as feeding nations could go so wrong.

It questions what is efficient about cramming millions of animals indoors, giving them antibiotics to survive, then spending vast sums transporting food to them, when they could be outside on grass.

It questions what is space-saving about a system that relies on millions of acres of fertile land to grow animal feed on estates often hundreds of thousands of kilometres from the farm.

It questions what is smart about having to remove mountains of manure from concrete floors and find a way to get rid of it, when if animals were in fields, their dung would return to the earth by itself, enriching the soil in the process, as nature designed.

It asks whether it makes any sense to encourage people to eat a lot of cheap chicken, pork and beef from animals specifically selected for their ability to grow so big that they produce fatty meat.

Finally, it begs the question, is the Farmageddon scenario – the death of our countryside, a scourge of disease and billions starving – inevitable? Through the eyes of the people and animals involved, this book sheds light on what they don't want you to know and asks, could there be a better way?

I
RUDE AWAKENINGS

Every now and then, something happens to shake the very foundations of how we view our food. 'Horsegate', the scandal of a horse-loving nation waking up to find it had been quietly devouring the object of its affection, was one of those moments. The fact that horsemeat was switched for beef in our food chain on such a widespread scale served to underline for many how little we know about our food: what's in it and how it is produced. There are fears of a gulf in understanding about the food on our plate; more than a third of young adults in Britain don't know that bacon comes from a pig, milk from a cow or eggs from a hen.[1]

Much of the meat on many supermarket shelves has a dirty secret that you won't find on the label – the way it was produced. It is a matter of convenience for producers that some consumers don't know that meat or milk comes from an animal that was once living and breathing, let alone understand how the animal was reared. Some producers go to great lengths to keep it that way; to keep the veil tightly drawn. Perhaps the most extreme example are moves in the US to introduce so-called 'ag-gag' laws, that would ban the taking of photos and film of intensive farming operations without permission, thereby making it harder to expose bad practices or wrongdoing. It raises the question, what have they got to hide?

A storybook vision of farming – of frolicking animals in pristine countryside – is all too often perpetuated. I was in my late teens when I started to find out that this wasn't always the case and that this vision was in fact far removed from the reality. This realisation changed the way I thought about what was on my plate – I'd like to think for the better. Thirty years on, I travelled the world to see the role that people, in often distant lands, play in the food we eat. Where better to start than California, the land of milk and honey? I decided to push beyond the seductive glitz and glamour of the Sunshine State's hotspots and headed to the dusty valleys that yield world-renowned harvests. It was through this *Alice in Wonderland* looking glass that I was to step into what seemed like another world. For me, it begged another question: was this a curious aberration or the future direction of farming?

I

California Girls

A vision of the future?

California, USA: according to the Beach Boys, home to the cutest girls in the world. I was there to get a sense of what our planet will look like if California-style farming becomes the norm around the world. Since the legendary Sixties hit, the population of females in the Golden State has soared, but the new arrivals share none of the glowing good health and athletic physiques of the sun-kissed babes in the song. They are milk cows and their purpose in life is to churn out supernatural quantities of milk before being turned into hamburgers.

Hollywood has dipped California in gold, drawing millions of visitors to the sun-soaked beaches, twinkling city lights and luscious Napa Valley vineyards. Few tourists ever see the real powerhouse: Central Valley, the fruit bowl of America, and home to perhaps the biggest concentration of mega-dairies in the world. One and three-quarter million dairy cows are reared in California,[1] crammed into barren pens on tiny patches of land that make a mockery of the vast potential space in this part of America. They pump out nearly 6 billion dollars' worth of milk every year,[2] and as much waste in the form of dung and urine as 90 million people.[3] Through a combination of selective breeding, concentrated diets and growth hormones designed to maximise milk production, they are pushed so grotesquely beyond their natural limits that they survive for just two or three years of milking before being sent to slaughter.

The US-style mega-dairy, often twenty to a hundred times the size of the average British dairy, is on the brink of migrating to the UK and other parts of the world. For this reason, and because it has embraced intensive farming with an enthusiasm and rigour unmatched anywhere in the world, Central Valley is a vision of how the countryside elsewhere could soon look. I wanted to see it.

It would not be my first trip to an American mega-dairy. During the successful battle to block plans for the UK's first mega-dairy, an 8,000-cow facility in Lincolnshire, England, I went out to Wisconsin to look at the farm on which that proposal was modelled. It was a flying visit, the result of a pledge I made on national radio at the height of the campaign. What I found was a soulless and highly tuned milk-production operation in which the cows might as well have been Kitchen Aid machines, designed to swallow up ingredients and spew them out in another form, keeping the process going twenty-four hours a day until they ran out of steam. What I found then was depressing and gave me some idea of what to expect in California, but nothing prepared me for the sheer scale of what I was to encounter this time.

Accompanied by a small team including Isabel Oakeshott and a camera crew, I flew to California in November, hoping to see as much of Central Valley as possible in just under a week. In the event, I nearly didn't make it – thanks to a chicken sandwich that my colleague had bought from a branch of Pret A Manger at Heathrow in case the food on the plane wasn't up to scratch. She didn't eat it there, but kept it in case there was nothing to offer in the motel at the other end.

We were at baggage reclaim at San Francisco airport, waiting to pick up our cases, when an eager sniffer dog bounded over. He'd smelled the sandwich in her hand luggage. Within seconds the dog handler was on the scene and we were the focus of a full-scale security alert, the sandwich inspected like an unexploded bomb, all my colleagues' bags examined and X-rayed.

The sandwich was subjected to further forensic examination before being carted off to be destroyed. At last we were sent on our way – but not before getting a long lecture about mad cow disease. It was a sharp reminder that while the UK has very deliberately relegated BSE to an embarrassing chapter in history, America has neither forgiven nor forgotten. The incident would seem even stranger later, when we witnessed the seemingly casual disregard for public health that goes hand in hand with mega-dairy farming in California.

We didn't hang around in San Francisco, but hit the road to Central Valley. It was a five-hour drive south through drab agricultural land. Approaching the valley, we noticed a curious yellowish-grey smog on the horizon. It looked like the sort of pollution that hangs over big cities, but there were none, only the dairies, which emit so much bad stuff that the surrounding air quality can be worse than Los Angeles on a smoggy day.[4]

As well as supporting an army of cows, Central Valley coughs up an incredible annual harvest of fruit, nuts and vegetables, despite having so little rainfall that it is technically classified as semi-desert. There are fields of pomegranates, pistachio orchards, grape vines and apricot trees. There are tomatoes and asparagus, and acres of red, pink and yellow rose bushes. There are miles of orange and lemon groves. There are also enough almond trees to provide four out of every five almonds consumed in the world. It sounds like the Garden of Eden. It isn't. It turned out to be a deeply disturbing place where not a blade of grass, no tree or hedgerow grows, except in private gardens and the ruth-lessly delineated fields.

The phenomenal output of fruit and veg is possible only thanks to a cocktail of chemicals and the plundering of the crystal-clear rivers that run down from the Sierra Nevada mountains. By remorselessly dousing the parched soil with fertilisers, insecticides, herbicides and fumigants, as well as diverting natural waterways, farmers have been able to pull off a multi-billion-dollar conjuring trick, extracting harvests from soil that is so depleted of natural matter it might as well be brown polystyrene.

All these chemicals make the air smell very strange. It caught in my throat and felt like it was creeping down into my lungs. In this breezeless bowl between the mountains and the coast, weed- and pest-killers from industrial sprayers can struggle to settle on crops, leaving a fine mist of toxins to hang in the air. Some days, when the temperature and air pressure combine in a certain way, clouds of chemicals can be seen hovering above the crops. All that fruit should be an irresistible draw for birds, bees, butterflies and other insects. We saw virtually none.

On this featureless chemical wasteland lie the mega-dairies: milk factories where animals are just machines that rapidly break down and are replaced. The dairies arrived in Central Valley in the 1990s, after being pushed out of the Los Angeles suburbs. Land on the city outskirts was becoming ever more valuable, and as the population expanded, farmers were finding it cost more and more to dispose of waste. Encouraged by realtors, many sold up and moved to the sticks, where they quickly discovered there was not much to stop them doing as they pleased.

At the time, agriculture was exempt from California's Clean Air Act. It was not until the late 1990s, when the cousins George and James Borba applied to build two 14,000-cow operations on adjacent properties in Kern County, in effect creating a 28,000-cow dairy, that serious attention began to be paid to the potential environmental and health impacts. The Borbas' plans were nodded through at first, but the sheer scale of the proposal galvanised people who were already worried by what they were observing in Central Valley. After a protracted legal battle, campaigners forced the authorities to undertake a full environmental impact assessment, the results of which were so alarming they changed the game for ever. Such assessments became standard procedure for subsequent planning applications, and California's exemption of agriculture from the Clean Air Act came increasingly under question.

The policy was finally changed in 2003, when a new system of permits was introduced. In theory, farmers now have to comply with tough air- and water-pollution regulations – a burden they bitterly deplore. In practice we heard – and saw – worrying evidence that many routinely flout the law while overstretched authorities turn a blind eye.

We spent several days driving around Fresno, Tulare and Kern, the three counties that generate the most agricultural produce in the whole of America. It was not long before we found our first mega-dairy, a series of towering corrugated-iron shelters set in mud pens. It was autumn 2011; the sun shone against a clear blue sky and the temperature was pleasant. Some cows jostled for space in the shade under the open-sided sheds. Others stood in the sun looking bored. There was no grass, just a deep pile carpet of earth and manure. From a distance it was hard to tell the difference between the cows and the stacks of rubber tyres in the yard.

We pulled up for a closer look. A large black beetle scuttled across the road, virtually the only non-flying insect we encountered in five days – though there were always plenty of black flies. They arrived, we were told, at the same time as the mega-dairies and are now a scourge, invading homes, schools and offices, forcing residents to install screens over windows and seals around doors.

Out of the car, we took in the scene. The stench of manure was overwhelming – not the faintly sweet, earthy smell of cowpat familiar in the English countryside, but a nauseous reek bearing no relation to digested grass. The cows moved very little, too engorged with milk. When they did walk, it was with a rocking gait, their legs splayed wide around their pink and grey beachball udders. We saw these farms every couple of kilometres, all with several thousand cows surrounded by mud, corrugated iron and concrete. Most of the shelters were rigged up with rusty fans, a pathetic defence against the searing summer sun. Between feed times and milking, there was little for the animals to do but wait – for food, for milking, perhaps for medication.

Near the town of Turlock we saw several mega-dairies next door to power stations, a grotesque inversion of the traditional concept of a farm. There was nothing rural about these locations: they were huge industrial estates, characterised by supersized animal feed and milk-processing plants belching out fumes. Alongside were railway tracks bearing freight trains half a kilometre long.

We had tried in vain to make an appointment for an official tour of a farm. It turned out to be unnecessary: most mega-dairies were located right by public roads. The only facilities we could not see were the milking parlours, often giant rotating carousels. Cows queue up, step onto the wheel, and are hooked to milking machines. When their udders are pumped dry they step off the merry-go-round to make way for the next animal, and return to their barren pen.

The landscape was so flat it was difficult to get a sense of scale, so we chartered a plane, a tiny four-seater Cessna, to look at the dairies from the air. It cost less than dinner for two at a London restaurant. At reception, the pilot's wife reassured us her husband had been flying for fifty years without incident. He was fond of his grandchildren, she said, and wouldn't take any risks. We showed our passports, paid up, and strolled out onto the runway with our aviation headsets. It was all strangely informal, pleasantly so. We might as well have been hopping on a city bus.

In the back of the aircraft there was barely room for two. I wedged myself behind the pilot's seat, put my earphones on and prepared for take-off. There were some cursory radio exchanges, the pilot revved the engines, and then we zoomed along the runway and were airborne. We levelled at around 600 metres. Squeezed into the front passenger seat with all his gear, our cameraman opened the window to take some aerial shots, letting in such a powerful blast of air that my colleague's contact lenses almost blew out.

When I peered out of the window it looked as if a vast steamroller had pummelled its way across the country, flattening every knoll and hillock and pulverising every plant and creature in its wake. All that remained was a vast empty canvas, carved into neat sections. Into these sanitised boxes were inserted crops, the oranges and almonds that make their way to kitchen tables all over the world. From the air, the dairies just looked like vast fields of filth peppered with black and white specks and the odd corrugated-iron roof. I could see the tops of silage mountains, their taut white plastic coats shimmering in the afternoon sun. By every dairy was a stinking pool of yellowish-brown slurry. Even up there, I could pick up the smell. What self-respecting bovine would want to be a California girl?

Back on the runway again, I stepped out feeling strangely elated – the flight had been fun. But that afternoon brought more depressing sights, including the worst mega-dairy of the lot. It was on an industrial estate, squeezed between a giant feed factory and an electrical substation. It looked run-down, the fences rusty and dilapidated, the corrugated-iron shelters unusually low, perhaps an effort to cut costs. Down the road was a huge battery chicken factory, and the acrid smell of bird droppings mixed with cow manure hung in the air.

Next door to the farm were a couple of run-down bungalows on a patch of scrubby wasteland. Piles of rusting scrap metal, old tyres and discarded plastic toys littered the yard. Just visible from the road were rows of suspicious-looking wooden crates. I went to investigate, and found that they contained around 200 cockerels. The Mexican owner spoke little English, but it was clear they were for cock fighting, an illegal but popular pastime among poor communities in these parts. I saw this neighbourhood at dusk, against a backdrop of darkening cumulus clouds. An exhausted cow stooped atop a mound of manure, silhouetted against the night sky. It was like a scene from Armageddon.

Of course that's not how the industry sees it. The California Milk Advisory Board (CMAB) claims that 443,000 full-time jobs in California are linked to the dairy industry and that it generates as much as $63 billion in economic activity a year. Apparently, a typical cow contributes $34,000. The organisation's website is a slick piece of PR, a collage of images of contented cows on green fields, wholesome-looking food, and all-American-looking farming families, their arms round each other as they pose for the camera, as if to say that being in this business makes for a happy, fulfilling life, which perhaps it does for some.

Since 1969 the CMAB has run an annual Dairy Princess competition, a beauty pageant for young women from dairy families. The twenty winners enjoy a one-year reign during which they are ambassadors for the industry, giving radio and TV interviews, talking to school assemblies and making appearances at meetings and agricultural fairs. It's all designed to put a gloss on what is a deeply dubious industry in terms of its impact on public health, the environment and animal welfare.

The mega-dairy was near the biggest single-site cheese factory in the world, a showpiece for the California industry. Above a luxurious tea shop plying a tempting array of attractively packaged cheeses, cakes and sweets is a visitor centre extolling the benefits of the mega-dairy industry. Through a window on an upper floor, you can look down at the factory floor, where workers in white coats and caps can be seen processing vast vats of bright orange cheese. Later I tried some. It was tasteless and rubbery.

Beside a display of 'educational' material about the industry, a TV screen displayed pictures of smiling farmers with their wives and children, sitting in hay-filled barns. There were touching images of farmers bottle-feeding newborn calves and cows being checked over by caring-looking vets. According to the spin, dairy farmers 'preserve water; care for the air; care for the land; care for the environment', as well as, of course, contributing to the local economy. Whoever put together the promotional film footage was not quite brazen enough to show any cows on grass.

Naturally, the visitor centre barely touched on the huge environmental and health issues presented by mega-dairies. There were a couple of information panels, with flow charts illustrating waste-disposal systems. They made it sound harmless; neat.

That is a gross distortion. A fifth of children in Central Valley are diagnosed with asthma – a consequence of air pollution linked at least in part to the mega-dairy industry. It is almost three times the national average paediatric asthma rate. Nearly a third of the 4-million-strong population of Central Valley are assessed as facing a high degree of environmental risk, both from toxic air and from water pollutants.

*

It's hardly surprising given the amount of muck being produced. One animal generates as much waste as fifty humans, meaning that a single mega-dairy of around 10,000 cows creates as much waste as a fair-sized UK city such as Bristol. According to the CMAB, as of November 2011 there were 1,620 dairy farms in California, housing a total of 1.75 million cows. Together they generate more excrement than the entire human population of the UK.

Finding somewhere to put all those cowpats is a huge headache. Most is channelled into vast lagoons attached to the farms. They let off noxious gases and leach into the ground. Even with clay liners, the lagoons are porous. The US authorities seem to accept that significant seepage is inevitable. I wanted to hear from people living near the farms about the pollution and health problems. We met Tom Frantz, a retired maths teacher who has lived in Kern County all his life. He lives in a cosy clapboard house among the almond orchards, with two energetic dogs and some noisy geese. Amid the sanitised acres of commercially produced trees his garden is a tiny oasis, a riot of brightly coloured flowers and verdant shrubs beneath towering palms.

He says that Kern County is where he will die – perhaps a decade before his time, if a recent scientific study into pollution in Central Valley is to be believed. 'I know I'll probably live ten, even fifteen years less than I would if I moved, but this is where I come from. I'm not going anywhere,' he told us matter-of-factly.

Lately Frantz has noticed he's developed a strange post-nasal drip. He is in no doubt about the cause. 'Living near mega-dairies is dangerous. We are looking at a potential health disaster. I can see a new strain of *E. coli*, some kind of plague, breaking out in Central Valley. That is the worst-case scenario. It sounds remote, but my worry is that it's just around the corner. Nobody will care, until it is too late.'

With his unkempt wiry hair and taste for reggae music, Frantz looked like an ageing hippy, but on the subject of mega-dairies and the environment he was extremely switched on, rattling off complex facts and figures about volatile organic compounds and nitrogen organic compounds with the authority of one who has spent years poring over the evidence. Over the past decade, he has become a thorn in the side of Kern County's dairy farmers, keeping their behaviour under relentless surveillance and calling them to account when he detects evidence that they are flouting environmental regulations. Where government agencies are too busy or lackadaisical to act, he starts to file official complaints, taking cases all the way to court if need be. It did not take him long to figure out that unless lawyers are involved people don't listen. This being America, you have to sue.

There are now ten dairies within thirteen kilometres of Frantz's home. The first arrived in 1994, the rest since 2002. Officially, they house a total of 70,000 cows. However, the real total is likely to be far higher: dairies are only required to submit the number of cows being milked at any one time, meaning they do not have to count animals that are being rested during their 'dry' (non-lactating) period. According to Frantz, what folk first noticed when the mega-dairies arrived was an influx of flies. Hardest hit was a school just over a kilometre from the first mega-dairy that opened. Teachers used to keep the doors and windows open in summer because the place had no air conditioning. These days that's out of the question. 'There were swarms of black flies in the classrooms. It was difficult for the kids to work with them buzzing around. That first year, they used rolls of sticky tape to catch them. Later, they installed screens on all the windows and sealed the doors.'

What bothers Frantz most though is air pollution. For much of the year, the smog is so thick and heavy that the Sierra Nevada mountains, 3,000 metres high, are invisible from the valley floor. Hardly anyone lives here, but air quality-wise, it might as well be Beijing. Of course there are many factors involved, from the exhaust fumes that pour from trucks carrying agricultural and other produce up and down the highway, to emissions from the industrial animal feed processing plants. They are a blatant source of pollution. The impact of mega-dairies is less visible but just as insidious, because of noxious gases from manure and silage.

Frantz and other concerned locals hold monthly meetings to discuss what to do. They work with sympathetic lawyers who take cases pro bono. Victory can mean multi-million-dollar fines for companies found responsible for environmental breaches. There's no money in it for the campaigners: the payouts are given to environmental groups.

For many small communities in Central Valley it is not air quality but water quality that is the most pressing issue. Studies have also shown a direct correlation between intensive dairy farming and contamination of water wells, especially with *E. coli* bacteria and nitrates.[5] During our trip, we spotted a well located within a few metres of the perimeter of a mega-dairy. It supplied local communities with their domestic water. Little wonder that so-called 'boil notices' – letters from local authorities ordering residents not to drink water out of their taps unless they have boiled it – are a way of life here.

Maria Herrera, a mother of four, runs the Community Water Center in the city of Visalia in the heart of the San Joaquin Valley. The organisation campaigns for universal access to safe drinking water. In a nation as rich as America, you'd have thought that people could take it for granted. Yet in parts of Central Valley it has become a luxury. In the small Hispanic communities that have sprung up to provide farm labour, it seems few dare to drink from the tap. Herrera told us:

> Ground water here is heavily polluted, mainly with nitrates, though there are also concerns about arsenic, which is linked to some ferti-lisers and is sometimes used as an additive in cattle feed. The meetings we hold with residents are always packed. The dairy farm-ers and their lobbyists come along and deny it has anything to do with them, but the evidence proves otherwise.

A permit system for farmers is supposed to keep water pollution within safe limits. In practice, it has proven impossible to prevent seepage from the Olympic swimming pool-sized lagoons of slurry attached to mega-farms. 'The lagoons aren't properly lined, and so the effluent leaks,' Herrera told us. 'It's crazy – we treat human waste so that it doesn't pollute water supplies; yet effluent from cows is allowed to get into the system. It makes no sense.' When nitrate levels in the water supply get too high, letters arrive from water companies advising folk not to drink from the tap. Herrera took us to meet some of the families affected who lived in a small trailer park surrounded by fields of citrus trees, a few kilometres outside Visalia. We stood amid the simple wooden bungalows, some so basic that the toilets were outside. Despite the obvious poverty, it was a welcoming place. The trailers and car ports were painted in pretty pastels; children and kittens scampered about; everyone knew everyone.

I was introduced to Luis Medellin, a handsome twenty-five-year-old who works at a mega-dairy but hates what the farms are doing to his community's water supply: 'To tell you the truth, our water here doesn't smell good. It looks cloudy and smells of chlorine. You can't trust it. From time to time, I do drink it right out of the tap, knowing it is contaminated, but mostly we buy bottled. I think there are Third World countries with safer water than we have here.' Every week, Medellin sets off with two giant plastic bottles the family keeps in the kitchen to replenish them with filtered water. It costs about four dollars a go, not a trivial sum for a family whose living conditions – an overcrowded trailer with little ventilation, lit by a few bare bulbs hanging from bare wires – would shock most well-heeled Americans.

According to the Community Water Center, there are at least six settlements within a few kilometres of Medellin's home where water quality is a serious issue. He has been campaigning for clean water since he was at high school. The irony that he earns his living from the very business that is responsible for polluting the supply does not escape him: 'I am not happy that I have to work there. The dairy has two huge lagoons full of crap – there's no other way to put it – and I am always thinking about what it's doing to the water.' He sticks it out because he is lucky enough to have a good boss, who looks after his workers and seems to care for the cows. His father has been working at the same dairy for fourteen years. It is the way of life here, but that doesn't mean people do not aspire to better or care about the damage the business is doing.

We were told about a growing body of scientific evidence about the health risks associated with living near mega-dairies. A recent study found that people living near factory farms have their life expectancy shortened by as much as a decade. We interviewed Kevin Hamilton, a blunt-talking registered respiratory therapist, at the Clinica Sierra Vista, in Fresno. His experience on the medical front line prompted him to become a committed activist against mega-dairies:

> We're talking about heart disease, birth defects, and stunted lung development among children who spent a lot of time outside playing sport. We're talking about high blood pressure and increased risk of stroke. We have the second-highest level of childhood asthma in the whole of the US. Fifteen years ago, I couldn't have said any of that with confidence. Now the evidence is overwhelming. It's terrifying.

The most vulnerable groups are children and pregnant women who can suffer long-term medical consequences as a result of even short exposures to dairy-related pollution. People over the age of sixty-five are also disproportionately affected. These high-risk groups make up a high percentage of Central Valley's population. Other groups heavily affected are the farm labourers who are exposed to noxious gases daily. 'They literally never escape the assault,' Hamilton said. 'They live in sub-standard houses, where the doors and windows are not sealed, and on the farms they are working very hard. The harder they breathe, the more stuff they take in. These workers are poor and disempowered. A lot of them are here illegally. Word gets round that if they make a fuss, they will be deported. It happens.'

Hamilton told us that one of his doctor colleagues recently moved away from the area because her son had developed bad asthma. 'She went to live in Colorado. I had a postcard from her six months ago, saying that since leaving this place, her son hadn't needed any medication.' Hamilton believes mega-dairies are a repellent symbol of a grotesquely unnatural agricultural food production system.

You have to use a phenomenal amount of chemicals to push multiple crops out of the soil we have here. If you look at the official data on the amount of pesticides applied per square mile, the figures would stun you. I don't think people have any idea. These pesticides are capable of penetrating the human body to genome level – meaning they can affect the very building blocks of the body.

My family is from Kansas, and we had one crop a year, and were really grateful for that. Some people would argue that CAFO [Concentrated Animal Feeding Operations] systems feed the world. I would argue that the world was feeding itself pretty well before we did all this. We've taken agriculture out of fields where it should be, into places where we have to farm against nature. If we have to have mega-dairies at all, why have them in a place like Central Valley, an airless bowl? We have loaded them in at a totally unsustainable rate. It's got way out of hand.

I left Hamilton's offices feeling drained and wondering who wins from this system. It would be easy to blame the farmers, but it's not as if they are all making fortunes. Most seem to feel under siege by environmental activists and regulation on a scale they never anticipated nor signed up for when they abandoned their small farms on the outskirts of LA.

Mega-dairy farming is a high-risk business exposed to global price hikes and price volatility. Evidence suggests they cannot weather recession as well as smaller pasture-based systems. Between 2011 and 2012, feed prices almost doubled, driving some dairies to the brink of financial collapse. According to one report, produced by the farming industry itself in the UK, mega-dairies only become more competitive in a hypothetical situation in which the price of milk is fixed for ten years. As this is impossible, the report states that pasture-based farms are more likely to turn a profit. It seems the system is not working brilliantly for anyone.

It was not until the last day of our trip that I fully appreciated what an unhappy business mega-dairying can be. We were in Turlock,

Stanislaus County, visiting a livestock market. I wanted to see what cows from mega-dairies look like at the end of their lives. I had a pretty good idea what to expect: sad black and white bags of bones with saggy, dried-up udders and the exhausted demeanour of animals five times their age.

Known as the 'Heart of the Valley', Turlock was a depressing place, just a sprawl of dilapidated houses and stores interspersed by gas stations, downmarket food joints and the odd tattoo parlour and palm-reading salon. The surrounding area was scarred by the towering concrete and steel apparatus of intensive agricultural production. Locating the market was a step too far for our satnav GPS, and I was twice forced to stop and ask for directions. There were few friendly faces and folk struggled to understand my accent. I wondered whether we would witness blatant cruelty and how the locals would react when we pitched up. Experience taught me they were likely to be suspicious. The Humane Society of the United States, America's biggest and most effective animal protection charity, has a high profile among those in the livestock industry, most of whom are primed for unwelcome visits. In the event it was okay. Though the owner Chuck Cozzi was wary, he was sufficiently intrigued by his unexpected English visitors to usher us in. Not only did he agree to show us round, he also allowed us to film, a highly unusual concession in this type of facility.

The auction house had the feel of a small-town football club, with a dog-eared office and a greasy-spoon café selling a daily breakfast special of two eggs any style, two bacon strips, two sausages and two pancakes, all for $3.99. The salesroom was like a little theatre, with rough wooden benches for customers, a small ring for the animals, and a raised kiosk at the back, where the auctioneer sat.

I sat down, feeling painfully out of place among the weather-beaten men in Stetsons. The sale began, and I listened in amazement. It was not the sight of the animals that was remarkable but the auctioneer's breathless sales patter which, to the untrained ear, was curiously captivating, like someone singing the popular country song 'Cotton Eyed Joe'. It turned out he had just qualified for the world livestock auctioneer championships.

I hung around until I'd seen a few depleted-looking dairy cows stagger into the ring. A pair of workers armed with plastic paddles swatted them out as the auctioneer sold them off for cheap meat to the slaughterhouse. Outside, a cowgirl on a paint horse cantered up and down corrals, herding animals towards the ring, her long blonde hair streaming out behind her. A prominent wooden sign in the lorry park warned farmers not to show up with animals too sick to walk. To my relief, there was no obvious sign of cruelty.

My film crew interviewed Cozzi, anticipating familiar complaints about low milk prices and the rocketing cost of animal feed. So it was completely unexpected when Cozzi, a 6ft 4in all-American guy, suddenly broke down in tears. He spoke of a close friend who owned a large dairy, but could not cope with the financial pressures. 'You know, he had enough. I think he shot himself, you know, leaving families behind. You know, kids. It's so sad.' He welled up and turned away from the crew, embarrassed. Composing himself, he said: 'You know I think maybe that guy was just so far in debt, he just gave up you know. Just got no more drive. I don't think anything could be that bad that somebody would want to do that, but it must have been for him.' It was a poignant moment. My crew and I felt hugely sympathetic to his account and thanked him for his honesty.

It is a reminder that it's not just California's milk cows who are suffering from this bizarre perversion of farming. Yes, the dairy cattle are dying young, but so all too often are the people who live and work with them. Everyone is struggling to survive – even the farmers who should be raking in their share from California's billions of dollars-worth of milk a year.

In the complex mesh of economic pressures and corporate interests that have given birth to factory farms, nothing is black and white. One thing does seem clear. In the land of the mega-dairy – a land that is inching perilously close to home – humans, cattle and the environment are dancing to a grim tune of extraction and depletion. Each is just an asset to be milked dry.

2

Henpecked

The truth behind the label

It was a perishing winter day, and I was pounding up the motorway to Northumberland in my old Renault hatchback to view some brand-new luxury homes.

According to the publicity, the development boasted cutting-edge fixtures and fittings and a wonderful location in Tyneside, on the rugged northeast English coast. I knew the area well – a place of stiff breezes, salty air, fish and chips, and cheerful folk. I'd often been there to visit the Farne Islands, a twitchers' paradise teeming with puffins and seabirds of all sorts. Out on the water, on birdwatching trips, they always provided a breathtaking interactive display, whistling past so close you could touch them. There would be a heavy smell of nesting birds and rotting fish, and you'd get the odd peck to the back of the head.

This time however, I was not going out in a boat. I was in the area to see birds of a different variety – hens – and more specifically their state-of-the-art accommodation. I'd been invited by a UK government advisory body, the Farm Animal Welfare Council (FAWC), to see an innovation: battery cages with perches. Officials were excited by the new features and wanted to know what I thought.

After hours on the road, I finally made it to the farm, where I was welcomed by a tall man with a thick Geordie accent. We shook hands, he offered me coffee, and I was introduced to the FAWC committee members who had also come to inspect the new cages. Among them was a government vet with thin greying hair, wearing a green anorak.

Coffee finished, we set off round the farm, a series of hangars crammed with battery cages: tiers of tiny, barren cages so small the hens couldn't even stretch their wings. Groups of five hens were usually crammed into each cage, where they could do little else but lay eggs and survive. The special ones we had come to see were really no different to the standard model – except that each contained a little wooden perch. We were told the hens loved the perches and that the new design made a huge difference to their lives. I resisted the temptation to roll my eyes. Anything was better than a bare wire cage, but it was hardly revolutionary.

Tour completed, we regrouped for a discussion. There was much debate about the ins and outs of what we'd seen. Eventually my moment came. I reeled off all the reasons why keeping hens in cages the size of an A4 piece of typing paper is dreadful, perch or no perch. Then I sat back and waited. The room was quiet. The government vet fixed me with a quizzical stare. 'Philip, I agree with much of what you've just said, but how can you say that hens *suffer*?' he asked incredulously. It sounds incredible, but back then in the early 1990s when my trip to Tyneside took place, the view that hens could not feel pain and suffer was commonplace. At the time, I had a fairly junior job with Compassion in World Farming. At first the work was mostly administrative, but soon I became involved in campaigns. As my knowledge grew, I was increasingly asked for my views by various organisations including government bodies. People often questioned whether farm animals could suffer. Along with colleagues, I spent a lot of time amassing scientific evidence to prove what seemed perfectly obvious to me: that hens and other farm animals are wired much the same as people when it comes to a nervous system, and can experience pain, fear, pleasure and excitement much in the way that we do.

As many as half a million people in Britain now keep hens in their garden, and I doubt that many of them would question this.[1] They can see that what happens to these complex creatures matters to them. They can see how much it means to them to feel the dust under their feathers, the sun on their wings, the soil beneath their feet. I am certain this isn't just sentimentality. People who keep chickens and other animals for pleasure, like our pet dogs and cats, often have an instinctive connection with their animals. I certainly don't feel the need to be anthropomorphic – to project human feelings onto animals, hens or others – because it's quite obvious that they are neither human nor automata, they're sentient creatures in their own right.

Our own little flock arrived one summer day in 2010, after my wife paid a visit to the local farm shop. We'd been talking about getting hens for months, but nothing had been agreed. In the end, I wasn't consulted. The assistant in the farm shop told Helen that their hens were soon going for slaughter. Without further ado, she grabbed a cardboard box, punched some air holes in it, popped three birds inside and brought them home. The first I knew of it was when I came home after a long day in the office and was relaxing outside with a cool drink when an unmistakable clucking broke out.

The hens spent a few days confined to barracks – a makeshift pen we constructed – before we released them into the garden. They burst out excitedly, running and flapping, and began rearranging our garden. They particularly liked pecking about underneath bushes, just as their ancestors must have done when they first evolved in the Asian jungle. It was obvious how much they valued scratching and foraging, being able to stretch their wings and feel the earth under their feet. I could see that each was an individual with her own personality, likes and dislikes. By winter, we had four and they had names: Hetty, Henna, Honey and Hope.

One particularly cold January afternoon, they were outside scrubbing around, when I heard an awful screech. It was a sound I recognised from childhood, when my mum and I kept bantams. Occasionally they were stalked by a cat and we'd hear a sudden cry of alarm, followed by an explosion of wings and frantic squawks. I would be sent to the rescue, by which time they were often in someone else's garden. This time the stalker wasn't a cat, but a fox. I felt a flash of panic. I rushed downstairs, shouting, arms flailing, and burst out of the back door, just in time to see the interloper making a run for it, hen in mouth. He tried to jump the 2-metre fence, then dived into the bushes, before tearing off towards the neighbour's shed, with me in hot pursuit.

At some point during his great escape, he dropped the hen, though I didn't know where. I was pretty sure she'd be dead, or fatally wounded. The rest of the hens had taken refuge by the house. They stood flicking and shaking their heads with stress. I counted them: one by the gate, a second by the table, a third at the back door . . . and the fourth? There she was, sitting on a wellington boot.

Hardly able to believe it, I counted again just to be sure, then scooped her up to check for injuries. To my surprise, apart from some superficial blood on her leg, she was in one piece. I attended to her wound and put her back on the grass. She hurried over to join the others. She seemed almost indignant at having been kept from her business of scratch and search. I felt ridiculously happy and relieved.

It was the first time I'd kept poultry since being at school. I never really liked classes. I was often told off for staring out of the window looking at birds when I should have been looking at the blackboard. However, one day I did sit up and pay attention. It was a Friday afternoon, I was eighteen years old and we'd all trooped into the lecture theatre to hear the latest in a long line of weekly visiting speakers.

It was the early 1980s and I remember that we were all excited because Kajagoogoo, a pop band from our town, had just made it into the charts. My neighbour Stuart Neale was on keyboards. Even back then, I was fascinated with birds and their freedom, and used to keep homing pigeons. I loved watching them fly over our neighbourhood, but sometimes they'd decide not to come back home, and would roost on Stuart's dad's guttering instead. I can see him now shaking his fist at them, as they peered down nonchalantly.

In the lecture theatre, I remember messing around until the teacher told us to shut up. Eventually, the hall went quiet and the speaker was introduced. He was a man called Chris Aston from Compassion in World Farming, and the teacher said he was going to tell us where our dinner came from. I remember being horrified at the pictures of pigs and calves in factory farms. I was particularly upset by the hens in battery cages and how they couldn't flap their wings, never mind fly. I thought about my pigeons and the wild birds that mesmerised me. The hens crammed into cages seemed a crime. It was a moment that changed my life for ever. I resolved to do something about it, which is how, a decade or so later, I found myself sitting in that farmhouse in Tyneside, listening to a government vet question whether hens could suffer and wondering why the Farm Animal Welfare Council existed, and why its members were interested in the perches, if they didn't think hens could feel anything anyway.

Fortunately, attitudes were about to change. In 1997, Labour swept to power after a landslide victory at the general election. A few weeks later I was in Amsterdam for a demonstration, the culmination of a long campaign by Compassion to change EU legislation. We wanted animals to be classified as 'sentient beings', an official recognition that they can feel pain and suffer. It was baking hot as we marched with colourful flags and banners, but the sweat was worth it, because the protest marked a watershed.

Later that day, European leaders agreed to our call, giving animals basic legal status as 'sentient beings'. We were now swimming with the tide. No longer were animals just 'goods' or 'agricultural products'. At last, it appeared their welfare was being taken seriously. The legal recognition won that day has gone on to become a dedicated 'Article', a core text to the European Treaty, giving it more weight. The wording states that 'since animals are sentient beings', the EU must 'pay full regard to [their] welfare requirements'.

Nothing much changed overnight. There was no sudden end to cruel practices like taking animals on long, unnecessary journeys for slaughter or caging them like units of production. Yet for those who cared about animal welfare, the new legal status changed the game for ever. Finally, EU law reflected both the scientific evidence and the message of common sense: that animals will suffer if they are badly treated, and experience a sense of well-being if they are well kept. It strengthened our calls for reform and meant I was unlikely to be hearing the sort of views expressed by that vet in Tyneside again – at least not from a government figure.

Two years later, in 1999, I was travelling all over Europe trying to persuade the EU to ban battery cages. It was an uphill struggle. Britain, the Netherlands, Germany, Austria and Sweden were in favour of reforms, but France and southern European states were deeply opposed. We had a battle on our hands, and we pulled out all the stops. We treated MPs and MEPs to breakfasts with free-range eggs; we delivered cakes made with cage-free eggs to embassies and governments on all sides; we locked celebrities and supporters in human-sized cages; and we organised marches all over Europe. Our protests were always colourful and carried out with a smile: we wanted people to join us, rather than turn away because they'd rather not know.

It was a huge effort, and right up until the last few days before crunch EU negotiations, we were staring at defeat. The southern member states simply would not budge. Then a young Italian called Adolfo Sansolini went on hunger strike against battery cages. It's not a tactic I'd advocate but it proved a turning point. Sansolini had contacts in the Italian government. Within days their position had changed; they supported the ban. It prompted a domino effect, resulting in one of the most significant victories in the history of animal welfare: barren battery cages were to be banned.

It takes a gargantuan effort to get production systems outlawed, especially when they dominate an entire industry providing a staple product. Yet we did it, by waving banners, writing letters, buying better eggs and articulating what was by then widespread public concern. I will never forget the overwhelming sense of elation when we heard the news. I had travelled to Luxembourg for the announcement, not knowing which way it would go. I remember standing on the steps of the European Council building, waiting for the UK minister to emerge and explain what had been agreed. It was a hugely proud moment for everyone who had worked so hard to make it happen.

The new law was far from perfect. For a start, there was a painfully long 'phase-in' period: twelve years for the industry to adapt. Furthermore, a clause allowed so-called 'enriched' cages: marginally bigger cages with perches and some pretty pathetic provisions for nesting and scratching in dust.

In summer 2011, my team was granted access to the facilities of the UK's largest egg supplier, Noble Foods, a company that introduced the new enriched cages well before the 2012 deadline. Getting permission was quite a process, involving extensive form filling and security checks. Having thoroughly checked our credentials and established that we were not animal rights protesters, the company eventually offered us a tour of their Nottinghamshire plant.

The Compassion team drove through lovely countryside – rolling fields of ripening wheat and shady lanes along which trundled lorries laden with fresh produce. The egg farm was a series of giant sheds clad in corrugated iron. Inside were a million hens. Throughout their short seventy-two-week lifespan (chickens can live eight to ten years), they would never see daylight. They lived in cages around 5 metres long, known in the business as 'colonies'. Suspended lights brightened and dimmed at particular times to create the impression of night and day, all geared toward regulating the egg-laying process.

Each colony cage contained four laying areas, with flaps to keep the hens' heads in the dark and a central scratching area for the birds to keep their claws short. Though the cages were less than half a metre high, the birds also had small perches. Their beaks had been severed by a laser to prevent them pecking each other, an almost universal feature of cage farms. The process appeared to have been carried out fairly clumsily: some beaks were of unequal length, and others cut diagonally. All perfectly legal – which raised the question of how far we've really come.

Nonetheless the ban on battery cages marked a sea change in attitudes, both public and corporate. Several of the world's biggest companies, including McDonald's in Europe, Sainsbury's, and well-known Unilever brands like Hellmann's mayonnaise in the EU, now sell or use only cage-free eggs. Yet while Europe has moved on, around 60 per cent of the world's egg-laying hens remain in barren cages. In the last decade I've travelled all over the world with my job and witnessed the unpalatable truth about how most eggs, meat and milk are produced.

A trip to Taiwan, which has 30 million laying hens, sticks in my mind. My fixer and companion was a Buddhist monk. He wore flowing white robes and had the classic shaved head. We would drive around together without a map, trying to find farms. Every now and then our local driver would stop to allow the monk to wind down the window and shout at a passer-by. The passer-by would shout back with equal fervour. We would drive on. This went on for a fortnight, as we made our way round the island, looking into how food is produced on this much-disputed territory. There were two key types of egg farm, one known as 'traditional' and the other known as 'controlled environment'. Sadly, on both types, the hens were caged. On the 'traditional' farm, row upon row of cages, covered by a tin roof, were otherwise open to the air. The wire cages were barely bigger than the hens crammed inside and were completely featureless. The hens had nothing, except access to a food trough and water, the very basics of life. Conditions on 'controlled environment' farms were little different, except that the barns were entirely enclosed, with temperature and ventilation controlled by computers and fans.

The spin about these sealed units, which exist all over the world, is that they are somehow safer and healthier because the birds are not exposed to agents of sickness and disease outside. The lie that this improves disease control was exposed in Britain years later when an intensive Bernard Matthews turkey plant was hit with highly pathogenic avian influenza.[2] It showed that merely to be 'closed' to the outside world and run by computer does not make a shed immune from the obvious laws of nature.

One traditional farm we visited in Taiwan was slightly different to the rest, in that it was open to the air. The birds were still crammed

into the most appalling tiny cages, but at least they got natural light
and felt a cooling breeze in what was often intense heat. I remember
watching as a woman in blue slacks, pink blouse and a wide-brimmed
wicker hat pushed a rusting cart along the rows of cages, collecting
the eggs.

I soon found that eggs weren't the only thing being collected. Two
perforated plastic bins stood by the cages, overflowing with chicken
corpses. Beside them was a sealed plastic bag. It appeared to contain
more bodies. Suddenly, something caught my eye: the white rubbish
bag moved. Horrified, I tore it open, and a chicken's head popped out.
She was panting for dear life. After a while the bird got to her feet.
Then she gingerly clambered out of the hole in the bag and crept away.
This, I learned, was how dying birds or those that didn't produce
enough eggs in Taiwan were all too often killed: they were simply put
in a bag and suffocated, then emptied into the bin.

I visited several farms in Taiwan's Miao Li province, and almost all
the birds I saw were caged. Some were crammed as badly as I'd seen
anywhere in the world, with a hand-span of space. It was a depressing
experience. I also visited what was in 2002 Taiwan's only 'organic' egg
farm. It wasn't what I expected. In Europe, organic production is
tightly regulated. Hens must have outdoor access and be kept in fairly
natural conditions. In Taiwan, far from seeing a few hens happily
scratching around small huts in the middle of a field, I found myself
looking at 300,000 birds, in cages stacked seven high, in four industrial buildings. The word 'organic' simply related to the feed they were
given; and I doubt even that would have withstood scrutiny.

There was a further shock in store. On most intensive farms in
Britain and Europe, laying hens are slaughtered after a year. Their
commercial lifespan is largely determined by the fact that they renew
their feathers every year, during which period they stop laying eggs for
a few weeks. In Taiwan, I discovered that the hens were typically kept
for two years. To minimise the drop in egg production linked to the
feather-renewal period, they were 'force-moulted': shocked into shed-
ding their feathers quickly. This involved starving them for ten days, a

treatment that appeared – rather surprisingly – to accelerate the moult-ing and refeathering process, and get them laying again as fast as possible. A year in a barren cage, then ten days without food, then another year before slaughter? I couldn't help feeling that the lucky ones were those released early through death and piled into those perforated black bins. Thankfully, 'forced moulting' is now banned in Europe, although it is still allowed in the United States.

My time in Taiwan is peppered with bitter-sweet memories. Though much of my visit was shocking and depressing, there was also a hint of change in the air. Interest was growing in healthier food from animals not routinely treated with antibiotics or fed growth-promoting drugs. When I delivered a lecture at a livestock research centre in Heng Chung, I was encouraged to find that both the goat and cattle farmers in the audience were keen to switch from rearing their livestock indoors to pasture-based systems.

Yet there is still a mountain to climb. A decade or so after my trip to Taiwan, I found myself in the middle of a cluster of battery chicken farms in Argentina. We were in South America to look at soya produc-tion and cattle. It was a sunny autumn day in Marcos Paz, about 50 kilometres outside Buenos Aires and a pleasant enough area. We drove past a centre for veterans of the Falklands war, still a highly sensitive issue in Argentina, particularly at the time of our visit, the thirtieth anniversary of the invasion. There was a children's playground with swings, a slide and a basketball hoop; a miniature football pitch; a brown horse in a paddock. We rumbled down a very bumpy lane where there was a sprawling bakery. It looked like any factory; you would not have known its business from the outside. Apparently it was where many of the eggs from the nearby battery operations ended up.

Soon we were among the farms. Set a little back from the road, there were seven or eight, looking much the same. We were told that once upon a time this had been an isolated area, but the town had grown, and the farms were now very near houses. Exasperated by the flies and smell, apparently locals had been trying to get them shut down for years. I got out of the car to explore. Though it was surrounded by a high fence, the gate to one of the farms was wide open. I walked in.

I noticed that there was a cottage right next to the barns. The family living there had strung up their washing between two trees on either side of the battery sheds. The air was thick with ammonia and dust – foul to breathe. Three small children and a dog peeped out of a window, eyeing me curiously. I hated to think what state their health must be in, living in such a place.

A worker, perhaps their father, approached. He wanted me off the property so I turned and left, he locked the gate behind me, and we went to try somewhere else. A number of the farms were deserted, the cages empty but for sad clumps of feathers caught in the wire. I walked into one open-sided barn, some 300 yards long, with empty cages stacked high from one end to another. Rusting ventilation fans were shrouded in thick cobwebs. A forlorn-looking pair of worn-out train-ers lay abandoned on a cage. The place was silent. It looked as if the hens had been cleared out fairly recently. Suddenly there was a shout. A chap in shorts and a baseball cap wanted me out. He picked up a large stone in case I didn't get the message. I didn't hang around.

I moved on to another farm. This time I was able to get very close to the birds. A teenager was pushing a galvanised metal trolley up and down between the rows of cages under a corrugated-iron roof. Each cage contained about seven birds, with barely enough room to stand. The boy was dispensing feed. The speed of the trolley flustered the hens into panicked squawks and attempts to flap. It was clear that in summer the heat would be unbearable. The hens had big blood-red 'combs' – crowns – making them look like cockerels. This is a way to deal with overheating. Some combs were so big they flopped down over one eye. The feathers on the hens' necks and backs were sparse. Bald patches were rife and their tails were like fistfuls of quills.

Their beaks were stumps, cut short in an attempt to stop them pull-ing each other's feathers out in frustration. A decaying corpse was trampled into the wire floor of one cage, rotting and grey. The survi-vors stood on it. They had no choice. Flies swarmed. Ash-grey droppings built up below. The birds cried with woe. I hated it. I focused on the eggs rolling off the back of the cages. Some people claim hens wouldn't lay 'if they weren't happy'. If ever there was proof of what nonsense that is, it was here. It's hard to see how anyone could witness birds in these conditions and still believe such a thing.

As I stood there among the cages, the stench catching in my throat, listening to the dismal squawking, I thought about the clean sanitised eggs, sold in nice little boxes, on supermarket shelves under 'fresh eggs'-type labels. They are so far removed from their filthy origins. From thousands of miles away, my wife was sending me cheerful text messages about what she was up to, in particular about how Hetty, Henna, Honey and Hope were doing. Compared with what I was seeing, the occasional threat of a fox seemed like nothing.

II
NATURE

Nestled among hedges and mature trees, the little village of Nocton in Lincolnshire seems an unlikely setting for a battle for the countryside. Life for the six hundred or so inhabitants revolves around a small post office, village hall and a primary school. There isn't even a pub, as an old by-law forbids such revelry. Those wanting a quiet pint must head down the country lanes to the nearest hostelry a kilometre or so away. Not so long ago, Douglas Hogg, Britain's former government agriculture minister, opened a guided trail of Nocton's historic features dating back to Roman times. Henry VIII apparently visited in the sixteenth century, planting a chestnut tree that still stands today.[1]

In 2009 this quiet hamlet was thrust into the media spotlight when it was chosen as the site for an 8,000-cow mega-dairy, planned to be the first of its kind in Britain and modelled on large-scale dairies in the USA. Angry locals were joined by MPs, environmentalists and a wide spectrum of campaigners worried about the impact of so many cows together, and all their dung, in such a small space. On local radio, the man behind the proposal suggested that: 'Cows do not belong in fields.'[2] His words came back to haunt him. Children from the local school drew pictures in protest; local buses carried advertisements insisting that cows belong in fields; questions were raised in the House of Commons. So loud was the public outcry that the plans were scrapped, the death-blow delivered by the Environment Agency and serious concerns about pollution. It was the first skirmish in what threatens to be a lengthy war.

The countryside is shaped and tended by the people who work the land. Three-quarters of the UK's land is farmed, providing a rich and varied landscape. The past sixty years have transformed it. Successive governments have presided over the uptake of industrial farming techniques often developed in the US. Now there are signs of damage. Sights once common are growing rare – the stunning aerobatics of huge starling flocks, the flutter of butterflies like the Brown Hairstreak, the buzz of previously abundant bumble bees.[3] Wild-flower meadows have vanished and areas of outstanding natural beauty are at risk as never before. Hedgerows that once criss-crossed the entire country have been disappearing along with woodlands. Large parts of the countryside are being built upon. Farms are getting fewer and bigger; farmers are struggling to survive.

The proposed Nocton mega-dairy may have been defeated, but the impetus behind it grinds on. Dairy cows are not the only farm animal affected in a new wave of industrialisation, this time conducted under the seductive language of 'sustainable intensification' and dubious justifications of feeding the world. It threatens to take the countryside, its wildlife and many of those who live in rural areas to breaking point.

3

Silent Spring

The birth of farming's chemical age

For more than three decades, I harboured a nagging sense of inadequacy, a puzzling feeling that I just couldn't see things in quite the same way as others. I remember how it started. It was the late 1970s when, as a young teenager, I read a book that would fire my imagination for the rest of my life. It led to years of gazing out of windows, a habit that got me into trouble with a string of teachers. It was a Sixties wildlife classic called *The Peregrine*, by J. A. Baker. I was enthralled, inspired, filled with a sense of nervous wonder. As I read his vivid and meticulous descriptions of the falcons he watched near his home in Essex, I dreamed of seeing them too. With eyes filled with awe, I scrutinised every kestrel I saw, just in case.

It would be some years before I saw my first peregrine falcon. Majestic, enthralling, and when they close their wings in a stoop, said to be the fastest animal on the planet, hurtling at speeds of 200 kilometres per hour. Since then I've seen many all over the world, as well as on my very doorstep at home in the South Downs of England. But that sense of inadequacy never quite went away. You see, after countless encounters, I've just not been able to see them so vividly, so close or for so long as Baker did. What was the matter with me? Was I doing something wrong? For thirty-five years that question remained unanswered; until now.

I was on my annual winter trip to North Wales with my wife, Helen. We love to call in at the wonderful nature reserve at Conwy. Situated on the banks of the estuary, with magnificent views of Snowdonia and Conwy castle, it is a spectacular place to while away an afternoon. The weather was foul, which meant more time in the shop and café surrounded by wildlife paraphernalia. Out of the bookshelves a title leapt out at me: *Silent Spring Revisited*, by Conor Mark Jameson. It explored the legacy of Rachel Carson, who first raised the alarm over the perils of pesticides sweeping across Britain and America and the demise of songbirds.

What happened next came as a bolt from the blue. Standing, book in hand, I started to read: 'In a book about the dawn chorus, about songbirds and birdsong, the Peregrine Falcon might not be the obvious place to start, but bear with me.' I had to agree, and was intrigued. Turns out both the writer and I had a love affair with *that* bird and *that* book by J. A. Baker. Both bird and Baker were described as having achieved 'almost mythic, prophet status'. Memories came flooding back of how much better, faster, closer the peregrines looked through the lens of Baker's eloquent prose. But then a twist: what I hadn't quite appreciated was that others were quietly questioning what Baker *really* saw. Through forensic analysis, the author suggested that Baker might not have been looking at wild peregrines, but instead at escaped falconers' birds, hence why they were so tame. Indeed, in some cases, they might not even have been peregrines at all, but some closely related escapee!

Okay, it's only a theory, but in that moment, decades of inadequate feelings, of wondering why I couldn't quite see what he saw, were lifted from me. The relief was so real that I ran round the shop telling anyone that would listen! You see, way back in the 1960s, when Baker was writing in southern Britain, peregrines had pretty much been wiped out. Agrochemical pesticides were poisoning wildlife, including those at the top of the food chain like birds of prey. People still remember how chemicals persisted in the food chain, accumulating in predators like falcons, causing nests to fail. What is less well remembered is how the countryside was littered with dead or dying birds. Foxes were affected by a mysterious illness where they lost their fear of people.[1] Those were extraordinary times in the countryside on both sides of the Atlantic; a time of dramatic demise.

When Rachel Carson published *Silent Spring* in 1962, it carried an introduction by Lord Shackleton, a member of the UK House of Lords, who declared: 'we in Britain have not yet been exposed to the same intensity of attack as in America, but here too there is a grim side to the story.' Things were bad in Britain, but worse in America. The US had given birth to techniques that treated the countryside like an industrial site, with unforeseen but devastating consequences. Half a century on, history is repeating itself; mega-farms are using the latest industrial practices pioneered in the US and now being exported to Britain and beyond.

I was in America and the car ride from our hotel was surprisingly short. Before I knew it, my cameraman, Brian, stopped the hire car and I was looking up at the childhood home of Rachel Carson. I had come to find out what inspired her to kick-start the environmental movement and how well we had heeded her warning. I stepped out into the chill air and reticent sunshine of the morning in suburban Springdale, Pennsylvania. A small black and white woodpecker clattered the branch above me. A simple white-boarded farmhouse stared imposingly toward the leafy street. It seemed isolated now that the once surrounding farm had gone. Instead, its natural appearance was bordered by the neatly manicured lawns of its neighbours. It was mid-April and the anniversary of Carson's untimely death less than two years after *Silent Spring*.

I walked up the raised driveway and banged on the back door. Peering in through the glass I could see Carson artefacts in an otherwise empty house. There was no answer, but then I was a bit early. The house was sheltered by mature trees: oak, maple, pine. A cluster of low wooden benches, set out theatre-style, where visitors would sit and learn. A sign welcomed the curious to the 'Wild Creatures Nature Trail', where

Carson began her lifelong fascination with the natural world. It was inscribed with her own words as a fourteen-year-old: 'The call of the trail on that dewy May morning was too strong to withstand . . . It was the sort of place that awes you by its majestic silence, interrupted only by the rustling breeze and the distant tumble of water.'

Back then, it clearly led to endless woods and fields. I followed the trail a short walk up the hill and peeked over the brow only to find an office and car park strewn with cars. I felt mildly disappointed, but then it has been over eighty years since the Carsons sold up.

Robert Pfaffman, a fifty-eight-year-old building architect, emerged welcoming and enthusiastic. He was on the board of the trust now running the Carson homestead as an official museum. He showed me round the five-room Pennsylvania German farmhouse which overlooks the Allegheny River valley. It came originally with sixty-five acres of orchard way back in 1900 when Carson's parents bought it, until bits of land got sold off to meet the expense of the children's education.

Pfaffman and I scaled the narrow staircase and entered a small bedroom complete with fireplace, low ceilings and pastel-pink walls. I felt a connection: the room was strangely similar to my own bedroom back home in my tiny eighteenth-century English cottage. This is where Rachel as a young kid would look out and daydream. I gazed out of her window to see the tree-covered beauty of the Allegheny valley, interrupted by two great chimney stacks dissolving smoke into the atmosphere. Coal-fired industry sits alongside agriculture in this community today, very much as it did during Carson's formative years. She was no stranger to the realities of life and landscapes. Perhaps this helped her to spot the blurring of lines between industry and agriculture that went on later in her lifetime. It was the insight I came for.

My plan was to drive from Pennsylvania and Rachel's former home and head across to historic Chesapeake Bay. I wanted to see whether the countryside would offer clues to Carson's legacy. Along the way, I called in to meet my fellow Brit Bill Sladen, ninety-two years old, a former professor at Johns Hopkins University in

Baltimore, where Carson studied zoology. He greeted me playfully with an impeccable old English accent: 'We've not met before have we? Why are you so bald?!'

Sladen is one of a distinguished band of 'bipolars': intrepid individuals who have travelled on icy journeys to both north and south poles. After training in the Second World War at London's Middlesex Hospital, Bill travelled to the Antarctic with the Falkland Islands Dependency Survey and United States Antarctic Research Program. It was during one of these expeditions that Bill discovered the first evidence of chemical pollution having gone global. During studies of Antarctica's Adélie penguin in 1959, the samples of six penguins and a crab-eating seal that he sent back for analysis in the US were found to have minute traces of DDT contamination, revealing that chemical pesticide pollution had reached all the way to the Antarctic.[2]

Sitting back, relaxed and reflective in a traditional leather chair, he told me of another of his ground-breaking projects: training goslings to learn historic migration routes by following a microlight plane. His long-standing dream was to re-establish the historic migration routes of the trumpeter swan to the nearby Chesapeake Bay, a vast estuary 320 kilometres long. But it was geese that grabbed the limelight, starring in a film, *Fly Away Home*, that documented the exploits of Sladen, his Canadian pal, Lichman and their microlight.[3]

I watched jauntily coloured American goldfinches and a wren behind him on the tasteful decking patio as he told me: 'Rachel was the pioneer of modern conservation. Her legacy is in having started a movement of concern for the environment.' But has it helped save Chesapeake Bay? I asked. He told me that the bay has declined, not just because of pesticide pollution, but perhaps even more because of manure run-off from the chicken industry. That was the clue I needed for the next part of my journey.

A farm tractor clanks along with what looks like thick red smoke belching from the back of a long green trailer and billowing across the adjacent road. Reddish-brown lumps spray out onto the field behind.

This is poultry manure being blown mechanically into the air and spread across the soil. 'The stuff along the ditches and field edges, if it rains, could run off and end up in Chesapeake Bay,' warns the local waterkeeper Kathy Phillips. 'The pungent smell of chicken manure being spread is a familiar part of spring here.'

Kathy has taken me on a 'muck safari' across Maryland. She moved here with her husband in the 1970s to live the beach life. After running for County Commissioner on a clean-water ticket, Kathy became local waterkeeper, charged with enforcing federal law protecting the coast of Assateague. 'CAFOs are everywhere in this area,' she told me, using her favoured acronym for Concentrated Animal Feeding Operations, better known as factory farms. 'They only grow corn and soya in these parts to support the area's poultry industry.'

Poultry manure is used as cheap fertiliser to spread on the fields growing the corn and soya that will end up as chicken feed. At first glance, it's a virtuous circle: the chickens eat the corn and their droppings replenish tired soils. The only flaw is the vast number of chickens in such a small area. Chicken manure is heavy in nitrogen and phosphorus, precious nutrients in the right amounts, but too much or at the wrong time and the rain washes it into waterways where it becomes a serious pollutant.

Kathy drives across a patchwork of woodlands and crop fields, tall pines and pasture, and pretty homes of pastel blue and white, painted using the 'coastal palette'. 'Smell that?' she asks, waving her arms. 'They're spreading manure out there.' This area used to be big on producing timber, orchard fruit and vegetables like tomatoes and cucumber. Now the dominant 'crop' is chickens. They try to portray massive industrial agriculture as family farms, she tells me, but it just won't wash. 'Family farms own and make all the decisions about the farm, that doesn't happen on these factory units.'

We cross the Pocomoke River, then drive past a great pile of manure, big and brown. It's not long before we pass another muck heap, then another. Some of the piles are so big that they look like small mountains. These fields have had manure newly spread on them; you can see clumps of it.' The smell filled the car. Then we found a 'muck shelter': sheds purpose-built to protect the manure from the elements before spreading. Kathy pointed out how local communities are often 'buffered' by lines of trees and shrubs to protect them from the emissions of dust and ammonia blown from the factory farms with their huge extractor fans.

I asked Kathy if she buys cheap chicken. 'No, I don't. Buying chicken in the supermarket is cheap food all right, but what people don't see is the hidden cost. It's their taxpayers' money that's being used to put up manure sheds, vegetative buffers around poultry houses to catch the dust and ammonia emissions. State money also goes into moving the manure out of the watershed.' We stop alongside the farm of a former secretary of agriculture. We can see a large 'muck shelter' with farm equipment standing inside; great piles of chicken manure stand outside. 'It's a great example,' Kathy says, 'of how Maryland's nutrient management programme isn't working, and it's in the Chesapeake watershed.'

We spent a morning seeing chicken sheds aplenty and mountains of manure, but not a single bird until we drove up behind a slow-moving articulated lorry stacked high with cages. Each one was crammed with hapless chickens on their way to the processing plant. Feathers flew from the battered and rusting crates. A couple of chickens managed to squeeze their heads through the bars, beaks open, eyes straining in distress.

Soon we were at Ocean City airport and keen to see the countryside from the air. The piping chatter of oystercatchers – noisy birds, all black and white with long carrot-like beaks – told me we were close to the beach. The quiet little airfield advertised scenic rides. A friendly-looking fellow in Hawaiian shirt and sun-strip cap sat inside the aerodrome talking into his Bluetooth headpiece while tapping into a computer. He spotted us and wrapped up his conversation. Neil Kaye, a medical doctor and helicopter pilot, had offered to take us up in the air as part of an initiative known as Lighthawk, specialising in helping environmental causes for free.

Soon we were strapping ourselves into Kaye's four-seater chopper. The rotors fired up, shaking the plane as they whirred faster and faster. I felt a little nervous. The chopper floated, swung round, tilted forward and my head went with it. We rose 150 metres, the same height as a passing bald eagle. It was like sitting in a tiny version of a London Eye capsule wobbling frantically side to side. We juddered along, following the river over expanses of woodland. We weren't up long before the forest gave way to fields. Then our first 'CAFO'. 'Twenty thousand birds per house in those,' Neil announces into my headset. We swoop over eight long, low-slung warehouse-like buildings, each accompanied by a huge feed silo. More sheds are scattered in other fields, perhaps run by different farmers. We see another two, then four more, then they are everywhere. Chickens are big business here.

The chopper tilts and we circle. 'This one disturbs me,' says Kathy over the headphones. 'Nine massive poultry houses, and directly to its right is a particularly sensitive tributary.' Any overflow from the meagre storm-water arrangements go straight into the creek. Kathy explained how it used to be a small farm, just crops. A CAFO like this takes as much ground as a small shopping centre. The difference is that a proposal for a shopping mall goes through a lot of planning processes. CAFOs, on the other hand, are treated a lot more lightly: 'Often the first you know about it is when it's being built.'

We fly on. I can see a mass of grey in the distance, like a lake surrounded by forest. As we get closer, I couldn't be more wrong. 'It's a big operation,' warns our pilot. Thirty huge factory-like sheds, each one covering perhaps 2,000 square metres, arranged in rows with massive extractor fans to keep alive the chickens crammed inside. Fast-growing trees shroud the site from the ground. The storm water from this CAFO, I'm told, runs into the Manokin River and then on to Chesapeake Bay.

It feels spooky to look down on what could be three-quarters of a million chickens and not see a single bird. I run the numbers and speculate: this site alone could produce upwards of 5 million birds a year. This is how 'premium fresh young chicken' is produced.

*

Chesapeake Bay is the largest estuary in the United States and has a rich history of battles. It is here that the British Royal Navy was defeated by the French in 1781 during the American War of Independence. A century of violent disputes over shellfish raged in the bay until the 1950s during the so-called Oyster Wars.

Today, the bay's vast watershed, which touches six states, is home to 17 million people as well as more than 3,000 species of plants and animals. Its skies are patrolled by the mighty bald eagle, while the beaches and mudflats swarm with some of the largest populations of shorebirds in the Western hemisphere.[4] However, the bay's latest and biggest battle is now against birds of another kind: chickens, and the muck they produce.

The sun is blazing on Shady Side, where I'm due to meet with people at the forefront of the battle to save Chesapeake Bay. A spectacular fish-eating osprey quarters the sky. Brown turns to white as the bird banks and hovers briefly before folding its wings to dive. A powerful plunge, a momentary pause, then broad, fingered wings lift bird and fish from the surface and away with a shake.

Betsy Nicholas, the executive director of Chesapeake Waterkeepers, waits by the jetty. We take a boat ride across the bay. A group of ruddy ducks with their comical plastic-blue bills bob ahead of us. A cobalt-blue belted kingfisher darts along the bank. A young bald eagle fixes us from afar with beady eyes. Chesapeake is well known for its oysters, but they're in trouble: less than 1 per cent of former numbers, thanks to overharvesting and pollution. 'The three biggest problems with pollution in this area are agriculture, agriculture, agriculture,' Betsy tells me. 'There are huge problems here with algal blooms, times when there are complete dead spots in the bay.' Folks are reluctant to address the pollution issue because they don't want to be seen to attack family farmers. 'But factory farms are a different matter. It's really become an industrial pollution source and really needs to be addressed as such.'

Betsy's main worry was the lack of accountability, the veil of secrecy that seems to be drawn around anything to do with agriculture. 'Until we can get through that, we can't address the problems, or separate the good from the bad actors.' Water-management targets have been agreed to try to reduce pollution from nitrogen and phosphorus run-off, but: 'I don't think we've done enough; no one wants to be the one responsible for making the changes.' The different interests and different states along the watershed would rather blame each other than take action.

In Baltimore I met Bob Martin, senior policy adviser at Johns Hopkins University. He echoes concerns about the extraordinary growth of the region's poultry industry and its effect on Chesapeake Bay: 'There are now nearly as many chickens being produced in the States surrounding the bay as there were across the entire country sixty years ago.' The vast majority of these are factory-farmed. Bob goes into greater detail about the effects on the local wildlife: 'The pollution from the chicken factory farms is causing a reduction in natural seabed grasses, making it harder for the oysters to grow. Local crabs are growing bigger and eating the oysters. Periodic fish kills involve thousands of dead and stricken creatures through nitrogen-induced oxygen starvation. Things are out of balance.'

Bob sees industrial agriculture as the biggest threat to environmental damage and public health in this area. 'I enjoy eating meat,' he confesses, 'but there needs to be some common sense in what we do.'

Farmland covers a quarter of the Chesapeake Bay watershed and is the largest single source of pollution, much of it relating to poultry manure.[5] The Bay lies on the eastern edge of the United States 'Broiler Belt' where many of America's chickens are reared. For want of a safe place to put it, excessive amounts of chicken waste are spread on the surrounding farmland, from there to wash into the Bay. Environmental authorities are slowly waking up to the crisis, and Chesapeake Bay has been placed on an unprecedented 'pollution diet'. Several states together with the Environmental Protection Agency (EPA) are working hard to reduce the nutrient flow into the water. However, despite best efforts, targets for reducing pollution have yet to be met. In environmental terms, more than half of the streams in the Chesapeake watershed are described as 'poor' or 'very poor', stripped of the snails, insects and other waterborne wildlife essential for a healthy aquatic environment.[6] Although agriculture isn't solely to blame, farming, and poultry litter disposal in particular, is the main culprit.[7] Nor will the problem go away in a hurry: the backlog of nutrients already built up will likely pollute the bay for years to come.

In *Silent Spring*, Rachel Carson focused on the overuse of chemical pesticides in the mid-twentieth century. I met with Ruth Berlin, executive director of the Maryland Pesticide Network, who told me that despite fifty years of reform, pesticide pollution is still a threat and found widely in Chesapeake Bay. The problem is far from unique to this area. Ruth told me how a cocktail of chemicals has been found in drinking water and fish samples. She sees agriculture as by far the biggest user of pesticides, affecting wildlife and implicated in serious public-health issues. Her verdict: pesticide overuse is probably 'a bigger problem now than fifty years ago'.

During my travels from her childhood home to the banks of Chesapeake Bay, I found people talking about the tremendous legacy of Rachel Carson in raising awareness. At the same time, there was a great deal of concern that the central message in *Silent Spring*, the threat from treating the countryside like just another industrial process, had been ignored. Yes, there have been some reforms – the banning of the infamous organochlorine insecticide DDT for farm use worldwide is a key achievement. The skies around my homeland are once again filled with buzzards, sparrowhawks and the like. Sometimes I also see the powerful pointed wings and black 'moustache' of the peregrine, thankfully restored to the skies over much of Britain and elsewhere. However, industrial agriculture is still linked to devastating effects, and Chesapeake Bay is far from unique. Wild bird populations are seen by the British government as an indicator of our quality of life, a yardstick for the health of the countryside. The sad fact is that half a century on from Carson's epic book, once common farmland birds still suffer heavy declines on both sides of the Atlantic.

4

Wildlife

The great disappearing act

NO FLY ZONE

It was the mid-1970s and the age of punk rock. I was eleven years old. When not hanging out with friends listening to the Sex Pistols, I was discovering birds. I liked the raw energy and anti-establishment lyrics of the bands at the time, but I was also drawn to the quiet of the countryside. It started when we went on a family holiday to a cottage in rural Norfolk, England. My grandad gave me a copy of the *Observer's Book of Birds*, and I remember thumbing through it trying to match the birds I could see out of the window to the pictures in the book. I was delighted when I figured out that the twittering dots in the sky were skylarks, and excited when two exquisite partridges came and sat on our window ledge and peered in at us.

It was the start of a lifelong passion. Back from holiday I joined the RSPB's Young Ornithologists Club, went to their local meetings and became an avid follower of nature programmes on TV. After school and at weekends, I would disappear into nearby woods and roam across farmland, inspecting the hedgerows and scanning the trees and skies for species I hadn't seen before.

One weekend, the Young Ornithologists Club organised a day trip to Tring Reservoirs in Hertfordshire. They were built in the early nineteenth century against the backdrop of the Chiltern Hills and act as a magnet for birds. Grebes dived, ducks bobbed and herons loafed in the reedbed. That's when I got really hooked. I spent pretty much every spare teenage moment cycling through the English Home Counties countryside, usually ending up beside the reservoirs. The bike ride would take me past farms and hedges full of yellowhammers and other birds. I used to love the sound of the corn buntings as they sat on telephone wires. I remember thinking they sounded like jangling keys.

Little did I know that just as my interest in birds was awakening, the objects of my fascination were disappearing almost before my eyes. Ten million breeding individuals of ten species of farmland birds disappeared from the British countryside between 1979 and the end of the twentieth century. And the problem was not confined to Britain. According to a group of scientists led by Oxford University, 116 species – a fifth of all the bird types in Europe – were at risk. Warning of a 'second Silent Spring', they described as 'damning' the evidence that agricultural intensification was the culprit.[1]

Nor was it only birds that disappeared – farm animals followed. The two are linked. About half a century ago, pigs, poultry and to some extent cows began vanishing from fields to be reared indoors on factory farms. Mixed farms, where crops and animals were rotated around hedge-lined fields – working with nature – began to be a thing of the past. Hedges disappeared – 100,000 kilometres of them between 1980 and 1994 – and farmers began concentrating on just one type of crop, using more and more chemicals to fertilise tired soil and wipe out pests. The industrial revolution had hit the countryside and it wasn't pleasant.

I remember the late Chris Mead of the British Trust for Ornithology speaking at our local bird club. He'd just published a book, *State of the Nation's Birds*, and talked of heavy declines in birds familiar from literature and folklore – skylarks, turtle doves, peewits. I remember him saying that intensive farming had created a 'wildlife desert', a view that resonated with me. His book highlighted the devastating loss of seeds on arable land – just a tenth remains of what birds could choose from half a century ago. Chemical insecticides do likewise for insectivorous birds.

Fast-forward a decade and the figures are shocking. According to the British Trust for Ornithology, the last forty years or so have seen the population of tree sparrows crash by 97 per cent; grey partridges by 90 per cent; turtle doves by 89 per cent; corn buntings by 86 per cent; skylarks by 61 per cent; yellowhammers by 56 per cent; even common species like starlings and song thrushes have fallen by 85 and 48 per cent respectively.[2] In 2010 the UK government published its own bird census which revealed that the population of farmland birds overall had more than halved since 1966. Their research revealed a sharp decline between 1976 and the late 1980s,[3] a period when farming in the UK was going through massive change, from traditional mixed farms to intensive operations.

This bleak picture is replicated in Europe and America. The European farmland bird census, which covers thirty-three species, found that the overall population fell by 44 per cent between 1980 and 2005. The sharpest drops once again took place between the late 1970s and early 1980s, a period of rapid agricultural intensification.[4]

In America, the 2011 edition of a multi-agency report called *The State of the Birds*, concluded that a quarter of the 1,000 bird species in the US were 'threatened, endangered, or of conservation concern'. In the States, farmland birds are often described as 'grassland' birds. The study noted that more than 97 per cent of publicly owned wild grassland in America had been lost, 'mostly because of conversion to agriculture', as a result of which grassland birds have declined far more than any other group of birds.[5] In 2009 it warned that some of the US landscape's most iconic species, like short-eared owls, which hunt during the day as well as at night, and the beautiful yellow-breasted meadowlark, which has a very distinctive melancholy whistle, are in deep trouble. It noted that ranchlands are 'often overgrazed, causing desertification' and that when migrant birds fly to South America for the winter, they often find conditions even worse, because grassland there is being converted to agricultural production.[6]

I spend a lot of time trying to convince governments and the food industry that factory farming is unsustainable. All sorts of arguments are put up to defend the system. Sometimes I'm told it's better to farm some land intensively so that we can save other areas for wildlife. An interesting idea – but in practice we don't save much land for wildlife.

Others try to claim that the blame lies elsewhere, usually abroad. I remember talking to a group of farmers on a rural leadership course who tried to blame the shocking decline of Britain's tree sparrows – the dapper and much rarer rural cousin of the city house sparrow – on factors in the bird's migratory wintering grounds. Just one problem: tree sparrows don't migrate.

Yet even the biggest apologists for intensive farming struggle to deny the impact. The evidence is just too overwhelming. In both the UK and Europe, farmland birds have declined more than those found in other habitats like woodlands or wetlands.[7] Tellingly, scientists have managed to reverse the decline of four species, at least on a local scale, by experimental changes in farming. Grey partridges – plump game-birds with cheerful orange faces – did better when pesticides were reduced and more nesting cover provided. Cirl buntings, which are pretty little European birds related to the yellowhammer and already have a tough time with the climate here, seemed to like it when farmers left fields as stubble in winter rather than immediately re-sowing, and left bigger grass margins around crops. The secretive little corncrake spends most of its time hiding in tall vegetation; it responded well to extra nesting cover and delayed harvesting of the grass it likes to forage in. Stone curlews, with their spindly yellow legs and big eyes, visitors to southern England in summer, seemed to appreciate mixed farmland, with spring crops planted next to grazed pasture.[8]

It's not all bad news: various initiatives by the EU and US federal government have helped slow the decline of some species. Under the EU Common Agricultural Policy, farm subsidies are no longer based simply on how much a farmer produces. There has been an enthusiastic uptake of various environmental stewardship schemes. Fertiliser application levels have dropped from around 150 kilograms per hectare in 1987 to just under 100 kilograms per hectare in 2009;[9] an indication that soils and the creatures that depend on them are being better protected. For example, artificial fertilisers tend to create acid conditions, which can be fatal to earthworms[10] – and quite a few British birds eat worms. In 2011, a study found that there are between double and quadruple the number of worms on organic farms as on conventional arable farms, probably because of the use of organic manure, as opposed to artificial fertilisers, and the lack of pesticides.[11]

In America, the US Conservation Reserve Program is a lifeline for millions of birds, because it pays farmers to leave some land out of production. However it is under threat from the government's decision to promote biofuels, which is likely to entice farmers to use their set-aside to grow corn for cars.[12]

Farming and wildlife can, and often do, go hand in hand. I talked to Richard Owen, a lifelong farmer now in his sixties, who works at Bickley Hall Farm in Cheshire. It's a demonstration farm to show that food production and nature can coexist to everyone's benefit. He runs 350 hectares of permanent pasture and hay meadows grazed by cattle and sheep. The talk here is of 'relearning traditional wisdom'; of farming sympathetically with the land, using fewer chemicals and rotating animals around the farm to preserve soil health. 'Farmers are often blinkered by what they're doing and the amount of money they owe,' Owen told me. 'Once they try this system of farming without inputs, with different breeds of livestock, that brings communities and people back to the farm, they become enthusiastic.'

Some of the local farmers have started to adopt this model. It's just one example of farming in tune with the countryside. As we talked, we walked up a steep bank overlooking a reed-fringed mere with lapwings nesting in an adjacent field, serenaded by skylarks. We peered into bushes and saw a blur of brown wings: tree sparrows; rare in Britain nowadays, but doing well here.

People care about birds – the Royal Society for the Protection of Birds (RSPB) has over one million members. That declines have been less steep in recent decades is probably due to growing public awareness of the damage done by industrial farming. Popular TV programmes like *Springwatch*, where birds are the soap-stars, can only help.

Almost every day I'm out with my binoculars scouring the South Downs, where I live. I've led birding expeditions to countries including Costa Rica, USA, Turkey, Israel and the Seychelles. I still 'twitch' when a really rare bird gets found in Britain, I've seen over 470 species so far, and have often driven overnight and sometimes squeezed into six-seater planes to chase rare birds on far-flung places like Scilly in the southwest and Fair Isle in Shetland.

I try to do my bit; feeding in the garden, supporting charity work and buying organic food when I can. One bird touched me like no other. I was at home in Hampshire one day when I got a phone call from a stranger asking for help. It was the kind of call I was used to, having been a volunteer 'ambulance' driver in my spare time for a local wildlife hospital. We'd just had a heavy storm, and the caller said they'd come across an injured bird. They didn't know what type it was. They gave me an address, and I jumped in my car and set off. On arrival, I found a young man and woman and a beautiful, if somewhat dishevelled, kestrel – a small falcon that you sometimes see hovering over fields and motorway verges. The bird had no obvious bone break but had clearly done something bad. I popped it in the back of my car and deposited it at the local Brent Lodge wildlife hospital, where experts were on hand.

A few weeks later I received a call to say the kestrel was ready for collection. The idea was to take him back to the place where he'd been found. It didn't go well. Weeks of rehab hadn't sorted him out. My heart sank when I released him, and he dropped to the floor. I tried hard to get him to fly, to no avail. With a combination of hopping and flapping though, he could move pretty fast. And he did; across the field, down a lane, up a bank and down again, all with me in hot, panicky pursuit. He might have been covering ground, but he would stand no chance against predators, and would most likely have met his end in the jaws of a fox. Eventually I snaffled him, and returned him to his box, and then to the wildlife hospital.

Months went by before I got the call to come and pick up him up again. This time I was told straight out that if he didn't 'go' when I released him, they'd put him down. I took him back to the field near to where he was first picked up, and wished him luck, urging him on with all my heart. I held out my hand and we both looked to the sky. He opened his wings, reluctantly launched himself forwards and then just dipped towards the floor, with my heart plummeting after him. Suddenly, as if by some guiding hand, his wings found strength. He flew in level flight away from me before gaining height and more height till he was above the treetops and away. Soon he was just a speck in the distance. As he disappeared, I leapt and cheered. I cannot describe the feeling. Even now as I write this, I can feel the surge of joy at having helped bring freedom and a new chapter to the life of a creature as amazing as a kestrel.

In the introduction to this book, I described Mao Zedong's war on sparrows as part of the Great Leap Forward in the late 1950s. Mao's war on nature was lost and people starved.[13] Looking back, it sounds absurd. Yet the same war is being waged – albeit in a far less visible fashion – every day in every country where vast swathes of land are given over to single types of chemical-soaked crops. It is rare to see a tree sparrow in the UK today: owing to changes to the countryside and intensive use of chemical sprays and fertilisers, they have effectively been wiped out, with less fanfare than Mao's war, but for the sparrows and other farmland birds, the result is the same.

BUZZED OFF

For fire chief Kenny Strandberg, 10 July 2011 started like any other Sunday in Island Park City, Idaho. Weekends in his part of the world are usually peaceful affairs, especially in summer, when the meadows are full of yellow and blue wild flowers, the sun shines against an azure sky, and for a few glorious months there is no need to battle the snow. It is a time for cowboys to hang up their boots and relax on the porch, taking in the mountain air. On Henry Fork of the Snake River, fly fishermen try their luck for trout, while elk and antelope roam through the high mountain pastures. The silence of the pine and aspen forests is rarely broken except by the occasional holidaymaker hiking out from a log cabin or guest ranch.

At the last census, Island Park had a population of just 215. It bears no likeness to a town, never mind a city, but back in the 1940s the owners of the many lodges along US Route 20, the highway that cuts across the state, somehow persuaded the authorities to call it a city so that they could dodge draconian liquor laws banning sales of alcohol outside city limits. Island Park now boasts the most extended and sparsely populated main street in America – a full 53 kilometres long.

As always, Strandberg was on call. Forest fires and car crashes are the usual trouble in these parts, but the day passed uneventfully until 4.30 pm, when his pager went off. There had been an accident on Route 20, and the sheriff's office was being bombarded with 911 calls. Details were sketchy, but Strandberg learned that an articulated lorry had veered off the road and partially overturned as the driver fought to steer his wheels back onto the highway. Apparently the vehicle had shed a large part of its load. The fire chief dropped everything and headed for the scene, where he met an extraordinary sight. Hundreds of beehives were strewn across the tarmac, and a giant swarm of bees – 14 million strong – was buzzing angrily overhead.

I tracked Strandberg down and managed to talk to him over the phone about what happened next.

'It was just a black cloud of bees,' he told me. 'You didn't dare open your window. You didn't dare get out. You didn't dare do anything.' The fire chief pulled alongside the driver's cabin, to check he was okay. 'He was suiting up. I didn't get out either. We just waited a few minutes until he got out. He grabbed the hose.' Clad in protective clothing, the driver and fire crews first tried spraying the bees with water to calm them down. Frantic calls were made to bee experts, who suggested dousing the swarm with fire foam to kill them, so that workers could approach the truck and begin the clean-up operation.

Strandberg was worried about something else: grizzly bears. A. A. Milne, author of *Winnie the Pooh*, was not wrong to claim that bears love honey. They find bees tasty too. The highway was covered in honey, and with many bees still buzzing around, it would offer an irresistible lure. Bears on the highway would have been pretty dangerous, so the crews worked all day to remove the stuff, receiving a number of stings for their pains. Strandberg admits that he and his team were ill prepared: 'We train for fires and extrication, but we have never trained for bees. We have learned a lot. Next time it happens, if it happens again – hopefully not – we will know what to do.'

Which is just as well, because he and his colleagues could easily find themselves fighting escaped bees again. At certain times of the year, three or four trucks carrying beehives rumble along Highway 20 every week. Their destination: California, where the bees are required for pollination services. During my time in California researching mega-dairies, I learned about an extraordinary consequence of intensive farming taken to extremes: industrialised pollination – a business that is rapidly expanding as the natural bee population collapses. In certain parts of the world, as a result of industrial farming, there are no longer enough bees to pollinate the crops. Farmers are forced to hire or rent them in.

Both wild bumble bees and domestic honey bees are under severe threat. In the UK, which has around twenty-four types of bumble bee, two species have gone extinct in the last seventy years. Six are considered seriously endangered,[14] and half the rest are considered at risk.[15] The British Beekeepers' Association fears the UK could lose all of its bees within the next decade.[16] In the United States, several species that were common as recently as the 1990s have disappeared. It's the same story in other parts of the world.

The potential implications of the decline are mind-boggling. Most fruit and vegetable crops are dependent on pollination by bees. It means the future of around a third of global agricultural produce is at stake. Governments have been slow to wake up to the problem. In 2007 the US House of Representatives held an emergency hearing on the status of pollinators in North America and set aside $5 million for honey bee research, but the amount was later cut by half.[17]

Farmers can't afford to sit and wait, so they are taking their own desperate measures, hiring commercially reared bees by the truckload, at vast expense. It's desperate measures for desperate times. Wild bees, so essential for natural pollination, have been driven out by chemical-soaked monocultures, industrial farming methods that have robbed them of the varied habitats they need to survive. In their place is the business of bees for hire, at least for now.

Some clue as to the scale of this bizarre new industry lies in the fact that the accident in Idaho was not unique. Just a year earlier, in another American state, there was a similar crash near Minneapolis, Minnesota, involving a flatbed lorry carrying 17 million bees in 7,000 beehives. Hundreds of hives were jolted free.[18]

Arriving at the scene, Lakeville fire chief Scott Nelson opened his door and immediately got stung in the face – 'It was a black haze. Never seen anything like that.' The driver of the lorry told reporters that he felt a bang, looked in his wing mirror and 'saw boxes and beehives exploding'. Rescuers were forced to stay inside emergency vehicles until the swarm dispersed. Three hours later, they were still trying to clear bees from the area.

Many beekeepers now make more money from pollination services than from honey production. The load that fell off the lorry in Island Park, Idaho was worth an estimated quarter of a million dollars. Charges for bee services are soaring. Prices have tripled since 2004, with hiring a hive now costing as much as $180.[19] These runaway prices have been enough to push some farmers out of business.

Nowhere is the problem more pressing than in California, where the almond industry is so in need of bees that vast numbers of US honey-bee colonies are joined by bees flown in from as far afield as Australia.[20] Here, the industrial production of a single crop, or mono-culture, has been taken to new extremes.

Every year, in late winter or early spring, some 3,000 trucks drive across the United States carrying around 40 billion bees to California's Central Valley, which houses more than 60 million almond trees. The orchards cover around 240,000 hectares of land, stretching the best part of 600 kilometres and producing 80 per cent of the world's almond crop – the largest pollination event in history.[21] Buying in these services is costly: Californian growers now spend $250 million a year on bees.[22] It is yet another sign of how nature's support systems are breaking down in the wake of unsustainable farming techniques.

China has a different answer to the same problem. Every spring, the pear orchards in Sichuan province blossom, transforming the hills into a snowy white fairyland. Each flower carries the promise of a juicy fruit. For centuries, farmers had only to sit back and wait until bees and the summer sun worked their magic, delivering a bumper crop in autumn. Fast-forward to the twenty-first century however, and nature has stopped providing these services free. Instead, thousands of villagers troop up to the orchards and perform the task themselves, armed with simple pollination sticks made of chicken feathers and cigarette filters. They climb the trees and dip their sticks into plastic bottles of pollen, then dab the pollen onto each individual blossom.[23] With a huge population willing to toil long hours for little reward, this may be a reasonable solution. It's hard to see such a labour-intensive operation proving viable in the West, where labour costs are much higher.

While I was in California investigating mega-dairies for this book, I came across a commercial beekeeper who told me what it's like being in this business. The plight of the cows in the mega-dairies and that of the wild bees are linked; both have become victims of the industrialisation of the countryside.

Mike Mulligan, a third-generation Californian beekeeper, and probably the most cheerful person I've ever met, invited me and the team to his home. Isabel, my camera crew of two and I set off for our breakfast invitation. From the moment we arrived at his farmhouse Mike was virtually hopping with excitement about his English visitors. Every now and again he would exclaim 'hot *dog*' and slap me on the back, grinning from ear to ear as he zipped around his kitchen preparing a magnificent breakfast feast. There was pumpkin pie and a plate piled high with freshly fried cinnamon bread. The kitchen counter heaved with bowls of fruit, supersized jars of Nutella, peanut butter and maple syrup. There were plates of biscuits and English breakfast tea. He was a born-again Christian and his company was called Glory Bee.

His home – a luxurious detached property, tastefully decorated, with a glorious garden amid the sanitised almond orchards – suggested that business was good. Behind the property was an extensive backyard with various open-sided barns storing the paraphernalia of beekeeping. On a patch of scrubland, bordered on one side by almond fields and on the other by a field of commercial rose bushes, stood a couple of hives. At very short notice – our visit had been arranged over the phone less than twenty-four hours earlier – Mulligan had somehow arranged the hives especially for us to see: he explained that his own bees were currently 'overwintering' on the central coast of California where conditions were warmer. It was November and they were effectively on holiday, conserving energy for the busy season ahead. They would stay on the coast until late January or early February. By Valentine's Day, they would be home earning their keep. After providing pollination services for almond farmers, they would be moved on to the orange groves and from there to pollinate buckwheat and sage in the so-called 'chaparral', a heathland scrub in the Californian hills. They return to Central Valley in mid-July, to work on cotton and alfalfa, before their annual holiday by the sea.

It all sounded rather lovely, but Mulligan was blunt. 'There's nothing romantic about it,' he told me. 'It is a difficult business.' Though his father and grandfather were both beekeepers, he is the first in the family to hire out bees for pollination services: previously, the money was in honey. He owns 6,000 hives, each containing between 30,000 and 50,000 bees. In 2011 he rented them out for between $150 and $160 a hive. It might sound a lot, but the bees are fragile, vulnerable to parasites and chemicals, and he has had some close shaves, like the time he took them to another state.

That year, there had been a drought in Central Valley, and business wasn't great. He'd heard there was a better market in the Midwest. Mulligan and his bees travelled by lorry to North Dakota, the heart of

American's honey-producing industry, where he rented a house. From the start, it was a nightmare. He and his staff arrived in early spring, and were dismayed to discover that there was still snow on the ground, dangerous conditions.

They had not brought the right equipment to care for the bees in cold weather, in particular the pollen and nectar supplements they need. Mulligan was worried they would go hungry. Slowly the weather improved, and in May he began unloading the hives into the wild-flower meadows. One by one, his bees died.

> Before going, I'd been having all these fantasies about getting rich out of honey production. I got quite carried away. Then my bees started dying – thousands and thousands of them. We had no idea what was going on. I was embarrassed to admit to my family that we were having problems – I felt a fool. But it got to the point I couldn't keep it from my wife – I had to tell her what was going on. I was almost crying, begging her to pray for me. I could see our whole livelihood going.

It turned out that the bees had been feeding from a musky flower called Zigadenus, otherwise known as death camas. It proved deadly to a large number of his bees. In the end Mulligan didn't lose them all and the trip wasn't the financial disaster he'd feared. However, by the time he returned to Central Valley he was deeply shaken, and behind schedule with an important annual health treatment for the bees. The 'varroa' mite, a parasite around the size of a pinhead, is one of many threats to his livelihood. If it gets inside the hives and reproduces, it's deadly. He says he has little choice but to spray the hives with chemicals. 'I don't like doing it, but I can't afford the risk. When you see the accumulation of pesticides our bees are exposed to, it is a worry.'

In fact commercial bees suffer a near-continuous chemical assault, and Mulligan worries about the consequences. 'It is possible that we could build up lethal doses of chemicals in our hives, without even knowing it,' he believes. His bees are also at continual risk from pesticides, to which they are highly sensitive. He explains that citrus trees are sprayed just after they have bloomed, and that bees can be severely damaged by the toxic material. His contracts with farmers specify time lags between any spraying and the installation of hives among the trees. His bees are also registered with Kern County authorities, who hold a copy of their proposed itinerary. In turn, his clients and other farmers who hire bees submit their spraying schedules to officials, theoretically enabling beekeepers to avoid sending bees into recently treated crops. It is an inexact science: in certain weather, pesticide spray drifts to neighbouring fields, affecting bees working there.

On the morning of our visit, Mulligan received 100 queen bees through the post. They cost $16 a head and came all the way from Hawaii housed in tiny cardboard boxes with mesh screens at both ends. Apparently the queens are particularly vulnerable to chemicals. 'The queen is exposed for months and months,' he told us. 'We are noticing that they are under stress – their economic life is being shortened.'

There is still some debate over the cause of collapse of the bee population, but most experts attribute it to agricultural intensification, particularly the use of chemical pesticides.[24] Research suggests that what has been dubbed Colony Collapse Disorder (CCD) is linked to the use of a group of pesticides called neonicotinoids. These are water-soluble, nicotine-like chemicals which, when sprayed onto the ground, are absorbed by the entire plant, turning it into what has been described as a 'poison factory'.[25] Plants grow extremely toxic to insects and, of course, to bees.

In 2013, the EU voted to ban the use of neonicotinoids on crops attractive to bees. The UK voted against the measure, arguing that the science was inconclusive. Speaking after the vote, EU Health Commissioner Tonio Borg welcomed the ban: 'I pledge to do my utmost to ensure that our bees, which are so vital to our ecosystem and contribute over 22 billion euros [£18.5bn; $29bn] annually to European agriculture, are protected.'[26]

Dave Goulson, a professor of biological sciences at Stirling University and a world expert on bumble bees, also believes that intensification of farming has been critical to their decline:

> Bumble bees are major contributors to pollination of crops and wild flowers throughout the temperate northern hemisphere. Many species have declined, contributing to fears that we might face a pollination crisis . . . In Europe, the primary driver is thought to be habitat loss and other changes associated with intensive farming. In the Americas, declines of some species are likely to be due to impacts of non-native diseases.[27]

Artificial nitrogen fertilisers mean there is no need for the old-fashioned rotation of crops, most importantly clover, that bees used to forage on, and herbicides have eliminated most of the wild alternatives. Their nesting sites have gone too. Some species live in dense grass above ground; others prefer underground cavities – typically abandoned rodents' nests. The removal of hedgerows and unploughed field margins has put paid directly to the upstairs bees and indirectly to the downstairs ones by starving out the voles and mice that create their homes. Any that do find nesting places are likely to have them smashed by farm machinery or zonked by pesticides.

Bees have become just as much a victim of agricultural intensification as the pigs, chickens and cows crammed into factory farms. The industrial rearing of animals goes hand in hand with intensive crop production, often grown to feed incarcerated animals.

In the developing world, where small producers cannot afford to hire hives or pay for hand pollination, there are fears that reduced yields will cause malnutrition. The UN estimates that around 70 per cent of crops that together provide 90 per cent of food supplies depend on bee pollination.[28]

Dr Parthiba Basu, an ecologist at the University of Calcutta, has carried out a number of studies that show a link between falling bee populations and poorer crop yields. Basu became an international bee expert almost by accident, after stumbling on the connection during a different project. He was originally investigating whether it would be commercially viable for farmers in India to turn their backs on intensification, a policy that is being aggressively promoted by the Indian government. The answer was a resounding yes. His research involved eighteen farms in sixteen different regions of India, from the Himalayan foothills in the far north to the swampy southern tip of the subcontinent. The farmers who took part previously focused on just one or two types of crop or livestock production and typically used high quantities of pesticides and fertilisers. They were persuaded to experiment with mixed farming, rearing several different types of live-stock and several different crops in the same location. Although they did not go 100 per cent organic, they were gradually weaned off their dependence on chemical inputs. At the end of the year-long experiment, almost every farm involved had made more money, partly because diversification enabled them to be productive all year round rather than only during the fertile monsoon season.

The results were exciting enough in themselves, confirming Basu's suspicions that Western-style agricultural intensification, championed by the Indian government as the key to riches, does not translate well to India's unique and highly varied ecological conditions. However, there was another interesting finding: crops that were highly dependent on bees for pollination were not doing as well as expected.

I met Basu while he was briefly in London in autumn 2011 to attend an international bee conference in Sheffield. Deeply worried about the bee crisis and its implications for poorer people in India, he told me: 'It was particularly striking in Bengal, where they grow a lot of a vegetable known as painted gold. It's a highly pollinator-dependent crop. They have been hand-pollinating it for a while now, because there are just not enough bees left. The plants are not very high and grow in dense rows, so you see a lot of children doing the job. They can squeeze between the rows and don't have to bend down to reach the flowers.'

India produces about 7.5 million tonnes of vegetables a year – about 14 per cent of the world's total – and has undergone an agricultural revolution since the 1960s. Half a century ago, the subcontinent was heavily dependent on food imports. Now it produces more than enough to feed its own population, though there are still distributional problems, and malnutrition remains rife. Buoyed by this success, the government is trying to go further faster. Recently, Basu has begun to look at bee problems in other developing countries and has found a link between declining populations and deforestation: 'Forest cover is essential for maintaining pollinator services. Bees need trees to survive. If you mix the two factors – pesticide use and loss of forest cover – you see a very strong link.'

He is in no doubt that agricultural intensification is to blame:

I had hoped that pollinator loss would not be nearly as serious in developing countries as it is in the West, but that does not seem to be the case. It is very sad. It is going to take a lot of effort to turn this around, but unfortunately the developing world is going down the opposite route right now, embracing Western-style intensification. That means more mono-cropping and chemical fertiliser and pesticide use, and more loss of the wilderness habitats on which bees depend.

Achim Steiner, director of the UN's Environment Programme, believes that as a society, we are labouring under the illusion that in the twenty-first century we have the technological prowess to be 'independent of nature'. In a world of close to 7 billion people, as he has put it, bees 'underline the reality that we are more, not less dependent on nature's services'.

THE ETERNAL SUN DANCERS

Catching butterflies sounds a cruel and old-fashioned hobby, but in the United States these days it's quite the thing. In the nineteenth century, the Victorians would snare the insects in delicate nets, then kill them with a drop of opium on the head. The drug kept the butterfly flexible long enough for the wings to be opened and set out neatly without breaking. The collector would then mount his specimen on a board with a pin through the thorax and display it proudly in a glass case.[29] These days, the aim is not to kill butterflies and turn them into works of art, but to attach a tiny tag to their wings, then set them free.

It's hard to imagine that a creature so fragile could survive being trapped and handled and still have the strength to fly away. Yet if they are treated gently, it can be done, and as thousands of proud American owners of butterfly nets will testify, it's not only worthwhile, but fun. This game is Monarch Watch, an extraordinary conservation programme involving enthusiasts of all ages in northern states in the US.[30] Every autumn, in gardens, schools, colleges and country parks, butterfly lovers head out with nets to try to catch and tag as many monarchs as possible. The aim is to track their unique annual migration, a phenomenon that is now under threat because of the industrialisation of farming.

The monarch is an emblematic species in America, the only type of butterfly that the average North American knows. More importantly, it is the star of one of nature's most spectacular shows – an international migration that takes it thousands of kilometres from the northern states of the US to Mexico and back again. En route, these delicate creatures with wings like orange and black stained-glass windows face incredible challenges. Fluttering along at around seven and a half miles an hour, they negotiate motorways and chimney stacks, railway lines and pylons, and power stations belching noxious smoke into their path. They navigate over rivers and mountain ranges, through towns and cities, weathering storms and downpours. Millions die en route, blown apart in the slipstream of aeroplanes, splatted on car windscreens, drowned in puddles and birdbaths, burnt in bonfires and trapped in webs. Against all the odds, millions of others make it, and the cycle continues.

Every year, tens of millions of the butterflies embark on this incredible journey, always starting on the same date: the autumn equinox, when night and day are the same duration.[31]

Every year, they end up in the same place, the high-altitude pine and fir forests of the Neovolcanic Plateau, some 390 kilometres from Mexico City. This is where they spend the winter months, surviving on little more than water. Not one who makes the journey across the continent has ever done it before; nor has a single one of those that return to America when spring comes ever previously travelled north. Each butterfly embarking on the migration in either direction is a member of a new generation. Yet they are guided by some kind of internal navigation system, and they all end up in the same place.

How they do it is one of the world's great natural mysteries that scientists have yet to grasp fully, and time may be running out to do so. Monarch butterflies are in decline and their migration is under threat. The population is currently the lowest on record. The finger points at pesticides.

Because monarchs all hibernate together in the same part of Mexico, scientists have been able to assess the strength of their population by measuring the area of forest covered by the colonies. According to the WWF (World Wildlife Fund) in Mexico, the current colony size is just 1.92 hectares, down from the previous low of 2.19 hectares in 2004 and far down from the recorded high of 21.6 hectares in 1995.[32] Though there is no danger of the monarch going extinct, experts fear the numbers are becoming so sparse that the migration will either disappear altogether or no longer bear any resemblance to the magnificent spectacle it is today.

At the heart of the problem is an unremarkable-looking plant called milkweed, with clusters of pale pink flowers. It used to be very common on arable farms. Monarchs love it, and won't lay eggs on anything else. Unfortunately, however, it is a weed, and farmers don't like it. The use of genetically modified crops, engineered to resist super-strength herbicides that destroy pretty much everything else, including milkweed, has all but wiped out the plant in many areas.

Dr Orley 'Chip' Taylor, an insect ecologist at the University of Kansas, is a director of Monarch Watch and a world-renowned expert on butterflies. I rang him to ask him what he believes is going on. He told me that the tagging programme run by Monarch Watch is now very large, but the number of butterflies volunteers can find to tag has dropped by almost a third in the last eight years. He is in no doubt that pesticides are to blame, describing the evidence as 'solid':

> Monarchs are dependent on milkweed plants, and milkweed plants are weeds. They used to be controlled by tillage – farmers would mechanically till to get through the weeds, and milkweed would survive the process to a certain extent. There would be a small number of plants per acre that would survive in corn and soybean fields. I've got pictures of those types of situations and you can see clumps of milkweed here and there within the corn and soybeans. These turned out to be extremely productive areas for monarchs. But in 1996 they developed a herbicide-tolerant soybean, and then they came up with herbicide-tolerant corn . . . Eventually what they did was eliminate the milkweed. At the same time as herbicide-tolerant corn and soybeans have been adopted, we've seen the decline of milkweed and a very strong decline in monarch populations.

Monarchs are the only butterfly species to migrate like this. Nobody even knew about their amazing journey until the 1970s, when the full wonder of their behaviour was revealed by a biologist who devoted his life to figuring out where they went.

Fred Urquhart began trying to track the movement of individual monarchs in 1937. At that stage very little was known about their migration, except that thousands headed south each September. Urquhart's day job was as a curator of insects at the Royal Ontario Museum, although he also had a position at the University of Toronto. At his home in the 1940s and 1950s, he and his wife Norah, who also worked at the university, reared monarchs by the thousand, and studied them in the hope of safeguarding their future. Their positions in the faculty of zoology gave them much-needed access to a laboratory and equipment to step up their research.

The Urquharts decided to try to track the monarchs on their autumn journey by marking and releasing individual butterflies, then plotting the distances and locations where they were seen or found. The couple spent several years experimenting with different tagging systems before they found something that worked, a sensitive adhesive label that would not come off in the rain. The tags were easy to apply by gently squeezing them to the wing. On each tag were the words 'Send to Zoology University Toronto Canada'.

In 1952 Norah wrote a magazine article appealing for volunteers to help with the tagging programme and twelve people responded. It was enough to start up an organisation, which they called the International Migration Association. The volunteers were given tags and instructions, and monarchs were soon being caught, tagged and released. Not long after, little boxes began arriving from all over Canada and the United States, containing tagged monarchs. It was proof that the creatures really could fly hundreds of kilometres.

By 1971, 600 volunteers had joined the association, with thousands more taking part every year. The Urquharts were now amassing a great deal of information, and were able to establish that the monarchs were migrating northeast to southwest. However, where they ended up for winter was still a mystery. Desperate to identify the butterflies' final destination, the couple went on numerous field trips to search for the hibernation site. Their travels took them from New England to the Californian coast, and from southern Canada to the Gulf of Mexico, with no luck.

The following year, 1972, Norah wrote to newspapers in Mexico about the project, appealing for volunteers to report sightings and help with tagging. It led to a breakthrough. A young American engineer called Ken Brugger who was working in Mexico City at the time saw the article. He didn't know much about monarchs but he was an amateur naturalist and keen to help. Over the next two years, he spent his weekends and holidays driving around the mountains of Mexico, searching for the monarchs' secret winter destination. In 1974, Brugger married a woman called Cathy, herself a butterfly lover, and together they continued the search, taking their clues from tattered monarch remains they came across along the roads in a certain area. When they showed some specimens to a group of Mexican loggers, they were told there was a place in the mountains where many such butterflies could be found.

On 9 January 1975, all was finally revealed. An overjoyed Brugger rang the Urquharts to say that he and Cathy had at last tracked down the butterflies' roosting site – 'We have found them – millions of monarchs!' He was breathless with excitement and no wonder, because the colonies are a truly amazing sight. Great clumps of butterflies coat entire tree trunks and branches in what looks like jagged orange and black fur. Most of the butterflies rest, wings folded, on the pines, but others swirl between the trees in dizzying masses. People who have witnessed this amazing sight say the biggest surprise is the noise of all the beating wings. It sounds like rushing water.

Understandably, the Urquharts were eager to see this extraordinary spectacle for themselves. They arranged to travel to Mexico the following winter. Despite being in their sixties, they were determined to climb the 'Mountain of Butterflies' – their reward for forty years of research.

Seeing the colonies of monarchs they had spent a lifetime seeking must have been an amazing moment, but there was one further magical surprise. While the Urquharts were staring in wonder at the fluttering masses, a pine branch collapsed under the weight of the insects, scattering a huge cloud of butterflies. Among a cluster that spilled at their feet, the Urquharts found one bearing a white tag. The butterfly had been tagged in Minnesota by one of their volunteers before setting off on its journey to Mexico. It was the final proof they needed. There could no longer be any question that the monarchs seen in North America were the same butterflies that spent the winter over three thousand kilometres away in Mexico.[33]

Since then, large monarch populations have been found at thirteen overwintering sites spread over five mountains. But the destruction of milkweed raises serious questions over their future, and it is only one of several threats they face. The other serious danger is illegal logging, which is destroying their winter habitat in Mexico.

Five hundred years ago, the Aztecs believed that the souls of the dead returned as butterflies. They called them 'eternal sun dancers', and treated them with care. Today, though the roosting sites are ecological reserves protected by the Mexican government, trees are still being felled. The pines and firs provide shelter for the butterflies, protecting them from fluctuations in temperature. If the canopy becomes too thin, they are exposed to storms and cold blasts that are often fatal.

The Mexican government has ploughed several million dollars into protecting the colonies, running campaigns to persuade locals in the 50,000-hectare area where the colonies spend the winter that the butterflies are worth far more as a tourist attraction than the trees are as wood. In the first year of the campaign, which began in 2007, the WWF reported a 48 per cent drop in illegal logging, but the decline of the colonies has continued.

When I spoke to Taylor about what is happening, he was deeply worried: 'Here we have one of the world's most magnificent biological phenomena, and to lose this would be absolutely drastic. I mean, this is one of the most spectacular events that happens on the planet.' He believes it is crucial to keep working with the Mexican authorities: 'Maintaining the conditions in Mexico is extremely important, and becoming more so as the population of butterflies goes down. These populations down there in Mexico can suffer seventy to eighty per cent mortality under extreme conditions as they overwinter, and if the population becomes really small, it will be tough coming back from a mortality event like that. Recovery is going to take several years.'

I was lucky enough to see the monarch migration myself while leading a wildlife tour in Cape May, New Jersey. We were there to watch the birds of prey that pass over those skies in September on their way back south. The bushes were full of tiny warblers who were also on the move, a marvellous sight. But the clouds of beautiful orange and black butterflies were simply breathtaking. A month later, I was on the Isles of Scilly off southwestern England. As I picked my way round a little airfield bordered by the Atlantic Ocean searching for migrants of the feathered variety, to my amazement I saw a monarch fluttering strongly over the grass. It must have come all the way from America, probably blown off-course. It is just mind-blowing to think that something so delicate could make it thousands of kilometres across the ocean.

Yet these superb migrants are under threat; and it's not just monarch butterflies that are suffering as a result of intensive farming. In the UK, farmland is the main habitat for three-quarters of butterflies, but according to a report by the Butterfly Conservation Trust, Natural England and the Farming and Wildlife Advisory Group, 'Typically, they now survive only on relatively small areas.' The organisations describe wild-flower meadows as vital. Such areas provide habitats for up to twenty-five different species, including those that are rare and declining. The advice is that no fertilisers or pesticides should be used in these pastures.[34]

In a recent report, the government-funded UK Butterfly Monitoring Scheme, which tracks the well-being of fifty-four types of butterfly, found that thirty had declined over the last ten years. A total of twenty-one species declined by more than a quarter, while twelve species declined by half or more. Victims include iconic British species such as the red admiral, painted lady and holly blue.[35]

One of the landmark extinctions in Britain when I first became interested in wildlife was that of the large blue butterfly. Decades of conservation effort came to a bitter end in 1979 when the species was declared extinct. Intensive farming was blamed for destroying much of its habitat.[36] It was particularly regrettable because the species is threatened globally. Not to be deterred, conservationists reintroduced the large blue to the British countryside in what was described as 'the world's largest-scale, longest-running successful conservation project involving an insect'. Twenty-five years later, its delicate pastel-blue wings can be seen again over more than thirty former sites.[37] It goes to show that declines can be reversed; but how much better to avoid them in the first place.

As for the monarch, there are large, non-migratory populations in other countries, so the future of the species seems assured. Yet its decline is symptomatic of a far deeper malaise. It would be devastating if the migration, a great natural wonder of the world, disappeared. As Taylor puts it:

> It's not that monarchs are important in themselves, it is that monarchs are symbolic of how we're doing. It's an iconic species that is symbolic of the biodiversity problems throughout the United States. If we can't support the monarch butterflies, it means we're not supporting a lot of things, because monarchs share habitats with virtually all pollinators in the US and they share habitats with a lot of small mammals and birds.

Like the proverbial canary in a coal mine, warning of impending danger, butterflies, like birds and bees, are the hallmarks of a healthy environment. Their plight is part of the complex web of life that underpins food and farming.

5

Fish

Farming takes to the water

PLAYING SCALES

It was one of those moments when everything seems just right. I was on the front deck of a glinting white catamaran, soaking up the sun with my new wife Helen. We were sailing through the turquoise seas off Mauritius, enjoying the tropical clear waters, seductive white beaches, cold drinks and cocktails. It was a heavenly experience and the honeymoon I'd dreamed of. The skipper turned the boat towards Ile aux Cerfs, a tiny island said to boast Mauritius's most beautiful beach. There was a party atmosphere. We were aboard with several other honeymooners, sipping beer and bottled water, sharing wedding stories and laughing about how you can spot male newly-weds because they fiddle compulsively with their wedding rings.

There was a splashing in the water. Dolphins! Everyone stopped talking and turned to look at the frolicking bottle-nosed dolphins swimming ahead of us. I was so excited I almost spilled my drink. Helen and I had been desperately hoping we'd see some on this trip. Just then, Helen spotted something strange in the distance: 'What's that over there?' she asked. I peered through my binoculars. It was a fish farm. In an instant, I was back in work mode, staring at the green pillars and white railings of the floating pontoons. They jarred on the landscape, a bum note in an otherwise perfect symphony. No doubt thousands of fish were caged in giant net-bags beneath the surface. I thought about the inevitable outpouring of muck from thousands of caged fish and how this might affect the pristine waters.

Mauritius has become a hot spot for fish farming, with twenty or more sites chosen for development by the island's Ministry of Environment.[1] Locals are all too aware of how this might affect the water. Mauritian environmentalists have voiced concerns that the feed and toxic chemicals used in intensive fish rearing will cause pollution, foster disease, and might even attract sharks, seriously jeopardising the all-important tourist industry.[2] There are also concerns for the health of coral reefs. Dr Deolall Daby, Professor of Marine and Environmental Sciences at the University of Mauritius, fears the effect could be fatal: 'We have to be very careful. It is because of corals that there are beaches. They are so important in the whole functioning and protection of the shores.'[3]

A tropicbird flew past our catamaran like a sharp-dressed gull, gleaming white with its long tail streaming like a ribbon stiff on the breeze. We'd had a wonderful day, one that I would never forget. But I couldn't suppress a nagging sense that the fish farms we'd seen were just plain wrong. The wildlife and environment in Mauritius have already taken a ferocious battering. The carpet of sugar cane throughout the countryside is a hint to why the native birdlife has its back to the wall. Some of the rarest birds on the planet are to be found here but are on the brink of dying out. The Mauritius Kestrel and Echo Parakeet, for example, are down to their last few hundred. That Mauritius still appears so idyllic is testimony to nature's resilience. I learned something on that otherwise perfect day: that industrial farming had taken to the seas in paradise. Until then, it was something that I associated much more with salmon farming on the remote coasts of Scotland.

In the shadow of snow-capped mountains, surrounded by bracken and Caledonian pines, lies Loch Maree Hotel, on the pebbly shores of one of the world's loveliest lakes. Ever since Queen Victoria stayed there in 1877, the hotel in Wester Ross, Scotland has been a popular destination for tourists drawn to the wild and lonely scenery.

The loch itself, the fourth-largest freshwater lake in Scotland, has long been thought to have special powers. In the late eighteenth and early nineteenth centuries, it was thought to hold some kind of magical cure for lunacy. The mentally ill were brought to the water as a last resort. The ritual involved tying the patient into a boat, which then set off for a small island in the loch called Isle Maree. The boat would circle the island three times – always clockwise – and on each lap, the unfortunate patient would be plunged into the freezing water. The vessel would then dock at the island, where there would be a further ritual by the remains of an ancient chapel, graveyard and holy well, thought to date back to Pictish times. It involved the patient drinking water from the well, and an offering made by nailing a rag or ribbon to an ancient tree. A visitor who witnessed the rites in 1772 told how the patient was forced to kneel before a crumbling altar and sip from the well, in a process that was repeated every day for several weeks in the desperate hope of a cure.[4]

In modern times, Loch Maree's star attraction has been something rather more tangible: fish. Every summer, thousands of adult sea trout and salmon used to swim back to the loch from the sea. They gathered in huge numbers in the bays, providing phenomenal angling. The loch developed a reputation as the world's finest sea-trout fishery,[5] a place where it was possible to land a trout weighing close to 9 kilograms. The record salmon landed in Loch Maree weighed in excess of 13 kilograms.[6] Loch Maree Hotel was a thriving business with ten boats and a team of Highland fishing guides or 'ghillies'. During high season, it employed as many as nine full-time ghillies. After sundown, a throng of anglers would gather round a roaring open fire in Ghillies' Bar, a glass of malt whisky in one hand, comparing notes about the day's catch. Now the hotel is deserted, because the fish have gone.

These days you'd be lucky to land a trout weighing more than one kilogram – as a recent expedition to test the state of wild fish stocks showed.[7] On 17 September 2010, around a month before the end of the fishing season, three boats set out from Loch Maree Hotel, each with an experienced ghillie, to assess a sample catch. According to a report by the Wester Ross Fisheries' Trust, which organised the trip: 'The day was cold, with showers and NW breeze. 21 trout were retained for measuring (all trout were subsequently released) and scale samples were taken to confirm identity. Andrew Ramsay caught the largest sea trout, a fish of 875g, nearly 2lb.'[8] In other words, tiddlers are all that remain. The sea-trout fishery in Loch Maree and nearby Ewe River has collapsed, leaving the owners of the hotel forced to find other ways of attracting guests, and the people whose livelihoods used to depend on the angling industry out of work.

Nick Thompson, who owns the hotel lease, says he barely gets one angler a week. The hotel is currently closed for renovation: his latest idea is to turn the place into a beauty spa. He's clear it has no future as a paradise for those who love the rod and line. When fish stocks vanish, the usual story is overexploitation, but at Loch Maree it seems anglers who got so much pleasure from landing one or two prime specimens every season are not to blame. The real culprit? Fish farms – there are two nearby.[9] One way or another, it appears these intensive fish-production operations have driven wild fish stocks to destruction. It is a picture that is being replicated all over the world.

Tucked along remote coastlines and river valleys, fish farms are the forgotten factory farms under the water, and one of the fastest-growing sectors of intensive animal rearing. Around 100 billion farmed fish are produced globally every year, 30 billion more than all the chickens, cows, pigs and other terrestrial farm animals reared worldwide.[10] Looking at it in terms of volume of flesh generated, in 2009 the world produced just over 80 million tonnes of chicken meat from an estimated 55 billion chickens; fish farming, known as aqua-culture, now produces about 70 per cent of that amount – using twice the number of animals.[11]

As natural fish stocks all over the world reach critical levels of deple-tion, there is a widespread assumption that fish farming is a happy solution. However, far from helping preserve wild fish stocks by taking pressure off the sea, fish farming requires ever more plundering of the oceans for the smaller fish fed to carnivorous forms of farmed fish, like salmon and trout. Bluntly, these kind of fish farms waste fish, rather than generate a net gain.[12] The figures speak for themselves. According to industry data, it takes between three tonnes[13] and five tonnes[14] of small fish to produce one tonne of farmed fish like trout and salmon. Supplies of wild fish are in danger of running out; according to the FAO (Food and Agriculture Organization), over half of all wild fish populations are already 'fully exploited' while almost another third have been overexploited.[15] In 2008, 23 per cent of the total global fish catch was small pelagic fish of the open ocean used to make fishmeal and fish oil, largely for farmed fish. In other words, around a fifth of the world's fish catch is effectively being wasted feeding other fish.[16]

Worryingly, a very significant proportion of this ground-up fish – around one-third of the total – is quietly being fed to other farmed animals,[17] usually chickens and pigs. So if your chicken nuggets or pork chops have a strange aroma, that could be why. Over the last few decades, the production of fishmeal and fish oil, mostly to feed farmed fish,[18] has removed around 20–30 million tonnes of fish from the sea.[19] It is usually made from pelagic fish like anchovies from the southeastern Pacific Ocean, and herring, mackerel and sprat species. Contrary to popular belief, fishmeal largely doesn't come from off-cuts and processing waste that isn't any good for anything else. In fact, most fishmeal is made from highly nutritious stuff that could happily be eaten by people. Indeed, the type of fish involved already provides half the fish for direct human consumption in over thirty countries.[20]

The use of pelagic fish for animal feed contributes to overfishing of other species, and thus to the current overfishing crisis in the oceans. It's not even as if the supply of pelagic fish is unlimited. According to the FAO, the two main stocks of anchoveta in the southeast Pacific, Alaska pollock in the north Pacific and blue whiting in the Atlantic are fully exploited. So are several types of Atlantic herring, while stocks of Japanese anchovy in the northwest Pacific and Chilean jack mackerel in the southeast Pacific are also running out.[21] Either way, it is clear that fish farming is a very inefficient use of limited resources, with a poor ratio of 'fish in:fish out'.

The industry is slowly beginning to recognise that it's unsustainable to continue feeding wild fish to farmed fish. Alarmingly, some people are suggesting the use of livestock meat and bonemeal as a possible alternative. Indeed, in areas where aquaculture is growing rapidly – China, Vietnam, Bangladesh and India for example – farmed fish are already being fed on 'cakes' that contain manure from intensive terrestrial animal feedlots.

At the Global Conference on Aquaculture in Thailand in 2010, scientists highlighted the threat to pelagic fish stocks and warned of a 'crisis in feed qualities' for farmed fish. Looking ahead to 2050, they predicted increasing use of material like chicken slaughterhouse waste and chicken manure to feed farmed fish.[22] In other words, the fish-farming industry may increasingly take to feeding fish with chicken muck. It is hard to see how this would be acceptable to Western consumers. I suppose it would depend how much they were allowed to know.

Like other factory farms, fish farms involve animals kept in intense confinement – up to 50,000 salmon in a single sea cage. Often suffering from blinding cataracts, fin and tail injuries and body deformities, and infested with parasites, they are forced to compete for space and oxygen. Salmon are reared in stocking densities equivalent to a single bathtub of water per 75-centimetre fish. Packed this tightly, these natural wanderers of the ocean swim as a group, or shoal, in incessant circles around the cage, in the same way that frustrated zoo animals pace their enclosures. Fins and tails are rubbed sore as the fish press against each other and the sides of the cage.

Life for farmed trout is even worse. Generally reared in freshwater 'raceways' or earth ponds, they can be kept at densities of 60 kilograms of fish per cubic metre of water. It's the equivalent of twenty-seven trout, each measuring the length of a ruler, being allocated a single bathtub of water. More normal rates are 30–40 kilograms of trout per cubic metre of water. The stress of crowding and confinement makes the fish highly susceptible to disease. In recent years, salmon farming has been hit by a number of serious outbreaks of illness, killing millions of fish. Mortality rates among farmed salmon are particularly alarming, with average losses in sea cages ranging from 10 to 30 per cent.[23] This kind of death rate would raise serious alarm bells in other farm animal species, but for some reason people care less about the welfare of fish. There is a sense that because they're not mammals or birds, it doesn't matter as much if they suffer.

I remember an experience that illustrated this attitude quite vividly. I was invited to Strasbourg, Alsace, to the Council of Europe, which promotes cooperation and human rights among European governments. Officials were drafting common standards for fish farming and I was there to observe. I spent a lovely couple of days in Strasbourg's narrow cobbled streets and half-timbered buildings. I marvelled at the famous Gothic cathedral that towers up in the city centre and enjoyed the varied local cuisine, making sure to avoid the ubiquitous foie gras made from force-fed geese. However, most of my time was spent inside the rather less aesthetically pleasing Council building, a modern whitish affair flanked by colourful flagpoles of member countries. On my way to and from the building, I liked to look out for the white storks in the nearby park.

The meetings were long and intense. European government officials in grey suits would pore over complicated texts, arguing over whether a sentence should read 'shall' rather than 'should'. 'Should' was a wishy-washy guideline, whereas 'shall' meant something had to be done. Sometimes, as 'expert' observer, I would talk into the microphone and get the right result: I wanted as many 'shalls' as I could get. That way, I felt we were at least beginning to get some protection for a kind of animal that frequently gets none at all.

Strasbourg sits within the Rhine Valley straddling the border with Germany in wonderful wine country. Our itinerary involved a day trip to look at examples of inland fish farms, presumably a rewarding break for staying the course of the interminable, bureaucratic discussions that occupied most of our time. I assumed EU officials in our party must surely have seen them many times before, or they wouldn't have been in a position to draw up legislation on the sector.

Along with thirty or forty government officials, advisers and vets, I boarded a coach for the day trip. Our first stop was a carp farm in the Rhine Valley. It had two types of pond, one for fish, the other for the tiny organisms – or zooplankton – that young fish like to eat. Apart from their early life, when the 'fry', as they are known, are packed tightly in tanks, the carp were generally reared in a fairly relaxed way, in unremarkable-looking earth ponds where they were left to grow by eating whatever they could find.

Our next stop was a farm rearing trout in rectangular concrete ponds surrounded by grass. The water came from a local spring and flowed through successive ponds, each one lower than the next. Each pond had bigger fish than the last. They seemed in reasonable condition, as did the water. You could see the bottom easily, though by the time it finished going through the farm, it was probably less pleasant. The fish were fed by hand twice a day every day except Sunday. I wondered whether the Sunday fast was for the benefit of the fish or the owner.

Our third and final stop was billed as a 'modern raceway site'. As I could probably have guessed from the euphemism, this was the factory-farm version of trout rearing. We were told that it came complete with automatic oxygen injection systems. It certainly needed them. Compared with the previous farm, three times as many fish were being reared in the same space. The water was a murky brown. The fish were barely visible, let alone the bottom.

I stood incredulous as government people, vets and other experts watched fish in obvious distress without a word of criticism or alarm. Lots of the fish had worn tails from rubbing the sides of the tanks and each other. Some had raw red flesh exposed above their tails; others had no tails at all. They were crowding round the water inlets, desperate for oxygen. The grids across the inlets churned the water, allowing oxygen to dissolve more readily. Schools of larger fish swam in tight lethargic circles, their backs reaching high out of the water. They were clearly not happy. One expert told me that this was probably due to 'clogged gills', which made for 'uncomfortable fish'.

I chatted with the farm owner, a tall amiable man in workaday wellington boots. He said he was happy to answer any questions and let me take photos. I took pictures of this stark factory-style place in the open air. It was very different to the two farms we'd seen earlier that day.

I saw fish with their eyes apparently popping out. A fish vet who was standing beside me said that they were probably suffering from 'impact damage' or infection. I took photos, much to the distaste of one of the other vets. He asked in a voice that barely concealed his anger whether I would use the photos publicly. He repeated the question a couple of times. I wasn't willing to give him any reassurances. He didn't like the idea that the pictures might be used to show what goes on in such places.

In the end, I never did use the photos. The water was so murky you couldn't really see anything. Yet ten years on, the memories of that day remain vivid. I can still see those fish in my mind's eye, with their missing tails and bulging eyes, swimming in appalling conditions. I can see a huddle of vets and legislators looking down, and saying nothing. I can remember how odd it felt getting back on the bus, with no one else concerned by what we'd seen. Vets are seen by many as the animals' champions; that some seemed unfazed by such poor rearing conditions really shocked me. I couldn't help wondering whether the veterinary profession was failing somehow to properly stand up for animal welfare, instead following and thereby supporting the status quo. In the complex moral maze presented by farming on land or water, it is easy to get lost in the profit bottom line: to lose sight of the patient – the animals – when dealing with the customer – the farmer. It's a theme we'll return to in chapter 6.

Since my trip to Strasbourg, the fish-farming industry has boomed. By 2009, nearly half of the total supply of fish on the global market came from fish farms. Between 2004 and 2009 alone, production tonnage increased by almost one-third. Regionally, Asia accounts for almost 90 per cent of global production, followed by Europe (4.5 per cent) and Latin America (3.3 per cent). In Asia, fish farms mostly produce carp and tilapia, though species of prawn and shrimp, as well as largemouth black bass, catfish and turbot, have also been introduced. In the West, the fish-farming industry is primarily driven by demand for salmon and trout.[24] Scotland's west coast is jam-packed with fish farms. It's a highly concentrated industry. Two-thirds of the market is reportedly controlled by the 'Big Four' fish-farm companies, the largest of which is Marine Harvest. Like its competitors, Scottish Sea Farms, Lighthouse Caledonia and Grieg/Hjaltland, it's owned by companies listed on the Oslo stock exchange.[25]

But for wild fish, such as salmon in Loch Maree, fish farms often pose a grave danger, primarily because they are hotbeds of sea lice.[26] These parasites, which look like tiny tadpoles, are one of the single biggest problems for salmon farms.[27] When they latch on to fish, they eat away at the skin and scales. The effect around the head can be so corrosive that the bone of the living fish's skull can be exposed – a condition known as the 'death crown'.

Wild salmon get rid of the lice naturally, because they drop off when the fish migrate into fresh water. Mature wild salmon also have a covering of mucus that repels the lice.[28] Farmed salmon don't have these advantages. Ridding them of lice often involves bathing them in powerful chemicals. Because farmed fish are kept in such high concentration, it doesn't take long for lice to multiply and spread. Young wild fish can tolerate brief exposure to sea lice, but when they migrate past salmon farms, they are often exposed to sea lice for several weeks, allowing the lice burden to accumulate to fatal levels.[29]

According to The FishSite, a fisheries industry news and information website, when wild salmon swim from the open ocean back inland to spawn, they have to pass through river mouths and inlets 'clogged with salmon farms'. Some wild salmon carry parasites, which don't bother them. However, when they swim past the cages, the parasites can jump ship. Due to the high density of salmon in the cages, once sea lice enter they reproduce like mad. Essentially, according to The FishSite, they become 'a reservoir of infestation'. The next generation of tiny smolts juveniles face this infestation with none of the mucus, scales or other armour against the sea lice of adult salmon, and many do not survive.[30]

According to the Atlantic Salmon Trust, in the Highlands and Islands of Scotland wild salmon are outnumbered by farmed salmon by a ratio of more than 700:1, creating more than 61 million potential hosts for parasites and pathogens.[31] Guy Linley Adams, an environmental lawyer employed by the Salmon and Trout Association, told me:

> The problem is when you've got a fish farm with 400,000 to 500,000 adult fish, all of which have a couple of female lice and all of which are producing eggs, then you can imagine the wild juvenile fish that are coming down the rivers have got to swim right past these cages, and are faced with this subsurface cloud of sea lice. And if you're a five- or six-inch-long wild smolt and you get ten lice on you, that's enough to kill you.

The evidence that sea lice from fish farms are to blame for the catastrophic decline of wild fish stocks is long established. Back in 1999, a study by the Scottish Environment Protection Agency (SEPA) concluded that in relation to salmon, the link should be 'accepted as beyond reasonable doubt'.[32]

A study carried out in British Columbia, published in 2006, looked at how lice affected juvenile wild fish on three different migration routes, each containing several fish farms. The farms were the primary source of lice throughout the entire migration route. The scientists concluded that if outbreaks of sea-lice infestation continued in British Columbia, 'local extinction of wild salmon is certain, and a 99% collapse in pink salmon population . . . is expected in four salmon generations'.[33] In 2009, the industry body, the British Columbia Pacific Salmon Forum, accepted that more needed to be done to protect wild salmon.[34]

A review for the industry that same year acknowledged that it would be hard to 'find anyone who would claim that lice from the farms *aren't* a factor in wild salmon survival' and that the industry needed to tackle the problem.[35] Peter Cunningham is a biologist at the Wester Ross Fisheries Trust, an organisation that is trying to restore the wild fisheries in Loch Maree and surrounding waters. I talked to him about what has happened to wild fish stocks and the role played by fish farms. 'There's been no fish over five pounds [2.3 kg] since I've been here, and in the old days, they were common in Loch Maree,' he told me. 'In the 1980s, something like 12 per cent of the sea trout over 1 pound [0.5 kg] were over 4 pounds [1.8 kg]. We see nothing like that nowadays. The fish aren't surviving as long at sea as they used to. There are a few one- or two-pounders around, but we find they have quite high numbers of sea lice on them most years.' Cunningham acknowledges that other factors have played their part in the demise of Loch Maree's natural fish population. However, Wester Ross Fisheries Trust believes that if the lice issue were resolved, there would be a 'substantial recovery' in the sea trout population.

As part of these efforts, some years ago the farms were moved a few kilometres further north up Loch Ewe. However, according to Cunningham it has not resolved the problem:

> It's helped a bit, but it doesn't seem to have made a difference yet in terms of getting the big sea trout back again. We're still getting some quite heavily affected sea trout when we sample them each year. Some of the fish may be going outside the Loch and picking up lice from other areas – it's hard to be specific about whether all the lice are coming from a single fish farm. There may be some drifting in from other areas.

Of course, fish farming is not just about trout and salmon. An article in the scientific journal *Nature* in 2008 pointed out that as the industry expands into other species, such as cod and halibut, 'the same concerns over disease and parasite transmission, and other impacts, will certainly apply to these species too'. The piece concluded that it is vital to assess the potential environmental costs and reduce them before the introduction of very large-scale farming of these new species.[36]

Fish farms pose another major threat to wild stocks: escapees. An incident in Scotland one icy November evening shows what can happen. As they were being hauled out of the water and off to market, 24,000 Scottish salmon had a lucky escape. The net ripped and the captives lost no time in breaking free and swimming off into the blue-black waters of Loch Ewe, never to be seen again.[37] For the escapees, it was a stroke of good fortune. But for wild fish in Loch Ewe and the surrounding area, the sudden influx of 24,000 farmed fish was a potentially deadly threat. As the company that owned the fish publicly acknowledged at the time, the great escape was deeply regrettable, and not just because of the wasted money.

Escapes from fish farms happen regularly as a result of careless handling or damage to nets and cages by storms and predators. Occasionally, farms deliberately release economically unviable young salmon. Farmed trout are also capable of mass breakouts – in 2000 there were six reported escape incidents from Scottish Rainbow Trout farms involving over 63,000 fish.[38]

Farmed salmon have been selectively bred since the 1970s to develop certain 'desirable' traits, such as faster growth rates and later sexual development. Large-scale escapes mean that former farm fish compete with wild stocks for limited food and places to spawn. Some anglers who accidentally catch farmed fish in Scottish rivers have discovered that the animals have salmon eggs and parr (young freshwater salmon) in their stomachs – suggesting they are preying on their wild counterparts.[39]

There are also worries about interbreeding. In 1991, the UK Salmon Advisory Committee expressed concern that interbreeds may result in offspring with a reduced ability to cope in the natural environment. This is because wild fish are genetically adapted to their surroundings. Even moving a salmon to a different river can reduce its ability to survive. This suggests that mixing wild and farmed salmon genes may make wild fish less robust. In 1999, SEPA declared that such 'genetic pollution' should be recognised as a 'real and present danger'.[40]

It certainly seems that farmed fish can be woefully ill-adapted to the real world. There have been a number of sightings of farmed salmon apparently stranded or caught out by the tide. Commenting on this odd state of affairs, the marine biologist James Mortimer told the *Orcadian* newspaper that it was 'unheard of for wild fish to be ignorant of the phenomenon known as tide'.[41]

Because of these concerns, fish-farm escapes are now closely monitored. Companies must immediately report accidents to the authorities. The Scottish government reported five escapes in 2009, in which a total of 88,000 fish broke free from seawater Atlantic salmon sites. The situation may be improving. For 2010, the Scottish Salmon Producers' Organisation, which represents 95 per cent of the industry, reported that around 15,000 fish had escaped, the lowest figure since statutory reporting began in 2002, and just 0.1 per cent of the total number of salmon farmed. Most of the escapes seem to have been due to 'human error or holes in the nets'.[42] It remains to be seen whether these far lower figures are just a blip, or the result of genuine improvements in the system.

The impact of the fish-farming industry on wild fish stocks might be less depressing if farmed fish were as tasty and wholesome as the organic alternative. Unfortunately that's not the case. Farmed salmon and trout, for example, contain significantly more fat than wild alternatives. Figures from the US Department of Agriculture show that farmed Atlantic salmon has more than twice the fat content of its wild cousin, and farmed rainbow trout is up to 79 per cent more fatty, while protein levels are comparable.[43] A study of chemicals in farmed fish led by Ronald Hites, a professor of environmental and analytical chemistry at Indiana University, in 2004 caused widespread alarm after it exposed worrying levels of contaminants. Hites's team analysed over two metric tonnes of farmed and wild salmon from all over the world and found that concentrations of chemicals were 'significantly higher' in the farmed samples. The conclusion was damning: 'Consumption of farmed Atlantic salmon may pose health risks that detract from the beneficial effects of fish consumption.'[44]

Chemicals are also used to make farmed fish an attractive colour. Wild salmon and trout eat crustaceans and algae, which make their flesh pink. To achieve this complexion in farmed fish, colourants are added, otherwise their flesh would be grey. Synthetic pigments – canthaxanthin and astaxanthin – are used, either individually or in combination.

Like other types of factory farming, the fish-farming business is designed to extract as much flesh as possible with minimum input. In the early 1990s scientists began exploring ways of getting round the fact that fish stop putting on as much flesh once they hit puberty. The problem was that they were maturing before they reached market size. The answer? A technique called triploidy, now widespread on trout farms. It's a procedure that involves creating fish that are sterile; technically neither male nor female.

Trevor Whyatt, an avid sea angler who runs a trout farm in Dorset, has been in the vanguard of triploidy since the 1990s and is a staunch defender of the practice: 'The problem is that maturing fish lose their appetite, and put all their energy into their reproductive system. The environmental benefits of fish that do not mature in this way are massive – you reduce your dress-out [gutting] loss from twenty-five to sixteen per cent. At the same time you gain back nine per cent of the food that would have been used.'

The process itself is relatively simple and can be performed by farm-hands. Eggs and milt (fish sperm) are collected from the trout and artificially fertilised under heat. The raised temperature prevents a natural process taking place in which a second set of chromosomes present in the egg disappears. Instead, that second set of female chromosomes is retained, along with a set of chromosomes from the male, resulting in offspring with three sets of chromosomes instead of two. These fish are infertile. It's a defect that sometimes occurs naturally; the technology is designed to guarantee it.

Whyatt believes that in some ways it's better for the fish, though he admits it makes them more vulnerable to temperature changes. 'They tend to be slightly more placid. It makes for a more harmonious population, which are easier to manage,' he said.

There is one distinct advantage: any escapees can't breed. For this reason, the Environment Agency has ruled that from 2015, all farmed brown trout in UK rivers must be triploid, to protect wild fish stocks. Whyatt, who is passionate about protecting wild fish stocks, has been pushing for this since the 1990s:

> I first saw the potential benefits of sterile fish a long time ago. There's a wonderful lake in the West Country with a great stock of brown trout, but there are only two small streams feeding into the lake. The problem was that all the fish wanted to spawn there at once, and the nests of those that spawned earlier were being destroyed, because the population was too dense. If half of the fish in the lake were sterile, they would not destroy the nests. Productivity would rise and mortality would fall.

He sees it as an opportunity to use a viable industry tool to enhance and restore the genetics of wild fish. It is always welcome when new technologies are used to solve environmental problems, rather than create them. However, there is evidence that triploid fish have higher levels of deformities, breathing difficulties, low blood haemoglobin levels and a lower ability to cope with stressful situations.[45]

The industry is all too often bad for other marine life, an issue highlighted when two decapitated seals were washed up on a beach on the Isle of Skye. It was a gruesome discovery that prompted a public outcry and the attention of the police.[46] Nigel Smith, a Skye businessman who runs award-winning wildlife-watching cruises in the area, found the carcasses – a pregnant female and a juvenile – as he was walking along the beach near a fish farm. 'Their heads had been blown off,' he told me. He has grown used to coming across grisly finds involving seals. The incident, in May 2008, threw the spotlight on one of the most unedifying and emotive aspects of the fish-farming industry: its impact on seals.

Huge numbers of fish in one place present an irresistible temptation to seals, as well as to birds, otters and other wildlife. Some farms see shooting these creatures as a legitimate way to protect their stock. In 2008, both common and grey seals were officially protected in Britain under the Conservation of Seals Act 1970. However, the legislation only applied during the breeding season and excluded fish farms, so that operators could kill seals all year round to protect their cage nets from damage.

For a long time, seal culling by fish farmers was largely unregulated. All that was required was a high-calibre rifle and a gun licence. Now, under new legislation, it is supposed to be a means of last resort, but there is considerable evidence that some in the industry continue to regard a bullet as the cheapest and easiest way to get rid of the problem.

John Robins, an animal welfare campaigner in Scotland, argues that many salmon farmers do not bother to invest in alternatives like protective nets to keep predators away from the fish:

> I first learned about this back in the 1980s when I got tip-offs from people in the industry. At that time, we worked out – and it had to be a rough estimate because there were no official figures – that roughly three thousand seals minimum were being killed by fish farmers every year. That was a guesstimate based on about ten seals per year per farm. I'd been told by one group of farm workers that they had killed over sixty seals at one unit in one year, and we had other information that made us think that 3,000 was probably a rather conservative figure.

In 2006, fish were given official recognition under the Animal Health and Welfare Scotland Act, a move that animal welfare campaigners such as Robins backed. It obliged fish farmers to take adequate steps to protect their stock from predators. Some in the industry have seen this as a green light to get out their guns. There is another way: anti-predator nets. Robins believes these are far more effective than bullets at keeping seals out: 'You'd have to be shooting seals twenty-four-seven to fully protect the fish. You really do need anti-predator nets.' The legislation makes it clear that shooting a seal should be the last resort.

The problem is that anti-seal nets are expensive to buy, install and maintain. Using the Freedom of Information Act, Robins has discovered that 80 per cent of Scottish fish farmers do not have anti-predator nets and even fewer use them. 'That does not make shooting a seal a last resort, it makes it very much a first or second resort,' he said. The Scottish fish-farming industry denies using shooting as a routine method of controlling seals, arguing that a single rogue seal can cause enormous suffering and kill thousands of fish.

The Scottish government claims that the number of seals being shot has fallen dramatically in recent years. However, in March 2011 it gave the go-ahead for the culling of more than a thousand seals, issuing sixty-five licences to kill up to 1,298 seals in total that year. The shooting goes on in remote areas, the government relying on the honesty of the shooters to provide accurate records of how many seals they kill. Robins believes: 'It's a true saying that when you buy a Scottish salmon you pay for bullets to shoot seals.'

PLUNDERING PERU

I was heading into the Peruvian desert when I discovered that my local guide and fixer for the week had a loaded gun in his pocket. Stefan Austermuhle, a German zoologist who runs wildlife expeditions and bird-watching tours out of Lima, turned out to have a rather more buccaneering lifestyle in mind: he was about to ditch the green front-line and become a gold prospector, at least for now.

His impending shift of status from penniless conservationist to potential kidnap target focused his mind on his personal security. Experience in this country had apparently taught him that carrying a pistol at all times was just basic self-protection. His investigations into environmental destruction in Peru had made him the target of death threats, a concern not lost on the police here, who gave Stefan a special gun licence and training to boot.

I did my best to hide my surprise as I made some rapid mental adjustments about both Stefan and Peru. He was clearly no mild-mannered ecologist, and my trip might be more of an adventure than I'd bargained for. It was the first of many unexpected developments on my expedition to investigate the impact of one of Peru's biggest businesses: exporting ground-up fish to China and Europe to be fed to farm animals.

Fishmeal is one of the filthiest secrets of the factory-farming industry, an environmental catastrophe that involves sucking millions of tonnes of small fish out of the sea and crushing them into fish oil and dry feed for farmed fish, pigs and chickens. The process deprives millions of larger wild fish, birds and marine mammals of their natural prey, drastically depleting stocks of important species. It also pumps vile fatty waste into ocean bays, creating 'dead zones'; pollutes the atmosphere around processing plants, causing widespread human health problems; and diverts what could be a highly valuable source of nutrition for people to industrially farmed animals.

There is nowhere better to see it in action than Peru, the second-largest fishing nation in the world after China. The country makes upwards of a billion pounds a year from a single species of fish, the anchoveta or anchovy, which is ground into fishmeal.[47] It has the biggest commercial fishing fleet in the world geared towards the exploitation of a single type of fish and exports more than a million metric tonnes of the stuff every year, making it the leading global fishmeal supplier.[48] The UK consumed 135,000 tonnes of fishmeal in 2010,[49] a third of it imported from Peru.[50]

Stefan was my fixer – a highly experienced biologist based in Peru for thirteen years. He had witnessed the impact of the fishmeal industry on marine life and birdlife for himself and seemed an ideal person to accompany me. He had worked with many international documentary makers and had the contacts I needed to delve beneath the surface

of the industry. As I was to discover, after more than a decade struggling to secure funding for conservation projects, he now wanted to
make serious money for his green campaigns by setting up what he
described as Peru's first ethical and environmentally friendly gold mine.
Until that was up and running though, he was available as a guide.

My primary destination in Peru was Chimbote, capital of the fishmeal industry and a port that lands more fish annually than the entire
Spanish fishing fleet.[51] I flew into Lima from Buenos Aires and spent
a few hours in the city getting my bearings. Perched on a cliff between
the desert and the Pacific Ocean, Lima seemed an odd mixture of a
place: ugly but cheerful, the city centre dominated by an entertaining
mixture of Sixties, Seventies and Eighties architecture.

I wandered downtown from our hotel, to find myself in a buzzing
backpacker district of cheap shops, money changers and juice bars. It
was Sunday and the vibe was relaxed, no one moving faster than a
stroll. The soundtrack was low-volume Europop from tinny old ster-
eos, interspersed with the odd 'poop' from battered-looking 1970s
local buses. As I made my way down to the seafront I found myself
amid concrete flyovers, multi-storey car parks and grey municipal
buildings. Overlooking the sea was a long strip of high-rise flats, their
blue- and black-tinted glass façades glinting in the sun. I stood at the
cliff edge and looked at the sea. There was no beach to speak of, just a
thin fringe of dark pebbles and grey-black sand. Some kilometres away
at the end of the curving bay was a barren hill shaped a bit like Cape
Town's Table Mountain, covered in pylons. I must admit that I was
disappointed. Peru had been high on my list of 'must-see' places, but
the capital was a let-down.

Before we left for Chimbote, Stefan offered to take me to Asia
Island, a rocky outcrop off Lima that once teemed with birds. Apparently the birdlife had been devastated by the anchoveta industry. He
told me it was a 'guano' island, uninhabited, with a dull white coat
consisting of thousands of years' worth of bird droppings. In some
places, the guano used to be 90 metres thick, evidence of the extraordinary number of seabirds that once lived there.[52]

In the nineteenth century, when cormorants, boobies and pelicans covered the rocks, the bird muck was a valuable commodity. By the 1850s it was Peru's biggest export and the government's principal earner. Some 20 million tons were shipped overseas between 1848 and 1875, half of it to Britain, where it was prized by farmers as a rich organic fertiliser.[53] It was an industry fuelled by abundant marine life, not least the anchoveta, which provided food for tens of millions of seabirds. Then the fish stocks were plundered; Stefan told me that most of the birds were now gone.

We picked up a 4x4 hire car in Lima and headed for Pucusana, a traditional fishing village not far away, where Stefan kept a motorboat. It's renowned for a good day out and its cebiche: salads of fresh raw fish. Dozens of little fishing boats were busy around the harbour, sorting their nets or unloading fish at the local market. People, dogs and birds wandered between market stalls and crates of newly landed fish. Pelicans flapped at the harbour edge looking for scraps.

I walked down the slipway, where the smell of seafood mingled with fumes from the boats. A wizened woman in a wide-brimmed hat stood under a blue gazebo, pulling huge lobster-coloured crabs out of a bucket. She and others were busy preparing food for sale: cockles, mussels, mackerel, a single tuna next to a small shark.

Stefan was worried that fog was rolling in and wanted to get going, so we clambered into his twin-engine motorcruiser and chugged out of the harbour. Almost at once we passed our first commercial fishing boat, about 30 metres long, its rusty black hull piled high with black nets. It was a purse-seine vessel, using a net to surround shoals of fish before the bottom is drawn shut like a purse. The mass of wriggling silver fish would then be lifted aboard by two elongated crane arms, which stood like idle sentries.

We passed independent fishermen bobbing their lines from rocks. The fog was closing in. I marvelled at the seabirds dancing above. Inca terns bounced on buoyant wings, their exquisite markings like sooty sea swallows. Through my binoculars I could see their waxy-red beaks, bright yellow at the base, and what looked like a curly-white pencil moustache. Gulls swirled about the boat. I saw two types of cormorant, and a rare Humboldt penguin standing high on some rocks. Gannets known as Peruvian boobies plunged and dived.

It thrilled me when a huge, shiny brown body, thick whiskers and bulldog-like face burst out of the water: a sea lion. In the mist, it all seemed other-worldly. A playful pod of bottle-nosed dolphins joined the boat, their back-combed fins glistening. They stayed close, swimming diagonally across our path, as if to say we weren't going fast enough. Stefan cranked up the speed and the dolphins kept the pace, four side by side, bow-riding; sometimes breaking the surface, sometimes dancing below. They were so close I could almost touch their smooth silver-grey bodies. I could see the scars and notches on their fins. Local biologists use pictures of these like fingerprints.

Small fish leaped, sprinkling the water like a handful of corn, probably trying to escape the dolphins. And then we were at Asia Island, now protected as a national nature reserve, uninhabited but for guards. As we approached, a warden on the rocks gestured at us to come closer. He'd received a tip-off of trespassers on the sanctuary's no-go zone and wanted help to investigate. As a thank you he allowed us to spend a few minutes ashore.

After the exhilaration of the boat ride, it was a depressing experience. The protected status had clearly come much too late: though there were still a few birds round the perimeter, it was unrecognisable from the place described in historical records. I saw rotting pelican carcasses, and a booby too ill to move. The strong smell of guano – dry, musty ammonia – was the only clue that it had once teemed with thousands of birds.

Today, guano is still harvested, but in much smaller quantities. The state-owned company that collects the deposits is said to be struggling to survive. Stefan told us: 'Around the middle of the last century, there used to be forty million seabirds across twenty-eight islands off the Peruvian coast. Now they have only 1.8 million birds. The decline in bird numbers correlates with the rise of the fishmeal industry. All these seabirds feed on anchovies. Since the anchovy industry, there has been a ninety-five per cent drop in bird numbers in sixty years.'

Seabird population crashes in this area are nothing new. Oceanic shifts in currents can have devastating effects on food sources for marine birdlife, leading to sharp but temporary declines. The waters on the Peruvian coast are not warm tropical seas but cool upwellings known as the Humboldt current, which are highly favourable to marine life.

During El Niño years however – a climatic pattern that occurs across the tropical Pacific – the ocean shifts and warmer waters reach the islands, causing the marine food chain to break down. Plankton disappears, along with the anchoveta that feed on it, and seabirds starve. After each natural crash, populations recover. They were hit hardest in 1957–8 when numbers declined by 70 per cent. Five years later, they'd pretty much sprung back. Lasting decline is something else. It seems that overfishing, primarily of anchoveta for fishmeal, is preventing recovery.[54] 'The island is now empty,' Stefan said. 'The guano workers of days gone by would lift clouds of birds into the air; so many, they would darken the sky for hours.' He described the current picture as a 'permanent El Niño low'.

Peru learned the perils of overfishing as long ago as 1970, when the industry collapsed. It prompted the government to introduce a quota system to ensure that some fish are left to spawn.[55] The quotas seem to be preventing anchovy stocks from further decline, but numbers remain at a base level compared with sixty years ago – hence the desperate state of marine birds. The Peruvian government would like the outside world to believe the anchovy fishery is now well regulated and sustainable. Others tell a different story.

Back in Lima after our boat trip, I met Victor Puente Arnao, a forty-three-year-old marine lawyer based in the capital. He told me that in 2010 his law office won a public contract to work on 12,500 cases of fishery violations. They had a combined fine value of US$200 million. I was not surprised when he added that the government ended up following up on less than half of these. Apparently the breaches all involved anchovy vessels. 'The authorities are not capable of regulating the quotas. The people that work in the ministry are involved in the fishing industry themselves. This leads to a very soft treatment of the companies concerned,' he alleged.

The 1970s fisheries crash also saw the start of dolphin hunting, a practice that has since been officially outlawed. However Stefan's own investigations suggest that between two and three thousand dolphins a year are still being killed illegally for human consumption.

Earlier in the day, while still on the boat, I had spotted a broiler-chicken farm on the mainland, one of many in Peru; the birds possibly being fed fishmeal. Whether it be here or oceans away, it was a reminder of the big picture: marine life sacrificed for factory farms.

Next morning we set off for Chimbote. I had been promised spec-tacular views along the 400-kilometre route, but it was a long time before they materialised. Lima's suburbs sprawled on and on along the traffic-choked highway. We ground along at a snail's pace through congestion worse than rush hour on the M25. All along the road, packed stiflingly close together, were flats and houses, many of them little more than shoebox-shaped brick sheds, stacked two or three high. Billboards with scantily clad women advertising who knows what loomed above them. Hawkers plied cigarettes, grapes and oranges. Children careered on rusty bikes. It was chaotic and colour-ful, exciting and slightly intimidating, teeming with life.

Finally we shrugged off the city and found ourselves in an arid semi-desert. Strings of pylons marched up scrubby hills. There was the odd hut, matchbox-small and rudimentary, though little sign of anyone at home. Some of the hills on the outskirts of the city had been turned into free advertising hoardings: scraped into the sand were names of politicians in huge letters, presumably carved out in the run-up to elections and never removed after polling day. A few kilo-metres beyond the city limits, we passed a new settlement, thousands of colourfully painted low brick huts, huddled close together. Stefan said it had sprung up within the last five years, a giant squat created by speculators. He explained that the inhabitants were bussed in by mafia types who illegally staked out the land and sold people 'leases'. For a while, he said, the settlement was entirely run by armed thugs who controlled entry and exit. It became a refuge for criminals on the run. Eventually the government moved in, built a police station and legitimised the place.

A few kilometres further along the highway we came across the first of scores of broiler-chicken factory farms. Rearing chickens for meat, from the outside at least they were are all the same: rows of long low sheds with saggy black or grey canvas roofs and black plastic tarpaulins slung haphazardly over the sides, presumably to provide shade. We didn't attempt to poke around: Stefan's gun was evidence enough that angry Peruvians can't always be relied on to waste time with words when they want you off their property. Further along the road we passed a field of burning sugar cane. Dark red and orange flames leapt 6 metres high, billowing yellow-black smoke over the plantation. The fire made a snapping, crackling sound. We passed through occasional small green valleys, heavily irrigated, with scrubby palm trees, tropical flowers and fields of maize.

And then, 200 kilometres out of Lima, we entered another world: the Peruvian desert, an extraordinary lunar landscape, desolate and endless. We were in the foothills of the Andes, but the mountains were made of sand. The different hues and textures were beautiful: no longer just barren brown scree, as outside the city, but undulating hulks in subtle shades. Some were great smooth dunes that looked as if they were made of soft brown sugar; others were more solid, rubbly, gunmetal-grey, pinkish-orange or ancient red. On the flanks were occasional hard ridges where rich seams could be found: of gold. It was a landscape unlike anything I'd seen before. We stopped to film. That's when I spotted some shells in the sand. The ocean was far away. Shells in the Andes: surreal.

Clinging to the road were occasional signs of life: the odd rustic shack, the walls just flimsy canvas, like those fold-up sun shelters that holidaymakers use on the beach, the most rudimentary of family dwellings. Now and again we passed roadworks, monitored by men clad in full-body protective clothes like chemical warfare suits, the labourers fully masked as they stood by the road, waving red or green flags. They were an eerie sight, but the heavy gear was vital in these conditions: every passing lorry threw up a wave of sand and grime in its wake.

As we drew nearer Chimbote, the dry mountains to the west fell away into smaller softer dunes, and suddenly we saw the South Pacific. The juxtaposition of desert and sea, the ocean stage-lit steely grey-blue by shafts of sunlight breaking through great puffy clouds, was a dramatic sight.

The journey through this ethereal scenery was exhilarating, which was just as well, because our destination most certainly was not. On reaching the outskirts of Chimbote, the largest town in the Ancash region of Peru, we were greeted by a polluted marshland strewn with litter and occasional pools of filthy stagnant water crusted with yellow algae. A sun-bleached sign over the main road welcomed us to the capital of the fishing industry. The air was humid, laced with the sharp stink of fish. My heart sank as we drove in.

The hotel where we would spend the next few nights looked quite promising from the outside. Built on the seafront, away from the bustle of the town centre, it had a cheerful pillar box-red façade, fancy white wrought-iron lamps and a mosaic porch, and it enjoyed spectacular views over the bay to a number of small guano islands. Behind the façade, however, was a much bleaker prospect. It was like a deserted government institution or student hall of residence, with long soulless corridors leading to desperately drab rooms of the kind you would expect in a backpacker hostel. Mine had dim strip lighting, brown curtains and dirty lino flooring. Its crowning glory was a portable fan which, when switched on, was so loud it sounded like an aircraft revving for take-off. A jaunty sticker on the stand said 'High Velocity Cozy Air'. It was no match for the suffocating heat. Night was falling.

I wandered down the corridor feeling uneasy. I seemed to have the whole floor to myself. Dark staircases led to more deserted accommodation. On the ground floor, several unlocked doors led to an unlit car

park. Stefan had made no secret of Chimbote's reputation as a dangerous place. However, the hotel was the best the town had to offer, so I
shut out negative thoughts and hunkered down for the night.

Next morning, after breakfast served by a middle-aged waiter who
appeared to be deaf and wore the expression of a man who had just
swallowed a rancid mussel, we headed to the harbour. We had
arranged to meet some local shellfish divers who were going to take
us out in their boat. They wanted to show us the contrast between
the rich pickings on the seabed a kilometre or so out in the bay, and
the toxic sludge towards the shore where the fish-processing factories
spew waste.

In the bay were perhaps a thousand fishing vessels of differing
shapes and sizes, all tied up. During the season, commercial vessels
typically head out to sea for several days at a time. Few if any have
refrigeration facilities on board, so that fish caught at the start of the
voyage are often rotting by the time the ship returns with its haul.
Boats dock at floating platforms equipped with pumping machinery
that sucks up the catch and pipes it over to the factories. The process
of turning it into fishmeal is complex. It involves steam-heating the
raw fish to sterilise it and squeeze out the liquid; separating out the
valuable oil from the rest of the fluid; and drying the heated fishmeal
into pressed cakes. The drying process is highly energy-intensive; the
waste – blood, guts, scales, fat – is repulsively discarded. The leftovers
are disposed of in rudimentary ways; typically it seems by pumping
the raw filth into the sea. The pipes through which the sludge flows
quickly clog up, like fatty arteries. Some processing factories are said
to use caustic soda to flush them out at the end of the season – a toxic
chemical that also ends up in the water. The effect of all this is a seabed
without oxygen: a dead zone.

At the harbour, a small market was in full swing, attracting a posse
of pelicans. The birds stood in a neat little row on the edge of a warehouse roof, their eyes fixed on some rusty blue barrels that contained
fish guts. Every time a trader flung offcuts into the bins there would
be a great flapping of wings. Out of the noisy melee a victor would

emerge with a slimy piece of flesh in its long beak and waddle round the corner to devour it.

We found the sea-clam divers, who led us to their boat, a basic affair with a small motor and no life jackets. On the bow was a huge fisherman's knife, so thick with rust that you could not tell it was once another colour. The benches were sprinkled with gull droppings. I sat on an empty fishmeal sack and we set off out of the harbour. The water was mercifully flat as we headed towards L'Isla Blanca, a guano island not far from the shore, where the seabed remains in reasonable condition.

Out in the bay, pelicans flew low over the boat, showing us their white bellies, and neotropical cormorants swooped and dived for fish. Through my binoculars I spotted a blue-footed booby, common on the Galapagos Islands but not so much around here, and an American oystercatcher. An empty fishmeal bag bobbed past us on the water. The midday sun was ferocious. There was no escape from its unforgiving rays.

As we approached the island, the divers cut the motor. I noticed a greasy film on the water. At first, I assumed it was pollution from our own boat, but the divers told us it was fat from the fishmeal factories. It shimmered gloopy yellow in the sun like the top of a bad takeaway meal. The diver, wearing an old black wetsuit, prepared to go overboard, carrying a small black bucket in which to collect what he could find on the seabed. He would get his oxygen through a plastic pipe leading to a motor on the boat. He plunged in, emerging a few minutes later with his bucket. In it were several dozen small crabs, clams and black snails, as well as a creature that looked like a rubbery yellow snooker ball coated in mucous slime. The diver poured them all onto the floor. The point was made: plenty of life on the seabed here. I signalled my appreciation, then scooped up the crabs and snails, managing to avoid being nipped, and dropped them back into the water. Trying not to wince, I also picked up the mucus-covered thing and sent it on its way too. We headed off towards the bay lined with fishmeal factories.

As we neared the shore, we came close to a number of deserted docking platforms and anchored commercial boats with names like *Susana* and *Mary Carmen*. The diver plunged back in, emerging this time with a bucket full of black gunge. I ran my hands through it, feeling the sludgy, slimy texture. It smelled of sulphur – rotten eggs. I poured it back. I was told that the muck is several metres deep here – too thick to allow life on the seabed any significant recovery when the boats are tied up. It was a total dead zone.

Until the 1990s, nobody bothered collating any data on this environmental disaster. The Peruvian National Fisheries Research Institute now carries out some research. An acoustic study of Ferrola that it undertook in 2008 concluded that there were around 54 million cubic metres of organic matter – toxic sludge – in the bay.

The National University of Santa in Chimbote also has an ongoing research programme. Romulo Aguilar, Dean of the Faculty of Biology at the university, has spent years studying the impact of effluent from fishmeal factories on marine life. We met him at his faculty, where he told us about his work:

> To the south of Chimbote is another bay, called Samanco, which is exactly the same shape and size as the Bay of Ferrola [lined by fishmeal factories]. Geologically, it is so similar as to be an excellent yardstick for how Ferrola could, or should, be. Our tests show that Samanco is still an extremely rich ocean system, with wonderful natural shellbanks and fish, crustaceans and algae. It is very high in biodiversity. In Ferrola, all this is gone.

Aguilar's team has measured what are termed oxygen-demand values in Ferrola Bay. It's a standard way of assessing water-pollution levels. The results showed oxygen-demand values of 56,000 micrograms per litre. I asked him what that meant. 'It shows pollution levels are off the scale,' he replied. Aguilar claims the bay is in such bad shape that even the tiny organisms normally seen as an indicator of high pollution cannot themselves survive. 'We see no sign of life, nothing,' he lamented.

The government has attempted to force the companies to clean up their act, to little effect. While a few are said to have improved their sewage-disposal technology, according to Aguilar most continue to operate much as before, discharging raw waste directly into the bay. Speaking about the use of caustic soda to clean out pipes at the end of each fishing season, he said: 'I have had several rows with the companies at meetings, where they have emphatically denied doing it. However our analysis of the water shows pH values at a level that clearly indicate it's being discharged. I am in no doubt it is going on.'

So thick and raw is the waste as it is pumped out to sea that it has spawned a cottage industry: fat collection. On a beach near some of the processing plants, desperadoes have made holes in the sewage pipes and can be seen siphoning the effluent into a network of vats. When the fat floats to the surface they skim it off and sell it for the equivalent of a few dollars. It sounded a foul and desperate way to make a living. We went to see.

Stefan and our cameraman had been to the beach before with a documentary team from Norway and had never forgotten the experience. They warned me that once I'd been there my boots were likely to be unwearable, so we went to a local market and bought the cheapest footwear we could find before setting off.

The approach to the beach was down a ramshackle lane in the suburbs. We took the car as far as we could and walked the final stretch, through hot sand strewn with bits of concrete, broken bricks and discarded tyres. To either side were fishmeal factories, hidden behind high walls with jagged broken glass along the top and watchtowers at each corner. As we reached the beach, the smell of decaying fish hit me like a punch in the stomach.

A gang of workers sat on a wall and eyed us warily. In front of them was a rudimentary system of tanks and vats, full of yellowy brown liquid. Every surface was coated in sticky black slime. A turkey vulture swooped low over the vats of fat. A pile of rotting fish lay abandoned on a bin bag. It was a squalid place.

The workers were hostile, wanting money. Officially what they were doing was illegal, but their facilities were well established. Far from anyone challenging them, it seemed to us that their operation was actively supported by the fishmeal companies. After lengthy negotiation the foreman agreed to talk to us. He told us that he and his team paid the companies for the right to 'treat' their water. He argued – with some justification – that he and his men were performing a public service by reducing the amount of fat pumped into the sea. He told us more than a thousand people were trying to make a living this way.

We asked whether fishing would be preferable to this dirty work and he laughed bitterly. 'Everything has gone. It's no longer possible.' He was pessimistic about the future, saying that he and his team would all leave Chimbote if they could. 'It is not possible to clean up the bay. This is all we can do. We need help to get out of here. Our children are suffering; they are being poisoned. The only reason we don't leave is because we can't afford to.' His claims of sickness and poisoning of children might sound exaggerated but are well supported by the evidence. Life expectancy in Chimbote is 20 per cent below the national average.

Dr Wilber Torres Chacon works for the department of health in the region. We met him at a nursery school not far from the beach, where he carried out health checks. It was a cute little place, with lots of bright wooden and plastic toys, colourful foam matting on the floor, and a lovely bed for naps under a spotless white mosquito net draped from a pretty hoop. The children wore coloured bibs according to their age. They looked healthy and happy, but Chacon told us all was not as it seemed: 'The fishmeal production activity here causes several health problems: severe respiratory infections, asthma, acute diarrhoea, malnutrition, parasitic diseases . . .' He examined some of the children, pointing out unusual-looking lesions on their skin. 'These children are constantly exposed to fumes from the factories. You can see the blemishes on their skin.'

Malnutrition is widespread in Peru, but rates in the Ancash region are particularly high, at between 20 and 30 per cent of the population.

Though the government has been attempting to tackle the problem for several decades, Chacon said progress was slow. In the last two years, despite increased efforts, rates of malnutrition have dropped by just 2 or 3 per cent.

Among other nutritional programmes, health chiefs and NGOs are trying to encourage locals to eat more fish – a habit that seems to have died with the rise of the fishmeal industry. As little as 1 per cent of the highly nutritious anchoveta caught off Chimbote is likely to end up on dinner plates,[56] a shocking fact, given how desperately the local population needs the protein. 'The population here consumes excessive amounts of carbohydrate, and not nearly enough protein,' the doctor informed us. 'Officials here on the coast are providing the infant population with fish to boost their nutrition, especially anchovies for the protein and non-saturated fats.' He told us that seven out of ten children in Chimbote suffer from skin conditions, with the figure rising to 90 per cent among those living next to fishmeal factories.

Such is the scale of concern about the health problems in Chimbote that some officials now believe the only solution is to move the entire fishmeal production industry out of the city. Chacon spoke of relocating all the factories to an industrial zone, 'far away from the population'. Chimbote's failure to provide adequate nutrition for its own population – a quarter of Peru's infants are malnourished,[57] while the country exports millions of tonnes of fish – is an irony not lost on everyone in the industry. Javier Castro Zabaleta, general secretary of the local fishing trade union, believes that the government needs to educate Peruvians in the value of eating fish instead of turning it into fishmeal for farm animals: 'We don't have a culture of eating fresh fish such as anchovy, even though it's so high in protein and we have it in such abundance. Peruvians are not used to eating darker flesh-coloured fish – we prefer white fish, chicken and other meat. We need to convert the anchoveta industry step by step into an industry for human consumption.'

We met Zabaleta at the union's headquarters on a main street in Chimbote. He sat behind an old wooden desk under a poster of Che

Guevara. Now in his sixties, Zabaleta was a weather-beaten man of the sea. Next to his desk was a ponderous bookcase that looked as if it might keel over at any moment. An old portable fan in the corner barely ruffled the air. It was hot in the room. A fisherman served us Coca-Cola in plastic cups and offered us biscuits.

Zabaleta told us his union represented 1,700 fishermen. In its heyday it had 5,000 members, but many big fishmeal companies now have in-house employee associations. He echoed what the fat collectors had told us about how hard it now was for independent fishermen to make a living: 'Before the fishmeal industry arrived, we used to fish for anchovy four or five miles off the coast. It was very productive and we did two trips a day. Now seventy to eighty per cent of the fishing here is industrial.' Because of overfishing, to find anything 'you need to fish two, three or even five hundred miles offshore. That is illegal for artisanal fishermen – they are only allowed to operate a maximum of fifteen miles offshore.'

He described illegal fishing as a persistent problem, though he said things were improving. In the last five years, the quantity of fish landed in breach of quotas had fallen from 3 million tonnes a year to 600,000 tonnes. 'The authorities don't enforce quotas. Some companies cover up the real weight of fish they unload. Either they don't register it when they bring it into port, or they fix the scales.' He claimed the most disreputable operators hired professionals to beat the system, deliberately underweighing catches by as much as 40 per cent.

Our last meeting in Chimbote was with a local heroine, Maria Elena Foronda Farro, who enjoys celebrity status among many ordinary folk in the town. She has campaigned against the corruption and environmental damage associated with the fishmeal industry in Chimbote since the 1990s. For her pains, she spent a terrifying spell in prison after being falsely accused of belonging to the Peruvian terrorist organisation Shining Path. Convicted and sentenced to twenty years behind bars in 1994, she was released after thirteen months following an international outcry spearheaded by Amnesty

International. Few doubt that her incarceration was due to her outspoken activism.

In 2003 she was awarded the prestigious Goldman Environmental Prize,[58] an international award for outstanding grassroots activists, but it took her two attempts to collect it: the first time she tried to get into America she was turned away at immigration because of her 'terrorism' conviction. Despite offers of asylum from eight countries she has never left Chimbote, where she was born and raised. Her home is two blocks away from a fishmeal company.

Now in her early fifties, Foronda continues to campaign through her organisation Natura to raise environmental standards in the fishmeal industry. She works on a shoestring budget out of a run-down little office in town, helped by a dedicated and loyal team of volunteers. I met her there and we went to a local restaurant to talk over lunch. 'Our first victory was forcing a number of the fishmeal companies to relocate out of the centre of Chimbote,' she told me. 'We managed to get an order against them in 2009, and in 2010 they began to move. It was a big thing – they had to build entirely new plants.' One of the biggest companies reportedly spent £27 million moving its premises. Foronda shrugged off the figure as insignificant relative to industry profits – 'In one season the combined fishmeal industry here makes 1,800 million US dollars profit. There are two fishing seasons a year, so annual profits are double that. Building a new plant is really no big deal. In any case, when they install new technology, they improve their yields.'

According to Foronda, twenty-six companies already located in the industrial area were given until the end of 2010 to clean up their technology, but as of March 2011 only eight had done so. She told me cheerfully that campaigning often meant two steps forward, one step – sometimes two – back. 'Just when we thought we were getting somewhere, the regional government started authorising the building of new low-grade fish-processing plants, arguing that it would create jobs. The national government turned a blind eye when factories opened that were dirtier than ever.'

Her hope is that the industry cleans up its act and begins to reinvest some of its huge revenue in the town and its citizens. 'The fishmeal business has left us without much to live on, plundering our natural resources and failing to put anything back into development in Chimbote. The bay used to be considered the pearl of the South Pacific. Look at it now.' As we talked, a cheerful procession was snaking its way through the streets outside the restaurant. At the front was a brass band, dressed in glorious red and black uniforms. Behind was a raggle-taggle bunch of people of all ages, carrying banners and balloons and tooting klaxons. It was a protest against a local oil company, accused of damaging fish stocks. In Chimbote, it seems, the fishmeal processors were not the only ones seen as villains.

I was so glad to see the back of the place that as we got back on the road to Lima later that afternoon, I was singing. We cranked up the volume on a CD of Eighties rock songs, and I was giving it gusto, only wishing I had my guitar. Chimbote symbolises all that is wrong with factory farming and its long tentacles that reach out across the world. I never wanted to set eyes on the town and its godforsaken industry again.

As dusk fell, we reached the desert. It looked so eerily beautiful it took my breath away. The sun slowly disappeared, turning the sandy mountains rose-purple. The moon hung in the sky. Soon there was nothing to see from the car window but inky black. I was monumentally relieved to be on the way home.

But Peru had one last surprise in store. On the outskirts of Lima, as we pulled away from a road tollbooth, there was a sudden jolt accompanied by a sickening crunch. Somebody had driven into the back of our hire car. Exhausted and overwrought after five hours dodging kamikaze lorry drivers in the dark on the winding desert road, Stefan was in no mood to be messed with. He leapt out of the car, flagged down the offending driver, and an angry confrontation ensued. It soon turned out that the other driver was drunk and refusing to supply us with his details or to stick around. For his part, Stefan was keen not

Selected examples of trade routes linked to industrial farming

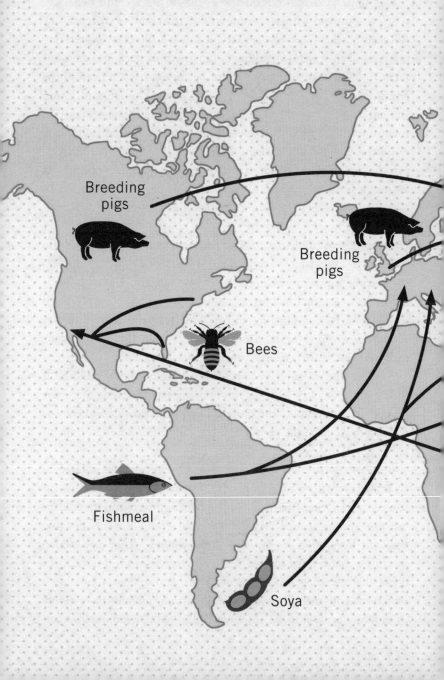

Breeding pigs

Breeding pigs

Bees

Fishmeal

Soya

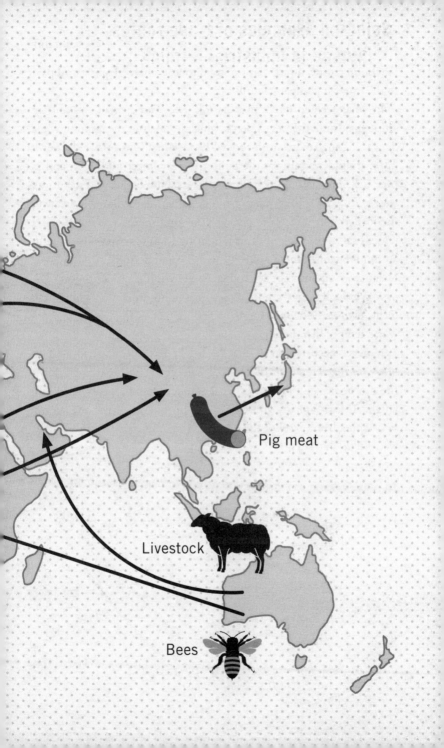

Pig meat

Livestock

Bees

to let him escape, aware that we stood to lose a hefty deposit we'd paid to the hire-car company, Europcar.

In what seemed like a masterstroke, our cameraman was able to swipe the drunk driver's car keys, which he'd left in the ignition while he was talking to Stefan. It looked like we had him cornered. We all hung around waiting for the police. It was almost midnight, and as time ticked by with no sign of the officers, both Stefan and the drunk were getting increasingly angry and impatient. I was horribly aware of the loaded gun in Stefan's pocket. The atmosphere was volatile and tense. I sat in the back of the car, figuring I was best keeping out of it. With no Spanish, there was not much I could do to help. Meanwhile Stefan was on his mobile, trying to get through to Europcar. The drunk, a small, heavy-set, unkempt man in his fifties, kept going back to his own car, disappearing for a few moments, then emerging in a worse temper than before.

It never occurred to us that he might have been attempting to hot-wire the vehicle, until he finally succeeded and roared off down the motorway in a puff of exhaust smoke. Despite it all, we had to laugh. At least Stefan's gun was still in his pocket. It was the last we saw of the drunk, and of the deposit. The only consolation was that we still had his keys.

6

Animal Care

What happened to the vet?

It started with James Herriot, the hero of the best-selling book and TV series *All Creatures Great and Small*. Everyone was charmed by the antics of the handsome young vet as he trundled about in his classic car tending to stricken animals in the Yorkshire Dales. The novels were written decades ago, but the public relations job they did for the veterinary profession was truly extraordinary. The image of the bright-eyed young man ready to battle through snow and floods, wade through manure, be kicked, knocked over and bitten by ungrateful animals and still be polite at the end of it all to their even less grateful owners stuck firmly in the public consciousness, and has never quite gone away.

Yet there is a darker side to the veterinary profession, which is rarely seen by those outside the industry. A growing army of vets do not work in small town surgeries microchipping pet dogs and patching up injured cats, but on farms that bear no resemblance to the picturesque homesteads of Herriot folklore. They often work in dimly lit sheds and in abattoirs, where they prop up the factory-farming system. In these grim settings, the aim is to keep animals alive long enough for profitable slaughter or ensure they continue churning out enough milk or eggs to justify their existence, then to dispatch them with as little ado as possible. The role played by highly respected and supposedly animal-loving professionals in this business reflects the less romantic side of the veterinary profession.

Few vets begin their careers intending to work in such environments, but slaughterhouse jobs have regular hours and the work involves more observation and inspection than physical intervention, meaning that it suits older practitioners and those with bad backs or other work-related ailments. Furthermore, there is a genuine chance to do good, an opportunity to minimise suffering for tens of thousands of animals in the hours before slaughter, a time when they are at risk of pain and suffering.

Jean-Claude Latife, a vet from Brittany, spent nine years working in UK abattoirs. He finally quit in 2010 after being threatened with a butcher's knife by a drug-crazed slaughterman who objected when he paused a production line because of a hygiene concern. I first met him at an animal welfare conference in Brussels. We chatted over dinner and my jaw dropped when I heard his story. Because the incident is subject to ongoing legal action, I'm not using his real name or any details that could identify where he worked. However, in his experience, UK slaughterhouses, widely assumed to boast the highest welfare standards in the world, are riddled with untrained workers who pitch up for work drunk or twitching for the next tea break when they can take another fix of drugs. Not surprisingly, vets are frequently pressurised or intimidated into turning a blind eye when they witness these dangerous men mistreating animals.

It's a far cry from what Latife imagined when he first moved to the UK. As a small-animal vet in France he had developed various medical problems and was advised by doctors to look for a less stressful job. He and his family had fallen in love with the English countryside on holidays, so when a position came up in a slaughterhouse in one of the most beautiful parts of the country, he jumped at the chance: 'The conditions were good – there was no work at the weekends, and there were decent career opportunities. It might sound like the last sort of place a vet would want to work, but I realised that caring for animals isn't just about looking after cats and dogs.'

By law, abattoirs in the UK must have a vet on-site. Their job involves witnessing animals arriving and being unloaded, checking that they are in a fit state to be transported, and that conditions on the lorry are acceptable. Abuses can be reported to the enforcement authorities.[1] They also keep an eye on how animals are handled before they are killed and whether they are properly stunned before slaughter. So-called 'missed stuns', which even in the UK can run at between 5 and 10 per cent at badly run abattoirs, are a huge source of suffering and can also be reported. The other part of the job involves leading a team of meat inspectors who monitor cleanliness and the processing of carcasses to rule out any risk to health and hygiene.[2]

Latife soon adapted to his new line of work in what he describes as a 'very very hostile environment'. 'You are surrounded by death, noise, shit and concrete, but it's something you get used to after a while. I felt I was playing an important role in a place where there is a huge risk of animals suffering, at every stage of the process.' It was the behaviour of rank-and-file workers that shocked him:

> It's the same in most slaughterhouses. Monday mornings are the worst. That's when new foreign workers arrive. Often they don't have a clue what an animal is, and have had no training at all for the job.
>
> As the vet, you have the power to stop the production line if you're not happy with anything you see, but it is a huge responsibility, and it's not easy to do. You might have fifty people working on the line, and if you stop it for some reason, delaying the kill for an hour or two, it's not just the workers who are upset, it's also the inspectors, the managers, even the supermarkets, who are waiting for their delivery of meat. The workers hate it because it means they lose money, or get a shorter break, and the bosses hate it because any delay in production has financial implications. The result is that you are under a lot of pressure and are pushed to accept situations that are far from ideal.

Latife grew quickly hardened to being insulted by workers frustrated when he questioned procedures and caused delays. He says he has a good sense of humour, realised there was no point in complaining, and just let it wash over him. However, when a slaughterman who had previously served six months behind bars threatened to kill him, he knew it was time to quit:

> In this particular slaughterhouse, around half the staff were British; the other half were eastern European, mostly Polish. The Polish workers were often drunk, and many of the British workers were on drugs. One day, a guy I normally got on well with went crazy. I knew he took drugs, and that day he took something during his tea break. When he came back, I noticed that there was not enough space between two carcasses, which raises a risk of cross-contamination. It meant stopping the line. When I told him we had to stop the line, he got very mad, I think because it delayed his next fix. He started insulting me and threatening me with a knife, and threatening to hurt my wife and children. At the same time another guy came at me with a knife. It was terrifying. I stopped the line anyway and reported the incident. I realised it was no longer a safe place for me to work.

Latife's account of the calibre of UK slaughterhouse workers, and the shameful way some incentivise staff to cut corners in animal welfare by paying them 'per kill' rather than hourly rates, is a damning insight into an aspect of meat eating that most people prefer not to think about: the way animals are treated when they come to be killed. What he experienced was not an isolated incident. Undercover filming in nine British slaughterhouses between 2009 and 2011 revealed evidence of cruelty and alleged lawbreaking in most of them, including animals being kicked, beaten and burned with cigarettes.[3] An exposé of twenty-five abattoirs in France uncovered rough handling and poor stunning practices.[4]

Given their professional status, vets are uniquely positioned to make a positive difference. Latife's experiences suggest that many are cowed into keeping their mouths shut about routine abuses, only intervening over the most flagrant breaches of standards. And the questionable role of vets in propping up factory farming is not confined to the end of the process. They play a vital part at every stage, often treating and preventing disease, but also acquiescing in a system in which suffering is inbuilt. Vets who are uncomfortable with factory farming and choose to have no part in it accuse colleagues who make a living on such premises of becoming institutionalised, seeing the animals they treat as little more than production units. These vets may have started their careers with ideals, but industrial farms are often big businesses where the bottom line is what counts. In time, some vets who work on such farms are affected by this culture, approaching sick farm animals as if they were faulty machines. A quick and dispassionate judgement is often made by farmer and vet together, much like car mechanics sizing up a broken-down car: either the machines are worth fixing, and they are treated; or they are not, and they are written off.

The harder farmers push animals beyond their natural limit, and the more closely animals are confined, often the greater the risk of disease and the heavier the reliance on vets to keep herds alive. Their weapon of choice is antibiotics. According to Dil Peeling, who qualified as a vet in the UK but spent much of his career working in developing countries:

> A vet's worth is now measured by his or her ability to deliver on production and animal health – not welfare. It is difficult to persuade vets who have invested so much of their careers in propping up intensive farming to turn their back on such systems. You're asking the high priests of the livestock ministry to reject everything they know. As far as they're concerned, this is how things have always been done.

Now working for Compassion in World Farming, Peeling believes that the industry is geared towards rewarding vets who focus on farm animal health in terms of how it relates to production, rather than seeing the welfare of the animals as an end in itself.

No vets ever produce scientific papers on anal glands in dogs, though one of the commonest things they see in their surgeries is dogs with itchy bottoms. It's not sexy and it isn't going to get them international recognition. Vets are scientists, and scientists are rewarded for writing scientific papers and presenting them at symposiums. Certain diseases are more fashionable than others, and the sexiest of all are diseases like foot-and-mouth disease, which impact on animal exports.

All seven of the UK's veterinary colleges incorporate animal welfare into the curriculum, as has been compulsory in the EU since the late 1970s.[5] The vast majority of vets who work on farms in this country genuinely care about the animals they work with and are upset when they witness suffering. They would alert the RSPCA to flagrant welfare issues.

Alastair Hayton is one of the UK's leading cattle vets and works on a variety of dairy farms in Somerset, some with over 1,000 cows. He insists that large farms do not necessarily go hand in hand with poor welfare and argues that the quality of management is more important than the scale of the operation when it comes to welfare. He kindly arranged to show my co-author, Isabel Oakeshott, around two very different dairy farms in Somerset where he works: one organic, the other one of the largest indoor dairies in the UK, with 1,000 cows, that I would later visit personally. He is happy with conditions on both: 'I am convinced there is no right or wrong system, only right and wrong management. Big is not necessarily bad. I have seen some very bad practice on small farms. The key is the individual farmer – how well they manage their farm, and whether their facilities are appropriate for the number of animals being kept.'

Hayton is adamant that most UK farmers genuinely care about their livestock, and have a strong vested interest in keeping them in prime condition: 'I have seen plenty of farmers cry when they've had to have a sick cow shot. A healthy herd is a productive herd.' However, he acknowledges that intensive farming is highly fine-tuned and that some farmers are increasing the size of indoor-reared herds without extending their barns:

> It's like these animals are on a mountain ridge. The higher the yield farmers are going for, the harder it is to keep the animals from falling off that ridge. A lot of farmers know that the housing they're providing is inadequate, but don't have the money to invest in something better. They know they need to sort it out, but financially it can be very difficult. But there are also welfare compromises with cows on grass.

I agree with Hayton that good husbandry and management is the key to getting the best out of any system. However, the reality is that some systems have only very limited scope for decent welfare. Take barren battery cages; they generally provide each hen with only enough space to stand on. They spend their entire lives standing on bare wire and can never flap their wings. The system is so intrinsically limited and restrictive that the birds are bound to suffer frustration and ill health. The fact is that no amount of stockmanship will create decent welfare in those cages. In short, it is a system with appalling low welfare potential.

By contrast, free-range systems for hens have high welfare potential. The birds have room to flap about; they feel the sun on their backs and the ground beneath their feet. The system provides scope for the hens to lead decent lives. Of course the human element is still important in unlocking the basic potential. If standards of husbandry and management are poor, or the birds are neglected, then they could still suffer. However, at least there's a chance for the animals to have a decent life.

I've spoken to vets who categorically reject the notion that colleagues working on factory farms in the UK perform a disservice to the animals they treat. However, others tell a different story. I recently talked to a young vet who decided to go into the profession after working on a small organic dairy farm as a teenager but was shocked by the realities of large-scale agriculture that she observed when she began her training:

I saw things that really upset me: calves castrated without local anaesthetic; animals prescribed antibiotics without proper diagnoses. A lot of the older vets, the men, have different attitudes that just aren't right today, but they're the people you're working under, and there's a lot of pressure to do things their way. Then there's the farmers who don't want to pay for blood tests or extra visits and might just keep a bottle [of pills] on the side . . .

She gave up practising as a small-animal vet two years after qualifying, in order to campaign for better farm animal welfare.

Katherine Jennings, a Norfolk vet who completed her studies at London's Royal Veterinary College a couple of years ago, agrees that the reality 'can be a bit grim'. She has dealt with pneumonia outbreaks in cramped, poorly ventilated veal sheds and flu epidemics among chickens, both with extremely high mortality rates, but doubts she has seen the worst of it. 'I don't go to the big farms because it's cheaper for them just to have their own vet on-site,' she told me.

There are other so-called experts working on factory farms whose role raises welfare issues, like the unqualified technicians many farmers use to perform routine tasks such as foot trimming and artificial insemination. They are cheaper than hiring a professional, are common in the US, and are becoming increasingly popular in the UK. Then there's the army of 'nutritionists' who ply their trade on factory farms. Although some are independent, most are linked to animal-feed companies, and have a close financial interest in recommending particular diets for livestock.

While there is an enduring tension between business-minded older vets and more animal welfare-conscious recent graduates, progress is slowly being made. Dr Anabela Pinto, a Research Associate in Animal Welfare at St Edmund's College, Cambridge, believes the demographics are helpful, but cautions that more vets are now coming to the UK from overseas. 'Vets have more compassion now. Older vets tend to be male and quite stiff-upper-lipped, but now the majority of veterinary students are women drawn to the job because they love animals. But there are foreign vets working in some of the more industrialised processes who won't necessarily have come from the same background.'

Some vets believe that animal welfare is hampered by the attitude of veterinary colleges towards organic farming. Some organic techniques could be used by non-organic farmers to reduce vet bills and dependency on drugs. Forward-thinking vets promote so-called herd health programmes, encouraging farmers to consider lighting, medication, rations, pasture and a range of other factors as part of a holistic approach to animal health. However, vet colleges do not appear to see any of this as a priority. Though students are taught not to overmedicate, they say lecturers often give the impression that organic farming is primitive. It seems there is an inbuilt bias towards intensive farming within the veterinary profession – hardly surprising, perhaps, when it provides so much employment.

Addressing the issue is as much about understanding what 'animal welfare' actually means as it is about changing the nature of farming or what it means to be a vet. It's not the same thing as simply animal health. Take battery hens again. Apologists often claim the birds wouldn't lay eggs if they weren't happy. Yet the truth is they're genetically programmed through selective breeding to lay about 300 eggs a year, and will do so whatever conditions they are kept in, so long as there's food and water. The mere fact that farm animals continue to function does not mean that they are 'happy' or well.

So where does farm size fit in? Does big have to mean bad? In principle, the answer is that it is not the scale that matters, but the nature of the operation. I've seen backyard farms in China that were just as awful as any mega-operation. Yet when farming becomes divorced from the land, problems are far more likely to arise.

Mega-dairies are a case in point. No farm animal works harder than the high-yielding dairy cow. In peak lactation, her body is being forced to work at a similar rate to a human jogging a marathon every day. No wonder their lives are so short. The average high-yielder often lives only for three lactations before being culled.

Thirty years ago, the average UK dairy cow was producing 5,000 litres of milk a year – no mean feat when you consider that beef cows, milking at a more natural level, produce only a thousand litres or so. However, average dairy cows today yield more like 7,000 litres a year. They need to be fed high-energy concentrated food to avoid 'milking off their own back' and growing weak and emaciated. Higher-yielding cows are under even more pressure. They churn out an incredible 10,000 litres or more a year. At this level, they simply can't survive on grass, can't possibly graze enough to keep up with the demands of such heavy lactation. So their diet is geared towards very high-energy food – grain, not grass.

It becomes unprofitable to turn cows out to pasture where they simply cannot take in nutrients fast enough. The result is the 'zero-grazing' system, in which cows are kept indoors for much of their lives. Once they're confined indoors, all their food is brought to them, and the system is disconnected from the land. If the herd size then rises from the UK average of about 100 animals to 'mega-dairy' proportions of thousands, this takes things beyond the carrying capacity of the land that the herd started out on. And from there, there's no way back.

In essence, the vets involved have become servants of the industrial farm machine, the technicians called in to fix things before they break down, to patch things up and keep the system going. Their job is to keep the animals healthy enough for long enough to be productive. They cannot afford to antagonise their clients by accusing them of cruelty, albeit institutional. To do so would be professional suicide. And so the system remains self-reinforcing.

A whole industry has grown up around factory farming, from breed companies developing the latest types of animal, to the feed companies with their latest formulas, the pharmaceutical companies with their wares advertised so prominently in farming journals, and the equipment manufacturers. Scientists are employed to build the evidence base to support the new way, often paid for by vested interests. I remember a leading member of the US egg producers' association beating his breast at a London conference, pronouncing that there were cabinets full of evidence to show that battery cages were best for the welfare of hens. It was a defensive statement made just after Europe had voted to ban the system due to animal welfare concerns. With so much behind it, it is not surprising that industrial farming has taken off with the momentum of a juggernaut.

Among the casualties are rank-and-file farmers who are often forced to produce more for lower prices. In any other industry, oversupply with plummeting prices would elicit a curb on production. In intensive farming, falling margins all too often trigger the next round in expansion; units get bigger and so too does the amount of suffering and damage to the environment, in the hope of maintaining income; and so the treadmill moves on, yet faster, often with greater debts to service. The inevitable outcome is that many farmers can't keep up and have gone out of business.

Alternative voices and approaches are often pigeonholed as being for niche markets, or they get shouted down. The status quo is further reinforced by the difficulty of making meaningful comparisons between the economics of different systems. After all, the industrial system is all too often supported by the distorting effect of agricultural policy, with its subsidies and incentives. And so the juggernaut hurtles on.

Fast-food chains and supermarkets have grown up and prospered in the wake of industrial agriculture: they sell cheap and plentiful ingredients from animals 'processed' through fewer but bigger slaughterhouses within a system increasingly geared around centralised supply and distribution. The industrial way has no doubt suited companies looking to trade at scale with standardised products. There can be no escaping their dominance in the marketplace; in Britain, five supermarket companies account for four out of every five pounds spent on food. The US retail giant Walmart is the world's number one supermarket, while Britain's biggest, most widespread store, Tesco, ranks third. In the restaurant business, McDonald's is the world's number one by a big margin. Yum! Brands, the owner of KFC and

Pizza Hut, is runner-up. Farmers are drawn to the massive markets that these companies represent, yet complain about their buying power and ability to drive a hard bargain.

How often have I heard commentators talking about the latest revelation in the food industry saying: 'I blame the supermarkets.' Yes, their business is commercial success – making profits for shareholders and investors. But that doesn't mean a company must, by definition, do bad things. John Mackey, CEO of Wholefoods Market, believes that most modern economists continue to see the purpose of business as being to transform factors of production into profit for the benefit of the investors. As he puts it: 'Corporations are probably the most influential institutions in the world today and yet many people do not believe that they can be trusted.'[6] Each time another scandal comes to light, be it in the oil, banking or food sectors, this feeling of mistrust is reinforced.

In the UK and Europe, the horsemeat scandal of 2013 saw consumer confidence and share prices hit hard. It also drove home the message that the world of business has become a much more sophisticated place, that businesses have to take account of interdependencies, be they societal or ecological. Writ large was the need to take the integrity of supply chains seriously if company bosses wanted to sleep well at night. Integrity doesn't stop at the basics: knowing where products come from and what they contain. It also requires being sure of the quality of products, be they meat, milk or fish; that there are no nasty surprises or questionable ethics lurking within and waiting to be found out. It is something that leading businesses are increasingly coming to recognise.

I saw this firsthand when Compassion in World Farming launched a new league table of food companies aimed at investors, based on how seriously they take farm animal welfare. As I rose to my feet in the heart of London's financial district, I wondered whether hard-nosed financial types would get it. Would they not ask what animal welfare has got to do with good investment decisions? I needn't have worried. Riding on the back of Horsegate, no one in the room needed convincing. They could see the danger in not addressing serious issues of public interest and transparency. The launch of the league table made headline news in the UK's *Financial Times* newspaper. Minds had been alerted to the perils of cutting corners.

The big supermarkets, restaurant chains and the like have become society's 'super-consumers'. They are seldom tied to some particular system or way of producing food. They themselves can be much more fleet of foot, able to choose what best suits their business and their customers. It is no accident that in the twenty-first century, 'super-consumer' companies are increasingly taking the lead by changing their policies for reasons like the environment or animal welfare. At Compassion in World Farming, I have made it a priority as CEO to assist businesses looking to do good things by changing policies. That is why we run an awards scheme that celebrates companies pledging to cut out whole swathes of cruelly produced food – committing to selling or using only cage-free eggs for example. I am proud that my organisation works with companies who have changed their buying habits in a way that brings better lives to hundreds of millions of animals. Rather than taking the one-dimensional view that business, big or small, is the cause of the problem, I see business as a key component in achieving change from the industrial model that has gone way too far down the track of putting profit before feeding people properly.

Governments will continue to have a role in shaping how things are done, be it through legislation or how they dole out public subsidies or incentives; but businesses can move far more quickly and decisively than governments. While leading businesses offer the scope to be part of the solution and governments in Europe have banned some of the worst forms of farm animal cruelty, the sad truth is that industrial agriculture still has a stranglehold in much of the developed world. And with industrial farming often comes an attitude that animals are mere units of production, a means to make a profit. James Herriot once said he hoped to help people realise how 'totally helpless animals are, how dependent on us, trusting as a child must that we will be kind and take care of their needs . . . [They] are an obligation put on us, a responsibility we have no right to neglect, nor to violate by cruelty.'[7] One can only wonder what he would make of what has happened to the British countryside in the last few decades, as pigs and chickens disappeared from fields into factory farms.

III
HEALTH

The Tower Wing at Guy's, London, is the world's tallest hospital building. Its thirty-four floors look down over a labyrinth of hospital wings and office blocks connected by long whitewashed corridors. The bustling reception area throngs with patients and staff. It is one of the country's oldest teaching hospitals: the in-house cafés teem with medical students relaxing after lectures.

Guy's Hospital stands in the heart of the capital like a monument to Aneurin Bevan's National Health Service, now a much-loved national institution, founded following the landslide Labour victory after the Second World War.[1] It is now the largest publicly funded health service on the planet and employs a staggering 1.7 million people, said to be second only to the massed ranks of employees of the Walmart supermarket chain, the Chinese army and the Indian railway. In 1948 it was launched with a budget of £437 million – equivalent to £9 billion at today's value. Today its budget is more than £100 billion a year. Since Bevan's day, life expectancy has increased dramatically,[2] and many deadly diseases can now be prevented or cured.

Yet new threats to health have emerged. Certain potentially fatal conditions – cancer, diabetes and cardiovascular disease – are far more prevalent. Over coffee in the shadow of the Tower Wing, Dr Michael Antoniou, a molecular biologist involved in developing gene medicines, speculated on the possible causes of this new wave of illness. A scientist with a wide-ranging interest in human health, he singled out cancer as a particularly interesting case. One in three people now get this awful disease. 'This is not just down to people living longer or to better detection,' he suggested. 'The incidence is going up due to environmental factors.'

So what has happened since the middle of the last century to trigger such a change? Antoniou believes that farming is partly – even largely – to blame, having become dominated by an industrial approach reliant on chemical pesticides and fertilisers.

Meanwhile there is an obesity crisis, epidemic in developed countries. Some blame diets high in saturated fats, such as cheap red meat, as well as lack of exercise. At the same time, more and more antibiotics are being deployed to prevent and treat disease in animals on factory farms. Evidence continues to emerge on the link between this dubious practice and the emergence of drug-resistant superbugs and new killer diseases.

The big question is: are factory farms making us sick?

7

Bugs 'n' Drugs

The threat to public health

PENICILLIN FOR PIGS

There was only a small audience for the health minister as he rose to his feet to tell the House of Commons about an exciting new discovery in America. It was midweek, 7.35 pm on 13 May 1953, and as usual at that time of day, most honourable members were otherwise engaged enjoying a pre-dinner tipple in the Houses of Parliament bars or smoking cigars on the Terrace. Winston Churchill was prime minister, Ian Fleming had just published his first James Bond novel, *Casino Royale*, and everyone was looking forward to the coronation of Queen Elizabeth II.

It's probably not surprising that a debate on a new piece of legislation, the Therapeutic Substances (Prevention of Misuse) Bill, attracted little interest among MPs. Aside from two government ministers, just six members spoke in the debate that night – probably the only ones who bothered turning up – and the bill was nodded through to its next stage after just fifty minutes. For health minister Iain Macleod, the poor turn-out was not unwelcome. He and his colleague George Nugent, a junior agriculture minister, didn't think the legislation they were presenting was particularly controversial, describing it as a 'little bill'.

The first part was designed to give the government more control over an array of promising but potentially dangerous new medicinal drugs that were being developed at the time. The part Macleod was more excited about was the second clause – dubbed 'penicillin for pigs'. It proposed to give farmers a new right to feed antibiotics to farm animals. He told MPs the Americans had discovered that putting a tiny amount of antibiotics in pig feed could have a 'most remarkable effect' on their growth. 'The amount of antibiotic is minute – a proportion between two and twenty parts in a million,' he said excitedly.

The health minister saw little to worry about. He had been assured by the Medical Research Council that there could be 'no adverse affect whatsoever upon human beings'. This was, he felt, 'mainly an agricultural matter'. Some of those present were more prescient. Hugh Linstead, the then MP for Putney, felt that allowing farmers to give their animals antibiotics to promote growth was 'entering into unknown country'.

> We have not been doing it long enough, I feel, to know what effect
> it will produce in the long term in herds and on meat, and indeed,
> on human beings who eat the meat . . . This drug has acquired a
> reputation as an efficient and magic cure-all, and particularly in the
> veterinary field, there is a real danger that if farmers can get hold of
> penicillin without having to pay the fee of a veterinary surgeon,
> they will be tempted to use it carelessly and in a widespread way on
> their flocks.

He wasn't the only politician to warn of the risks associated with giving one of the most important and powerful medicines available for human use to farm animals. Dr Barnett Stross, MP for Stoke-on-Trent Central, saw the potential for disaster:

> We are really treading into strange country. If pigs are fed in this
> way, new types of bacteria may evolve and thrive which are resistant
> to the penicillin, which the pigs are eating regularly in their
> food . . . Should that arise, it would mean first that we should lose
> the benefits that we are now about to gain . . . if there be migration

of the bacteria to humans we may find ourselves in trouble. I do not want to frighten anybody, but these are matters we may look at.

He sounded a further alarm bell:

> We should remember an experiment in America which gave the lead to so much of this work nowadays. They found that another chemical substance, an oestrogen – a type of ovarian hormone – could be used to fatten table poultry. It produced birds with large breasts which were very succulent and they made a much greater delicacy on the table. What they did not know, of course, when they offered those birds to people in expensive restaurants, and no doubt to Senators and Congressmen, was that the oestrogen remains in the breast of the chicken and causes, in men only I am glad to say, sterility, which is a very serious matter.

The official parliamentary record, Hansard, records that Stross's warning was greeted with guffaws.[1] What a pity that the MPs who attended the debate did not take him more seriously. Fast-forward sixty years, and events have unfolded almost exactly as Linstead and Stross predicted. Antibiotics, the 'wonder drug' of the medicine cabinet, are now so widely used and abused on farms that they are losing their potency in human medicine. And just as Stross foresaw, science is struggling to keep pace with the speed at which bacteria are adapting to resist them.

In his annual report in 2008, Sir Liam Donaldson, the UK's then chief medical officer, warned that bacteria were becoming so resistant to antibiotics that 'In some diseases . . . the last line of defence has been reached.'[2] So acute is the situation that the Director General of the World Health Organization (WHO), Dr Margaret Chan, warned on World Health Day in 2011 of a 'post-antibiotic era, in which many common infections will no longer have a cure and once again, kill unabated'.[3] In such a scenario there would be no effective treatment for a range of killer diseases, such as typhoid, tuberculosis, pneumonia, meningitis, tetanus, diphtheria, syphilis and gonorrhoea. Of course, the misuse of antibiotics in human medicine is a big part of the problem. Nevertheless, the excessive use of antibiotics on farms is helping to bring this medical Armageddon closer.

Penicillin, the first mass-produced antibiotic, was first used experimentally in farm animals in 1942, before it was widely available to doctors. Studies showed that hens fed low doses of the drug laid more eggs and sows produced more surviving piglets.[4] No wonder farmers were so keen. For a while, it seemed there was no dark side to the magic. The first warning sign cropped up in the 1960s, when there were several serious outbreaks of salmonella. Thousands of people were hospitalised and at least four children died. It was the world's first recorded 'superbug' resistant to a range of drugs.

It was clear by now that every dose of antibiotics given to a person or animal was a chance for resistant bacteria to develop. The greatest risk came when people or animals were given low doses of the drug, as was happening on farms. These were the ideal conditions for bacteria to fine-tune their resistance, and there was every probability that the powerful new bacteria were being transferred from animals to people.

In 1968 the UK government launched an official inquiry into the problem, led by a distinguished biologist, Professor Michael Swann.[5] His committee clearly wanted to ban all non-essential use of medically important antibiotics in agriculture.[6] Little came of it. The then government succumbed to industry pressure and allowed farmers to carry on as before. Swann's committee also called for a ban on advertising antibiotics to farmers, pointing out that if drug companies could persuade farmers that their livestock would perform better with certain pills, farmers would put pressure on vets to prescribe the drugs. That recommendation was also blocked by industry lobbying. Four decades on, the UK remains the only country in the EU to allow direct advertising of antibiotics to farmers.[7]

People have been sounding the alarm about this for years. The world's public-health experts, from the EU, the US and the WHO, agree that resistant bacteria from food animals are being transmitted

to people. High-level public-health authorities including the
European Medicines Agency and the European Food Safety Authority
have issued stark warnings about the danger, arguing that it is essential
to curb antibiotic use in farming before it is too late. Yet the veterinary
drugs industry and factory-farming lobby continue to dispute the
science and resist any clampdown.

The quantity of antibiotics used by the livestock industry is enormous. By the turn of the century, approximately half of all antibiotics
produced in the world were destined for food animals.[8] According to
one estimate, 80 per cent of antibiotic use in America is on farms, 70 per
cent of the total to boost growth or prevent disease rather than to treat
it.[9] In theory, the use of antibiotic growth promoters is now banned in
the EU – though some farmers have found a simple way round the law,
using low doses to ward off disease, whilst also boosting growth. In
America and elsewhere the practice stays legal and widespread.

Of course antibiotics and other veterinary medicines should be
used to treat genuine illness in farm animals; few would argue with
that. But the fact is that precious antibiotics are being squandered to
prop up an inherently bad and disease-ridden system. Intensive farms
are breeding grounds for disease, because they keep legions of animals
in very close proximity. The European Medicines Agency has described
factory farms as places that provide 'favourable conditions for selection, spread and persistence of antimicrobial-resistant bacteria'.[10]

Factory farms actually promote disease. When pathogens – both
bacteria and viruses – can find an endless supply of hosts to infect
among crowded animals, they do not die out. Viruses can mutate as
they infect a succession of different animals, growing more virulent,
and potentially developing the ability to infect people and be trans-
missible between people. To make matters worse, because of the
miserable conditions in which they are reared, factory-farmed animals
are typically stressed, depressing their immune systems. Being trucked
and shipped around heightens the stress. Studies have shown that
when animals are moved, they shed increasing amounts of bacteria
and virus particles,[11] meaning that more are infected at the end of the

journey than were infected at the outset. If the journey is to a slaughterhouse, the pathogens can migrate to meat.

As chickens, cows and pigs are housed in ever-closer confinement and pushed further beyond their natural capabilities, farmers have grown more and more reliant on antibiotics to prop up the system. This usually involves giving low, sub-therapeutic doses of drugs to animals through their feed or water. Treatment often lasts for several weeks. The aim is to compensate for the sickly environment the animals are kept in.

Dairy cows are prime targets. In a practice known as 'dry cow therapy', they are routinely given antibiotic infusions in their udders to prevent them getting mastitis. 'Blanket' dry cow therapy, in which all dairy cows are treated this way during the two-month period that they stop producing milk ahead of calving again, is widespread in Europe, with the exception of organic operations. Dairy cows receive an average of two courses of treatment a year.

Pigs receive similar treatment. On factory farms, they are typically separated from their mothers at four weeks old, so that the sow can conceive again quickly. Under more natural conditions, sows would wean their young at three to four months of age, when their immune systems are more robust. Pigs weaned at one month old, the minimum age under EU law, are far more vulnerable to serious infection, so many intensive pig farmers start adding antibiotics to their feed once they are weaned. Often they continue to be given medicated feed of one sort or another at intervals throughout their six-month lives.

Not all livestock farms are hooked on antibiotics, and a strong lobby within the community itself argues for drug-free farming. I talked to Richard Young, a farmer with an encyclopedic knowledge of antibiotics and the perils of not using them wisely. His farm is in one of the loveliest parts of the Cotswolds. His cattle graze on rich pasture under oak trees over six hundred years old. He is in no doubt that farming can kick its drug dependency:

There has been little public scrutiny of farm antibiotic use for over a decade, yet during that time we have seen farmers dramatically increase their use of antibiotics . . . critically important in human medicine, and we have also seen the development of several serious antibiotic-resistant bugs in farm animals which are passing to humans on food and in other ways. It is high time that the government took this problem seriously. Organic farmers have shown it is entirely possible to raise healthy animals with minimal use of antibiotics. We could immediately start a Europe-wide programme of change to look after animals in ways that naturally keep them healthy.

Where antibiotics are used on farms to treat sick animals, this usually involves administering high doses of drugs for relatively short periods. On many intensive farms, if a few animals are ill, the whole flock or herd gets the drugs, sick or not. The cross-over between the classes of antibiotics used on farms and those used by doctors has serious implications for public health – as those who catch so-called 'superbugs' discover.

Simon Sparrow was just seventeen months old when he died in the US in 2004 of the killer superbug MRSA. Unlike many victims of methicillin-resistant *Staphylococcus aureus*, he didn't contract the bug during a hospital stay. The day before he passed away, the toddler woke up with a fever and was disorientated. His parents took him to the paediatric emergency room at their local hospital to have him checked over, but the doctors weren't that worried.[12] According to his mother, Dr Everly Macario, they ran a few standard tests and thought he might be asthmatic.[13] She was not reassured:

I could tell something was really wrong given how irritable Simon was . . . he truly was inconsolable. When my husband came to pick us up, [he] noticed Simon's lips were blue . . . We went back in and pointed this out to the doctors. They once again measured his oxygen level and informed us he was within normal range. We went home and gave Simon some albuterol [a treatment for asthma] via an inhaler. When we did this, Simon's eyes rolled back in his head in a way that really alarmed us. But we said to ourselves:

'He'll be fine – he's just sick like any other kid his age gets sick; he'll be fine . . .'[14]

When Simon's breathing began to change, Dr Macario, herself a Harvard-educated public-health expert,[15] called a paediatrician friend, holding the phone to her son's nose and mouth so that the friend could hear. Her friend told her to call the emergency number, 911, straight away. Simon was rushed back to hospital. 'As soon as Simon was wheeled into the ER, doctors hooked him up to everything imaginable. And I kept hearing: "Your child is very, very sick. Your child is very, very sick." At this point, I became almost hysterical,' Dr Macario recalled.[16]

Simon's condition deteriorated fast. His heart raced, his blood pressure crashed, and his lungs filled with fluid. He was pronounced dead at 12.45 pm the following day. No cause of death was given.

Two months later, an autopsy informed Dr Macario and her husband that their son had died from MRSA, probably 'community acquired'. In her words: 'It seems unfathomable that a healthy, hearty and beautiful little boy could have such a bacterium – one that attacked his organs by releasing lethal toxins – and in less than twenty-four hours was gone. MRSA took my son swiftly and totally.'

Following Simon's death, Dr Macario launched a campaign for action to combat the spread of MRSA. She is now the public face of Moms for Antibiotic Awareness, an American-based pressure group, as well as co-founder of the MRSA Research Centre at the University of Chicago. Moms for Antibiotic Awareness is sponsored by the Pew Charitable Trust, an organisation that has highlighted the damage done by factory farms, including their role in rendering antibiotics less effective.[17]

Every year 25,000 people die in the EU from infections caused by drug-resistant micro-organisms. The European Commission estimates it costs the EU economy at least 1.5 billion euros a year. The Commission has described antibiotic resistance as 'an important, largely unresolved issue in public health'.[18]

If a disease is antibiotic-resistant, it means initial treatment is less likely to be effective. The consequence is more severe illness, more hospitalisation and higher death rates. Doctors are forced to use more expensive and complicated drugs that may have worse side effects. Young children are particularly vulnerable.

It is easy to imagine how antibiotic-resistant bacteria can be passed from animals to humans through their meat, but that is only one of the ways it happens. The bacteria can be passed on to people who work with infected animals, through manure or even in airborne particles. The bacteria can then spread further from person to person.

Of the so-called superbugs, MRSA is the most notorious. Until a few years ago, it was found almost exclusively in hospitals. They have responded vigorously to the threat through better hygiene, and infection rates on wards are now falling. However, the bug is now striking people who have had no contact with hospitals, like baby Simon. There have been outbreaks of so-called 'community acquired' MRSA in many countries, including the UK, US, France, Germany, Switzerland, the Netherlands, America, Canada, Australia, New Zealand and Japan.[19]

A Soil Association report tells how in 2004 a previously unknown strain of MRSA, called MRSA ST398 (or NT-MRSA), was identified in pigs and began to spread to people. The first recorded cases of human colonisation by 'pig' MRSA were in a Dutch baby girl and her parents, who were pig farmers. Now half of all Dutch pig farmers are thought to carry the new strain – 760 times the average rate in the wider population.[20] Tests suggest MRSA is often present in raw meat. It was in 35 per cent of turkey samples inspected in the Netherlands, and at least 10 per cent of chicken, pigmeat and beef.[21]

It's no surprise that factory-farmed pigs are generating superbugs. They get more frequent doses of antibiotics than any other farm animal. The European Food Safety Authority believes MRSA ST398 is probably 'widespread' in the food-animal population, 'most likely in all Member States with intensive animal production'.[22] Dutch scientists and government officials have no hesitation in pointing the finger at the intensive pig industry and its reliance on antibiotics for the rise and rapid spread of farm-animal MRSA.

In June 2011, the respected medical journal *The Lancet Infectious Diseases* published findings of the first-ever documented cases of MRSA in British farm animals. Scientists found fifteen cases of a completely new type of MRSA in milk from English dairy farms. It's already infecting people in England and Scotland – though not from drinking milk, as the pasteurisation process kills bacteria.[23]

MRSA is just one threat to human health created by antibiotic use on factory farms. Rising numbers of food-poisoning cases are now antibiotic-resistant, which means that catching these bugs can threaten life. Just how many people fall victim to food poisoning every year is hard to assess; as most people don't bother reporting it, official statistics represent only a fraction of the true number of cases. The UK government department Defra (Department for Environment, Food and Rural Affairs) estimates that there are between five and nine times as many cases as officially recorded.[24] In 2009, there were almost 200,000 reported cases of campylobacter in the EU and around 109,000 cases of salmonella.[25] The main sources were poultry meat, pigmeat and eggs. The real annual figure for campylobacter and salmonella cases in the EU may be more than 2 million.

Across the EU there were also 3,573 cases of dangerous toxic strains of *Escherichia coli* (*E. coli*), such as *E. coli* 0157. Defra has said that *E. coli* 0157 is 'widespread in cattle in the UK'.[26] Other types of *E. coli* now offer greater resistance to antibiotics; in the UK, 12 per cent of all cases of blood poisoning caused by *E. coli* are resistant to almost all antibiotics.[27]

In the US there are around 9.4 million cases of food poisoning every year. Of those leading to hospitalisation around a third are due to salmonella and 15 per cent to campylobacter.[28] Between March and September 2011, ground turkey processed and distributed by the meat production giant Cargill and contaminated with salmonella was reported to have infected 119 people in forty-two states.[29] Cargill recalled over 16 thousand tonnes of the ground turkey produced at one of its plants. The strain of salmonella involved was multi-drug-resistant.[30]

The peak of the scandal over salmonella in eggs in Britain was in the late 1980s, when over 90 per cent of British hens were kept in cages[31] and the cabinet minister Edwina Currie famously lost her job for claiming that 'most' British eggs were infected. Now, almost 50 per cent of laying hens are either free-range or organic,[32] and salmonella rates have plummeted to 0.25 per cent.[33] The reduction may be partly due to testing, vaccination, and culling of infected flocks, but it may well also be linked to more free-range conditions. A study of salmonella in British laying hen flocks published in 2010 found that smaller flocks and non-caged flocks were much less likely to carry the bug: over 18 per cent of caged flocks tested positive for *Salmonella enteritidis*, the most common strain causing food poisoning, compared with less than 3 per cent of non-caged flocks. The largest flocks, of 30,000 birds or more, were seven times more likely to carry salmonella than the smallest flocks of 3,000 hens or less.[34]

It has been estimated that two-thirds of bacteria, viruses or other micro-organisms that can cause human disease are zoonotic, meaning they originate from animals.[35] The aggressive viral diseases like bird flu and swine flu have strong links with intensified farming. Intensive animal production has provided a new route for the likes of swine flu and bird flu to develop and spread. Highly pathogenic bird flu such as the H5N1 virus emerged during a time of massive expansion of the poultry industry in the Far East. It was first spotted in Hong Kong's live-bird markets and chicken farms in 1997. Six people died. From 2003 it spread across east Asia, at exactly the time when the poultry population was soaring and poultry production growing more intensive. China reared three times as many meat chickens in 2005, when bird flu was rampant, as it did in 1990.[36]

The H5N1 virus spread across Asia, the Middle East, Europe and Africa. It has been found on chicken, goose and turkey farms, and in some wild birds, mainly swans and geese. By August 2011, 564 people were confirmed to have been infected, of whom 330 died – a fatality rate of almost 59 per cent.[37]

Most people who caught it lived cheek by jowl with chickens or were involved in killing them. However, an outbreak of a different strain of bird flu, H7N7, in the Netherlands in 2003, during which a vet died, showed that such viruses can be transmitted from poultry workers to other people. Tests found that eighty-six poultry industry workers and three family contacts were infected with the disease during the outbreak, and around thirty more family contacts were also probably infected, although tests were inconclusive.[38]

Factory farming was always bound to cause a disease backlash by pushing nature way beyond its limits. Keeping massive numbers of poultry on intensive farms worldwide appears to have come back to bite us in the form of potentially fatal strains of avian influenza. Luckily, at the moment H5N1 is not easily transmitted between people. The big fear is that every time someone is infected, the virus gains more ground to mutate in, raising the prospect that it will become far more contagious. This could lead to a global epidemic.

Recently, scientists have shown that just a few mutations would allow H5N1 to become as infectious as seasonal flu.[39] An editorial in the *New Scientist* described the risk of a pandemic as 'fact, not fiction'.[40] Public-health experts writing in *The Lancet* have estimated that such a flu pandemic could kill as many as 62 million people, mostly in developing countries.[41]

I remember Peter Roberts, former dairy farmer and founder of Compassion in World Farming, sharing with me his conviction that factory farming would have to end because of the vast tide of disease coming from it. The fact that pigs, humans and birds can exchange flu viruses or elements of viruses raises the nightmare prospect of a highly contagious and lethal flu strain that starts out in animals transferring to people. These worrying new mixes of human, bird and pig viruses are likely to be very hard to treat.

A key element in the spread of diseases originating on farms is long-distance transport of animals for fattening and slaughter. As well as being a major welfare issue, it allows diseases to 'hitchhike' their way to new places and populations. I was reminded about this issue when

I went to Ghana in 2011 for the Commonwealth Veterinary Association conference in Accra, where there were representatives from the UN, the European Commission and the World Organisation for Animal Health. Several eminent speakers talked about the role of global travel and international trade in agricultural products, including live animals, in the emergence of disease. The slogan 'one world, one health' was widely bandied around by vets in the conference hall, summing up the general view that animal and human health are intertwined. It seems to highlight the folly of the live-export trade, in which millions of live animals are trucked and shipped long distances across the world, just to be slaughtered at the end of the journey.

Many scientists still seem reluctant to speak out about the role of intensive farming in antibiotic resistance and the rise of aggressive strains of viral disease. The science is complex, pressures from the industry are intense, and it takes a long time to track specific antibiotic-resistance problems all the way through the food chain. It can be hard to be precise, because the bacteria usually changes slightly during the process. Naturally businesses benefiting from the system seize on any holes in the evidence. Some have even tried to suggest that the wire or slatted floors typical on factory farms reduce disease and increase food safety for consumers, because the animals are separated from each other's faeces and isolated from infections circulating in the outside world. There may be some truth in this, but any protection farm animals get from being reared indoors is far outweighed by the health risks associated with being in such a confined and artificial environment.

Disturbingly, some have tried to scapegoat wild birds as responsible for avian influenza and use this as an excuse to support greater farm intensification. They claim that keeping poultry inside protects them from wild birds carrying disease. What this argument conveniently overlooks is that low-level avian flu is a perfectly natural disease in wild birds. It's only when it enters the pressure-cooker environment of an intensive farm that the disease tends to mutate dangerously. Once a virus gets into an intensive poultry shed it can move quickly through the flock, constantly replicating itself. Any 'errors' or changes to the genetic code during replication don't get repaired: this is how the virus mutates and new variant strains emerge. The tragedy is that while intensive farms provide ideal conditions for the emergence of new aggressive disease strains, wild birds can then become infected too.

Experience from the 2005 outbreak of highly pathogenic avian influenza (AI) H5N1 suggests that the disease is more likely spread along major road and rail routes than on the flight routes of migratory birds. Also, the overwhelming majority of wild birds found infected with H5N1 were dead, preventing them from carrying the virus over long distances. When H5N1 hit a Bernard Matthews turkey farm in Suffolk in 2007, there was no evidence of highly pathogenic AI in wild bird populations in Britain. Defra reported at the time that over 4,000 wild birds had been tested over the previous six months; only 0.4 per cent were found to be infected with AI, of which none were highly pathogenic strains.

Blaming wild birds is an excuse for doing nothing about the dominant source of the problem: the factory-farming system itself. Dr Aysha Akhtar, a neurologist and public-health specialist and Fellow of the Oxford Centre for Animal Ethics, sums it up like this: 'By confining billions of animals on factory farms, we have created a worldwide natural laboratory for the rapid development of a deadly and highly infectious virus.'[42] Akhtar works for the Office of Counterterrorism and Emerging Threats of the US Food and Drugs Administration. She points out that human terrorists don't have a monopoly on killing and causing chaos. Factory farming, she fears, has as much potential as, and probably more than, any terrorist to do the job.

MEXICO'S 'GROUND ZERO' FOR SWINE FLU

High in the mountains of southeast Mexico is an unassuming little town with a very unfortunate reputation. La Gloria in the Perote Valley is a cheerful place where children play volleyball on the dusty roads and the streets are decorated with yellow bunting. Lying

8 kilometres from one of the biggest concentrations of pig farms in the world, it was at the epicentre of one of the biggest-ever health scares linked to factory farming. It was here early in 2009 that people first started falling sick with a previously unknown disease, a new and highly contagious virus containing genetic material from a mix of pig, bird and human influenza.

Soon the virus had ripped across the world, leaving thousands of people gravely ill and La Gloria notorious as the 'ground zero' of swine flu. The spotlight has long moved away from this place and the surrounding area, where a company called Granjas Carroll de Mexico (GCM) owns a network of pig farms. Now the immediate panic is over, most people have stopped worrying about catching swine flu, just as they have stopped worrying about contracting bird flu. Yet there is every chance that the combination of conditions that created these strange new viruses will sooner or later throw up another pathological monster. In the meantime the daily battle goes on between communities like La Gloria and industrial farms on their doorstep.

I visited Perote Valley in November 2011 just before the annual festival to mark Revolution Day, commemorating a long armed struggle that began in 1910 and resulted in the overthrow of Mexico's autocratic ruler and the creation of the Constitution of 1917. I wanted to see the supposed birthplace of swine flu and find out whether what happened was still affecting people there.

Starting our journey in the bustling city of Xalapa, capital of the state of Veracruz, my camera crew and I hired a car and headed for the hills through uplands that could have been English, except that the light was so bright the grass looked greener than I'd ever seen. Small groups of black and white cows dotted a rolling landscape. We were on the lower slopes of an ancient volcano looking towards the Sierra Madre mountain range, shrouded in the morning mist. In the distance, a city winked in the sunlight. After an hour or two talking with dairy farmers in the area, we pressed on to the high mountain plateau, at 2,500 metres. Here we found an entirely different landscape – a scrubby desert littered with great fat cacti. The occasional dusty village sat amidst vast open terrain on which people strove to grow anything that could survive on such dusty land. Growing crops here was clearly a struggle – though there was a fair amount of maize – but factory farms were another matter.

In hangars scattered over a few square kilometres in the valleys of Perote and Guadalupe, GCM (a partnership between a Mexican animal-feed corporation and US-based Smithfield, the world's largest pork producer) rears more than a million pigs a year here.[43] As with the mega-piggeries we saw in China, the animals spend their whole lives indoors in cramped concrete pens, their manure flushed into open-air lagoons.

The first time we thought we'd spotted one of GMC's establishments it turned out to be a prison – factory farms and jails can look very similar. It wasn't long before we came across what we were looking for: a huge industrial piggery made up of eighteen long corrugated-iron hangars. Next to each one was a giant steel bin, most likely an automated feeding cylinder. The farm was surrounded by an electric fence topped with barbed wire. From the outside there was nothing else to see: no pigs, no farmers, no signs of life; just a series of grim metal structures that looked as if they'd been more or less dumped at random on the countryside. With its forbidding locked gates and high fences, the perimeter screamed 'Keep Out!' and so we did. There was no one around to ask if we could see inside, and it's hardly likely they'd have let us, so we drove on through the valley.

Our first stop was the village of La Tlalconteno, where a vibrant market was in full swing. Shoppers and customers mingled under bright awnings that gave some respite from the intense sun. Standing among the brimming stalls was a man holding a colander piled high with crackling golden-brown snacks: fried pig skin. He tipped them onto a tray and piled more on top, till the stack nearly spilled onto the floor. Eager hands reached out and the pile was gone. A small boy danced to the music as his dad watched proudly.

The boy was a picture of health, with a cheeky smile and glossy black hair, his little feet shuffling to the beat in tiny buff-coloured cowboy boots. His name was Alan, and he was not quite five years old. His father offered him a toy from a nearby stall. He told us he was thankful for this day, because back in 2007 he almost lost his son. Alan was less than one year old when he fell sick with a mystery illness. Doctors offered different diagnoses, ranging from a nasty cold to pneumonia, but it was no ordinary sniffle. Alan developed severe breathing difficulties and became so sick that he was hospitalised for almost three weeks, a long time for an infant.

His father, Gerardo Praxedis Serrano Diaz, firmly believes that the child's illness was linked to pollution from nearby pig farms. His symptoms were uncannily similar to those suffered by victims of swine flu, but the family will never know. At the time, Diaz was asked by doctors if he kept pigs at home. He doesn't. However, he lives less than 100 metres from a slaughterhouse.

He was keen to show us more piggeries, so we piled into the car and set off. We passed several livestock transporters being cleaned in an area just off the road. Diaz explained that this was where the lorries were usually disinfected. Past a hump in the road, the valley opened up, industrial pig farms strewn in all directions. We passed an access road with a pig-farm sign; then another and another. In all we could see a dozen or more pig farms, each with a row of factory-like sheds and giant metallic feed bins. Our car stank of slurry.

We pulled up next to one of the farms. The smell was overwhelming. Around us was semi-arid landscape. Diaz told us that dead pigs from the valley were taken to La Gloria for disposal. GCM claims that carcasses from pigs that have died from 'stress or old age' are disposed of using biodigesters or composting,[44] but there are persistent complaints from locals that carcasses are left lying around, attracting wild dogs.

We drove on, and could see more farms: another fifteen. They looked like clusters of low aircraft hangars or military installations. The steel, concrete and electrified boundary fences warned us off. Articulated lorries trundled in and out, ferrying huge quantities of feed. The finished product, fattened pigs, would be loaded onto huge transporters for their final journey to slaughterhouses nearer the city.

I knew from past experiences that these closed farms rarely welcome casual visitors; being in a place once caught in a media hurricane made it even less likely that we'd be allowed through the firmly shut gates with our cameras. My suspicions of sensitivities were confirmed when we passed a pickup truck, presumably from one of the pig farms. The driver wound down the window and asked what we were doing in the area. No sooner had we started to move off than he was on his walkie-talkie, most likely alerting others to our presence.

We parked near a lagoon beside the road. Not a picturesque lake, but a slurry lagoon full of liquid effluent from the pig farms. I scrambled up a bank for a better view, doing my best to avoid being skewered by cacti. The smell told its own tale. Just behind it was another piggery. A figure in white overalls walked between the barren buildings. I could hear pigs squealing. It was all surreal: the heat, the dust, barbed wire over a 3-metre boundary fence. None of it bore any resemblance to the traditional image of a farm.

Diaz wanted us to talk to his father, who lived in La Gloria and was a key figure in a local protest group opposed to the expansion of industrial pig farms. On the outskirts of the village we passed a graveyard; a colourful riot of stone and wooden crosses and miniature monuments to loved ones. Bright pink and orange ribbons, ornately arranged into crucifix shapes, stood proud over graves. A gecko wiggled up a tomb, casting a tiny shadow in the morning sun.

The village of La Gloria has a population of less than 3,000. It is sheltered by tree-lined mountains that bring water to the area. The wind swept dusty streets, children scampered in the road, and a red scooter drifted by with a megaphone booming a recorded voice to advertise *tamales*, a traditional maize mixture wrapped in banana leaves. The bunting was up for Revolution Day; the walls of a hacienda were peppered with bullet marks from the historic battles of that time. A century on, a different type of civil war continues: between people and pigs.

Diaz's father, Guadalupe Gaspar, a charismatic man with a deep tan and a white Stetson, was sixty-eight years old but seemed younger. He was a farmer and a leading member of the local campaign group Pueblos Unidos (People United). He welcomed us warmly into his modest home. Next to the blue front door, a little shrine stood in a bay window. The sitting room inside was painted mauve, with quintessentially Mexican light blue and pink curtains. By an empty cabinet, a collection of statuettes attested his religious faith. Gaspar seemed a proud man of strong beliefs.

Sitting with his feet tapping the tiled floor, he explained that La Gloria had long been a farming community, growing corn, beans, potatoes, barley and wheat. He told us that as a child he had spent happy days walking in the countryside with his dad. In those days, he claimed, there were no water shortages in the area. Now it is a different story. He blames deforestation in the mountains and the mega-piggeries. Locals must now compete with swine for scarce water in the dusty valley. He talked of a lack of jobs, a real concern for families, particularly young people, many of whom are moving away. It is already said to be common for villagers to commute to work in Mexico City, 200 kilometres away, returning at weekends.

The arrival of multiple industrial pig farms could have been the answer to villagers' prayers, providing jobs and opportunities. In practice, Gaspar claimed, few locals were being employed. Most farms seemed to be staffed by outsiders. We heard the familiar story of pollution, strong smells, contaminated drinking water and flies. Since few locals have any economic stake in the piggeries, it was little wonder they were seen as the enemy. Gaspar told me he felt cheated. 'We were deceived. They said some companies were coming that were going to give people jobs. They never said anything about coming here to pollute.'

His account may not be entirely fair. According to GCM's website, it is the main source of employment in the Perote Valley, generating more than 3,000 direct and indirect jobs[45] – though that is not to say that the people it employs come from local communities. The company points out that it also pays a lot of tax, contributes 2 per cent of its payroll to support Veracruz's infrastructure, and has funded reforestation and irrigation. It also says it provides free medical care to eighteen communities in the states of Puebla and Veracruz.

Whatever the true costs and benefits, soon after the first pig farm arrived near La Gloria, villagers say they noticed a change in the quality of the groundwater. There was widespread concern and they began writing letters of complaint to the federal government, the state authorities and the town council. In 2007, when they heard another farm was planned, closer to La Gloria than the first, they organised a protest. Several hundred demonstrators tried to close the road, a common form of protest in Mexico. They managed to partially block traffic onto the federal highway and annoy the authorities.[46] One morning, around a month after the protest, there was a knock at Gaspar's door. He opened it to find three federal policemen, together with a man in civilian clothes. He was told he was being charged with launching an 'attack' on a federal highway. 'They caught me and arrested me as if I was a drug smuggler.' He claims he was forced to sell his corn crop to defend himself, and that every fifteen days for a year he was forced to make a two-hour round trip to Puebla, the nearest big town, to check in with the authorities.[47]

Two years after these early protests, more than a quarter of the population of La Gloria – 28 per cent – went down with swine flu or something like it.[48] Mexico's government first reported the outbreak in La Gloria to world health authorities in April 2009. A week later, clusters of pneumonia were reported in Mexico City. At the same time, California identified a novel influenza virus. On 23 April, Mexico confirmed cases of infection with the same virus, newly named as H1N1 influenza A. It spread far more quickly than anyone anticipated: within a week, ten countries were affected; 180 countries by the end of August.[49] Within a year, according to the World Health Organization, the virus was linked to over 18,000 deaths worldwide.[50] It was traced back to La Gloria, where a young boy, Edgar Hernandez, was the first confirmed case. He later became known as 'kid zero' or 'patient zero'.[51]

In a small park in the middle of La Gloria is a lifesize statue of Edgar on a podium surrounded by a strange waterless well. The paint around it has begun to peel, but the statue makes it difficult for anyone in La Gloria to forget what happened.

What was unusual about the viral strain was that it had a mix of genetic material from two different swine influenza viruses as well as from human and avian strains of flu. It was suspected of having been spread by people arriving from affected areas in Mexico.[52] Since then, question marks have been raised over the scientific link between the emergence of the new virus in La Gloria and the neighbouring piggeries.

Naturally GCM has highlighted doubts over the evidence. It says that Mexican public-health authorities inspected its facilities, particularly farm 113B, the nearest to La Gloria, and certified the absence of 'any sign associated with the flu in our animals'. Their website[53] states that 'the company requested [the authorities to] collect new samples to certify the state of health of our swine.' The tests were conducted by the National Health Services, the authority responsible for animal health. The website states: 'There is no scientific evidence that links GCM with the A/H1N1 flu virus.'[54] It goes on to say that these reassuring findings were endorsed by another government agency, the Federal Commission for Protection Against Health Risks. Apparently they carried out an extensive study in La Gloria, which concluded that the firm had 'no diseased pigs or sick people, nor any respiratory or diarrheic problems'.[55]

Maybe it's just a coincidence that a novel strain of flu, a key component of which was two different strains of swine flu, was first diagnosed right by some mega-piggeries? Either way, the focus of local protests about the farms shifted from pollution, smells and flies, to disease. 'When we realised what was happening, all the villages came together to demonstrate,' Gaspar recalled. 'People were dying. That is when we got up and started to fight; to make demands, so that the farms knew we didn't want them here.' They marched along the main road to make their point to the authorities, but according to Gaspar they were 'treated like troublemakers'. He claims he has effectively lost everything as a result of speaking out, and that many locals are too frightened to stick their necks out or have been bought off.

At the time of my visit, it was almost three years since the swine flu outbreak, but the anger and fear in the community burned as powerfully as ever. Gaspar said it felt like 'living in a time bomb'. 'We don't know when something else bad is going to happen to us. The government must get rid of the farms because while they remain, the pollution will continue and I am sure there will be more new diseases.' Locals say that while La Gloria was in the global spotlight, they were quietly assured that no new farms would be built. Now the world is no longer watching, and they doubt the pledge will be honoured. Indeed, there are already signs of new pig farms arriving.

Having said goodbye to Gaspar, we went to meet a local elected official who had agreed to talk to us on condition of anonymity. He took us in our car down a dusty, cactus-lined track. We passed through fields of maize growing in corn-dolly pyramids, stopping 2 kilometres outside La Gloria. The sixty-year-old official was taking us to see what he claimed was the future site of a new industrial pig farm.

The first signs of development were already in place: concrete foundation posts wrapped in barbed wire and a well for drawing water. A white cross, 6 metres tall, towered over the plot. It was not a religious symbol but one of a line of identical posts that stretched as far as the eye could see, erected to carry power or communications to the place.

The anonymous official told me he was sad and angry at the way companies had moved in offering jobs when so few people in the community benefited. It would be much better for the community if money were invested in traditional agriculture, which generates more jobs. As a civil servant, his particular concern is for the future of the valley's young people. 'So many are leaving. We need job creation here in the countryside so they aren't forced to go. If more farms are built, they are going to cause problems for our grandchildren and great-grandchildren in years to come.' He wants communities faced with similar proposals to act quickly. 'Don't allow them to build more farms, because we will never rid ourselves of the pollution,' was his message to others.

We returned to La Gloria, where we went to speak to a thirty-year-old mother of two girls, who gave her name as Hortensia. When we arrived she was standing in the doorway of her brightly coloured home. She beckoned us inside, where she felt she could talk more freely. It was more evidence that people were uneasy about speaking out. Hortensia told me she had lived in La Gloria nearly all her life. She sat underneath a painting of a child wearing a hat with sad blue eyes and tears rolling down his cheeks. It was a strange picture to choose for your wall, but seemed symbolic of the sadness and fears of the community. Hortensia was one of the protesters against planned expansions of the industrial pig-farming area. She described the people who took part in blocking the road as ordinary folk, worried about the environment and their children. She said strong winds blew bad smells over the village.

I asked her about polluted water, and she looked resigned:

People have got used to it. They say that once these farms are established, there's nothing they can do. Even if they complain, nothing will happen. The water that supplies our village comes from the forest, but sadly it is being cut down illegally, so you can imagine the concern we feel, knowing that it is being felled, and a business is going to be established that uses thousands and thousands of litres of water every day. It is a huge worry.

Like others we talked to, she was sceptical that any new pig farms, like the one apparently already rising just out of town, would create jobs for locals: 'They have promised, as they always do, to provide jobs for the men here. It is the same every time. I don't believe it.' She thinks the new farm will be the biggest yet. From what we observed, the company had already bought the land and dug wells.

I asked her if the village would benefit from the pigmeat produced. She told me that it was normal for people in La Gloria to have their own pigs at home, which they feed on leftovers. They don't need pork from a mega-farm. It turns out that the pigs from these industrial farms are sold live for slaughter in Mexico City and other mainly urban places.

I asked about the swine flu episode. She told me that the first villagers knew of it was when journalists arrived and then it was all over the television. She was frightened for her children and relatives. Her mother went down with the illness and the family feared everyone would catch it. She said it was blamed on the pigs at first, but acknowledged that nothing has been proven.

I left her house and walked past the hacienda. I remember thinking that the pockmarked walls of the crumbling citadel fortress were no match for this new type of enemy. It doesn't have to be like this – as we could see when we visited a very different type of farm in the same area. We stopped for a breather outside a village called Acajete which had beautiful views over the Sierra Madre mountain range. The slopes were shrouded in morning mist. A farmhand was walking up a hill carrying three white buckets. He spotted us, waved in a friendly sort of way, and beckoned us over. We explained what we were doing in the area, and were treated to an impromptu tour of the farm.

The farmer, Ana Maria Frauzoni Hernandez, also a vet, took us round, first through a cluster of modest flat-roofed buildings – the farmhouse and dairy – and then into the fields. There was a faintly pleasant smell in the air of baby-sweet dairy mixed with a slight hint of manure. Hernandez explained that the farm belonged to her brother. She talked about respecting the cow as a noble animal. We walked past a few trees and found twenty calves loafing in the sun. I stood in the inevitable cowpat.

The place turned out to be one of thirty-four farms in a local dairy cooperative. It was a pretty big farm by European standards: 500 cows in all, but you wouldn't have known it. They were well dispersed, grazing happily on the hillside. We watched as forty Friesians were milked out there in the fields. The animals and a couple of farmhands stood among a cluster of silver milk-churns. Hernandez explained that they were milked twice a day. Her father used to milk them three times a day but the cows got stressed. When milking was over, a horse carried the churns up the hill. The cows followed along behind. It was wonderful to see them walking naturally, without the bloated, bulging udders and splayed back legs we'd seen in California's mega-dairies.

The cows on this particular farm were kept outdoors all year round. No chemicals, preventative antibiotics or hormones were used, though a bit of supplementary feed was offered when the grass was short. Hernandez told us they had an average lifespan of twenty years – around four times the life expectancy of cows in intensive dairies.

Towards the end of the tour, she started talking about the battle to get a decent price for the milk. It's a familiar theme on both sides of the Atlantic, with farming systems big and small. The milk from this farm was sold under the name 'Joyalat' – *joyal* means jewel in Spanish. Hernandez said she saw the milk as 'white gold'. She shared some customer feedback about how good the product is, apparently because the cows graze naturally on grass full of nutrients.

Before we left, I tasted the yoghurt. I found it full of flavour, very smooth, with no hint of sharpness: delicious. A poster in the dairy window proudly proclaimed that 'The best milk in the world is produced in Mexico.' Looking around me, at that moment, I had to agree.

I went to southeast Mexico looking for answers. I hoped to trace swine flu to its origins, and find out whether the virus was really linked to the piggeries. I left with no definitive answer: nobody seemed to know. There is speculation that the virus may have been circulating in the US pig industry long before it showed up in remote, dusty La Gloria. Standing in an arid valley surrounded by many thousands of pigs, without seeing a single animal, was an eerie experience. What I did gain was an insight into the reality of life for communities near factory farms. It's an issue that would have received far less public attention had it not been for swine flu. However, as always, once the immediate panic was over, the media caravan moved on, leaving the community to fight on alone.

8

Expanding Waistlines

Food quality takes a nose-dive

The fattest man in the world lives in a small ground-floor flat in north London, and spends his life sleeping or sitting propped up in his reinforced bed. Keith Martin has not ventured outside for more than a decade, except for the odd visit to hospital, and the time he moved house, which involved an uncomfortable journey in the back of a van.[1] At 368 kilograms (812 pounds), the forty-two-year-old is too large to go anywhere or do anything other than eat, drink, read, play on his games console or watch his enormous plasma TV. He is so fat he can't even roll over in bed by himself. His girth is greater than his height.

Every day, seven carers working in shifts arrive to wash and change him. On alternate days he is visited by two nurses who tend his bedsores. Last time he left his flat to go to hospital after falling over, it took a specialist team and a £90,000 ambulance designed for morbidly obese patients to transport him. He had to be dragged across the floor to the vehicle in a special bag. That's nothing compared with the practicalities involved in taking the world's fattest teenager to hospital. Also from the UK, Georgina Davies weighs 400 kilograms (880 pounds), and getting her out of the house involved pulling down two walls.

Martin began to binge at the age of sixteen, when his mother Alma died. He left school that year, with poor qualifications, and went to work as a warehouseman and labourer until he got too fat to do the job. He continued to overeat and every year his weight ballooned: 'I let myself go. I just didn't care. I got so bloated on sausages, bacon and roast dinners. I just ate whatever I felt like.' Now he is desperately trying to change the habits of a lifetime, knowing that his weight will soon kill him. Doctors have warned him that unless he succeeds, he will be dead by the age of fifty.

For years, he has started the day with eight hot dogs and four slices of bread, or a stack of ham sandwiches washed down with sugary coffee. For lunch he gorges on chocolate bars, cakes and biscuits, while a typical dinner has been two entire roast dinners with all the trimmings, or sixteen sausages, plus a family-sized bag of oven chips. His target is to slash his daily calorie intake from 9,000 to 2,500, the recommended amount for men.[2] To stand a chance, he will have to overcome his addiction to cheap processed meat. The sausages and hot dogs he loves provide little nutrition for the calories.

Martin is an extreme example, but on current trends, by 2030 half of American adults will be obese. The projections for the UK population are almost as bad. Annual obesity-related health costs are expected to soar over the same period by $48 billion in the US, and nearly £1.25 billion in the UK.[3] The availability of low-cost factory-farmed food has played a key role in the global obesity epidemic. Scoffing roast chicken and sausages might seem healthier than guzzling cake, but evidence shows that factory farming has stripped away much of the nutritional value of the meat on offer in supermarkets and fast-food joints. At the same time the fat content has soared. Intensification has had so severe an impact on meat quality that some scientists claim you'd have to eat four entire factory-farmed chickens to benefit from the same level of some nutrients as you would have got from a single organic chicken in the 1970s.

'The intensification of animal farming has virtually destroyed the nutritional quality of our food,' according to Professor Michael Crawford of the Institute of Brain Chemistry and Nutrition in London. Crawford visited my office in Godalming, Surrey, on a lovely summer day in June. As we talked about his decades of research into this issue, we strolled along the banks of the river Wey, which meanders through the town. He pointed out the strange appearance of the trees on the grazing marsh, which had been browsed by cattle that graze there in summer. The leaves had been chomped to exactly cattle-head height, giving them a sort of bobbed-haircut look. The professor predicted that the meat from those cattle would taste good as well as being nutritious. His research suggests that the best-quality meat comes from animals allowed to forage for food as nature intended, grazing on grass and browsing on trees and hedges. Confinement to 'improved' pasture – grassland with little variety – reduces the quality. Feeding intensively on grain, an unnatural diet for ruminants, makes it even worse.

In a presentation to my staff team at Compassion in World Farming, Crawford talked about his study published in *The Lancet* more than half a century ago highlighting the huge difference between the fat content of farmed animals and their 'wild' counterparts. The research revealed that the ratio of 'bad' to 'good' fats in farmed animals was 50:1, compared with less than 3:1 in their wild counterparts. Since then, the picture has dramatically deteriorated.

Crawford describes modern industrial chicken-rearing as 'fat production, not meat production'. He argues that intensively reared farm animals are effectively 'selected for obesity' and get virtually no exercise. The result is meat 'marbled' with excess fat. 'If you eat obesity, you become obese' is the way he puts it. According to UK government nutritional advice, most people eat far too much saturated fat, a large proportion of which comes from fatty cuts of meat and products like sausages and pies. Diets high in saturated fat are associated with high cholesterol and heart disease. According to one study, if the UK population reduced its intake of saturated fat from animal sources by 30 per cent, rates of coronary heart disease would fall by 15 per cent, and the number of premature deaths would fall considerably.[4]

The science on this subject is complex and mainly relates to the balance of 'good' and 'bad' fats in factory-farmed meat, specifically amounts of polyunsaturated fats, known as omega-3 and omega-6 (good), versus saturated fats (bad). The life-or-death difference between 'good' and 'bad' fats in the human diet was first identified in the 1970s, when Danish doctors noticed that Greenland Inuit had exceptionally low rates of heart disease and arthritis, despite an apparently high-fat diet. Their good health was traced back to high levels of omega-3 in fish, which was their staple diet.[5] Scientists believe that people evolved to eat roughly the same amount of omega-6 fat as omega-3.[6] Current recommendations allow for a bit more of a tilt: up to four times as much omega-6 as omega-3. Yet the average Western diet ranges from ten to twenty-five times more omega-6 than omega-3.[7] There is compelling evidence that this is linked to the seismic shift from natural grazing on farmland to rearing animals on grain.[8] Unlike grain, grass is full of omega-3.

The link between the nutritional value of meat and the animal's diet is now well documented. According to an American study published in 2010, fresh forages (plants, grass and leaves eaten by grazing farm animals) have ten to twelve times more ALA – an important building block for omega-3 fatty acids – than grain. The study showed that when the amount of grain given to animals on a basic grass diet was increased, the concentration of omega-3 in the meat fell. It concluded that 'grass-finished beef consistently produces a higher concentration of n-3 FAs [omega-3 fatty acids], resulting in a more favourable n 6:n 3 ratio. The amount of total fat found in a serving of meat is highly dependent upon the feeding regimen.'[9]

A Bristol University review of grass feeding versus grain feeding in cattle found that beef cattle given a diet of fresh grass between the ages of fourteen and nineteen months had much higher levels of omega-3 than steers given grass silage – and very much higher levels than steers fed on concentrated diets that include high levels of grain or soya, as happens on feedlots (cattle factory farms).[10]

A major review of the literature found strong evidence to show that animals kept in higher-welfare conditions – not reared in highly intensive factory farms – provide more nutritious food. It looked at data from over seventy-six studies and found that meat, milk and eggs from higher-welfare farms often contain less fat and higher levels of key nutrients than their factory-farmed alternatives. As compared with factory-farmed produce, pasture-reared beef has 25–50 per cent less fat, and free-range and organic chicken up to 50 per cent less fat.[11]

The difference in omega-3 levels is striking: compared with factory-farmed produce, pasture-reared beef is on average 2.7 times higher in these essential nutrients; higher-welfare chicken is from 20 per cent to five times higher; meat from higher-welfare pigs is 40 per cent higher; free-range eggs are 30 per cent higher; and milk from pasture-raised cows is 100 per cent higher. These are important health benefits given that the chronic lack of omega-3 in the modern diet is linked to heart disease and cancer.[12]

Antioxidants in our diet are vital to good health and help fight diseases like cancer. Keeping animals in better conditions has been found to provide benefits here too. Free-range eggs can have up to double the amount of vitamin E, a powerful antioxidant, vital in fighting disease. It may play a role in preventing cancer by blocking the formation of certain carcinogens and boosting the immune system. It is also thought to prevent cataracts. Keeping hens so that they are free to roam outside also produces eggs with nearly three times as much beta-carotene. The human body converts beta-carotene into vitamin A, key to maintaining healthy eyesight, bone growth, reproduction and cells. It also helps promote healthy skin and a strong immune system. Whether it is beef, pigmeat or milk, keeping the animals in better conditions where they have a more natural diet has been found to produce better-quality food; free-range pigmeat has, on average, 60 per cent more vitamin E; higher welfare milk has 180 per cent more beta-carotene.[13]

Perversely, some nutritionists and health experts seem to think that the solution is to supplement animals' diets with additives and continue to feed them grain, rather than the more obvious and simpler option of letting them live and eat as nature evolved them.

There is a widespread assumption that chicken is a high-protein, low-fat choice. Recipes involving steamed or poached varieties frequently feature in weight-loss programmes. But factory farming has dramatically changed the nutritional quality of chicken: it is no longer as healthy as it should be. Up to a fifth of the weight of broiler chickens is now fat.[14] It's partly genetic – decades of selective breeding for plump birds – and partly diet. In any case, factory-farmed chicken contains around 40 per cent more fat than protein.

In 2005, Crawford's team published the results of an analysis of modern chicken meat. His study revealed that a typical supermarket chicken today contains almost three times more fat than a typical chicken in 1970, and a third less protein.[15] All this means that a portion of chicken today contains 50 per cent more calories than it did in 1970. Crawford also discovered that modern meat chickens contain only a fifth as much DHA – another of the omega-3 fatty acids – as is found in 'wild' chickens.[16]

Crawford attributed these huge changes in the nutritional content of chicken to factory farming, pointing out that traditionally reared chickens used to be active and eat vegetation and seeds, whereas modern, intensively reared birds are fed on high-energy foods and can barely move: 'Such chickens are no longer a protein-rich food, but a fat-rich food. The explanation is simple, namely that they are fed largely on cereals.'

That fast-food restaurants risk peddling obesity and ill health every time they sell extra-large burgers made with factory-farmed beef and supersize portions of chicken nuggets made from factory-farmed broilers is so glaringly obvious that they find themselves taken to court more and more often. In 2010 the former manager of a McDonald's outlet in Brazil was reported to have successfully sued the fast-food giant after claiming they made him put on 20 kilograms (45 pounds) during the twelve years he worked there. The thirty-two-year-old man, whose name was kept anonymous, ballooned from 70 kilograms (155 pounds) to 105 kilograms (231 pounds) while McDonald's employed him. He blamed it on the company's policy of mandatory food sampling and free lunches of burgers, fries and ice cream. He was awarded $17,500.[17]

Other cases have been more complicated, and some have yet to conclude. In 2002, a man from the Bronx, Caesar Barber, filed a class-action lawsuit in New York against a number of fast-food companies, claiming that eating regularly at the restaurants had made him obese. In the first action of its kind, he sued the companies for failing to warn him of the health risks associated with regularly eating their food, blaming them for his two heart attacks and diabetes. The fifty-six-year-old weighed 120 kilograms (266 pounds) and had been eating fast food up to five times a week. He claimed that until a doctor pointed out the dangers of his diet, he was misled by deceptive advertising suggesting the stuff was healthy. 'Those people in the advertisements don't tell you what's in the food. It's all fat, fat, and more fat. Now I'm obese. The fast-food industry has wrecked my life. They said 100 per cent beef. I thought that meant it was good for you,' the *Guardian* reported.[18]

Barber's lawyer Samuel Hirsch claimed that the main aim of this lawsuit was to 'get the chains to inform customers that their food is guilty of expanding their waistlines'. Hirsch subsequently dropped the case. According to some reports, he felt he stood more chance with a case involving children.[19] He later took on a nineteen-year-old client, Jazlyn Bradley, who, at just 1.70 metres (5 foot 7 inches) tall, weighed 122 kilograms (270 pounds); and a fourteen-year-old, Ashley Pelman, who weighed 77 kilograms (170 pounds) despite being only 1.47 metres (4 foot 10 inches) tall. This particular litigation was aimed solely at McDonald's, and originally included several other teenagers, all of whom, according to *Time* magazine, claim that 'as [a] result' of eating Happy Meals, McMuffins and Big Macs for several years, they became obese and developed diabetes, coronary heart disease and high blood pressure.[20]

It was this case that inspired Morgan Spurlock to make his famous 2004 film *Supersize Me*. It charted what happened to his physical and mental health when he spent thirty days eating only at McDonald's. During the experiment, he dined at the chain three times a day, trying every item on the menu. His average daily calorie consumption was 5,000 – double the recommended amount. The result was that he piled on almost 11 kilograms (24.5 pounds); his body mass increased by 13 per cent; his cholesterol level rose and he experienced mood swings, sexual dysfunction and fat accumulation in his liver. It took him fourteen months to lose the weight.

Hirsch argued that McDonald's child-focused advertising and toy promotions portrayed the restaurant as child-friendly, making his clients think it was fine to eat there regularly – sometimes as often as two or three times a day.[21] In September 2003, New York District Judge Robert Sweet dismissed the suit, saying the allegations were 'vague' and 'insufficient'.[22] But litigation has continued.[23]

It's not just Americans with less access to education and food choices whose health is being compromised by high-fat diets involving too much meat. There is now abundant evidence from public-health scientists that what has become a 'normal' level of meat consumption in industrial countries is excessive in terms of individual health. An overview of the global obesity pandemic published in *The Lancet* blamed the public-health crisis on 'changes in the global food system',[24] a reference to factory farming, which has made animal products, especially animal fats, too cheap. Average meat consumption in rich countries is currently around 200–300 grams per person per day. According to a group of public-health experts from Cambridge University in the UK, the London School of Hygiene, the Australian National University and the University of Chile, this needs reducing to a global average of 90 grams per person per day, for both environmental and public-health reasons. These scientists argue that limiting meat consumption to 90 grams a day would bring 'important gains to health' for people who currently eat more than that, including a likely reduction in the risk of colorectal cancer, breast cancer and heart disease, as well as the poor health linked with overweight and obesity. They argue that the reductions in heart disease would be mainly due to reducing the consumption of saturated fat in meat.[25]

Factory farming is behind the very high level of meat consumption in the most industrialised countries these days. Grain or other specialised high-protein and high-energy feed is used to raise yield per animal and maximise 'food-conversion efficiency' – in other words, the amount of meat produced relative to the amount of feed used in rearing the animal. Animals are kept in highly concentrated and confined conditions to reduce labour and feed costs. These conditions limit the amount of energy the animals 'waste' by limiting their movement; reduce the effects of temperature fluctuations by keeping them indoors; and cut the amount of time and energy 'wasted' by farm labourers who do not work in open fields. Specialised high-yielding livestock strains are used, selected because of their ability to grow and mature faster, put on more muscle or produce more milk, and requiring specialised feed, veterinary drugs and housing in order to reach maximum productivity. These methods meet – and arguably fuel – the huge demand for meat.

However, the conversion of potential human food like grain into meat in factory farms remains fundamentally inefficient: more calories go into the farm animal than come out in the form of meat, milk or eggs. When farm animals were first domesticated, ruminants like cows and sheep would eat grass that people couldn't eat and turn it into food. The conversion rate didn't matter because the animals weren't competing with people for food. Similarly, pigs and poultry were kept to eat scraps and leftovers and to forage; again, providing a useful service where they didn't compete with people for food. On the factory farm, potential human food – grain – is now absurdly fed to farm animals in a process that produces unhealthy meat and other products.

There are encouraging signs that meat consumption in the West may already have peaked or be on the verge of doing so. Euromonitor International, the world leader in consumer market research, predicts that sales of alternatives to meat will rise by 15 per cent in value between 2010 and 2015, and has highlighted the 'massive potential' for growth in demand for non-meat-based food of all types.[26] In September 2011, Rabobank, the international agricultural bank based in the Netherlands, released a report called *Where's the Beef?* claiming that meat consumption per head in America 'appears to have peaked' and that the industry should no longer rely on increasing domestic demand to get them through 'over-production situations'.[27]

In December 2011, the Values Institute at DGWB, a California advertising and communications company, predicted that one of five 'health and wellness trends' Americans are most likely to embrace in 2012 is 'flexitarianism' – the reduction of meat consumption for health reasons, without eliminating it altogether. One example of this trend is the growing popularity and social media following of so-called 'Meatless Monday', an initiative promoted by Johns Hopkins Bloomberg School of Public Health.[28] However, the trend is exactly the opposite in rapidly developing countries, particularly China. Rabobank predicts continuing rising global demand for meat.[29]

IV
MUCK

The old saying 'Where there's muck, there's brass' meant that there was money to be made from dirty jobs. In the twenty-first century, livestock farming has turned this adage on its head: disposing of mountains of animal muck is costing the earth.

In 2002 there was an outcry among UK farmers when they were banned from dumping manure wherever they wanted. New EU legislation was introduced to tackle a mounting environmental crisis linked to excessive quantities of animal waste and fertiliser spread on fields leaching into ponds, rivers and lakes. Desperate to overturn the new measures, the agricultural industry launched a scare campaign, suggesting that up to 10,000 farmers could be forced to cart millions of tons of manure across the country. They stoked up fears of congestion and disease, not to mention the inevitable stink. 'There will be manure criss-crossing the countryside from livestock areas to arable areas,' said a consultant to the National Farmers Union. The NFU's president wrote to the government claiming that the restrictions on spreading manure and fertiliser in sensitive areas would cost farmers about £100 million.[1]

The row highlighted just how serious the consequences of separating animals from the land had become. As animals have been moved off fields into barns, the age-old nutrient cycle of manure from grazing cattle and pigs replenishing tired soils has been broken. When too much manure is produced for any given areas, as happens on megafarms, rivers and lakes are all too often polluted.

Britain's farm animals produce 80 million tons of muck a year. An average-sized dairy herd of a hundred cows can produce as much effluent as a town of 5,000 people. Across the country, there is a total of 1.8 million dairy cows,[2] not to mention many millions of pigs, chickens and other farm animals. Disposing of all this manure has become a phenomenal and increasingly costly battle. The muck presents the single gravest threat to our waterways, as Britain's farms account for the lion's share of water pollutants like nitrates and sediments.[3] Fertilisers and manure spread on the land can be washed into waterways; the excessive nutrients suck oxygen out of the water, choking aquatic life. Eventually, the water can become a dead zone, in which nothing can live.

The trend toward intensification of livestock farming, with more animals on fewer farms, is likely to make things worse. Manure, once a valuable commodity in the natural process of growing food from farm to fork, is now a problem. From the beaches of Brittany to the rivers of North Carolina, we find stark warnings of the perils from mountains of muck. But is anyone listening?

9

Happy as a Pig

Tales of pollution

BRITTANY IN BLOOM

Minutes before they collapsed, Vincent Petit and his horse had been happily galloping along the beach. When they set out for a long ride along Brittany's rugged northern coast, both the young man and his fifteen-year-old mount had seemed in perfect health. As a vet, Petit would have been better placed than most to spot anything wrong with his thoroughbred, and there had been nothing to trouble him as they thundered along the sand.

Coming to a rocky stretch, Petit pulled the horse up and dismounted, to walk the animal along a rough beachside road. It was here that they slipped, plunging into a hidden pool of algae that had accumulated on the shore. The sludge was so deep it reached to the horse's withers. Panic-stricken, Petit yelled for help. Struggling not to sink further himself, he battled to hold his horse's head out of the sludge, but within seconds the animal was dead. 'I cried for a man on a tractor to throw a rope, and then I looked at my horse and saw that his nose was falling into the sludge. I held his head up for him, but a few seconds later, he went into respiratory arrest, without even a fight. It was incredibly fast,' Petit recalled.

The horse was not the only victim. Moments later, the vet too lost consciousness. He was rescued by passers-by, who prevented him from drowning in the algae. Initial veterinary reports suggested that the horse had died from asphyxiation. An autopsy told a different story. The horse hadn't drowned: it was poisoned. Both animal and rider were overcome by toxic gases from the algae.

The tragedy happened in summer 2009, casting the spotlight on the environmental disaster unfolding on Brittany's beaches. Once upon a time Saint-Michel-en-Grève, the quaint seaside resort where the accident took place, attracted thousands of tourists. Now the beaches in the area are no-go zones, and hotels are struggling. Sunbathers and sightseers have deserted the picturesque inlet and coves. Holidaymakers have been replaced by workers in bulldozers battling to clear the unsightly and potentially lethal algae that have been washing up on the shore.

Commonly known as sea lettuce, the weed-like algae, resembling lettuce leaves, are naturally present in small quantities all along the littoral. But in certain conditions, when the warm summer sun hits the calm shallow waters of some of the prettiest bays in the region, they proliferate, spurred by an excess of nitrogen carried downstream by polluted rivers and waterways. The algae sweep in with the tides in what is described by locals as mighty green swells and amass in stinking piles on beaches. As they dry up and begin to decay, they release hydrogen sulphide, otherwise known as 'sewer gas', and other toxic fumes that get trapped in pockets under a fine white crust of residue.

Apart from their reek (they smell of rotten eggs as they putrefy), the algae were long believed by locals to be harmless. According to Yves-Marie Le Lay, a philosopher who heads up a local conservation group, though dogs occasionally went missing on the contaminated beaches, the owners simply assumed they'd drowned.

However, in recent years a series of fatal accidents, beginning with the death of two large dogs on the Saint Maurice beach in 2008, have left little doubt that the algal blooms pose a threat to all forms of life in the water and on the shore. In summer 2011, the corpses of thirty-six wild boars, a badger and a river rat were found washed up on the Saint Maurice beach near the mouth of the Gouessant River, all within days of each other. Autopsies revealed lethal doses of hydrogen sulphide in the tissues of all but one of the animals. Despite the weight of evidence, it was nearly a month before the authorities confirmed publicly that the animals had been killed by the noxious algae.[1] 'It must not be said that the green tides kill,' Le Lay told us when we went to Brittany to investigate. He was referring to the way that this crisis is being hushed up by local government, for fear of what it could mean for tourism.

The Saint Maurice beach has since been closed; a sign in French blocking the way simply reads 'Access Prohibited'. On the brushy headland next to the beach, a small chapel bears witness to the monster on the beach. Local activists have nicknamed it 'Notre dame des algues vertes' (Our Lady of the Green Algae).

As life ebbs from areas contaminated with the green algae along Brittany's northern shores, elsewhere in the region business is booming. For just beyond the windswept vistas of the Côtes d'Armor and quaint cobbled summer resorts that draw in tourists from around the world lies a very different kind of landscape: an ugly vista of industrial pig farms. Not far from the bay of Saint Brieuc is a dense network of highways and roads that link together a thriving industrial hub. There are warehouses, factories, processing plants and superstores, all connected to the region's most important economic sector: the production of pork, eggs and milk.

Brittany produces 14 million pigs a year.[2] The region is France's top pork producer: over half of the nation's pigs are reared here.[3] There are as many pigs produced annually in these parts as there are people in London. Despite the number, you'd be hard pressed to spot a single one rooting in a field or farmyard. Here, pig farming is heavy industry, driven by modern machinery, cheap labour, biotechnology and metrics that demand the highest output at the lowest cost.[4] In France, this type of farming is known as '*hors sol*' or off-land.

Ask a local environmentalist about the green algae crisis and you'll invariably get a history lesson that begins just after the Second World War. It was then that '*les Américains*', helping to rebuild the country after the war and keen to develop new export markets for their own agro-industry, brought factory farming to France. As André Ollivro, head of a local environmental campaign group, told us when we visited, no one could have anticipated the consequences. 'A farming model was developed but no modelling or trials were ever carried out,' he said gloomily.

Brittany's landscape was transformed. Le Lay describes how massive monoculture cereal plantations replaced the graceful rolling pastures of clover or alfalfa that animals once grazed on. Farm animals were sent indoors into feedlots, industrial triumphs of productivity that measure their performance with terms like 'stocking density'. According to Le Lay, cereal crops now harvested on Breton lands are processed into animal feed, alongside other crops sourced from around the world, including soya, often from deforested land in Brazil. Access to the global commodities market means there is never a shortage of food for the ever-growing herds.

And the pigs themselves contribute to their own nourishment. They are housed on perforated concrete floors above massive containers that collect their waste. The liquid manure is then spread heavily as agricultural fertiliser, largely on the corn plantations that now cover most Breton farmland. Corn is a resilient crop which can withstand doses of pig slurry as fertiliser that would flatten other crops such as wheat. The fields are said to be doused in treble or quadruple the amount needed to optimise yields. 'Where else are they to put it?' Le Lay told us with a French shrug.

According to André Pochon, a pioneer of sustainable farming in the region, Brittany's corn fields have become a 'dumping ground' for pig manure: 'Local authorities have aggravated the situation by allowing farmers to spread the slurry when the land is bare [of crops]. When it rains, the slurry drains into the rivers and aquifers. It's nonsensical; these practices defy the most elementary laws of agronomy.' The result is that massive quantities of phosphates and nitrates sweep into the region's aquifers and waterways, ultimately draining into the Atlantic. There they feed the spread of algal blooms that deplete the water's oxygen level and suffocate fish and marine life. In the end, there are dead zones that can no longer sustain life. 'Nothing lives in the bay of St Brieuc any more,' said Le Lay. 'Everything has died. We used to be able to fish periwinkles and clams from the rocks. The sand is black now; it's saturated with hydrogen sulphide.'

When the weeds finally make their way through the tides onto dry land, they pose a potentially fatal health risk. Thierry Morfoisse worked for one of the local companies charged with cleaning some of the beaches contaminated with green algae. In 2009, having unloaded a final load of seaweed at a treatment centre, he passed out and was found dead by the foot of his truck at the roadside. Paramedics took a blood sample onsite; his body, its darkened skin showing signs of suffocation, was taken straight to a funeral home. His family was told he had died of natural causes.

The affair might have been kept quiet were it not for the horse that died days later on the beach in Saint Michel-en-Grève, raising alarm bells at the Elysée and prompting a visit to the region from the then French prime minister François Fillon. An autopsy carried out two months after Morfoisse's burial, unknown to his parents and against their wishes, revealed that the forty-eight-year-old algae worker had died of cardiac arrest,[5] possibly resulting from a pulmonary oedema, the often fatal signature of hydrogen sulphide poisoning.

Understandably, the authorities appear to have been all too well aware of the public alarm that this was likely to cause. Apparently, Morfoisse's death was reported to have been linked to basic poor health and his smoking habit. His family is now suing for involuntary manslaughter. They believe his employers were negligent, failing to protect him from the poisonous fumes. 'He was well and truly taken to the gas chambers,' is how his father Claude put it. They are still waiting for the case to be heard before Parisian courts:

There were no gas masks, no metering devices [used to detect the presence of hydrogen sulphide], no automated boxes [to unload the algae], nothing at all. My son had to come out of his truck to unload the containers; he was caught between two fires – the gases at the treatment centre and those coming from the truck. I keep saying, I don't want my son's death to be in vain. Something has got to be done.

His wife Jeanne, a small, gentle woman, speaks of her concern for the children playing on the beaches. She sees them scrambling on the rocky shores of the Côtes d'Armor, where perhaps her own son once played. They are rough-and-tumble playgrounds where seaweeds naturally accumulate, beyond the reach of the bulldozers sent to clear the beaches. 'If a child dies on these beaches, there will be no other name to call them but murderers,' she told us, referring to the local authorities.

Experts agree that the green algae crisis is just the most visible aspect of the pollution from pig farms. Dr Claude Lesne, who specialises in pollution at the University of Rennes, told us that the pig manure contains traces of pesticide and cadmium, a carcinogen that can disrupt the endocrine system and is added to zinc supplements to speed up growth, as well as residues of the antibiotics routinely given to the pigs. There are many public-health implications when these wash off the fields into the water supply. In the dark, rank sheds where the animals are reared, the air is filled with toxic gases, faecal germs and bacteria. Both animals and the farmers in this area frequently suffer from chronic respiratory disorders and other health problems.

Were it not for the powerful ventilation system required to keep the animals, to the untrained eye the pig farms would be fairly indistinguishable from any other warehouse. According to Thierry Dereux of the pressure group Côtes d'Armor Nature Environment, the reach of the gases is so extensive that residents of Lamballe, a small city in the hinterland of the Côtes d'Armor, are forced to breathe higher than normal doses of ammonia.

Few things in the region seem untouched by the industry. When we visited in August 2011, locals were celebrating the annual mussel festival. The celebration will soon become history. We were told that the water pollution is so bad that harmful bacteria including *E. coli* are rife, and by 2015 the EU is expected to impose a ban on the sale of mussels from the area. People are doing their best to put a brave face on it. There is a solidarity that local activists prefer to call denial.

'*Dieu merci*, there will always be smokers,' a woman who runs a small tobacco shop and news-stand told us as we scanned the latest headlines about the wild boars that died on the beach. She voiced fears that the dead animals would finish off many local businesses. Le Lay believes most folk in the area are still in denial. 'The stakes are too high,' he said. Tackling the problem 'would mean radically revisiting the agricultural model. But when thirty-eight animals die, it's time for emergency measures.'

Campaigners in the area have modest demands: an information campaign to raise public awareness of the risks and a more thorough clean-up of the shores. I couldn't help feeling that such measures would be no more than a sticking plaster over a gaping wound. It's clear that the problem won't go away until it is stopped at source. This means starting to restore the balance between land and animals, and recognising that there are now far more pigs than the land can sustain.

There are about fifty pig farms in the region now using what is called the Pochon system, a more sustainable form of pig farming where the animals are fed mainly from what is grown on the farm and live happily on straw beds.[6] Alternatives to industrial farming exist and are successfully being practised by some farmers, but they are still considered mavericks by many of their peers in these parts.

PINK LAGOONS IN AMERICA

After giving a quarter of a century of service to his country in the United States Marine Corps, Rick Dove was looking forward to a gentler way of life as a fisherman on the river Neuse near his home in North Carolina. Snaking its way right across the state, the waterway, 442 kilometres long, has a proud history. It is said to have been named by two of Sir Walter Raleigh's scouts, Arthur Barlowe and Philip Amadas, who found it while exploring the New World in 1584. They named it after the American Indian tribe known as Neusiok, who lived along its reaches.

Dove's childhood dream was to make a living on the river, and when he left the forces in 1987 he bought a boat and set to work. For a while it was all that he hoped it would be. He caught enough to make a good living, and his son went into business with him. Between them they owned three boats. They crabbed and net-fished, selling the catch both retail and wholesale. They opened their own seafood store in the local town.

Just a few years later, however, father and son were forced to abandon the business after the fish began to die and they themselves fell seriously ill. Their livelihood and health were under threat from a deadly organism. The second element of its name, *Pfiesteria piscicida* (PP), is the Latin for 'fish killer'. It's often referred to as the 'cell from hell'. Between 1991, when it was first detected in the Neuse, and 1999 when it peaked, this single-cell organism is thought to have killed more than a billion fish in North Carolina. It also made many fishermen sick.[7] Its arrival in the Neuse was traced to pollution from pig manure.

The commercial hog industry in North Carolina is huge. It is worth around $2 billion a year to the economy,[8] and at any one time there are around 10 million pigs being raised on factory farms in the state.[9] That number of pigs makes mountains of poo: in one North Carolina county alone, 2.2 million pigs generate as much untreated manure as central New York City creates sewage.[10] Like the manure produced on every factory farm, it has to be disposed of somewhere – and all too often it ends up in the wrong place.

The state of North Carolina is no stranger to factory farm-related disasters. In 1995 it was the scene of a catastrophic spill when the dike of a 10,800-square-metre lagoon filled with liquid manure belonging to a pig-farming company ruptured. A torrent of pig waste – 117 million litres – poured into the headwaters of the New River. At the time it was the biggest environmental spill in US history, more than twice the size of the *Exxon Valdez* oil spill six years earlier.[11] Almost two decades on, environmental legislation in North Carolina is tighter but the potential for manure-related disasters persists. Huge lagoons of pig sewage, upwards of 3,000 of them,[12] dot the North Carolina landscape. Each can hold vast quantities of faecal matter, urine and other obnoxious by-products: blood, excrement, afterbirths, even still-born piglets. The liquid is pinkish brown. Anyone who falls in is almost certainly doomed.

In an exposé of the pig-farming industry some years ago, *Rolling Stone* magazine described the perils of trying to save a person who falls in. The article cited a case in Michigan involving a worker who passed out while repairing a lagoon, overcome by the fumes, and toppled in. His fifteen-year-old nephew dived in to save him but was likewise overcome. The worker's cousin went in to save the teenager and suffered the same fate. Almost unbelievably, two further relatives died the same way that day trying to rescue the others.[13] It sounds too incredible to be true, but pig manure is ten times more water-polluting than untreated domestic sewage.[14]

Inevitably the lagoons sometimes overflow, allowing the noxious soup to flood over fields and seep into groundwater. Campaigners claim that major floods turn entire counties into 'pigshit bays'.

To stop this happening, farmhands sometimes reduce the level of the lagoons by pumping out some of the stuff and spraying it over nearby farmland. Unfortunately the process is not always carried out in moderation, which brings what the industry calls 'over-application'. The land becomes saturated with pig waste, which festers in stagnant toxic pools. Scientists attribute to this the devastating attack of *Pfiesteria piscicida* in the Neuse. It appears to have caused eutrophication of the river, a process in which phosphorus and nitrogen (high concentrations are found in livestock waste) over-enrich water, distorting the ecosystem. The excessive nutrients create the ideal conditions for algae like PP to flourish, sapping oxygen levels in the water till eventually there is too little oxygen to support any other life, and leading to so-called 'dead zones'.

Dove shudders at the memory: 'Between 1991 and 1999, we lost billions of fish – certainly a billion and a half. I know that because I was on the river. I watched. These fish had huge holes right through their body. They'd get sores, these bleeding, ulcerating sores, and fishermen get the same sores.' He said the problem was unknown in North Carolina before the swine industry arrived. It was not long before Dove himself got sick, as did his son. As they would no longer eat fish from the river themselves, he says they couldn't possibly have continued to sell it to others. He described PP as a 'vampire' because of the way it eats away at fish flesh. In humans, prolonged exposure has been linked to brain scarring, central respiratory problems and memory loss. Dove's memory was impaired, his respiratory system was damaged, and his immune system remains weak.

Dr JoAnn Burkholder, the biologist from North Carolina State University who first identified PP, experienced its corrosive effects herself. One day early in her research, while pouring the algae into a flask, she recalls how she became disorientated and began to suffer from stomach cramps. Her eyes became so bloodshot that for a few hours she could hardly see. She also suffered short-term memory loss. The research was moved to a secure room, but not before another colleague had to be hospitalised.

After he gave up fishing Dove did something else for a while, but he pined for the water. When an opportunity arose to be river keeper on the Neuse in the mid-1990s, he leapt at the chance. That's when he began investigating who was to blame for wrecking the river. 'I've been trying to pay them back ever since,' he said. Now retired, Dove regularly takes to the skies above North Carolina in a beat-up old Cessna. From the air, he documents the environmental impact of large-scale industrial pig farming. Over the last sixteen years he has compiled a dossier of more than 80,000 pictures and hundreds of hours of film footage exposing the pollution caused by the farms, most of it linked to manure.[15] He uses the material he has gathered to fight for better standards, particularly from the meat giant Smithfield, the largest pork conglomerate in the world. The company owns most of the pigs in North Carolina, but as Dove explains, likes to claim it doesn't own their poo: 'Smithfield tries to contend that when the poop comes out the back end of the pig they no longer own that – that belongs to the grower who's going to grow the pigs for them.'

After years of legal battles, things are looking up. There are new regulations in North Carolina governing the disposal of pig waste. Pricey Harrison, North Carolina's Democratic member of the House of Representatives, has also introduced a bill to make all hog operations compliant with new air and water-quality standards by 2016.

Yet the environmental and public-health nightmare created by the seas of manure from factory farms is not unique to this place. Wherever there are factory farms, there are all too often problems with disposing of the manure they produce. For fly-by-night operators, the temptation is huge to cut costs by 'accidentally' allowing lagoons to leak, or by breaking regulations on the quantity that can be spread on fields. Even conscientious operators make mistakes. On a typical factory farm, the amount of manure is simply so enormous that disaster is rarely more than a cowpat away. According to the US Government Accountability Office, an independent non-partisan agency that works for Congress, a very large hog farm rearing 800,000 hogs – of which there are at least two in America – could generate more than 1.6 million tons of manure annually, or more than 150 per cent of the urine and faeces produced by the 1.5 million residents of the city of Philadelphia, Pennsylvania.[16]

On traditional mixed farms, manure is a valuable fertiliser. The trouble with factory farms is they produce far too much, too far away from land that could benefit. All too often, it ends up in the wrong place. According to a review by American and Canadian environmental scientists, releases of phosphate into the environment now exceed 'planetary boundaries' – they have overloaded nature's means to cope.[17] The result is widespread eutrophication of lakes and rivers, something I witnessed myself in China. There are now dead zones across the world, devastating fishing and tourist industries as well as contaminating drinking water.

Repairing the damage is costly. In Kansas alone, scientists have put the cost to taxpayers at $56 million. As Kansas is not one of America's major dairy or pig-producing states, the bill in other states must be far higher. Based on data put together by the Union of Concerned Scientists, a not-for-profit environmental pressure group, a rough estimate of the total cost of cleaning up after American pig and dairy factory farms could approach $4.1 billion.[18] No wonder factory farmers are so careless about the waste they produce. Disposing of it properly involves big money.

The US Department of Agriculture believes it would cost around $1.16 billion a year to spread manure over farmland in an environmentally friendly way.[19] The trouble is that many arable farmers would refuse to accept it as fertiliser anyway – even if it were free. One reason is the stench. The large tanks and lagoons in which it is usually stored create chemical conditions that make it smell far worse than it does when it initially emerges from the animal – like rotten eggs.

The scene is little better in Europe. Under EU law, large pig and poultry facilities are now classified as 'industrial installations' for pollution purposes. It means they are bound by the so-called 'Pollution Prevention and Control regime'. This requires each 'installation' to have a permit setting out emission limits and various other conditions on the release of pollutants. In theory, the rules imply a minimum 'spreadable land area' available per animal. A 1991 EU directive on nitrates set out certain limits to the amount of manure that can be applied to land and when it can be spread. However, Brussels admits the effects have been limited, saying recently that member states need to 'step up their efforts regarding monitoring, identifying pollution hot spots, and tougher action programmes'. The European Commission has noted that 'storage capacity' for manure remains a 'frequent problem', suggesting that factory farmers are failing to cater for periods when applying manure to land is banned or impossible because of bad weather.[20]

Throughout my travels, the muck issue has always come up, from Lake Taihu in China, where pollution has ruined the drinking water, to California, where rivers have been stripped of fish. It's probably no coincidence that a proposal for Britain's first US-style mega-dairy was dealt a death blow by the Environment Agency's concerns about water pollution. Near and far, the effects are being felt.

10

Southern Discomfort

The rise of the industrial chicken

After an idyllic childhood in rural Georgia, USA, Janisse Ray never doubted what she would do for a living. As a little girl, she spent endless happy days with her grandparents Arthur and Beulah on their farm, roaming around, building dens and eating crab apples, pomegranates and muscadines, a delicious type of purple grape. She remembers her grandfather's mules, her grandmother's chickens, fields of vegetables and sprawling watermelon vines, and granaries full of corn. During her pre-school years, her grandmother milked a cow.

The post-Second World War years brought the so-called Green Revolution in the USA, and agribusiness began to woo farmers towards a new dawn. It marked the start of a long slide from agrarian to industrial life. American farmers began using heavy machinery and fertilisers; the face of the countryside was transformed.

When Ray was six, her grandfather died. Soon after, farmers in the area began to grow subsidised tobacco and douse their crops with Roundup weedkiller. 'Then came monster combines, terrible erosion and the invasion of privet,' Ray recalls. The cane grinder was sold, the smokehouses fell, the last hen wasn't even eaten. Her grandmother sent the milk cow to the livestock auction. Ray remembers the final pea patch. On that same farm where she spent her childhood, the fields are now sown with genetically modified soyabeans. The fences have been torn down, the wax myrtle has gone, and the wild cherry trees have been felled. The mockingbirds and cardinals have fallen silent. The sassafras tree her grandfather carefully skirted with his harrows is long dead.

In the neighbours' field are two industrial chicken houses, lit day and night, where tens of thousands of birds are fed grain with antibiotics until they are fat enough for slaughter. Whoever is in charge is not a farmer but a 'contract grower', rearing the birds for some distant corporation. Sometimes the smell of burning chicken corpses fills the air. Sometimes the air is acrid with the smell of chicken litter being spread on fields. Everyone hates it, but something must be done with the muck.[1]

Today Ray lives in a beautiful old farmhouse with forty-six acres of land in southern Tattnall County Georgia, in the delta of the Altamaha and Ohoopee Rivers. Despite the ugly transformation of the agricultural industry in Georgia, she never gave up her dream of becoming a farmer and now makes a living growing organic vegetables, pecans, fruit and seeds, as well as producing meat and eggs from grass-fed animals free from growth-promoting antibiotics and hormones. Her farm shop sells homemade sodas – birch beer, ginger ale, root beer, raspberry and cream – as well as jams, jellies, woodcut note cards, granola and cakes. She also runs workshops on cheese making, fermentation, backyard chickens and other modern homesteading and sustainability skills.[2]

It could not be more different from the dominant type of farming in the state of Georgia today: industrial chicken production. Chickens are the most populous farm animal on the planet. Worldwide, 55 billion chickens are reared for meat each year; nearly three-quarters are factory-farmed.[3] The US produces nearly 9 billion broiler chickens for meat a year on some 27,000 farms.[4] A typical farm will raise 600,000 birds a year.[5] And with that number of chickens comes an awful lot of chicken muck. The chickens in Maryland and Delaware alone generate 42 million cubic feet of litter – enough to fill the US Capitol dome nearly fifty times over.

Down the years, I've visited many chicken farms rearing broiler birds for meat. From Britain to Beijing, they're all much of a muchness. Commercial chicken production is dominated globally by just two or three breed companies responsible for supplying the genetic strains of an estimated 80 per cent of the chicks. The white-coloured fast-growing broiler chicken, together with its health and welfare problems, has become a standardised global product.

In hotter countries like the Philippines, there are often slightly fewer birds packed into each building and the sheds are frequently open-sided with natural light. In the EU, the birds are most often kept in fully enclosed industrial buildings. In the Peruvian desert, they were housed in what looked like long low tents. These differences are mainly cosmetic: the conditions inside the buildings are all along the same industrial lines. Europe recently passed new legislation designed to protect the welfare of meat chickens, but it did little to help. In fact, Britain's chickens can be packed even more tightly into sheds than under government guidelines in the 1990s.

Georgia is the largest producer of meat chickens in the United States, rearing 1.4 billion meat chickens every year.[6] If it were a country, it would be the sixth-largest poultry producer in the world.[7] The heart of the industry is in the foothills of the Blue Ridge mountains in a place called Gainesville. It enjoys a dubious status as 'poultry capital of the world'. Georgia's chicken industry started to evolve as early as the Great Depression, when Jesse Jewell, a feed salesman from Gainesville, was struggling to keep his business afloat. Millions of Americans were out of work, some on the brink of starvation. Bread lines were a common sight. Farmers were desperate and had little money to buy feed or chickens. Jewell came up with a plan to sustain sales: he started selling baby chicks to Georgia farmers on credit. The farmers raised the chicks and then sold the fully grown birds back to him for a profit. Eventually Jewell had enough farmers producing broilers for him that he was able to invest in his own processing plant and hatchery.[8] He was a pioneer of so-called 'contract farming'.

In 1939, there were fewer than sixty chicken farms in Hall County. With the onset of the Second World War, the poultry industry in Georgia began to grow. The US War Food Administration, which administered food reserves during the conflict, snapped up all the chicken processed in North Georgia, giving poultry farmers a guaranteed buyer. The industry boomed. Within a decade, Hall County had over 1,000 chicken farms.[9]

Jewell's own business expanded. In 1954 he added a feed mill and rendering plant.[10] Eventually, J. D. Jewell Incorporated had the resources to manage every phase of chicken production, from hatching to processing, distribution and marketing. His 'vertically integrated' production model, in which he controlled a chain of companies managing every step of the process, set a new industry standard, as did his trademark frozen chicken. His approach to hiring was also innovative: he was one of the first employers in Gainesville to take on black workers.[11]

For the next couple of decades, broiler production in Hall County continued to thrive. Sensing a quick buck, everyone piled in. Competition drove down prices, but through the 1970s and 1980s Americans were acquiring a new taste for poultry over red meat, and there was no shortage of demand. During the 1990s the number of companies involved in the business fell, but not because there was any less appetite for chicken. It was simply that the industry consolidated, with fewer and bigger operators, serviced by an army of contract growers, who continued to expand. Within less than a decade, the number of chicken farms fell by half.[12] Today, just a few companies dominate the industry. In 2006, Gold Kist, an Atlanta-based company founded during the Great Depression, merged with Pilgrim's Pride Corporation to form the world's largest poultry company.[13] These food giants are vertically integrated and continue Jewell's practice of contracting out to farmers the business of turning chicks into broilers.

In 2012, Compassion in World Farming expanded its work beyond Europe into the US. One of its first initiatives, led by US director Leah Garces, was to set up a coalition of concerned organisations under the banner 'Georgians for Pastured Poultry'. A detailed study was commissioned of the broiler chicken business in Georgia, where conditions and practices are fairly typical of those in the industry worldwide. The findings exposed the grim realities of this type of farming, not only for the birds themselves, but for many of those who make a living from the industry.[14]

For the big companies who pull the strings, it's a business that continues to bring in megabucks, but for the farmers – or rather 'contract growers' – it's a stressful line of work to be in, characterised by the constant threat of being cut loose by their sole customers: the companies who employ them to fatten the birds. The chickens remain the property of the firms, who set out their expectations in legally binding contracts and pay according to targets and performance league tables.[15] Growers have little negotiating power, and anti-competitive behaviour by those who hold the purse strings appears commonplace.

Carole Morison, an award-winning broiler farmer whose contract was terminated after twenty-three years, has told how she was constantly bullied by the companies she worked for. 'I can't count the many, many times that I have heard in one shape or form that our contract was going to be terminated if we did such and such. That's no way to communicate with people who are your business partner,' she told a public workshop.[16] This is the way that 99 per cent of broiler chickens in America are reared,[17] not by farmers on homesteads using generations of acquired knowledge about livestock and land, but by producers at the beck and call of company bosses in far-off offices who issue instructions and targets.

For the chickens themselves, it's a grim existence, with virtually no legal protection. All farm animals are exempt from America's Federal Animal Welfare Act. Despite comprising 95 per cent of farm animals reared in the US, for some reason chickens are not even protected by the Humane Methods of Slaughter Act.

No other farm animal has been as selectively bred as the meat chicken to reach an unnatural size so quickly. Over the last fifty years, growth rates have quadrupled.[18] Getting the birds to target weight now takes no more than seven weeks.[19] At this age, they are still very

young; chickens don't reach puberty and start laying eggs, for example, until about eighteen weeks old. This rapid growth has allowed for
mass production of cheap meat, but the chickens pay a heavy price,
from leg problems to heart disease and a condition colourfully dubbed
'flip-over syndrome', characterised by sudden frantic wing-flapping,
shaking and loss of balance, sometimes accompanied by a shrill clucking. Within a minute or so, birds fall onto their backs or sides and die.

Like other factory-farmed animals they are kept in extremely
crowded conditions. The typical 'grow house' for a flock in Georgia is
reported to be 15 metres wide and 150 metres long,[20] and contains
more than 30,000 chickens. Each bird has what has become the
customary floor space, the size of a sheet of standard typing paper.

Because of cramped conditions and obesity, leg disorders are
common. In fast-growing breeds, the development of birds' bodies
cannot keep pace with their weight gain, making walking painful.
Most move around only when absolutely necessary to reach food or
water. Towards the end of what the industry calls their 'growth cycle',
or just before slaughter, they spend most of the time sitting or lying
down. Some are in such poor shape they can barely walk.

Leg problems are just one of many health problems associated with
the system. Heart and lung disorders are another, linked to the sheer
size of the birds' bodies. Not surprisingly, an estimated 42 million
chickens in Georgia every year die before they get to slaughter weight.[21]
Because it makes them eat more, and therefore grow faster, they are
typically kept in continuous, or near-continuous, light. This has
become a major welfare issue, disrupting their natural bodily rhythms.

When they are ready for slaughter, chickens are manually caught
for taking to the slaughterhouse. Catchers grab up to seven chickens
at a time – three in one hand and four in the other – before pushing
them into crates for loading onto trucks. It's an unpleasant process,
usually carried out in the dark in an attempt to reduce stress. According to the Southern Poverty Law Center, a civil rights organisation
based in Alabama, supervisors require catchers to grab and crate birds
at the almost unbelievable rate of around 1,000 an hour.[22] On arrival

at the slaughterhouse, where as many as 200,000 birds a day are processed,[23] they are typically taken out of their crates, loaded onto conveyor belts and shackled upside down by their legs. They may be stunned – rendered unconscious – before having their throats cut, but unlike in the EU, it is not a legal requirement in the US.

On welfare grounds alone, there is clearly a case for changing the way meat chickens are reared, but it isn't only the birds who suffer. According to a 2010 report by the US Bureau of Labor Statistics, poultry processing is among the industries with the highest rate of non-fatal occupational illness.[24]

Georgia offers a depressing insight into the working lives of the 47,000 people employed in the industry, a high proportion of them female and Latino.[25] Many of these workers, especially those in the lowest-paid roles, face daily hazards and have few rights and protections.[26] Being a catcher is particularly unpleasant. Workers are contracted by poultry companies to round up birds for transportation to slaughter. They are typically paid per truckload filled, at rates that work out at less than the minimum wage.

Tom Fritzsche, from the Southern Poverty Law Center, has interviewed many catchers. 'One thing that is really noticeable about many chicken catchers is how their hands look,' he reported. 'They are swollen to double the size of a normal hand and some workers even shrink back from shaking hands, because it's so painful. There's even a condition that people refer to as "claw hand", that some chicken catchers develop from gripping so many chickens so tightly over the years.'

Catchers tend to be divided into crews of seven or eight workers, and have limited contact with the companies that ultimately pay their wages. It's a fly-by-night business. Investigations have suggested as few as 5 per cent of workers have clear employment conditions and many travel from job to job in unlicensed, unsafe vehicles. Fewer than 15 per cent of crews keep proper records to ensure that workers are paid properly for the hours they do.[27]

Jobs at processing plants fall into three broad categories: slaughtering, deboning and packaging. Specific roles include live hanging, a task that involves removing birds from cages when they arrive and hoisting them by their feet onto moving shackle lines; wing folding – twisting and tying carcass wings into position for cutting; wing cutting – using scissors to remove chicken wings from carcasses; and deboning – cutting meat from dead birds. There are also many unpleasant cleaning jobs.

Of these jobs, live hanging is among the worst. To keep the birds calm, they stand on hard floors in almost complete darkness, lifting each bird by its legs to hang it on hooks, at shoulder or head level, attached to moving conveyor belts. They may hang up to twenty-six birds per minute, a process that often strains workers' upper shoulders and necks.[28] Here's how one worker described the experience:

> I hung the live birds on the line. Grab, reach, lift, jerk. Without stopping for hours every day. Only young, strong guys can do it. But after a time, you see what happens. Your arms stick out and your hands are frozen. Look at me now. I'm twenty-two years old, and I feel like an old man.[29]

According to official figures, around one in seven workers is injured on the job, more than double the average for all private industries. Poultry-line workers are forced to keep pace with speeding conveyor belts, repeating the same finger, hand, wrist, arm and shoulder movements as many as 20,000 or even 30,000 times a day.[30] The result is that they are fourteen times more likely than other workers to suffer repetitive strain injuries.[31] The speed of the conveyor belt is directly correlated with company profit. Workers are under constant pressure to perform faster, and fear of retaliation pervades processing plants.

I wish I could say that US-style industrial meat-chicken farming was confined to North America. Sadly, it's how four out of five chickens are reared in Britain too. The system is dominant in Europe and has spread across the world.

Some years ago I was invited to be an after-dinner speaker at a gathering of poultry producers. I was pretty sure the entertainment value would be in me entering the lions' den, rather than anything I was going to say. And once I'd finished, they didn't hold back. Some of the remarks from the floor are unprintable. Others were very defensive. I was asked: 'How can you criticise my farm if you've not seen it?' It was only a matter of time before someone invited me over. Their tone was confrontational: 'I challenge you, Philip, to visit my farm . . .' To this day, I'm still not clear why that was seen as such a big deal. I accepted the invitation without a flicker of hesitation. After all, I'd been to plenty of chicken factory farms before then, and many more since.

When the meeting was over and I was sportingly given a case of English wine as a thank you, the chairman had the task of organising my 'Challenge Philip' visits. It was not straightforward. It turned out that the ones reluctant to let me come were the very individuals who'd thrown down the gauntlet in the first place. It hadn't taken them long to have second thoughts. To avoid the club's embarrassment, Andrew Maunder and David Lanning of a company called Lloyd Maunder filled the breach and the visit was arranged. I'll never forget Maunder's face when I told him bluntly that modern broiler chickens were genetically selected to suffer. We were sitting in his farmhouse kitchen, having a chat. He looked surprised – almost wounded. We spent the day looking round farms, standing in huge sheds thick with top-heavy white birds. I'm sure he'd already seen the potential in moving away from the intensive system. A few years later, his business changed the way they reared their chickens, giving them more space and attaching far greater priority to welfare.

The state of industrial chicken farming in Britain was exposed in 2008 by an investigation commissioned by Compassion in World Farming. Grainy footage from inside dimly lit warehouse-like sheds showed tens of thousands of birds, packed so close together they looked like thick white carpets. They would have more room in the oven. Some were limping or lifeless. Outside were dustbins full of dead chicks; the ones that didn't make it to six weeks old. We didn't know where the chickens were bound. What we did know was that this was the kind of production system supplying major supermarkets with cheap chicken. At the time, 95 per cent of Britain's 800 million meat chickens were kept like this.[32]

The subsequent outcry, spearheaded by the celebrity chefs Hugh Fearnley-Whittingstall and Jamie Oliver, made so-called 'battery' chicken a high-street talking point. People in unprecedented numbers started demanding higher-welfare chicken. By 2010, despite the economic recession, nearly a quarter of fresh chicken sales were from farms keeping their birds in better conditions: free-range, organic or RSPCA Freedom Food standard.[33] Some supermarkets responded strongly to customer concerns. The volume retailer Sainsbury's is moving all its fresh chicken to a new higher-welfare standard, dumping what the media have dubbed the 'battery' bird. That one commitment alone means that nearly 100 million birds a year face the prospect of better lives. The Co-operative, Marks & Spencer and Waitrose have committed to similar things.

On a recent trip to America, I found myself in a retail park in midtown Atlanta, Georgia, home of civil rights leader Martin Luther King. Slap bang in the centre was a Whole Foods Market store, a food emporium if ever there was one. On chilled shelves lay neat rows of chicken products complete with animal welfare ratings. Signs extolled the virtues of the produce – 'Raised with care', or 'Great tasting meat from healthy animals' – and a chalkboard listed local farmers supported by the store. I recognised some of the names. They were part of a new group, Georgians for Pasture Reared Poultry, which recently exposed the dark side of the state's industrial chicken business. Even here, in the epicentre of the global industrial chicken phenomenon, the signs of a better way are starting to show.

V
SHRINKING PLANET

A police motorcade with outrider bikes and flashing lights sweeps into the East Anglian town of Ipswich, making for the docks. A helicopter buzzes overhead, keeping a watchful eye on the crowd gathering below. It's a blustery September evening. In the fading light, a horde of cameramen scramble for a picture. Yet this is no celebrity exit from the country. Flanked by police vehicles are six livestock transporters carrying sheep destined for slaughter on the Continent. It's the latest skirmish in a twenty-year war of attrition between live-animal exporters and opponents of the trade.

Banner-waving protesters yell into the night as the four-tier juggernauts disappear into the gloom. A former Soviet tank carrier waits by the dockside for its live cargo. It's the first shipment of live animals for slaughter out of Britain since the exporters were thrown out of another port following an incident involving two arrests and the death of over forty sheep.

Many people, myself among them, have spent much energy trying to stop animals being transported on horrifically long journeys simply to be slaughtered at their destination. Today this notorious trade is but a stubborn rump of its original shape: fewer than 100,000 animals exported live from Britain every year, compared with 2.5 million two decades ago.

Yet in ports across the globe, another trade linked to the questionable treatment of animals is booming. It's the import and export of grain and soya for animal feed, much of it destined to feed animals on factory farms. It takes place day and night, without fanfare or protest, and in many ways is just as insidious as live animal exports. As I discovered on a personal journey to bear witness, producing and transporting all this soya and cereal to feed animals is a dirty, thirsty business, ruining many lives.

Over the coming decades, the world's livestock population is set to near-double in the face of rising global demand for meat. It comes at a price. Much of the predicted increase is from industrial farms that use vast quantities of water, oil and land to churn out meat, milk and eggs of dubious quality. The resources required are in diminishing supply. Many of the world's best brains are now focused on solving an increasingly vexed question: the mismatch between growing global demand for food and the world's ability to deliver on a shrinking planet.

Land

How factory farms use more, not less

NO MAN'S LAND

Deep in the forests of northeastern Argentina live the last survivors of a once proud tribe. For centuries, the American Indian Toba Qom were feared and revered by outsiders. So remote and inaccessible was their terrain that it was known as the Impenetrable Forest. According to early Spanish settlers, the nomadic hunter-gatherers were fierce and unforgiving, hardened by their inhospitable surroundings. Today just a handful of Toba Qom remain in the Chaco and Formosa provinces. Theirs is a hot, humid lowland, stretching from the banks of the Paraná River to the Andean foothills and across international borders into Paraguay and Bolivia.[1] Once it was a rich and spacious hunting ground with all the natural resources needed to support their primitive lifestyle. Now, far from being impenetrable, much of the ancient woodland has disappeared, the elegant quebracho trees felled for their fabulous hardwood timber and the land carved up by anonymous speculators whose business is growing genetically modified soy, some of it for biofuels, some destined for factory farms, in the UK and elsewhere.

For the hard core of Qom who have doggedly refused to leave the Chaco, life today is a bitter struggle. Shockingly, in a country whose economy has boomed, in recent years some are reported to have starved to death.[2] Without land on which to hunt, grow and forage for fruit, gnawing hunger has become the norm. The traditional cotton cultivation in Chaco, which provided seasonal work, has disappeared to make way for the less labour-intensive soya production. There are few if any jobs left. The Qom manage as best they can, eking out a living from the little land they have left and turning to witchdoctors when they fall sick, but herbal remedies and rituals are proving no match for the scourge of malnutrition.

In 2011, five Qom made a drastic decision. Carrying little more than the clothes they were wearing, they left the Chaco and embarked on a 1,000-kilometre journey to Buenos Aires, where they set up camp at the intersection of two of the capital's main avenues and began to stage a hunger strike. The aim was to secure an audience with government officials, to demand the 'immediate return' of land they claim is rightfully theirs and has been appropriated by the provincial government. A few months earlier, two Qom had been killed in a violent clash with police in Formosa, after they blockaded a road in protest at the disappearance of their land. They manned it for four months, during which tensions with local authorities rose. Finally it seems both sides opened fire, with tragic consequences. The Qom were not the only victims: a police officer was also killed.[3] As each year goes by and the Qom feel more oppressed, the stakes for the last survivors rise. Many have abandoned what appears to be an ever more hopeless battle to preserve their heritage and, against their instincts, left Chaco for a new life in the city.

The Qom are just one group among many who have become victims of factory farming's appetite for land. In a disturbing form of modern colonialism, land in developing countries is being carved up by richer nations to guarantee a steady supply of cheap meat. The flow can only be maintained if there is enough grain and soya on which to fatten the animals, no matter how far it is grown from where they are reared.

The notion that factory farms save land is a widespread misconception, yet an argument commonly deployed in their defence. I remember a prominent figure in the poultry industry insisting factory farms were space-saving and scoffing at the idea that all chickens could be reared free-range. 'There wouldn't be a mountain in Scotland or a valley in Wales or any spare business land in any town, city of village that would not have a chicken on it,' he claimed. Free-range for all chickens was, in his words, 'absolutely ludicrous and impossible to achieve'.

In fact, if all chickens in the UK were reared for meat free-range, they would take up an area around a third of the size of the Isle of Wight – not so ridiculous after all. The entire global population of meat chickens – about 55 billion birds – could be kept free-range on an estate the size of Hawaii.[4] And though it is hard to imagine all 70 billion of the world's factory-farmed animals being reared in such a way, the idea becomes less ludicrous when you consider how much land they already require, including a third of the world's entire cropland.

Intensifying farming, getting more crops or animals out of the same land, may seem like a space-saving idea, but in practice it encourages more land to be used, not less.[5] Scientists from Argentina, the US, Canada, France and Belgium found that agricultural intensification between 1970 and 2005 was accompanied by cropland expansion. Part of the reason is that the intensive farming of animals is fuelled by the industrial production of animal feed crops. So much land is used for industrially farmed animal feed that if it were all in one field, that field would cover the entire surface of the European Union, or half the surface of the United States.[6] A third of the entire global cereal harvest goes to livestock; even more – 70 per cent – in rich countries.[7] In addition, about 90 per cent of the world's soya meal is destined for industrial livestock. If that food wasn't diverted to feed animals in factory farms, the land could be used for something else. Every year, an area of forest equivalent to half the UK is cleared, largely to grow animal feed and for cattle ranching.

So factory farming actually uses a vast amount of land or 'ghost acres' – it's just that the animals themselves are often no longer on it. The concept of ghost acres was invented in the 1960s by Georg Borgstrom, a professor of food science and geography at the University of Michigan.[8] He used the expression to describe the difference between the amount of food a country consumes and the amount it grows on its own land, a gap filled by imports. Half a century ago, he was warning that this gap was too big. Factory farming has dramatically increased it. According to one calculation, the amount of land used worldwide for agriculture has increased by almost 500 million hectares – ten times the size of France – in the last forty years.[9]

This might not matter if the supply of fertile land was infinite, but it is not. Estimates vary widely of the amount of remaining virgin land in the world that is suitable for cultivation. The most optimistic figure is 15.6 million square kilometres of rain-fed land – roughly equal to the existing cultivated land area.[10] But this is likely to be a gross over-estimate of what might be realistically available, because much of it is forested. The World Bank has come up with a far more modest figure that excludes forested areas: 4.45 million square kilometres.[11]

It may sound like plenty. The trouble is that rising population, urbanisation and erosion are swallowing up fertile land at a ferocious pace. And much of it is in Africa and Latin America, which means that wildlife will likely be bulldozed to make way. The result is a global land grab on a phenomenal scale. It is the twenty-first-century equivalent of the scramble for Africa in Queen Victoria's reign. Though it is being driven by a number of factors, particularly the global demand for biofuels, the insatiable demand for animal feed to supply factory farms bears much of the blame.

In Argentina, 200,000 hectares of woodland are believed to be lost each year to make way for soy.[12] Genetically modified soya now covers at least 19 million hectares[13] of the country – 65 per cent of the entire farmland[14] – much of it for export. The level of soy planting is causing disquiet as yields stagnate due to high inflation and lack of crop rotation needed to maintain healthy soils.[15]

Carlos Vicente works for GRAIN, a small international organisation that campaigns to safeguard land for local communities to manage their own food production. I met him at his home outside Buenos

Aires, where he told me about the devastating impact of 'land grabbing' for soya production in his country:

> What we're looking at is a new breed of food colonialists. In Argentina, the stark reality is that we have stopped producing food for ourselves. Fruit and vegetable production has fallen drastically – they now have to bring it into Buenos Aires from a thousand kilometres away. More than fifteen thousand dairy-producing farms have gone out of business in the last fifteen years. Milk production has declined so dramatically that at times there's been talk of having to import it from Uruguay.

I asked him how all this had affected ordinary people.

> The single-crop model does not really need people. In Argentina, the authorities themselves have said that it takes only one person to work five hundred hectares of soya. So we can produce twenty million hectares of soya with a few thousand individuals, and the rest of the people – rural folk who used to be farmers – are simply left over.
>
> The impact on biodiversity and public health has also been huge. For example, we have had outbreaks of dengue fever, caused by mosquitoes. They have been extremely precisely mapped and occur where there are soya plantations. Why? Because those areas are now monocultures, where the natural habitats of the animals that controlled mosquitoes have been destroyed.

Land deals in Argentina and other parts of South America are often highly complex, involving multiple countries based in different parts of the world. Speculative buyers pool finance with hedge funds, investment funds and pension funds. 'It's impossible to get a clear picture of who is buying the land to hoard and produce food,' Vicente said, 'though we know Saudi and China are both trying to strike these deals with the government here.'

Naturally there are winners from the system, like Gustavo Grobo-copatel, CEO of the agriculture giant Los Grobo. In Argentina, he is known as '*el rey de la soja*' – 'the soya king'. His company controls more than 250,000 hectares of soya.[16] 'He boasts about being the most successful farmer with no land,' Vicente told me.

During a trip to Argentina I went to talk to some Qom people who had lost their land to soya. It was a sad and troubling experience. Most of those who abandoned the Chaco now live in a poor suburb of Rosario, an industrial town 300 kilometres northwest of Buenos Aires. Arranging an audience with them was a delicate task and involved lengthy negotiations. Instinctively suspicious of outsiders, they had grown cynical about sharing their story, feeling that they had often given their time and energy for no immediate tangible return.

My co-author, Isabel, and I enlisted a local anthropologist, Laura Prol, to speak on our behalf. Over a period of weeks, she talked to the Qom about our work, gently persuading them of the case for drawing wider attention to their experiences. A meeting of community leaders was held to discuss our request. Finally they agreed to see us, but it was clear it had not been a straightforward nor a unanimous decision. We were keenly aware of the sensitivities, anxious to make it clear that we came neither as voyeurs nor as wealthy benefactors bearing material gifts.

We found their neighbourhood in a rough suburb of Rosario, down a bumpy dirt road lined with weeping willow trees. We had been warned that the area was a hotbed of crime; rife with car-jacking and theft at gunpoint. We left our valuables in our hotel.

The Qom houses were solid and weatherproof but extremely basic, with perhaps two rooms for sleeping and eating, and outdoor toilets. Out in yards, families cooked over real fires and chickens pecked in the dust. Stray dogs roamed the streets and children played in the dirt. Our meeting was at a community centre. We arrived to find about ten men sitting around a table in a small, poorly lit room. It was full of antiquated IBM computers – like the old Amstrads of the 1980s – which had been donated by well-wishers. Though none of the machines were switched on, we were told they all worked and were used for computer literacy classes. The men – there were no women – were drinking maté, a bitter herbal tea made from the yerba plant. As is traditional in Argentina, it was served in a communal cup through a curved nickel straw, and offered to everyone present. They eyed us warily.

It was a difficult few hours, complicated by the fact that the Qom have their own language, an additional challenge for our Spanish–English translator. The secretary of the group introduced himself as Abel Paredes, a name he described as having been 'imposed' on him by white people. What he meant was that it was not his indigenous name, which he did not seem to want to give. He wore a sporty fleece and cargo trousers and looked in his late forties. As he told his story he became emotional, his voice dropping and breaking:

> I came to Rosario with my brother and cousin, who live in a shanty town. My parents stayed in Chaco and are now dead. Over the centuries, our people have been pushed into smaller and smaller territories. Then some years ago a multinational came in and bought our land. The provincial government sold our land, with us included in the price, because we happened to be there. We had no value of course. We used to live in the forest; hunting for alligators, honey, iguanas, fish, and we tilled the land, growing vegetables and cotton. When the multinational arrived to grow soya, they fenced off the land and installed armed guards. What could we do? We are peaceful people, and so we decided we would try our luck here in Rosario.

For Paredes and his family, it had clearly been a desperate decision. 'We didn't want to come here and live on the edges, as beggars, but we were driven out,' he said. The displacement of the Qom from their ancestral land began long before soya cultivation took off. However, the huge sums to be made from growing soyabeans have accelerated the process.

Paredes said that when they first arrived in the city, they were treated as foreigners. They faced a bitter fight for the most basic services, including education for their children, and were ostracised by the local community, who accused them of bringing parasites and diseases like tuberculosis to the city. Now they have their own school in Rosario, where pupils are taught in the mother tongue. However, they claim that their access to healthcare services and other public services remains extremely limited.

Others had similar stories. The vice president of the community centre, Domingo Lassaro, whose elderly parents are still in Chaco, said he came to Rosario five years ago, after a company bought 40,000 hectares of his ancestral land. They fenced it off and installed sophisticated surveillance systems, leaving his family destitute.

We asked the men what their message would be to the outside world, a question that prompted lengthy debate. It boiled down to a demand for the same rights as the rest of the population, particularly in terms of access to healthcare and secondary and adult education, as well as reservations for Qom still in the Chaco. 'We don't want handouts, just recognition of our rights,' Paredes said finally.

> There is a healthcare centre just a block away from here, but all it is, is a building, with a sign outside. There was an election recently, and the candidates promised they would get it up and running, but it is still empty. We want to preserve our identity as indigenous people, but we also want to get on in this city. It would be good to have a government teacher to help us with literacy. Even our houses here are not our own – we could be evicted at any point.

We left the community centre feeling drained. It was a truly moving experience. We owed the Qom something for talking to us, but it was hard to see what we could deliver, given the scale of the difficulties they face. They show how the poorest and most disempowered people in the world are being cast aside to ensure a steady supply of artificially cheap, poor-quality chicken, pork and beef for people thousands of miles away. For that is where much of the soya meal ends up: feed for industrial livestock.

Just as it was in the nineteenth century, Africa is the other target for rich land-grabbers. A 2009 UN study of land transfers in five sub-Saharan countries – Ethiopia, Ghana, Madagascar, Mali and Sudan – found that 2.4 million hectares of land had changed hands in such deals since 2004.[17] A separate study the same year by the International Food Policy Research Institute put the figure for Africa between 2006 and 2009 at 9 million hectares, about three-quarters of the total amount 'grabbed' worldwide.[18]

More up-to-date studies dwarf these figures. A report by the World Bank found evidence of 'large scale land acquisitions' of 45 million hectares in 2009 alone. Many of these ventures have yet to get off the ground – the WB described more than half as in 'initial development' stage – but the general (upward) trend is undisputed.[19] An international conference on Global Land-Grabbing held in London in 2011 estimated that more than 80 million hectares of land were acquired in this way in 2008–9[20] – an area nearly twice the size of California, the USA's third-largest state.

It's a phenomenon driven by speculators, such as the UK investment management company Chayton Africa. Glancing at its website, you might think its business is luxury travel. There are glorious photos of pink sunsets over fields of sunflowers; rainbows over rushing waterfalls. In other images a patchwork carpet of luscious green pasture rolls out under a vast skyscape and a huge sprinkler quenches fields of ripening corn.[21] It could be France but in fact it is Zambia, and the firm's line of work is making money for Western investors from rich African soil.

The firm's founding partner, Neil Crowder, a former Goldman Sachs man, believes there are substantial returns to be made from growing food crops in this part of the world and has snapped up thousands of acres in Zambia for maize, soya and wheat. He is using Zambia as a guinea pig before launching similar ventures across sub-Saharan Africa.[22] 'We have identified Botswana, Malawi, Mozambique and Tanzania as target markets for our investments,' he has said, boasting that once his firm's huge agribusiness in Zambia has taken root, it will have 'numerous competitive advantages'.

For the hordes of international investors now piling into agriculture in Africa, this sort of financial adventure is little more than an experiment with what Crowder calls 'a new asset class'.[23] So far, there is no evidence that his project has upset or disadvantaged the local population. However, if it does not it will be the exception rather than the rule. In Africa there are multiple examples of both warlords and legitimate governments blithely selling land to foreign companies, at the expense of local people.

The Ethiopian government appears to be a particularly enthusiastic seller and has set aside 3 million hectares to lease to foreign and local companies for food crops.[24] In 2008, it signed a deal with an Indian-based multinational, Karuturi, involving 311,000 hectares in a remote part of the west of the country – a tract of land larger than Luxembourg. Karuturi's usual line of business is cut flowers, mostly roses grown in giant greenhouses in Ethiopia, Kenya and India.[25] Every year the company produces 555 million stems which are shipped across the world to luxury markets in places like Hong Kong, Muscat, Dubai and Japan. Now they are diversifying, as their website says, into 'agri-business' and have taken up cultivation in Ethiopia 'on a mega scale'.[26]

The Ethiopian government believes that partnerships with companies like Karuturi are good for the people, 85 per cent of whom are smallholder farmers. Berhanu Kebede, Ethiopia's ambassador to the UK, has dismissed the amount of land involved as 'relatively small' compared with the 76 million hectares of arable land in the country, saying that investors such as Karuturi bring 'huge benefits, not just the jobs, houses, schools, clinics and other infrastructure, but knowledge transfer, skills training, tax revenue and other benefits to the workers and country as a whole'.[27]

However, by their own admission officials have sometimes been too hasty to sign deals. Karuturi's original concession ran right through a national wildlife reserve, which had been created in 1974 to protect an endangered species of antelope. The Ethiopian government belatedly realised that Karuturi's venture was located dangerously close to the antelopes' migration route, and the boundaries had to be redrawn.

Officials later admitted they had 'signed an agreement without any coordinates or delineation of the land . . . a mistake'.[28]

Meanwhile the catalogue of benefits for the people has been slow to materialise. Previously, locals earned a living by farming or fishing. More recently, they have been reported as being employees of Karuturi, earning less than the World Bank's $1.25 per day poverty threshold.[29] The Ethiopian press has carried various reports of exploitation, and child labour appears rife. Local people say they were not consulted about the deal. Karuturi says that things will get better, and that it pays at least Ethiopia's minimum wage. Bosses claim they are 'very very cognizant' of the fact that the company is dealing with people who are 'easily exploitable', and have grand plans to build a hospital, cinema, school and a daycare centre in the settlement.[30] Yet the reality is that locals have lost control of their own land and with it their destinies, perhaps for the sake of cheap burgers thousands of miles away.

The near-miss for Ethiopia's rare form of antelope underscores the future threat of a world where wildlife is forced out. There is already intense pressure for land, bringing wildlife into competition with people. Over the next few decades, climate change is likely to increase this competition. Witness what has happened at Titchwell RSPB bird reserve in Norfolk. One of my earliest memories of this magical place was a night I spent on the reserve when I was eighteen years old. It was 3 am on a warm spring morning and I was sitting in a makeshift dugout looking out over an expanse of marshland glowing silver in the moonlight. Every now and then the gulls would spook themselves, taking to the air and filling the hush with harsh cries. The glimmering white of a barn owl sometimes hovered overhead like a ghost. But this was no ordinary birdwatch. I was a volunteer on the night watch. My vigil was to protect the nest of the rare wading bird the avocet, famous as the emblem of the RSPB, and returning from the brink of extinction in Britain. It was nesting for the first time on Titchwell's brackish marsh, a wonderful shallow lagoon where fresh water and salt water meet.

These days, the primary threat to birds is habitat destruction, but anachronistic as it seems, rare species still face a threat from egg collectors. It's a strange throwback to a bygone age when it was commonplace for schoolboys to collect birds' eggs. Although it is now illegal, a small number of offenders persist. They will go to great lengths to steal the eggs of target species, and the avocet was definitely on their hit list. The warden of Titchwell marsh, a nature reserve owned by the RSPB, was very worried. 'Eggers', as we called them, had been reported at other sites along the coast, so my fellow volunteers and I organised a rota to keep watch.

When it came to my turn, I sat through the night in my dugout a mile or two from the nearest help. It was long before the days of mobile phones. It was a privilege to see the natural world at this hour. I felt drunk through lack of sleep but purposeful and resolute. Every now and then, I wondered what I'd do if the eggers came. And then the moment I'd been dreading happened. The wind suddenly whipped up, the gulls spooked, and against their eerie cries, I could hear what sounded like boots on gravel – 'scrunch, scrunch, scrunch . . .'

The sound was distant at first, but it grew louder. My heart pounded as I considered my options. Grabbing a stick, I leapt out of my dugout to do my duty – only to discover that my foe was water lapping on gravel. The rest of the night passed without incident and I felt a bit foolish. Several days later as dawn broke on another night watch, I peered out to catch sight of avocet chicks. The first Titchwell avocets had hatched, and on my watch! I was elated. All the efforts of the warden, us volunteers and the support of people far and wide had paid off.

That was 1984 and the early years of the RSPB's tenure at Titchwell Marsh. It has since gone on to become one of the most popular and best-loved nature reserves in the country. Some 90,000 visitors flock there every year. It is famous for its breeding bitterns, bearded tits and marsh harriers, the latter among my favourite birds.

Now the reserve is under threat from rising sea levels. The brackish marsh that I watched over as a volunteer all those years ago is to be sacrificed to save the rest of the reserve. The RSPB wants to save the freshwater haunts of the bitterns and harriers, and that means realigning coastal defences, at the expense of the marsh. The magical expanse on which those first avocets hatched will slowly be claimed by the sea. New islands have been created to tempt the avocets to other parts of the reserve.

While the threat from the sea at Titchwell is part of a natural process of coastal erosion under way for many years, climate change is making things worse. The work to strengthen sea defences in this part of Norfolk should save these important freshwater marsh habitats for the next fifty years, but I can't help wondering what happens after that – especially as the reserve is hemmed in by farmland and village. Will we face a future where wildlife has nowhere to go?

The submerging of the Norfolk marshes is a reminder that climate change is not a theoretical problem for faraway countries to worry about. Its effects in places like the UK are very real. By the middle of this century, wildlife and many millions of people are likely to be displaced by rising sea levels. This will intensify competition for land, and is likely to cause profound difficulties, particularly as so many major cities are close to the sea.

Experts differ over the extent to which sea levels will rise over the next century. Unless the scientific consensus is completely wrong – probably a forlorn hope – then over the next few decades, millions of people will be forced to move away from the coast onto land previously used for agriculture. Planet Earth has reached 'peak land', the point in present history where land will begin to disappear as sea levels rise. It seems that a two-degree temperature rise this century is fairly inevitable; it would bring with it a one-metre rise in sea level.[31] This would have profound consequences as the rising tide laps at coastal cities and farmland alike.

A more modest sea-level rise of 0.7 metres this century would affect around 150 million people living in low-lying coastal areas, including some of the world's largest cities. Almost half of the world's top fifty cities are at risk, including Tokyo, Shanghai, Hong Kong, Mumbai, Calcutta, Karachi, Buenos Aires, St Petersburg, New York, Miami and London.[32]

The UN predicts that the growing appetite for meat, particularly in developing countries, will lead to a near-doubling of the current global livestock population. If this happens on an industrial basis, more land will also be needed to produce extra feed crops. It seems that human demand for land and other resources could well be on a collision course with global warming. Sea-level rises threaten to swallow up land, affect crop yields and cause widespread disruption.[33] To keep pace with the food demands of a growing population, experts suggest that an additional 2 million square kilometres of land will be needed by 2030.[34] That's the same amount of land that would be lost to the sea if the planet warms by two degrees.[35]

SCORCHED EARTH

Sundown on a cattle ranch, Argentina. Three gauchos in cowboy boots and berets leap and lunge, swinging their lassos at galloping calves. From a distance it's a curiously beautiful dusty dance, the animals bucking and ducking, their rusty hides rich red in the setting sun. Their hoofs kick up clods of mud as they hop and swerve to dodge the ropes. Hobbled, they crash to the ground, legs flailing, the thud of rumps throwing up great puffs of dirt. On the corral fence there's a box of Philip Morris cigarettes, the classic cowboy vice.

In a slick air-conditioned office, a rancher is mulling over whether to let us take a closer look. He is a lean fellow with a quirky moustache and the rich complexion of a man who has spent most of his life in the sun. He wears expensive-looking shoes, tight-fitting black trousers, a luxury watch, and a pair of Ray-Bans hooked into the top of his crisp white open-necked shirt.

He sits at the head of a long boardroom table. Behind him are a dozen silver-framed photos: three generations of his family. Before him is a neatly arranged stack of business cards, a sheaf of invoices and a large calculator.

On a couple of dozen acres in the Argentine pampas, the rancher's twenty-five-strong team of workers rear and fatten 4,000 beef cattle in conditions designed to maximise profit by getting the animals to market weight in the shortest possible time span. He has invested US$10 million in this feedlot facility, and by the looks of it he's making a handsome return. Soon he hopes to double the size of his beef herd. 'There are no limits. There's no roof out there,' he shrugs, gesturing at the corrals outside. This rancher is a typical Argentine feedlot operator. He also owns about 3,000 hectares southwest of Buenos Aires. For generations, this land would have been used for cattle grazing. Today only a fraction of the estate is for livestock. Like many other landowners in Argentina, the rancher now uses much of his estate for something else: cultivating genetically modified soya beans. His cattle live on little more than a patch of mud, while his most fertile fields are devoted to crops. 'That's where the real money is,' he told us with a knowing smile.

Soya production is a little-known but vital part of the factory-farming system, meeting the colossal demand for high-protein animal feed for industrially reared animals. Much of the soya produced on the pampas, the plains of Argentina, is ground into soya meal and shipped to Europe and the UK, as well as China, where most is used to fatten pigs and chickens.

Though the destructive effects of soya production for biofuels have been widely documented, particularly in relation to deforestation, the devastating environmental, social and health impact of soya production for animal feed has largely escaped the spotlight. In terms of overall production, Argentina grows nearly a fifth of the world's soya beans, behind only the USA (35 per cent) and Brazil (27 per cent).[36] Most of Argentina's land used for arable crops – 18 million hectares of a total of 32 million – has now been turned over to genetically modified (GM) soya.[37] Cultivators churn out almost 50 million tonnes of the stuff every year,[38] much of it for the overseas market.

Argentina has established itself as the soya meal export capital of the world, accounting for nearly half of global exports.[39] Soya meal is the high-protein powder that is separated out from the oil when the beans are crushed. Although soya can make perfectly good nutritious food for people, much of it is destined for industrial animal feed. The UK is an eager customer – nearly half the soya meal imports used on British farms come from Argentina.[40] For Argentina, it's big business. Exports of soya meal alone were worth over US$7 billion in 2008.[41] With a 32 per cent export tax, it's a major money spinner for the government. For a handful of landowners and foreign speculators too, the returns have been spectacular.

Meanwhile the impact on the cattle-ranching industry has been far-reaching, with animals pushed off the land they used to graze and forced into intensive fattening systems: 'feedlots'. Corralled into grassless pens carpeted in manure, they are often sustained on a diet of concentrated feed and antibiotics and swiftly transformed into beef.

So Argentina offers a double window on the world of industrial farming. On the one hand, it is literally feeding factory farms in the UK and other parts of Europe, via shiploads of ground soya. On the other, it is a showcase for one of the most intensive farming systems in the world: feedlot farming, or what some call 'battery' beef. Already rife in South America and the US, these cattle feedlots could soon make their way to the UK and other parts of Europe. Meanwhile, in restaurants across the EU, diners continue to tuck into bargain-basement beef from Argentina thinking it is some kind of luxury product. Most are blissfully ignorant of the conditions in which it was produced, and know little of the true quality of the meat.

*

My journey to look at the transformation of Argentina's countryside had begun in the northeastern city of Rosario. With a population of 1.7 million, it is the third-largest city in the country. It lies at the heart of a major industrial corridor and is the hub of Argentina's soy industry, its sprawling grey port the dispatch point for soya exports.

The city is a 300-kilometre drive northwest from Buenos Aires through the pampas. With my journalist companion and camera crew in tow, my schedule and budget was tight. We landed in Buenos Aires mid-morning local time after an overnight flight from London, and with only a coffee and a rather leathery croissant from the airport, we picked up a hire car, met our local fixer and got straight on the road.

I was looking forward to the journey to Rosario and imagined a lush and varied landscape of eucalyptus trees, horses and haciendas. My heart sank as we broke out of the Buenos Aires suburbs and found ourselves in a pancake-flat landscape of endless soya. Though the iconic pampas grass still grew in abundance by the roadside there was precious little else to see from the car window but field after field of the scrubby little soya crop, either green or a dreary yellowish brown, depending on the ripeness. Farmers had turned the perimeters of their fields into rolling adverts for seed and pesticide companies, hanging ugly plastic posters from fences. There was little to break up the monotony: no villages, no churches, no farmhouses, and little sign of life except the odd stray dog lolloping along the verge.

It was an uninspiring introduction. Rosario wore a concrete tutu of slaughterhouses, meat-rendering operations, factories and soya-processing plants. All this industrial paraphernalia seemed to throttle what remained of the old town. A handful of fine nineteenth-century colonial-style houses and municipal buildings were eclipsed by more modern blocks of flats, offices and shops.

Overlooking the river was a clutch of sparkling new high-rise flats, apparently built with fortunes made from soy. Argentina remains haunted by the economic trauma of 1989, when inflation rates rocketed, and many entrepreneurs are still deeply mistrustful of leaving their money in the bank, preferring the security of bricks and mortar.

Rosario owes its status as an agro-economic hub to its location on the Paraná, the second-longest river after the Amazon in South America. It flows almost 5,000 kilometres through Brazil, Paraguay and Argentina before emptying into the Atlantic. In Tupi, a native Brazilian language that is now extinct, Paraná means 'as big as the sea', and at Rosario it is as wide as an estuary.

Since 1997, governments in the five countries of the Rio de la Plata Basin – Argentina, Brazil, Bolivia, Paraguay and Uruguay – have been plotting to transform the Paraná into an industrial shipping canal in a highly controversial scheme known as the Hidrovia project.[42] The river alterations outlined in the original plan, backed by the Inter-American Development Bank and the United Nations Development programme, included dredging, rock removal and structural channelling at hundreds of sites along the way, according to the organisation International Rivers. It could have had a devastating effect on the Pantanal, the world's largest tropical wetlands. The project was thrown out following an outcry by a wide coalition of environmental, social and indigenous organisations and various independent technical critiques, but is now back on the cards. The Andean Development Corporation is reported to have given the governments involved almost US$1 million for new studies, and there are fears that a revived scheme will involve even more intensive dredging and rock removal to guarantee the passage of barge convoys through twenty-three 'critical' river passes.[43]

If it goes ahead, the project will ease the flow of agricultural and other commodities through South America, to China and Europe – but at Rosario, it didn't look as if the export industry needed any help.

We arrived mid-afternoon, exhausted and bleary-eyed from the international flight and long drive. Next day we hired a boat to look at Rosario's many soya-processing factories from the water. At the small boatyard on the outskirts of the city there was a sickly meaty smell. The river was vast and appeared to be heavily silted, a great milk-chocolate flow punctuated by enormous tankers with romantic names like *Crimson Venus*, *Storm Ranger* and *Sea Honest*, flagged to Panama, Nassau and Limassol. There was nothing dreamy about the voyages they were about to undertake – no cargoes of coffee, wine, brightly coloured ponchos or other exotic Latin American goods. Just one product: soya, be it soya meal or soya beans. The vessels were magnificent in scale if not purpose, great hulking things the length of a small-town high street. There was little sign of life on board, though at one point a lone figure in an orange boiler suit appeared on a deck. Seeing us filming, he gestured at us to leave.

Soon it would be Easter, and the odd fisherman was trying his luck on the mud banks. On the riverbank, there were no trees or vegetation, only cracked brown mud, above which reared the dreary edifices of the soy-processing industry: enormous tangles of grey and brown metal, cranes and winches, funnels and chutes, pipes and ladder-like structures that looked like extended fire escapes. Huge pipes carrying waste water from the mills spewed their contents into the river.

The soya meal market is controlled by a handful of multinational operators. We sailed past a plant and watched an avalanche of soy pouring from a funnel jutting out from the factory into a ship container. Dust billowed over the boat: we were being sandblasted by particles of soya meal. The tiny beige flakes settled on our clothes.

On the other side of the river were verdant islands. Apparently they stretch for 70,000 square kilometres along the Paraná. During El Niño years, when the water is low enough, we heard that farmers drive cattle over the river to graze – but not before felling the trees, destroying a fragile ecosystem. The fact that cattle are on these islands at all has been blamed on the encroachment of soya, eating into grazing land. In 2009, the military were called in to help ranchers evacuate a million cattle off the islands for fear of the kind of flooding that claimed the lives of 300,000 cattle only two years earlier.[44]

The boat trip gave a stark impression of the sheer scale of the soya industry in Argentina: gargantuan factories and ships, industrial behemoths that feed factory farms all over the world. Who would live in such a place? The answer is: people who lived there before the industry arrived, and are stuck.

Rosario's soya factories have sprung up in the last decade or so, devastating what used to be peaceful suburbs. At a community centre in the neighbourhood of San Lorenzo, near the river, we met

Lillian Ober, a mother of two, in her forties. She moved to the area fifteen years ago after the old industries moved out, thinking it would be a nice place to bring up children. Today, no fewer than 1,000 trucks a day – she's watched and counted – thunder through her neighbourhood en route to the processing plants we saw from the water. It's a relentless, noisy, polluting caravan that is claimed to whip up coughs, asthma, and perhaps more sinister conditions.

A study by the Italiano Garibaldi Hospital in Rosario showed that in six towns in the region, the incidence of testicular and gastric cancer in males was three times higher than the national average; the incidence of liver cancer ten times higher; the number of cases of pancreas and lung cancer two times higher.[45] People who live in San Lorenzo believe it is the site of a cancer cluster. The dust is always bad, but what really gets them is the chemical spraying of soya as it arrives at the plants. To ensure the beans are insect-free when they are shipped overseas, we learned that the wagons are doused with chemicals before they are unloaded. The soy is sprayed again when it's dumped into storage containers, and sprayed a third time before being loaded onto the ship. I was told of local newspapers reporting that in the last decade as many as sixty truck drivers have dropped dead after being overcome by chemical fumes. It may only be anecdotal, but it's not hard to believe.

'When we moved to this area, my husband and I had no idea what was coming,' Ober told us. 'We thought it would be a lovely area to raise a family.' The community wants a proper assessment of the public-health risks associated with the industry. They have done their own informal research. It suggests that 90 per cent of health problems in the neighbourhood are respiratory or allergies. 'Everyone is complaining of the same sorts of conditions. We want an investigation. It seems obvious to people living here that the health pressures are linked to the soya industry, but we would like to see if there is scientific evidence.'

Six years ago, when a giant milling company decided to expand its plant in Rosario, Lillian and her husband thought about upping sticks and taking their two daughters, now eleven and fifteen, somewhere less polluted. After much soul-searching they decided to stay. They often wonder whether they did the right thing.

> My husband's elderly parents and family live here. You can't uproot an old tree, so while they are still alive, we won't move. It was also a principle for us – we wanted to teach our children that if you hit a problem, you don't just run away. You face up to it; tackle it. We can't change what's happened here, but now we are fighting to ensure the agro industry doesn't expand one more inch into San Lorenzo. But I do worry about my daughters' health.

Daniel Pablo, another local activist, in his early fifties, told us he had lost several members of his family to cancer. Perhaps it's just bad luck – after all, one in three people develop cancer at some point in life – but most of his relatives were relatively young, suggesting that something more is afoot. He told me:

> There's no history of cancer in my family. Yet since we moved here, my sister-in-law got a tumour; my brother got a brain tumour that killed him; and my brother-in-law got testicular cancer, for which he was successfully treated. Two of my cousins had premature babies, and my mother got a glandular condition. Two of my best friends died suddenly. All of them lived in this area. For the last four years, I've been asthmatic. Everyone here has allergies, particularly the kids. Soya is a business for the few and an epidemic for the masses.

The team and I left the community centre to look at the road where 1,000 or so trucks a day trundle to and from the factories. We stood on the verge, watching the wagons rumble along and churn up clouds of dust. Outside the plant nearest San Lorenzo a car park full of heavy vehicles waited to set off again. Behind barbed-wire fencing and a large sign saying *Prohibido* (Keep out) we saw the usual giant mesh of grey and brown metal: one factory, the size of a village.

With its apparent cancer clusters, nauseating smells and ugly industrial skirt, it was hard to see how anyone with a choice would hang around in Rosario. The following day we headed to San Jorge, 180 kilometres north, in the heart of the soya plantations. In 1996, Argentina became the first South American country to allow the use of genetically modified varieties.[46] Now the entire soya crop in Argentina is GM.[47] Initially, the new technology was a bonanza for farmers, increasing yields by 173 per cent.[48] It did not last. The weeds grow increasingly resistant to the chemicals, and farmers are having to use ever-stronger chemicals to produce the same harvest.[49]

The figures are startling. In 1990, before the arrival of GM in Argentina, 35 million litres of chemicals were used on crops per year. In 1996, the figure rocketed to 98 million litres. By 2000 it had soared again, to 145 million litres; and by 2010 the figure reached 300 million litres – almost ten times the amount of pesticides and herbicides used pre-GM.[50]

According to an organisation of concerned Argentinian doctors, known as Physicians in the Crop-Sprayed Towns, 12 million Argentinians are affected by the agrotoxins every year, as aerial and land-based sprayers douse houses, schools, parks, water sources and work areas. Incidences of serious public-health issues are on the rise, including rates of birth defects and stillbirths in such areas.[51]

One area that has been particularly badly affected is the poor Urquiza neighbourhood of San Jorge. Cancer rates in the suburb are reported to have spiked 30 per cent since 2000.[52] Our first stop was a press conference on chemical spraying at the municipal centre. A local politician, Esteban Roglich, was there along with Daniel Vezenasi, a doctor and one of Argentina's leading specialists in respiratory conditions. Roglich was unveiling a new bill he'd drafted that would create buffer zones around residential areas, over which aerial spraying would be banned. Local campaigners were also demanding that the few existing regulations governing pesticide spraying be properly enforced.

The doctor gave a powerful account of the health case. He talked of higher rates of miscarriage and birth defects in areas exposed to chemical spraying. He spoke of a research project he carried out with medical colleagues, involving eight communities with populations of under 10,000 in areas affected by crop spraying. In all, he and his colleagues surveyed a total of 45,000 patients. The research revealed that hypothyroidism is one of the top health issues in such areas – in stark contrast to the rest of the country.

Afterwards we went to visit a woman who had played a key role in exposing the human health risks associated with pesticide spraying. Viviana Peralta's story was sobering. The mother of six told us that she had been cooking tortillas at home in Urquiza when her baby nearly died. It had started out an ordinary day. Her husband was at work, selling furniture for a local carpenter, the older children were at school, and she was keeping an eye on the little ones while doing domestic chores. In the fields, soy cultivators were aerial-spraying the crops again. Then her four-year-old daughter Michaela burst in. 'Mum! Come quickly!' she shouted. 'The baby's not breathing. She's gone purple.' In her cradle, seven-month-old Aileen was having some sort of seizure. Peralta rushed the infant to the local medical centre, from where she was transferred to hospital, fighting for life. Later doctors told her that they had feared that a tracheotomy – radical throat surgery – would be the only way to save the infant.

Aileen, who had suffered from respiratory problems from birth, pulled through. The medics were in little doubt what was causing her problems: the remorseless aerial crop spraying in Urquiza. Peralta was advised to move out of the area and seek compensation.

So began a long campaign for justice: a quest not for money but for laws to stop the spraying. The day Peralta brought her baby back from hospital, she heard the familiar sound of a crop sprayer – this one a terrestrial 'mosquito' control vehicle – grinding its way towards the fields by Urquiza. She ran out of the house and stood in its path. The driver refused to turn back, so she hurled a brick at his machine. She chose to fight, not to move: she told us she and her husband had invested everything in their home.

'We built this place with our own hands and worked so hard for everything in it. It didn't seem right to me that we should be forced out, while the soy cultivators carry on their work.' Though her home was modest, there were signs of how hard she and her husband had worked: a fitted kitchen; a large TV and DVD player and an impressive hardwood dining table, large enough to seat twelve, with matching heavy wooden chairs. Outside was a pleasant yard where the children could play.

With Roglich's help, Peralta found a sympathetic lawyer in Santa Fe and took her fight to the provincial courts, having compiled enough medical evidence to prove beyond reasonable doubt the link between the pesticides and her daughter's near-fatal breathing difficulties. Local soy cultivators offered her money to shut up, and various other deals that she dismissed. 'At one point they offered to take me and my children to a hotel every time they were spraying. I didn't even consider it. They are so cynical.'

Eventually, her campaign paid off, but it was only a partial victory. In March 2009 a provincial judge banned cultivators from aerial crop spraying over her home, but absurdly the order only applied to her property. It means that theoretically the rest of Urquiza remains exposed. Aileen is now four and a half years old, an exceptionally cute little girl with long golden-brown hair swept up in a pony tail, dancing brown eyes and a cheeky smile. She made a full recovery from the seizure, and her health problems have largely disappeared since the spraying stopped, though she has been hospitalised on a number of occasions since with breathing difficulties. They appear to have been linked to drift from chemical spraying in the area. Some of Viviana's other children continue to suffer from respiratory problems. Her campaign goes on.

Our next stop was a provincial town outside Buenos Aires, where we hoped to find out more about how Argentinian beef is produced. It meant another long road trip. On the face of it the motorway network in Argentina is impressive: superhighways with toll booths, the road surface in impeccable condition. When the going's good, it can be great, as there are so few cars. The problem is grain lorries: there are so many of them that if there's an accident or roadworks, the motorway very quickly turns into a car park.

As we began a long drive southwest we found ourselves stuck on a slipway because of an overturned soya load. We were forced to leave the highway and take a scenic route. There was little to see. The countryside was featureless: nondescript soya fields, dusty villages, and the trappings of the agro industry. Somewhere along the way we passed a soy-processing complex the size of a town. Rows of rusty, dusty wagons were parked nose to tail in forty numbered lanes as if they were waiting to board a fleet of car ferries. Bored drivers loitered by their cabs. Beyond the rows of trucks we could see the plant itself, a grim grey monster. To the front were three enormous dark grey warehouses like aircraft hangars; behind, the usual giant tangle of funnels and chutes.

It turned out to be the edge of an agro-industrial estate. Half a kilometre away was another equally gargantuan plant, this one a huge grey block like a nuclear power station. Around the corner was a Monsanto factory with billboards outside advertising Roundup, the company's best-selling weedkiller.[53] From outside, with its innocuous green and red logos, it could have been a garden centre selling shrubs and plastic tables. Inside, brilliant minds were no doubt poring over new chemical solutions to nature's attempts to compete with monoculture.

The first feedlot we passed housed around 400 calves. By the road were tall conifers and clumps of trees, beyond which we could see maize fields. The countryside was richer, more interesting, than the usual soy, but in the corrals there was nothing for the animals to do except pick listlessly at their food. The temperature was pleasant, about 27 degrees. There was no shade for the cattle – not a serious problem in autumn but surely unbearable in summer.

I've seen many smaller versions of this – beef cattle crowded in one space – but this was something else. It was on a massive scale and very similar to the mega-dairies of California. I scrutinised the cattle: they were very young, and bored. Our arrival provided a brief distraction. The herd played grandmother's footsteps with us; as we stood looking at them from the road, they would inch towards us when we were still, retreating the second we advanced in their direction. I thought about the grass-fed cattle I was familiar with in places like the UK and felt glad this type of mega-farm has yet to arrive.

The town turned out to be a wealthy little place of coffee bars and plane-tree boulevards and what could almost be described as a boutique hotel, all mood lighting, low-slung sofas and wi-fi in the lobby. Its clientele seemed to be middle-aged businessmen from Buenos Aires, local landowners now rich enough to live in the city and pay someone else to do the hard work. After the grimness of Rosario, it was all quite pleasant.

We dumped our bags at the hotel and went to meet Pablo and Pilar Guerra (whose names have been changed to protect their identity), a couple who live by a feedlot on the outskirts of the town. Their home was a decent-sized, solid house with thick walls keeping it cool during the day and an extensive backyard. By the front door a skinny white cat lounged in a flowerbed, soaking up the afternoon sun. She was no ordinary family pet but one of a team of rat catchers, much needed around the Guerras' home since the feedlot was built ten years ago.

The couple's livelihood was a farm shop next to the house, where they sold home-cooked pastries, chicken and pork. They'd had the business since they bought the house twenty-six years ago. At that time it was in a lovely spot. Guerra told us he reckoned that the takings at his shop had halved since the feedlot arrived. The problem was the smell and the flies. Fighting to get the farmer to do something had proved hopeless.

> For five years, I did everything I could to get the feedlot to clean up, or compensate us. I took my complaints everywhere I could – the local, provincial and national legislators; even the national watch-dog. We demanded that either the problems were tackled – the flies, the rats, the smells – or they closed down.

I organised community meetings – we would get fifty or sixty
people turning up, all unhappy about the feedlot – which is a very
good turnout for a small place like this.

Despairing, Guerra and a handful of his neighbours sued the operators of the feedlot, winning a judicial order in 2008 to get the place
shut down. It remained open pending an appeal, which the owners
won. 'Gradually I realised I was knocking my head against a brick
wall. Over the years, I must have discussed what was going on with
150 different politicians. They all fobbed me off with platitudes. Only
one guy was honest. He told me: "This is your fate. Too bad. Accept
it."' Guerra seemed a broken man.

Though they gave up the battle years ago, he and his wife had not
shrugged off their reputation as troublemakers in the eyes of local
farming magnates. He told us that when he recently visited the local
authority to renew his shop licence, out of the blue the official
suddenly demanded to see the original plans for his house and shop.
Guerra had never had them, and in twenty-six years had never
needed them. Now, apparently there had been a 'complaint' about
his property. 'Someone is making trouble for me. They'd be glad if I
left the area,' he said sadly.

When we met, he still had not obtained his permit and did not
know how the situation was going to evolve. To make matters worse,
he and his wife had just found out that a mega-piggery was being built
over the road. Work had already begun. We saw the early signs of what
was being built, all shiny aluminium. On the front was a bright orange
sign saying 'Mega'.

The value of Guerra's property had already fallen dramatically
because of the feedlot. It must have felt like the last straw. The couple
had spent the morning of our visit discussing whether to sell up, but
like so many of the people we have interviewed for this book who have
found themselves living next to factory farms, they seemed deeply
reluctant to uproot.

Later that day we met Miguel Martinez (whose name is changed for legal reasons), another man with a sad story about the feedlots. Approaching retirement, he was a civil servant in Buenos Aires, but spent weekends out here, where he grew up. He still had a little place near the town and some land, and once or twice a year he liked to head into the pampas to hunt wild boar. Dressed in his khaki hunting gear, he was clearly no sentimental animal lover. Yet he hated the feedlots, and what they had done to this area. He took us onto one of the fields he inherited from his grandparents, where he once hoped to build a house to retire. He abandoned the project when he discovered the neighbouring field had become a feedlot. 'I was going to build it right here, with a veranda looking out west,' he told us, gesturing to the plot he had in mind. 'There was a lovely little duck pond just over the hill, and woods behind. There was a fig tree my mum used to pick fruit from to make jam.'

The duck pond and fig tree had gone. In their place were thousands of cattle on mud. Around us was GM soy, interspersed by a lot of weeds and more mosquitoes than any of us had ever seen in one place. On Martinez's beige cap alone were at least eighty of the insects. They swarmed around our faces, attacking any bare flesh they could find. There was no stagnant water nearby; nothing that would obviously attract such a shockingly high concentration of flies. It seemed clear we were in the midst of an ecosystem out of balance.

Martinez seemed unfazed by the mosquitoes, but remembering his childhood and his retirement dream made him emotional:

I was three months old when I first came here. This place is part of my life. It's where I learned to ride horses, and later to shoot, enough for dinner, never any more. There was always a strong sense of community; neighbours helped each other. My grandparents owned a thousand hectares, four hundred of which was divided between us grandchildren when they died. The rest was sold. Until the 1990s, all the farmers round here were local. They knew the area; had a connection to the place. Now it's just investors. One of the big land-owners round here owns car parks in central Buenos Aires. When I

came back here in 2005, after a while away, all the gates to properties in the area were padlocked. It was a shock. Some blighter who
bought the land from my grandparents' estate set up this feedlot.
When I saw it, there was the same feeling of terrible loss that I experienced when I had to clear out my grandparents' place, and when I
got divorced.

His reflections might have been dismissed as the misty-eyed nostalgia
of an old man, but Martinez made it clear that he did not simply
yearn for a bygone era:

> I understand that business is business. I am not against progress or
> anti-development. I don't mind living among cattle and horses. But
> four thousand cows on one field is too many, and the consequences
> are far-reaching. Small farmers cannot compete; the water supply is
> polluted; the air is bad. For investors, it's just money, but farming is
> not the same as working in the car industry; or owning parking lots.
> It's an art, and a responsibility.

The feedlot next to Martinez's land was owned by a successful
rancher. We knocked on his office door and told him about our
research. We handed over our business cards and after lengthy
consideration the rancher seemed satisfied, first ushering us inside
for an interview, and then accompanying us on a guided tour of the
feedlot. He told us that although he was brought up in this area,
these days he prefers the bright lights of Buenos Aires. In common
with many feedlot farmers who have made a success of their businesses, he now spends just three days a week on the farm; his home
is in the metropolis.

He was clearly proud of his business and spoke compellingly of his
drive to build a future for his children and grandchildren. He told us
he cared about his cattle. As we walked around the facility, I had spotted a dead cow abandoned at the edge of one of the corrals, swarming
with flies. It was a worrying sign, not least raising questions about

Equivalent area of global arable land dedicated to growing feed crops for farm animals

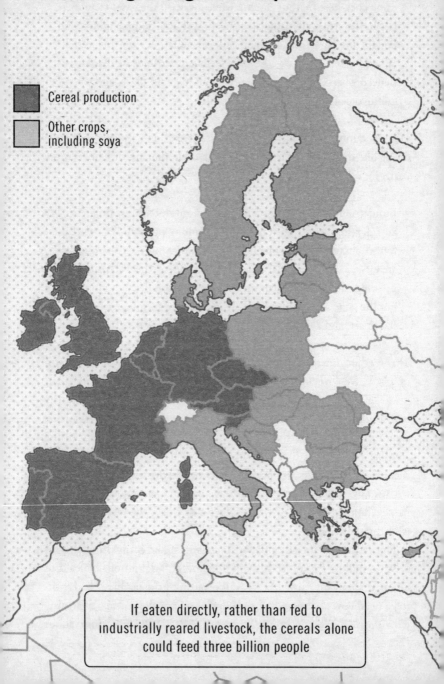

Cereal production

Other crops, including soya

If eaten directly, rather than fed to industrially reared livestock, the cereals alone could feed three billion people

How growing feed for industrial livestock increases pressure on finite arable land

Total land area suitable for rain-fed crop production worldwide: circa 30 million km²

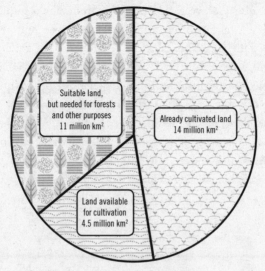

Suitable land, but needed for forests and other purposes 11 million km²

Already cultivated land 14 million km²

Land available for cultivation 4.5 million km²

'Ghost acres' farmed elsewhere to produce feed for industrially reared chickens

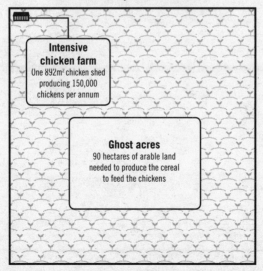

Intensive chicken farm
One 892m² chicken shed producing 150,000 chickens per annum

Ghost acres
90 hectares of arable land needed to produce the cereal to feed the chickens

First illustration based on OECD-FAO, *Agricultural Outlook 2009–2018. Highlights*, 2009 and K. Deininger, D. Byerlee, et al., *Rising Global Interest in Farmland: can it yield sustainable and equitable benefits?*, World Bank, 2010.

whether the rancher had the facilities to deal with the high mortality rate on feedlots.

Later that night, our last in the area, a local environmental group showed us film footage they'd taken of around two hundred dead cattle from a local ranch operation. They had been dumped, apparently using a digger, just outside the perimeter of the ranch, on what was thought to be public land. The corpses were in varying states of decay, some already little more than skeletons; others quite fresh, suggesting this was not some sort of one-off catastrophe.

I don't think the rancher we met was a bad man. Indeed, he was very courteous. Like so many of the people I have met in the factory-farming business, he struck me as just another enterprising but ultimately misguided individual trying to make money from a bad system.

12

Thicker than Water

Draining rivers, lakes and oil wells

TURNING OIL INTO FOOD

Northern Alaska is perhaps the most beautiful part of the United States. It is vast pristine tundra, illuminated at night by the dancing pinks, blues and greens of the Northern Lights and seared through the days by the dazzling Arctic sun. Almost a million caribou roam over this vast expanse, while the ice-bound coast is habitat for seals, polar bears and whales that fish beneath its freezing waters. This epic snowscape is also home to millennia-old native communities whose property is beginning to sag into the melting permafrost while their icy hunting grounds recede further offshore through global warming. Beyond the region's living beauty are treasures of our planet's evolutionary heritage, unique fossils of the dinosaurs and early mammals that first adapted to live in the cold.

However, the natural riches of this Arctic idyll go beyond what you see on the surface. The US Department of Energy estimates that about 10 billion barrels of oil lie beneath the tundra and frozen waters of Alaska's Arctic National Wildlife Refuge (ANWR) and its adjacent districts,[1] enough to feed US and EU domestic consumption for a year.[2] There is also an as yet unquantified natural gas reserve in the shale beneath whose rocks much of the oil is also trapped, complicating the process of extraction and increasing the potential for damage to the environment.[3]

Since 1980 there has been a Congressional moratorium on drilling in the area, reissued each year despite efforts by George W. Bush
to have it overturned during his presidency. In the wake of the 1989
Exxon Valdez disaster in Alaska's Prince William Sound, which was
the worst oil spill on record until the *Deepwater Horizon* catastrophe
in the Gulf of Mexico, it had seemed that the ANWR might be
spared the plundering of its natural resources. Indeed, President
Obama opposed exploiting Alaska's environmentally sensitive areas
during his first presidential election campaign, saying that he wanted
to reduce oil consumption and promote renewable energy.[4] His then
opponent, Senator John McCain, stated: 'As far as ANWR is
concerned, I don't want to drill in the Grand Canyon and I don't
want to drill in the Everglades. This is one of the most pristine and
beautiful parts of the world.'[5]

However, as high fuel prices continued to hurt the US consumer,
the president changed his position, announcing the issue of new leases
to oil companies in a vast region directly bordering the ANWR. Meanwhile Shell, which had shelved its Arctic drilling plans in the face of
opposition from environmental groups and regulatory hurdles, has
announced it is now considering pushing ahead again following positive signals from the government.[6]

Several other oil giants are heading in a similar direction. Total
wants to develop the Shtokman natural gas field off Russia, the largest
single potential offshore Arctic project, 560 kilometres into the
Russian-controlled part of the Barents Sea.[7] Meanwhile BP is planning to join a £10 billion venture to develop onshore oilfields in the
Yamal-Nenets Autonomous Area of Russia, in the Polar Circle.[8]

Such is the global hunger for oil that it seems no potential new
source, wherever it lies, remains off-limits to energy giants for long.
Factory farms are a big driver of the insatiable demand. While golden
fields of corn and cows in sheds may seem a world away from ugly
power stations and the plunder of precious habitats in the relentless
quest for more black stuff, farmers are among the fossil-fuel industry's
most important customers. Whereas traditional farms relied on

manual labour, modern agriculture is heavily dependent on oil- and gas-guzzling machinery and huge quantities of petrochemicals. That's why farmers are so often at the forefront of fuel protests. When oil prices rise, it hits them hard.

Most academic research into the amount of fossil fuel used by industrial farms has been carried out in America, where a study led by Professor David Pimentel, a renowned ecologist based at Cornell University in New York State, found that conventional crop production in America swallows up the equivalent of 6.3 barrels of oil per hectare. Of this, two-thirds is used for petrochemicals like fertilisers, pesticides and other inputs.[9]

One tonne of US maize, a staple feed crop for intensive livestock, takes a barrel of oil to produce,[10] while modern farming methods globally use two barrels of oil on average to produce enough fertiliser and pesticide for one hectare of crops.[11] Agriculture for food production is such a big oil eater that it accounts for 7 per cent of America's entire energy usage.[12] Oil and industrial farming are so tightly connected that Albert Bartlett, a physics professor at the University of Colorado, has described modern agriculture as 'using land to convert petroleum into food'.[13] This process is likely to become increasingly expensive as the era of cheap, easy oil draws to a close.

Conventional sources of crude oil and gas are running out and there is widespread consensus that energy is only going to get more expensive in future. This will have a knock-on effect on food prices. A doubling of oil prices, for example, could lead to US grain becoming about 20 per cent more expensive.[14] This would mean people paying more for staples like cereals. It would also increase the price of factory-farmed meat, for which feedstuffs like cereals form two-thirds of the production cost.

In 2007, a meeting of the National Petroleum Council in America heard evidence that 28 per cent of existing oilfields were in decline by 2005 and 40 per cent would be in decline by 2008–9.[15] This bleak forecast seems to have been right. In 2010, the International Energy Agency published figures showing that production from crude oilfields currently operational peaked sometime between 2006 and 2008. The IEA predicts that output from these fields will decline by as much as a third of 2010 production levels by 2035. At a press conference in London in November 2010 to unveil a report on the future of world energy supplies, the organisation declared: 'The age of cheap oil is over,' though policy action could bring lower international prices than would otherwise be the case.[16]

The Oil Depletion Analysis Centre believes the world has now consumed almost half the total amount of conventional oil that most experts estimate will ever be available for recovery. 'A dozen recent independent analyses all show global production reaching a natural peak within the coming decade [i.e. up to 2020],' the organisation has said.[17]

So-called 'peak oil' – the point at which the maximum rate of petrol extraction is reached, after which production falls into decline – was a big talking point a few years ago, but has now largely been replaced by discussion about a less drastic-sounding scenario – 'oil depletion'. The new language may have been adopted because oil companies are reluctant to admit they may not be able to meet demand.[18] Persistently high oil prices have made it worthwhile for companies to exploit more expensive and 'unconventional' fossil energy sources, such as shale gas, oil from deepwater platforms (like *Deepwater Horizon*), Canadian and Venezuelan tar sands and coalbed methane in Australia. 'Fracking', the hydraulic fracturing of rock to release natural gas, has become popular despite protests by those worried about the environmental consequences. Either way, demand for oil shows no sign of letting up, oil companies are having to work harder to maintain supplies, and the upward trend in prices is expected to continue, with serious implications for farmers and consumers.

The oil companies themselves share this view. A 2011 report by Shell International on future energy scenarios suggested that global demand for energy could triple in the next four decades compared with 2000 levels. They warned that this could leave a gap between supply and demand equal to the size of the whole industry in 2000.[19]

Higher prices are expected whether or not fossil fuels continue to be heavily used, because of the higher cost of exploiting unconventional sources and exploring new geographical fields. Against this backdrop, farmers and policymakers need to think about how to produce food in ways that use less of this more and more costly resource.

One town in Britain is already doing just that. I came across it by accident while I was on an August holiday. My wife and I were renting a cottage in the English countryside, and took a day trip to the pictur-esque town of Totnes, northeast of Plymouth. The town was in full-on tourist mode, with a medieval market under way. There were women wearing blue, beige and rust-coloured Elizabethan period costumes manning stalls, and it was all very cheerful. We walked along the steep high street and past the Norman castle built by invaders centuries ago as a supreme lookout. It's a place with a strong sense of history – a sign boasts that it has more listed buildings per head of population than any other town in the UK.

But it also seems to be a place with a strong sense of the future. People have been thinking seriously about how to prepare for a time when fossil fuels are in shorter supply. As we sat having lunch in a high-street café, something pinned on the noticeboard caught my eye. There was a poster about 'Transition town Totnes'. I discovered that Totnes is the first UK town to be preparing for peak oil – all the more impressive because it is a community-led initiative, not something forced on locals by a forward-thinking local authority.

The poster urged people to think about the amount of energy they use, adding that, properly planned, a town using much less energy would be 'more resilient, more abundant and more pleasurable than the present'. Among leaflets promoting Latin dance, a local auction and a myriad of local attractions, was an invitation to visit homes in Totnes and the nearby village of Dartington already making the tran-sition to lower-energy households.

Looking up at the cattle grazing on the hillside above the high street, it was difficult to imagine that much of our farming too needs to consider a future without cheap, easy oil. When animals are farmed as I could see them around Totnes, on land rather than inside industrial farms, they almost always use energy and other resources more efficiently. I was struck by a recent comment in the *Times* newspaper, which put it this way: 'Scientific agriculture has led us to a point where many times more energy goes into a field in the form of fuel, heavy machinery, pesticides and chemical fertilisers than is harvested from it.'[20] There is a range of alternatives to oil-hungry industrial agriculture. Perhaps the best-known is organic. Organic farming is considerably less oil-dependent. Researchers at Cornell University found that producing organic maize takes 31 per cent less energy than producing conventional crops. In a report published in 2006, they claim that if 10 per cent of all US maize (corn) were grown organically, it could save the USA the equivalent of 4.6 million barrels of oil a year.[21] And the huge quantity of oil used to produce crops pales into insignificance compared with the amount used to produce factory-farmed meat. Cornell has looked at the input of energy taken to produce maize and wheat relative to the output (calories) in the end product. Pimentel and his team carried out the same calculations for the production of beef, pigs and chicken meat, based on the amount of grain and forage consumed by the animals and the calories of animal protein they eventually produce.

The ratio for harvested maize was 4:1 and for harvested wheat 2:1, whereas for beef production it was 40:1, for pig production 14:1, and for chicken-meat production 4:1.[22] In other words, growing wheat and maize – and probably vegetables – is far more energy-efficient than producing meat.

In the UK, the Soil Association has studied the amount of fuel used in intensive meat production relative to the amount used to produce it organically. They found that organic meat production was generally far more energy-efficient, though organic chicken and eggs required more, because the birds live much longer and are allowed space to exercise. Their calculations, based on official government statistics on energy use in farming, are that organic milk production uses 38 per cent less energy than non-organic; organic beef uses 35 per cent less; organic lamb 20 per cent less, and organic pigmeat 13 per cent less.

From what I've learned, there are even greater energy savings to be had by moving away from the practice of feeding livestock copious cereals and soya – feeding them food waste for example.

The Soil Association also looked at crops and found that organic crops like wheat, oilseed rape and carrots use about a quarter less energy compared with conventional growing methods.[23]

At the moment, most businesses seem to assume that the answer to the challenge of peak oil simply lies in finding more. In April 2012, Lloyds of London, the world's biggest insurance market, became the first major business organisation to break ranks, voicing concerns about the huge potential environmental damage from oil drilling in the Arctic. The City institution estimates that US$100 billion of new investment is heading to the far north over the next decade. They have warned that the cost of cleaning up any spills in this remote and spec-tacular part of the world could be 'significantly greater' than anywhere else. A spokesman said: 'Given the Arctic's fragile ecosystem, it is vital that policy-makers and businesses take into consideration worst-case scenarios. Migration patterns of caribou and whales in offshore areas may be affected,'[24] he declared, and reeled off 'multiple ways in which ecosystems could be disturbed'.[25]

When a company like this voices such concerns, it does raise my hopes. What seems obvious is that we are reaching a tipping point.

INTENSIVE FARMING IS THIRSTY WORK

When equatorial winds sweep heavy rain clouds across the Pacific Ocean, they blow liquid gold to Fiji. As the clouds burst over the Yaqara Valley, drenching the tropical rainforest, they replenish the supply of water hundreds of feet beneath the canopy of leaves. After passing through layers of ancient volcanic sediment it eventually settles in a vast bed of porous rock – from where it is sucked up into bottles and transformed into one of the most sought-after and expensive drinks in the world. 'Untouched by man', according to the advertising blurb, Fiji Water is marketed as the purest mineral water on earth.[26]

The business is owned by Lynda and Stewart Resnick, entrepreneurs from California who shrewdly saw the way the world was going. The water is bottled in a totally sealed facility built directly above the aquifer in the rainforest. Apparently no human hand is allowed to touch it before it reaches the customer who breaks the seal. The company boasts that the taste is 'as unique as its creation'. They say their product has a 'unique mineral profile', and thanks to its special qualities and a highly sophisticated marketing campaign involving product placement on hit American TV shows like *Friends* and *Desperate Housewives*, it has become a celebrity favourite.[27]

In the UK it is highly sought after by well-heeled shoppers at Harrods and Selfridges, as well as reputedly being drunk by the Beckhams.[28] How much its extraordinary success relates to its supposedly distinctive taste (in ad-speak, its 'soft, smooth mouthfeel'[29]) is another question. At the end of the day, it's water: H_2O.

Over the last forty years, bottled water, once a business that nobody took seriously, has become a multi-billion-dollar global industry. Seduced by clever marketing and beautiful packaging, wealthy consumers pay handsomely for a product that looks identical to what comes out of the tap and generally tastes pretty much the same. In what was derided at the time as a caricature of capitalism, in 2007 Claridge's Hotel in London introduced a water menu consisting of more than thirty varieties of bottled water from around the world. The Four Seasons Hotel in Sydney offers something similar – some two dozen options of still or sparkling – some of which are said to be shipped to the Antipodes all the way from the UK. It is difficult to square this booming industry with the vast quantity of water that is casually wasted, often by the same consumers who spend so much buying it in plastic bottles. For while people are prepared to pay top dollar for packaged water, ordinary water is undervalued and squandered wherever it is not obviously running out, especially by industrial farming.

Make no mistake, rearing animals is a thirsty business. Worldwide, around a quarter of freshwater use relates to producing meat and dairy.[30] On average meat needs around ten times the amount of water per calorie to produce as vegetables and other plants.[31] Just producing one kilo of beef requires the equivalent of nearly ninety bathtubs of water, while around thirty-three bathtubs are needed to produce a kilo of pork, and twenty-four tubs for a kilo of chicken.[32]

The water 'footprint' of the concentrated feed given to industrial livestock is more than five times greater than the footprint of the kind of food eaten by animals kept on pasture, such as the grass itself and silage.[33] In essence, factory farms are draining lakes and rivers to irrigate crops to feed to animals. Reared indoors in barns, the animals themselves drink and the sheds need to be hosed down. When animals are genuinely kept on a grazing system, much of the water is simple rain on grass – a natural process. In industrial rearing, water is often drawn from rivers and aquifers, thereby diverted from other useful purposes. Experts have calculated that grain-based animal feeds use forty-three times more irrigation water than pasture-based feeds.[34]

The perversities of the system are symbolised by an extraordinary business located deep in the Arabian desert. In the heat of the midday sun, temperatures in this part of the world reach a blistering 50 degrees Centigrade. There is precious little life save for a few scrubby bushes. But one species is defying the laws of nature: the American Holstein cow. Thanks to a complex irrigation system that pumps millions of gallons of water from deep under the sands, Al Safi mega-dairy in Saudi Arabia is flourishing.

In this most hostile environment, 29,000 listless cows are reportedly kept alive and functioning in open-sided sheds fitted with fans that waft a fine spray of water over the animals. The cooling equipment is imported from Arizona and enables them to survive the heat. According to a visiting journalist in 2002, they produce a phenomenal 549,000 litres of milk per day.[35] Of course the cows need to be fed, and Al Safi has found a way to make the desert oblige. Where there was once just hot sand, luxuriant green alfalfa now grows at the rate of a crop every three weeks. Fields are drenched by pivot sprinklers that suck water from sources up to a mile underground. It comes out close to boiling. Between drinking and cooling, each cow is said to use around 135 litres of water a day.[36]

The dairy uses sophisticated computer programming to monitor the milk output of every animal. Performance details are sent to a central database, where they are analysed and used to identify cows that are not coming up to scratch. The minute they run dry, it's off to the slaughterhouse.

Unpleasant though it is to see cows living like this – the *New York Times* reporter Craig Smith described the animals as 'so skinny that they look like a row of piebald dropcloths draped over scaffolding'[37] – this surreal operation is all the more disturbing for the vast amount of water it uses. After all, fresh water is a finite and diminishing resource, and the soaring global demand for meat is putting already dwindling supplies under increasing pressure. Factory farms in deserts expose the worst excesses of the system.

Many countries are already taking water from aquifers at well above the rate that nature can replenish them.[38] The UN warns that farming is already by far the dominant cause of water depletion globally.[39] And that simple equation, of taking more water out of the ground than the rain puts back in, is causing the oceans to rise, accounting for about a quarter of current sea-level rise.[40] Scientists report that global stocks of groundwater were shrinking more than twice as fast by the turn of the century as in 1960. If the USA's Great Lakes were being siphoned at the same rate, they would run dry after about eighty years.[41]

In wealthy countries, there is little public awareness of the real pressure on global water resources. Cynics often claim that water can't really be wasted because it is never destroyed: it evaporates, forms clouds, falls as rain, runs off mountains in rivers or seeps into the earth and emerges as springs that in turn feed rivers and lakes, in an endless cycle that never changes. But while the same amount of water exists on Earth today as it did millions of years ago – calculations put the total at roughly 1,620 quintillion litres – almost all those quintillions (97 per cent) are in the sea, where they're useless unless the salt can be removed.[42]

The hydrological engineer Michal Kravcik, who has studied water systems in Slovakia and neighbouring countries, has described how fresh water is being lost to the sea. According to the book *Blue Gold*, he describes how a drop of water first evaporates from a plant, earth surface, swamp, river, lake or sea, then falls back down to earth as rain. If the drop of water falls back onto a forest, lake, blade of grass, meadow or field, it 'cooperates with nature' to return to the hydrologic cycle.

However, if the earth's surface is paved over, stripped of forests and fields, or drained of natural springs and creeks, the drop will not go into rivers and lakes to be used by people and animals, but will head out to sea, where it will be stored. 'It's as if the rain is falling onto a huge, low-lying roof, or umbrella, of pavement and treeless areas: everything underneath stays dry, and the water runs off to the perimeter.' The result is rising sea levels.[43]

There is little doubt that rising pressure on global water supplies from agriculture, climate change and population growth, to name a few factors, is leading to critical shortages. Water is heavy and phenomenally expensive to transport, meaning that while parts of the world suffer catastrophic floods, there is no simple way of channelling their extra water to places that don't have enough. Up to 2 billion people in the world already suffer from water shortages, and scientists predict that the number will at least double, perhaps affecting as many as 7 billion people, the majority of the world's population, by 2050.[44] Since 70 per cent of the world's freshwater supply is used for agriculture,[45] any debate over water shortages must be a debate about the future shape of farming.

Many people mistakenly assume that this is only an issue for poor countries. This fallacy was exposed in 2012 when the UK government declared that London and the densely populated southeast of England were facing a drought. Reservoirs were said to be at an all-time low and a hosepipe ban was announced. A few days later it started raining – and barely stopped for six weeks. April 2012 was the wettest recorded since 1910, followed by the second-wettest summer for a hundred years.[46] The supposed drought became a national joke. Yet the underlying supply problem was real, and it was not the first time the government was forced to sound the alarm.

In 2006, water shortages became so serious in London that authorities considered bringing in supplies from Scotland and Scandinavia by sea tanker.[47] Other desperate measures contemplated included towing icebergs from the Arctic up the Thames Estuary, then breaking them up to provide slush to pump into the water supply. Thames Water, the company responsible for the capital's water supply, was also reportedly looking into so-called cloud seeding, the creation of rain clouds by dropping particles of silver iodide into the air.[48]

In the end they came up with a longer-term solution: the creation of a £270 million desalination plant. Described by the city's former mayor Ken Livingstone as 'technology more appropriate for the desert', it is Britain's first foray into desalination, and opened in 2010.[49] It is capable of providing enough water for 400,000 homes a day in 'seriously water-stressed' London, but given the vast amount of energy required to strip salt water from the sea, will only be used in an emergency.[50] Worldwide, there are now more than 15,000 major desalination plants, mostly in the Middle East. They are becoming increasingly common in Spain and America.

Bitter political battles are already being fought over dwindling freshwater supplies in other apparently rich countries. In some cases, the combination of shortages caused by excessive draining of rivers and the drastic measures taken by politicians to prevent the crisis escalating is forcing people off their land and robbing them of homes and livelihoods.

In Australia, the desperate state of the Murray–Darling, a major river basin that drains a seventh of the continental land mass, dominates politics. It is a hugely important source of water, irrigating farmland that accounts for 30 per cent of the nation's agriculture. For decades, the river was a free-for-all: dammed, diverted and piped off all along its 3,375-kilometre course by whoever could afford to. Nobody gave a thought to the consequences downstream as the river was plundered to supply towns and cities and thousands of farms. The party went on until the early 1990s, when the catastrophic consequences of this haphazard overexploitation became all too evident. By 1994, human activity was consuming 77 per cent of the river's average annual flow and the mouth of the river was beginning to silt up.

Two developments propelled the issue to the top of the political agenda, where it has remained. City dwellers whose water comes from the Murray–Darling started noticing that it tasted salty. Around the same time a vile toxic bloom sprung up along a 1,000-kilometre stretch of the river, a result both of the dwindling flow and of pollution from farms. The hideous sight and smell killed fish, repelled tourists and devastated hundreds of businesses. People in Adelaide began to worry that their taps would run dry. Meanwhile farmers continued to siphon off unsustainable volumes, wasting as much as a third of what they took through leaks, seeps and evaporation.

Ever since there has been an almighty row over what to do, focused largely on the role of farmers. Dozens of contentious rescue plans and initiatives have been launched, some of which have helped but none of which have proven sufficiently robust, long-lasting or effective. Not until 2008 was a single intergovernmental body set up to monitor and manage the river. This Murray–Darling Basin Authority (MDBA) has ploughed hundreds of millions of taxpayer dollars into various schemes, capping the amount of water that could be taken from the river and keeping clear public records of water-use rights. A water-trading system was launched, under which farmers were given annual water allocations that they could sell. The reforms made a significant difference – the salty taste in tap water disappeared, fish began to return, and water levels in the river rose – but serious problems remain, and disputes over potential solutions are more bitter than ever. There is evidence that smaller farmers are slowly being driven out of business, as big operators flex their muscles to commandeer heavily restricted water supplies. In 2007, the late Peter Cullen, who was an academic and member of the Australian government's National Water Commission, predicted that investment and water will continue to gravitate towards big, professionally managed farms – ironically, the sort of operations that typically waste water – while small family-owned farms where animals are pasture-reared will decline.[51]

At the time of writing the MDBA was about to release yet another plan to save the river, after a previous scheme collapsed in acrimony. The original plan involved huge cuts in farmers' allocations but was shelved following furious protests by farmers and rural communities. The MDBA was ordered to come up with something else. Tellingly, some scientists walked out of the talks in disgust, believing that politicians were shirking the hard decisions needed to preserve the river.[52]

The plight of the Murray–Darling is far from unique. According to the World Economic Forum's 2009 report on water, seventy major rivers around the world are near maximum extraction levels, including the Colorado, the Ganges, the Nile and the Tigris-Euphrates.[53] In a joint report, the United Nations Environment Programme, the World Bank and the World Resources Institute have warned that water is likely to become 'one of the most pressing resource issues of the 21st Century'.[54]

In the Middle East, North Africa, parts of Asia and parts of Europe, water is being sucked out of aquifers at a far faster rate than they can be refilled. The problem is accelerating. Scientists have warned that there may soon come a point when groundwater levels are so low that 'a regular farmer with his technology cannot reach it anymore'.[55] According to the UN Food and Agriculture Organisation (FAO), water use for agriculture – arable farming as well as livestock – is almost entirely to blame. The World Economic Forum has described the livestock sector as a 'key player' in increasing water use, mostly for the irrigation of feed crops.[56]

With demand for food expected to soar between now and 2050, there needs to be a new approach. According to a 2009 joint report by the International Water Management Institute (IWMI) and the FAO on the future of irrigation in Asia, 'there are clear limits in most places on the amount of additional water that can be used for agriculture'.[57] IWMI's projections show that on current trends, South Asia is going to need 57 per cent more water for agriculture, and East Asia 70 per cent more, to meet demand, 'Given the existing scarcity of land and water, and growing water needs of cities, such a scenario is untenable,' the report concludes.[58]

A key factor in the mounting pressure on water in Asia is the region's greater taste for meat. Since the early 1960s, rising meat consumption has resulted in a 3.4-fold increase in the amount of water needed per person for food in China. Some experts have described this as the 'main cause of the worsening water scarcity' there. According to a contribution to the scientific journal *Nature*: 'If other developing countries follow China's trend towards protein-rich Western diets, the global water shortage will become still more severe.'[59]

Of course it's not only industrial agriculture that is to blame. Other sectors play their part – even innocuous-sounding businesses like growing flowers. Yet it is the industrialisation of farming in step with population growth that is likely to place the greatest strain on water supplies over the next four decades. And not only human population growth, but also the burgeoning number of factory-farmed animals. On UN FAO predictions the livestock population will almost double by 2050,[60] which is likely to mean a population explosion from 70 billion animals slaughtered a year to 120 billion. The greatest proportion of this increase is predicted in thirsty industrial systems reliant on vast areas given over to growing feed. Growing livestock feed is not a sustainable use of scarce water resources. It takes up good rainfed cropland, which itself is globally in short supply; and where it does not, it increases still further the use of irrigation. Both would be better served growing food directly for people.

A reduction in meat consumption, coupled with a shift to pasture-based rearing of livestock, could make a big difference. Outdoor grazing systems generally use far less fresh water than industrial systems. This is particularly true for genuinely pasture-based systems such as grass-fed cattle rearing, where the animals are not fed concentrates based on grain and soya. The disparity in water use soars where the crops used to make concentrated feed are irrigated and grown using large doses of fertiliser, which is common practice. Including the amount of water that is polluted by agrochemicals and waste products, the water footprint of concentrate feeds like cereals and soya can be up to sixty times higher.[61]

In places like Britain, water tends to be taken for granted. We're obsessed by the weather, and for us there's always too much rain. At least, that's how it's long been seen. But things are changing. I laughed along with everyone else at being drenched every day for weeks during the so-called drought. I stood mouth agape as the river that normally trickles past our front door swept by us in torrents, and cursed as every dog walk was accompanied by a thorough soaking. But warnings of drought alongside persistent rain show just how precious water really is, and how scarce it is becoming.

13

Hundred-dollar Hamburger

The illusion of cheap food

With just £23 a week to spend on food for herself and her husband, Annabel Abram has learned how to make every pound stretch. She's discovered that if she goes to the local supermarket after 8 pm there are bargains to be found as store managers discount fish and vegetables they failed to shift during the day. She buys 'just add water' dried mashed potato to thicken soups and make them more filling, though the price recently shot up from 18p to 28p a packet, and she makes good use of her 10 per cent discount at the café next door to the Peace and Justice Centre in Edinburgh where she works as a volunteer.

If it weren't for her husband, Annabel would probably be vegetarian, but David, formerly a long-distance lorry driver, likes his meat, so she occasionally buys it, especially as sausages and burgers can be cheaper than fresh vegetables and fruit. On the face of it, this couple are exactly the sort of people who should benefit from factory farming. They live in a council flat on a deprived estate in Edinburgh, and with David off work due to a heart condition, they depend on state benefits. They have to watch every penny, so should appreciate the opportunity to buy bargain-basement meat. David is particularly fond of sausages, bacon and ham, and doesn't really care about the conditions in which the pigs were reared.

Yet Annabel is as discerning as her budget allows, and would no sooner eat a £2 chicken than a tin of cat food. She is an intelligent woman with an interest in animal welfare, and the price tag would ring alarm bells. 'I would definitely think there was something suspect about it,' she says, and she is right to be suspicious, because that price tag is an illusion. It might only cost supermarket shoppers £2, but somewhere else in the world, people even poorer than her and her husband are paying the price.

Defenders of 'pile them high, sell them cheap' farming systems always wheel out economic arguments, claiming that it produces affordable meat for the masses. They talk as if industrial farming is some kind of driver of equality and suggest it's unreasonable – immoral even – for people who can afford the luxury of food produced in a gentler, kinder and healthier way to question the system.

So successfully have they propagated this myth that few people in rich countries ever pause to question it. When it's possible to buy an entire chicken – plucked, packaged and ready to pop into the oven – for £2 in a Tesco price-war promotion,[1] or snap up two litres of milk for a quid at Poundland, or get a packet of no-frills rump steak for less than the price of a cup of tea in a greasy spoon café, perhaps it's not surprising that consumers are taken in by this logic. Such bargains seem to show that factory farming delivers spectacular value.

Yet the truth is that factory farming drives up food prices, because of the vast quantity of grain and soya required to feed the animals. This is tipping the delicate balance of the supply-and-demand see-saw. So while factory farms may deliver low-cost (and lower-quality) meat for shoppers in developed countries, it is at the expense of people else-where. The irony is that the people who pay the real price for the production of the bargain chicken or steak have no choices and no welfare state to rely on. The result is that many starve.

Around the world, food prices are rocketing as global food produc-tion fails to keep pace with soaring demand. There is a direct link between this phenomenon and factory farming. In the UK, the cost of a typical basket of shopping rose by 6 per cent between 2010 and 2011,[2] but the price of certain products – such as pasta, butter, coffee – shot up by many times that figure. Food-distribution charities in Britain have reported a huge increase in the number of people going hungry.[3] As the economic downturn dragged on and the number of people out of work increased, the government instructed job centres to refer poor and hungry people to charity food banks for emergency food parcels.[4] Worldwide, the UN estimates higher food prices pushed a further 100 million people into poverty between 2007 and 2008, and nearly 50 million in the latter half of 2010.[5]

Weather conditions have played a role in this disaster. A blisteringly hot summer in 2010 ravaged 2 million square kilometres of wheat and cereal crops across Russia, Ukraine and Kazakhstan, sending world prices spiralling upwards. In 2011 France and Germany, Europe's biggest cereal exporters, suffered the driest spring for almost a century, affecting yields. This added to inflationary pressures. Droughts hit Australia, America and Africa. In Lusaka, Zambia, the price of bread soared by 75 per cent between September 2010 and April 2011.[6]

Extreme weather events that damage harvests, rising energy prices and financial speculation all play a role in food-price hikes. Yet the surging demand for cereal crops, corn (maize) and wheat, propelled by factory farming, and to a growing extent by biofuels, is exerting further pressure. The insatiable demand for grain and soya beans to feed the billions of animals being reared on industrial farms means that vast tracts of prime agricultural land are being used to grow feed instead of crops for human consumption. A third of the world's entire cereal harvest and 90 per cent of its soya now feeds industrially reared live-stock.[7] At the same time, millions of hectares of fertile land are being turned over to biofuels. Together, these pressures are limiting supply of cereal crops for human consumption, pushing prices beyond the reach of many. There is now a dangerous competition between crops for people, crops for industrial farms, and crops for cars.

Multiple organisations and experts have acknowledged this link. In 2009 an Oxfam report on sustainable UK consumption warned that increased demand for grain to feed livestock is likely to push food prices 'beyond the limits of affordability' for the world's poorest people: 'The recent rises in food prices have already caused misery for millions, but future price rises and pressures on food supplies are likely to be increasingly compounded by, perhaps even driven by, rising global demand for meat and dairy products.'[8]

It's a theme echoed by many other agencies. The 2011 report on agriculture produced by the Foresight unit at the UK government's Department for Business, Innovation and Skills warned that major jumps in meat consumption – 'particularly grain-fed meat' – will have serious implications for competition for land, water and other resources. It warned that soaring demand for grain-fed meat is likely to lead to 'substantially higher food prices'.

The outlook is grim. Most economists and agricultural experts who have examined the evidence believe that the era of food surpluses is over and that high food prices are here to stay. The Foresight report says there is a 'significant likelihood' that prices of major crops will rise, 'perhaps dramatically', over the next forty years.[9]

It's no exaggeration to say that the consequences of these food-price rises are already proving catastrophic. Hunger played a key role in the Arab Spring of 2011, with initial unrest in some places sparked by the rising price of bread.[10] Bread is a staple food in Arab nations, the key source of low-cost sustenance. In Egypt it is known as *aish*, meaning 'life', and the flat round variety called *baladi* is heavily subsidised by the government. Any dwindling of supply or sudden price hike raises the spectre of civil unrest.[11] According to *The Economist*, the food-price spike of 2008–10 was 'the final nail in the coffin of regimes that were failing to deliver their side of the social contract'. Bread riots in Bahrain, Yemen, Jordan, Egypt and Morocco in 2008 were followed by political uprisings three years later[12] and the fall of President Mubarak in Egypt. It seems clear that food scarcity will increasingly shape modern politics; even trigger wars.

Food-price rises threaten to negate progress made on the millennium goals on child poverty and malnutrition. Scientists at the International Food Policy Research Institute have warned that on current trends, the number of malnourished children in sub-Saharan Africa will be nearly the same in 2050 as it was in 2000 (even though it would be a smaller proportion of the overall population).[13]

Were all the cereals alone currently fed to factory-farmed animals offered directly to people instead of being converted into meat, it could feed a mind-boggling number – as many as 3 billion folk.[14] It would certainly be a far more efficient use of resources, considering just how much plant protein is required to create chicken, pork and beef: an average of 6 kg of plant protein such as cereals to make 1 kg of high-quality meat.[15] But not all this 'meat' is truly fit for people. Calculations by the University of Manitoba in Canada suggest that producing 1 kg of genuinely edible beef by industrial methods requires as much as 20 kg of feed. For pigmeat and chicken, it requires 7.3 kg and 4.5 kg of feed respectively.[16]

Chandran Nair, an Asian environmentalist who runs a think tank called the Global Institute for Tomorrow, believes that drastic measures are needed in relation to food pricing to avert a human and environmental catastrophe over the next four decades. He has voiced an argument that is almost impossible for politicians to make: that the price of factory-farmed meat will have to rise dramatically to correct the huge discrepancy between what consumers pay and the real cost of production. He argues that the true economic cost of a US$4 burger, if you factor in externalities (such as the cost of converting grain to meat, water and energy use), is 'something like US$100'.

Nair even goes further, bravely suggesting that the burgeoning populations of China and India will never be able to live like Americans 'simply because there isn't enough to go around'. 'A sort of intellectual subservience on the part of many Asian policymakers and economists has resulted in a denial of the scientific-based evidence that five billion Asians in 2050 cannot live like the average American,' he told the BBC. In the same interview he pointed out that Americans currently consume around 9 billion poultry a year. Asia, with around ten times America's population, currently gets through just under twice that. If Asian meat consumption increases as expected, in 2050 Asians will consume around 200 billion birds. 'This again is not going to be possible . . . [at] that level of consumption, we will see . . . collapse in ecological systems that we are very much dependent on,' Nair has said.[17]

At the moment, factory farms keep growing as if none of this matters. The economics of these outfits force the companies involved onto a desperate treadmill, on which they must run faster and faster just to stand still. If the only reason people buy your chicken as opposed to someone else's is that yours is cheaper, then you have to keep the price low or lose sales. As production costs rise because of energy prices and pressure on other resources, your profit margin per animal likely falls. Often, the only way to keep making money is to rear more animals. And so these farms expand and expand, multiplying the suffering, waste and environmental damage they cause.

That has not stopped developing countries trying to emulate the system. Impressed by the way factory farming has delivered artificially cheap meat in the West, governments in many developing countries are actively encouraging small farmers to adopt mass-production techniques. They're egged on by multinational food and biotech companies, often by organisations like the World Bank, and any number of others who spread the false gospel that intensive farming, with all its chemicals, its waste, its hideous animal-rearing methods, fills hungry bellies and makes farmers rich.

Yet the evidence suggests that factory farming is a dangerous export. In a number of countries, attempts to adopt Western-style industrial livestock production systems and arable farming have been little short of disastrous. It is becoming painfully apparent that factory farming is not the solution to poverty in the developing world.

Nowhere is this clearer than in India. The tragedy that has been quietly unfolding on the subcontinent offers a stark example of what can happen when governments and multinational companies try to foist intensive agriculture on farmers in developing countries. The story is encapsulated in the fate of an ordinary man called Shankara Mandaukar, reported in the British press.[18] The weapon he used to kill himself was a tool he once hoped would help make his farm more profitable.

For years, the Indian father-of-two had struggled to make ends meet as his crops fell victim to bad weather, weeds and pests. He was easy pickings for the biotech company salesman who dropped by one day with an enticing offer of 'magic' seeds with special powers to resist bugs. The price was high but Shankara was desperate and the salesman promised huge yields. With the help of the village moneylender, he scraped together the cash. But the genetically modified seeds failed twice running. Bankrupt and about to lose his farm to the loan sharks, Shankara could see no way out. One afternoon, he downed a bottle of insecticide, and died in agony in the dust outside his home. Though a crowd of villagers looked on as he moaned and vomited, they knew there was no point summoning medical help. They had seen it all before and knew he would be dead within an hour.[19]

In the state of Maharashtra, such scenes have become commonplace. Hundreds of farmers kill themselves every year, as they do in many other parts of the subcontinent. The statistics almost defy belief. Since 1995 more than a quarter of a million Indian farmers are reported to have taken this desperate step, leaving hundreds of thousands of wives and children destitute. It's equivalent to one suicide every thirty minutes, and is believed to be the largest wave of recorded suicides in human history.[20] And despite a number of interventions by the Indian government, there is little sign that the problem is going away. Maharashtra state, the area of the country worst affected, has become known as 'the suicide belt', consistently recording the highest number of cases.

It's impossible to imagine such a tragedy occurring in Europe or America without the whole world knowing. Yet this ghastly spectacle has played out against a backdrop of near-silence from the international community. Few people outside India are aware of the crisis, though the Prince of Wales briefly drew it to public attention in 2008 when he condemned the 'truly appalling' death rate and highlighted the role of GM crops in tipping thousands of small farmers over the edge.[21]

Every suicide has multiple causes, but a number of common themes have emerged from these deaths. Most of the suicides appear to be triggered by financial difficulties linked to attempts to adopt new farming methods, typically involving high quantities of expensive fertilisers, pesticides, herbicides or GM seeds. Palagummi Sainath, an award-winning journalist on *The Hindu* newspaper who has investigated the crisis in depth, has written about the 'predatory commercialisation of the countryside' in India, and the 'corporate hijack of every major sector of agriculture, including, and especially, seed'.[22]

A high proportion of deaths afflict cotton farmers, lured into borrowing money to pay for expensive GM varieties that have failed to yield the promised returns, leaving them heavily in debt and at the mercy of moneylenders. Yet agents working for biotech companies persuade farmers who can ill afford it to find the money. So far from proving pest-proof, a number of varieties of GM seed appear to have succumbed to parasites, leaving poor farmers with nothing. There are also reports of salesmen failing to warn farmers that the so-called magic seeds require up to double the amount of watering as traditional varieties – a life-or-death matter in areas prey to drought.

In 2005 the government of India commissioned a report from the Tata Institute of Social Sciences into the suicide crisis. The study was small – it looked in detail at just thirty suicide cases in the region – but it revealed the shameful vulnerability of struggling farmers to sales patter by biotech agents or government officials. Tata's report found that many of the farmers depended on fertiliser and pesticide company agents for advice on seeds and crop care. It concluded that they were peddled a 'false perception of prosperity', prompting them to take 'serious risks' such as abandoning traditional crop rotation. They were unable to meet the rising costs of production, spreading on yet more fertilisers even while crop performance was declining.[23]

Professor Joan Mencher has pointed out that as well as examining the circumstances under which farmers have killed themselves, it's also useful to look at the circumstances and practices of those farmers who have not: 'On the whole, they tend not to have taken such large loans for chemical pesticides, herbicides and seeds. They tend to continue to grow food crops to be eaten not only by their own families, but for local markets and nearby cities. They inter-crop, rotate their crops, and have a backup in case one particular crop fails.' She concludes that: 'Excessive commercialisation usually means growing for distant markets and ignoring soil regeneration.'[24]

Dil Peeling, whom we met in chapter 6, has spent most of his career working in developing countries and has seen how unhelpful it can be to promote agricultural intensification in poor places. It's a lesson he learned right at the start of his career, when he was working with Voluntary Service Overseas in Indonesia. He recalls trying – and failing – to teach Western production methods to farmers on the tropical island of Sulawesi.

'My Indonesian wasn't good,' he remembers. 'I was bellowing instructions through a megaphone. They were nodding respectfully and pretending to listen, but I knew they weren't taking much notice.' His audience was arranged by status – the richest farmers at the front, the poorer standing behind them, a straggle of women at the back. In front was a posse of government officials from the ministry of agriculture. They sat round a collapsible wooden table and oozed superiority.

Armed with his loudhailer, Peeling was doing his best to convey the received wisdom that if they moved into intensive livestock production they would get richer. His heart wasn't in it, and not just because he found the condescending attitude of the local government officials distasteful. He'd noticed that the standard line of impressing Western intensification on farmers in the developing world wasn't working and that his audience would go home and ignore everything he'd said. Most of the frustrated NGOs and government agricultural officials seemed to think this was because the farmers on Sulawesi were just too stupid. They could not understand why the men seemed so resistant to the message. Most likely, it was because they simply knew better.

Peeling stayed in Indonesia for a while, but a few months later, two of his colleagues who had been working with the UK's Overseas

Development Administration (ODA) returned to London. During their time in Indonesia, Sarah Holden and Peter Bazeley had likewise observed that foreign aid-funded schemes designed to help the poor by promoting livestock intensification weren't working. They wanted to get to the bottom of it. Once a Whitehall backwater, ODA was being transformed by Tony Blair's new Labour government. It had a shiny name – the Department for International Development (DFID) – and it was an exciting place to be. The new administration wanted to do foreign aid differently. Old norms were being torn up; Clare Short, the secretary of state at the time, was determined that taxpayers did not throw money at stuff that did not work.

Together with the agricultural economist Steve Ashley, Holden and Bazeley secured public funding to see whether the limitations of the projects they'd observed in Indonesia were evident elsewhere. They reviewed several hundred aid projects in developing countries geared towards moving poor farmers away from traditional practices.[25] Most of the livestock projects aimed to provide poor farmers with new technology and services designed to boost production. The results were damning: the projects were not helping. Peeling, who kept in close contact with his old friends while they were carrying out the review, told me:

> By encouraging poor farmers to increase production, projects were intended to help the poor by a number of routes. Firstly, it was assumed that increased production meant more access to meat for the producers. But the problem is that the poor generally chase calories. They may have access to more meat, but high levels of protein are an unaffordable luxury when your family is underfed, so they sell it for something that gives more energy, like sorghum.
>
> Secondly, the projects were based on a false assumption that if you intensify, you will boost jobs. That's not true. Whereas arable farming tends to be labour-intensive, livestock production is capital- and energy-intensive. Ironically, intensifying actually pushes people out of jobs – the employer needs money to buy stock and feed, and can't afford to pay staff.

Thirdly, the projects hoped to increase the supply of meat for the hungry, by boosting production, and therefore bringing prices down for consumers. It didn't work because it ignored two key facts. The first is that hunger is still, in the main, a rural phenomenon. Even if highly intensive production could generate plentiful and cheap animal protein at source, delivering those perishable products to remote and dispersed rural populations through poor transport links in hot countries is simply economically untenable.

When it comes to Third World poverty, Peeling believes that factory farming is the wrong answer to the wrong question. You have to ask why people are going hungry in the first place:

> People are hungry because they are poor. People who are not poor are not hungry. So the problem is one of access, and frustrated demand, in that the poor cannot access food even if it exists. But the prescription that governments and aid agencies have provided to address hunger focuses on the supply side – by stimulating increased production. Yet eighty per cent of the world's malnourished children live in countries that produce an agricultural surplus. So the key question is not 'How do we produce more food?' but 'Who produces it?'

Having spent most of his career working with small-scale farmers in developing countries, often for the UN, his experience suggests that it is women who tend to suffer most, because they have significantly less access to resources.

> In many societies, keeping poultry, goats and sheep remains one of the few areas where productive assets are controlled by women. This type of small-scale livestock rearing provides a rare and precious opportunity to target women and children through development initiatives. The frustration is that where production has been intensified and commercialised, men have tended to take over and a golden opportunity for poverty reduction is lost.

Peeling feels that many people who are pushing for developing countries to adopt Western farming techniques misunderstand the way the poor use livestock. Out of 1.4 billion people in the world who are classified by the UN as 'extremely poor' (surviving on less than $1 a day), the vast majority (between 700 million and 1 billion) live in rural areas[26] and rely on their animals for a living.

> If you ask them what their best chance of getting out of poverty is, they say livestock. But if you ask them what their biggest source of income is, livestock only comes fifth or sixth on the list. Why the discrepancy? Because people in developing countries use livestock for things other than the market.
>
> There's an old Swahili expression that a man without a donkey is a donkey. People use their livestock as an asset, which reproduces faster than interest in a bank; or they use it for labour. They keep it as an insurance policy, perhaps to buy the right for their daughters to marry into a wealthy family. They keep their animals if they possibly can and only sell when they absolutely have to – something which goes some way to explaining the indifference of those Sulawesi farmers to my highly educated, strident but totally irrelevant advice on how they could commercialise their production.

In any case, deeply ingrained cultural values based on socio-economic realities in poor rural communities are not possible to overturn or shift quickly. The non-market services provided by farm animals are not just a social construct: they underpin rural economies. Food poverty is largely found in rural areas, whereas factory farming is geared toward feeding cities. In developing countries, it presents the poor with a double whammy, failing to provide them with affordable food and denying them the opportunity to grow food to sell to their urban cousins.

A number of factors are now starting to work against the approach of rolling out industrial grain-fed livestock agriculture across the world. These include falling water tables, degraded soil and rising temperatures. Good arable land and the aquifers that produce irrigation water are both in decline. While the world's food production continues to increase, the rate of increase has slowed. From 1970 to 1990, world grain production grew by 64 per cent. From 1990 to 2009, it increased by only 24 per cent.[27] Since 2002, world grain stocks (measured in days of available consumption) have been lower than at any time since the early 1970s.[28] A growing human population is in a furious competition for food with a burgeoning farm animal population. The number of animals is on course to near-double between now and 2050, mainly on factory farms. The global livestock industry already contributes 14.5 per cent of human-produced greenhouse gas emissions[29] – more than all our cars, planes and trains put together.

As the effects of global warming take hold, the sea is likely to rise and flood farmland. In some of the poorest tropical regions crop yields will decrease as a result of heat, drought and salination of soil.[30] For food production alone, at least 2 million square kilometres of additional land will be needed by 2030[31] – an area eight times the size of the UK. But a similar area of land could be flooded by the middle or end of this century.

When world food prices were rising in 2008, the UK government warned: 'In the developing world, food price rises threaten to throw millions of people back into poverty and increase the number going hungry.'[32] Yet few governments are willing to tackle the hidden costs of 'cheap' industrial food exemplified by the hundred-dollar hamburger.

VI
TOMORROW'S MENU

It was 2011, the year that *The King's Speech* reigned over the British Academy Film Awards, winning seven BAFTAs, including Best Film and Leading Actor for Colin Firth. Five million people tuned in to watch Firth receive his award. At the British equivalent of the Oscars, Natalie Portman won Leading Actress for her portrayal of a tortured ballerina in *Black Swan* and J. K. Rowling was awarded a special BAFTA for her contribution to British films.

So it was with some excitement that I accepted an invitation to lunch at the prestigious BAFTA building in the heart of London's West End. The meal was hosted by two distinguished chefs, Paul Merrett and Anton Manganaro, who had taken on the challenge of imagining what the restaurant menu of the future might look like. Still dressed in their chefs' aprons, they weren't claiming to have the answers, but were hoping the meal would start a conversation about how, in a world with a rapidly growing population and diminishing resources, our choice of food might need to differ from a typical meal today. Would the answer prove tasty and enjoyable to eat?

That day, we dined on a choice that included apple and parsley salad, parsnip soup and cobnuts, local line-caught fish, slow-baked Cornish lamb or roasted vegetable crumble with sheep's milk and cheese dauphinoise, and beetroot carpaccio with spelt. The gathered throng agreed it was delicious!

The following chapters are inspired by that meal, and the challenge faced by a future world of more than 9 billion people. The UN has warned that global food supply needs to increase by 70–100 per cent by 2050. Yet today as much as half the food produced worldwide is squandered – binned, left to rot or fed to farm animals.

How do we tackle this situation to everyone's benefit, including the animals'? What might the future hold for technologies like GM? How will pressure on the global food system change as China establishes itself as a superpower, and people in other rapidly developing countries in Asia and South America demand a slice of the meat- and dairy-rich diets long taken for granted in the West? What new ingredients might we sample, and is there a recipe for how we might feed everyone in a way that is kind to animals and saves the planet too. That day at BAFTA, Merrett and Manganaro declared on their menu: 'It is time for change. It is time to think differently.' I hope these suggestions will give you a taste of the future. *Bon appétit.*

14

GM

Feeding people or factory farms?

GOLDEN RICE

Ingo Potrykus is elderly now, but is determined to live for at least another two years. Happily he is in robust health and he has a huge incentive to stay that way – because soon he expects to see his life's work fulfilled. The German scientist has spent more than a decade campaigning for his extraordinary invention to be delivered to the world's poorest children.[1] During the agonising wait, he believes that millions of people have died who would still be alive today if only they had been able to take advantage of his breakthrough.

It's called Golden Rice, and it is perhaps the best advert for genetically modified food. It is a variety of rice that has been adapted so that it contains high quantities of the nutrient beta-carotene, which the body converts into Vitamin A. It's a substance not found in ordinary rice. Potrykus's genetically modified version of this staple food looks as ordinary and appetising as the saffron-coloured rice served in Indian restaurants. Its supporters, including the billionaires Bill and Melinda Gates, the Rockefeller Foundation and the US Agency for International Development, believe it has the potential to prevent between 1 and 2 million deaths a year in the developing world, and to save as many as 500,000 children from going blind.[2]

Around 124 million people in 118 developing countries suffer from Vitamin A deficiency (VAD) and its potentially fatal side effects. In southeast Asia alone, more than 90 million children suffer from the condition.[3] By eating just one bowl of Golden Rice a day, there is a decent chance these people could be saved.[4]

Yet since 1999, when Potrykus and his collaborator Peter Beyer worked out how to alter rice genes so that they produce beta-carotene in the part of the plant eaten by people, rather than just in the husk, where it is found naturally, Golden Rice has been stuck in a laboratory. Potrykus, Beyer and many humanitarian organisations who support their work have spent the best part of this time campaigning for permission from governments all over the world to start production and distribution. However, the long wait has not been entirely worthless – the pair and their collaborators have also busied themselves engineering an even better version with much more beta-carotene than the original creation. It will mean people don't have to eat nearly as much to get the benefits.

Now, finally, Golden Rice is on the brink of clearing all its regulatory hurdles in a number of developing countries, a moment that promises to be the proudest of Potrykus's life. I rang him in Switzerland, where he lives these days, and he told me it would be the culmination of his life's work. 'I have devoted everything to Golden Rice. It has taken an awfully long time. I just need to live another two years, and then I will see it finally saving lives.'

Golden Rice raises hope that the answers to world hunger can be developed in science labs. Many people believe that if science can solve malnutrition, we have an absolute moral responsibility to let it get to work. Even the Vatican is on-side. In 2010, after a study week by a group of scientists, the Holy See put out a statement declaring that governments have a duty to make GM crops more readily available and to remove barriers to their development: 'There is a moral imperative to make the benefits of [this] technology available on a larger scale to poor and vulnerable populations who want them.' Supporters of Golden Rice even invoke human rights to their cause.

They cite Article 27 of the Universal Declaration of Human Rights, which states that everybody is entitled to 'share in scientific advancement and its benefits'.

That's certainly how Potrykus sees it. What's particularly attractive about Golden Rice is that it was developed for humanitarian reasons, and Potrykus, who is in his late seventies, remains adamant that there will be no role for profiteering. He and Beyer could have made a fortune by selling the patent, but they have given it away free. 'We have donated the technology,' he told me. 'Neither of us was ever tempted to make money from it. My entire career has been dedicated to this. It is all about food security, and nobody will have financial benefit.'

Throughout the long years in his laboratory and the battle to win regulatory approval, he has been spurred by the memory of what it's like to go hungry. Growing up in post-war Germany, he and his brother had to steal and forage for food. 'That's what it was like for me, between the ages of twelve and fourteen. It was a very difficult time. I was fighting to survive. That has been my motivation with Golden Rice.'

The first country likely to authorise Golden Rice is expected to be the Philippines, shortly followed by Bangladesh, then probably by India, Vietnam, China and Indonesia. Though he is philosophical now, perhaps because the end is in sight, Potrykus never imagined it would take so long. Yet there are good reasons for the delay. A number of scientists have raised questions about the safety of Golden Rice and regulators have been reluctant to give it the go-ahead until these questions have been answered.

Most of the critics of Golden Rice cite general fears relating to the unintended consequences of messing around with the genetic makeup of edible crops. The vagueness of these concerns has been a major source of frustration for the international organisation that campaigns for the adoption of Golden Rice.[5] Yet questions remain about whether it will achieve everything Potrykus believes. There are fears that it could actually exacerbate malnutrition by discouraging poorer people from taking their own steps to improve their diets by eating more leafy greens.

Professor Dr Klaus Becker, of the University of Hohenheim in Germany, has warned that it could 'encourage a diet based on a single industrial staple food rather than upon the reintroduction of many vitamin-rich food plants with high nutritional value that are cheap and readily available'.[6] Other scientists believe that very malnourished people may not be able to make use of the beta-carotene because their bodies do not contain enough fat and iron to absorb it. There are also concerns that compounds linked to beta-carotene could cause birth defects. The World Health Organization (WHO) argues that there are simpler solutions to vitamin A deficiency. Their strategy focuses on promoting breastfeeding (breast milk is a natural source of vitamin A) and supplying high-dose vitamin A supplements for deficient children, a policy the organisation says has proven a 'simple' and 'low-cost' intervention that has had 'remarkable results' – reducing mortality by 23 per cent overall. The WHO is also working to encourage poor families to grow vitamin A-rich vegetables in their gardens.[7]

DO GM CROPS DELIVER?

It seems strange and incongruous that Golden Rice has taken so long to reach the table. After all, other genetically engineered crops are now widespread. In 2008, 13.3 million farmers in twenty-five countries – 90 per cent of them smallholders in developing countries – were growing genetically modified crops, covering 125 million hectares.[8] The most common modifications are herbicide tolerance and insect resistance. By tinkering with the genetic makeup of crops it is possible to give them new traits that enable them to thrive despite being doused in powerful weed- or pest-killers. The trouble is that these 'broad-spectrum' pesticides are indiscriminate, wiping out helpful insects that prey on the bugs that destroy crops in the first place. In other words, they destroy natural biological controls that can work very well to prevent blight.

GM crops may seem like a quick-fix solution to challenges farmers have faced for millennia, but they create a new set of problems, if not for the farmer, for everyone else. Research into the environmental impact of herbicide-tolerant crops carried out in the UK and published by the government department Defra in 2003–5 showed that GM crops were linked to a dramatic fall in the number of butterflies, bees, weeds and seeds compared with traditional intensive production using conventional weedkillers and insecticides.[9] An obvious criticism of this study is that it compared one environmentally damaging method with an even worse one. A 2009 University of California report into the ecological impact of GM crops and conventional fertilisers and weedkillers concluded: 'The massive use of transgenic [GM] crops and agrochemical inputs, mainly fertilisers and herbicides . . . pose grave environmental problems.'[10]

Outside the European Union, which remains wary of the technology, GM crops are commonplace. Proponents argue that they are a vital weapon in the battle to feed the world. In truth, the crops that are most likely to be GM are also those most often destined for animal feed. Over the last two decades, the technology has become a vital component of the industrial farming juggernaut, boosting production of feed crops for factory farms. The 'big four' GM crops – corn (maize), soya, cotton and rapeseed – are all used for animal feed, with corn and soya far the most important. About half of the world's corn (maize),[11] and over 90 per cent of soya beans,[12] in the form of soya meal, are used for animal feed. GM feed production is highest in North America, the global leader in industrial animal production. The US feeds almost 40 per cent of its corn to farm animals,[13] and 85 per cent of it is GM.[14] Similarly, the vast majority of the world's soya harvest is GM.[15] The uptake of GM corn and soya is high among developing countries where factory farming, or growing associated feedstuffs, has taken off.[16]

The hype about GM feeding the world appears little more than rhetoric, especially given that factory farms waste more food than they make, GM or otherwise. Only about 30 per cent of the calories in terms of corn or soya used to feed incarcerated farm animals are returned in the form of meat or other livestock products.[17] In other words, 'world hunger' is a convenient cover for what can be a deeply dubious business, where profit comes before the efficient production of food for people.

Saving seeds is intrinsic to farming worldwide: seeds are sown, the resulting crops are harvested and the seeds are extracted for the next season's planting. That's how it's always been. But in 1980 the US Supreme Court agreed that GM seeds could be patented – a move that has distorted this natural cycle. It is now illegal for American farmers growing GM crops to save seeds: they have to buy a new set every year or run the risk of prosecution.

Eagleville, Missouri, population 321, is hardly a crime hotspot. A cluster of houses huddled around two church spires and a water tower, it boasts a sleepy town square and the quiet, white picket-fenced lanes of any American village. Its general store, The Square Deal, supplies groceries and other bits and pieces to the villagers and farmers. Grateful to be spared the drive to neighbouring towns, locals pop in to buy everything from hunting kit to ice cream, and perhaps swap some gossip with the store's owner, Gary Rinehart.

One day in 2002 a stranger marched into the shop and asked for Rinehart by name. His tone was officious. He was there to make a serious allegation: planting genetically modified soya-bean seeds in violation of a patent held by the designer, the US agribusiness giant Monsanto. Rinehart was puzzled. Of course he'd heard of Monsanto – their 'Round-up' herbicide is used all over America – but he personally didn't plant or even deal in soy seeds, and his only contact with farming was via the sharecropping his brother did in a few family fields.

Monsanto's private investigator was insistent, accusing him of breaking the law and warning that he would be made to pay. There were other customers in the shop. Angered, Rinehart refused to answer any more questions and the Monsanto man walked out. He had been in the shop for less than two minutes.

The scenes that took place in Eagleville are repeated daily across North America as Monsanto and its competitors dispatch their legions of private detectives to protect their revenues. Such is the culture of persecution that American farmers now talk of a climate of fear. The way firms like Monsanto see it, farmers who save seeds from expensive genetically modified crops and use them to grow another harvest for nothing are simply thieves, and that is how they treated Rinehart. After going to his shop and openly accusing him of breaking the law, Monsanto hit him with a federal lawsuit, accusing him of 'knowingly, intentionally and willfully' planting GM seeds in violation of the company's patent right.[18]

In the months that followed, Rinehart learned that Monsanto had drafted in a St Louis-based investigative agency called McDowell and Associates to put local farmers under surveillance after receiving an anonymous tip-off that someone was illegally saving patented seeds. According to documents filed to the courts, a private detective named Jeffrey Moore had seen Rinehart planting soya-bean seeds from plain brown bags. According to Monsanto, when he had finished he tossed the empty bags into a ditch and drove off. Moore retrieved the bags and sent a handful of beans left in the bottom for analysis. The results showed they were 'Roundup Ready'.

After much toing and froing during which Rinehart continued to protest his innocence, Monsanto finally dropped the lawsuit. It had been a case of mistaken identity. The real culprit turned out to be his nephew Tim. The case was finally settled when the younger man apparently agreed to pay for the seed he'd saved. Both sides remain sore about what happened. According to Monsanto's website, they have 'not collected one cent'.[19]

It didn't take biotech companies long to figure out that the legal hassle associated with pursuing farmers who save seeds could be avoided if there was a way of making saved seed useless. The result is so-called 'terminator seeds', designed to yield only one crop. And so a millennial model of sustainability, in which farmers plant their own seeds, is coming to an end in America – and where America leads, others follow.

It might not seem unreasonable for companies to find innovative ways of stopping people using goods they have not paid for, but consider the implications for farmers who cannot afford GM seed, or simply prefer to do things the old-fashioned way. If their property borders fields sown with GM seeds, a phenomenon known as 'genetic drift' can occur, resulting in the traditional farmer's fields being contaminated with plants from patented seeds. This effect could have devastating consequences for farmers in developing countries like India, which has begun to import and experiment with terminator seeds. What if poor farmers who are used to sustainably replanting their own seed have their crops contaminated and rendered sterile?

Growing a means to survive has never been easy. Droughts, insects and crop blights have caused countless famines and continue to kill millions despite technological advances. Naturally farmers are always looking for ways to reduce these risks – and companies like Monsanto seem to offer an easy solution. Around 80 per cent of all food consumed in America now contains material from GM crops. Monsanto argues its GM technology can double yields at the same time as reducing the resources required to complete the cycle from seed to harvest. The company claims it helps farmers 'produce more while conserving more'[20] – as long as that doesn't mean conserving Monsanto seeds, of course.

Beyond Eagleville lie North America's plains, vast fields of swaying crops that stretch as far as the eye can see. In the old days, these endless plains were sown with a variety of crops. Now large tracts of land are cultivated with just one crop, from GM seed owned by one company, engineered to be resistant to herbicide supplied by that company. It's known as mono-cropping. Inevitably, it means more power to the seed company and less to the farmers.

The inevitable damage to wildlife, biodiversity and farmers who prefer to do things differently would not jar so much if GM seed dramatically increased production, saved land, or reduced food prices for ordinary people. But there's not much evidence it does any of these things. Disturbing reports are emerging that in some places Monsanto's

GM corn has stopped doing what it's supposed to do. Bugs appear to be developing a resistance. There are serious concerns that the same will happen with weeds, creating an even worse problem for farmers than the one they faced in the first place.[21]

There are also concerns about Monsanto's Bt-corn, designed to kill a bug called rootworm. Many farmers in America's cornbelt consider this creature their number one enemy, especially in Minnesota, where the corn harvest is worth $7 billion. So there was an enthusiastic uptake of Monsanto's genetically engineered variety, which promised to eliminate the problem.

For a long time it worked wonders. Then farmers began noticing some of their plants were toppling over – the worm was back. The first scientific confirmation of a problem came in August 2011, when Iowa State University released the results of a study.[22] The university described the reappearance of rootworm as very significant, dismissing the notion that there were just a few isolated cases. Now, in half a dozen states stretching from Illinois to South Dakota, the bug appears to be outwitting the men in lab coats. Michael Gray, an agricultural entomologist at the University of Illinois in Urbana, says that replanting GM crops year after year increases the danger that the insects will develop a resistance to the substance designed to finish them off. Monsanto's share price dropped 4.5 per cent on the news.[23]

It seems that GM technology doesn't even reduce food prices. At a time when more GM maize and soy is being grown than ever before, cereal and soy prices are at a record high. And high prices are now begging serious questions for governments worried about reliable food supplies.[24] As a way to solve world hunger, it doesn't stack up.

Britain's dairy cows had a close shave with GM when Monsanto tried to get a licence to sell a GM milk-boosting hormone to farmers in the UK. The product, widely used in the US, involves giving cows regular jabs with an artificial version of natural growth hormones to boost milk production by 10–20 per cent. The trouble is that high-yielding dairy cows are already pushed to their limits, producing ten times more milk than the calf would naturally drink. The product, called Bovine Somatotropin (BST), is associated with serious health risks including painful udder infections, digestive disorders and lameness. At the time, I was working as campaigner for Compassion in World Farming. We led a campaign to get it banned on animal welfare grounds. We wrote reports, badgered Westminster and Brussels and kicked up a real fuss. My team went on tour with a white coat and giant syringe to jab a lifelike model cow up the back end in high streets up and down the country. The resulting press did its job. After twelve years of hard campaigning, BST, known in the US as bovine growth hormone, was banned for both sale and use across the European Union.

The EU has adopted a highly cautious approach to GM. The only GM crop grown commercially by member states is Bt maize (MON810). In 2009, France and Germany both suspended approval for the crop, which is also banned in Austria, Hungary and Greece.[25] By contrast, in 1992 the US government decided to classify GM crops as 'GRAS' – Generally Recognised As Safe. This means GM food no longer has to carry any special label. No particular tests are required before it is sold to consumers. Companies can even introduce new GM food to the market without telling the authorities.[26]

Yet we still know little of the potential health risks. Research by the Institute of Ecology and Evolution of the Russian Academy of Sciences and the National Association for Gene Security found that rodents fed GM-soy lost the ability to reproduce within three generations.[27] An international team of scientists found that rats fed GM corn eat more and grow fatter than those on a non-GM diet. The effects were repeated when rats were fed fish which, in turn, had eaten GM food. Talking about these findings, Professor Åshild Krogdahl, of the Norwegian School of Veterinary Science, asked the obvious question: 'If the same effect applies to humans, how would it impact on people eating this type of corn over a number of years, or even eating meat from animals feeding on this corn?'[28]

Meanwhile, though GM crops are banned in the UK, there is no law against giving farm animals GM feed. Since there's no requirement to say anything about that on meat and dairy product labels, shoppers simply never know.

CLONING

Like millions of pet owners, Duane Kraemer and his wife Shirley dote on their cat. Pretty brown and white CC and her mate Smokey are so pampered they have a custom-made two-storey kitty house in the yard with a screened front porch, air conditioning, heating, catwalks, lofts and an enclosed outdoor play area. Care has been taken over the interior decor, with a framed photograph of CC as a kitten hanging on the wall. CC's favourite spot is at an upstairs window where she spends much of the day sitting looking out for visitors while her offspring Tess, Tim and Zip scamper around. Those who come to see her are often surprised by how ordinary she looks. For CC – or Carbon Copy as she is officially known – was the world's first cloned cat and her birth on 22 December 2001 in a Texas laboratory made international headlines.[29]

She spent the first two months of her life at the College of Veterinary Medicine, Texas A&M University, being paraded for the cameras and television crews and monitored by scientists for any sign of abnormality or ill health. Then she was adopted by the Kraemers. After they took her home she largely disappeared from the public eye, but in 2011 she was wheeled out again to mark her tenth birthday.[30] She certainly seemed in excellent health. Despite her unique origins, she is said to have lived a pretty normal life, producing a litter of kittens with Smokey in 2006. Apparently, she proved a great mother, and still is, though she takes life gently these days.

Like CC, the Kraemers are also beginning to slow down a little, but though Duane is now in his seventies, he is far from ready to hang up his white lab coat. He was one of the research team that created CC and has devoted the last decade of his career to exploring how cloning technology can be put to use – particularly by the farming industry.[31] Ironically Kraemer himself originally planned to be a farmer. He grew up on a dairy in Wisconsin and left school intending to learn animal husbandry and take over the family business. But as a young student at Texas A&M, the sixth-largest university in the United States, he was bitten by the research bug. To his father's disappointment he decided on an academic career instead.

The research project that produced CC was known as Operation Copy Cat and was part of a project called Missyplicity which has all the makings of a Hollywood script.[32] Bankrolled by an eccentric American multi-millionaire named John Sperling,[33] the scheme was inspired by the creation of the world's first cloned mammal, Dolly the Sheep, born at the Roslin Institute in Scotland in 1996 to a worldwide fanfare.

Sperling and his partner Joan Hawthorne were fascinated by what the Roslin Institute had achieved and saw an immediate application for the groundbreaking technology in their home – creating a carbon copy of their beloved pet dog. Missy, a border collie and husky cross, was getting on a bit and the couple were prepared to spend whatever it took to immortalise her by creating a genetically identical replica who would make them feel as if she was still with them after she'd gone to the great kennel in the skies.

News of their unusual project spread quickly and they were soon contacted by wealthy people all over the world who also liked the idea of immortalising their four-footed friends. Scenting a business opportunity, they set up a company called Genetic Savings and Clone with a view to providing an international pet-cloning service. The company opened for business in February 2000, but cloning dogs proved trickier than anticipated. For a while, some of their resources were directed at cloning cats, which is how CC came to be.[34] By the time scientists cracked the technology for cloning dogs, Missy had died, but her DNA had been gene-banked for future research. In 2005, Korean scientists created the world's first cloned dog, named Snuppy,[35] and two years later Missy's owners had her DNA flown out to Seoul, where it was used to create three carbon copies of their late pet.

In the meantime, Kraemer and his team were moving on to other species of clone. Kraemer doesn't see a problem with rich people investing their money to create copies of their pets, though the technology can't guarantee that the clones look identical to the original donor or 'blueprint'. (CC's donor, Rainbow, who is now dead, was a different colour – white, brown and ginger – to CC, who has no orange markings.) Perhaps that's partly why it has never really taken off, though the huge sums of money involved must be the main reason.

Since CC was developed, a handful of individuals have used the technology to clone pets, including a colourful character from California called Bernann McKinney who spent US$50,000 commissioning a lab in South Korea to create identical copies of her dead pit bull terrier Booger.[36] She went to collect the five puppies in person, but refused to let them travel in the cargo space of the plane home. The airline would only let her carry one pup on her lap at a time, so she spent another US$20,000 making five separate trips to ferry each puppy in style.[37]

However, predictions of a vast commercial market for resurrected pets through cloning fell flat. BioArts, the leading US pet-cloning company, which was set up by Joan Hawthorne's son Lou, closed in 2009 after concluding that the market was too limited. And there was another, uglier, problem: the number of deformed animals that were being produced. Lou Hawthorne has described how one clone that was supposed to be black and white had 'greenish yellow' bits; others had skeletal malformations, which he has described as 'generally not crippling, though sometimes serious, and always worrisome'.[38]

However, all the scientific money and effort that had gone into developing the technology had not been wasted. While there were too few pet owners who were willing or able to spend tens of thousands of pounds cloning their dogs, the technology pioneered by Kraemer and other so-called embryo transfer scientists had already found another type of customer: agricultural animal breeders.

Cloning is now an important tool in the deeply dubious and growing business of developing ever more unnaturally high-yielding beef cattle, dairy cows and pigs.

A superficially attractive aspect of this new branch of science is the creation of animals that are resistant to certain diseases. After helping to create CC, Kraemer took the technology one step further, using it to clone a bull resistant to a bunch of nasty conditions common in cattle: brucellosis, tuberculosis and salmonellosis. Nicknamed Bruce, the bull was the first animal in the world cloned specifically for disease resistance. He was created using genetic material from a long-dead animal named Bull 86, who had been identified as naturally resistant to those illnesses. At the time, Kraemer's team described the implications of their achievement as 'potentially monumental'.[39] Though the three cattle diseases have been virtually eradicated in the USA and Canada, brucellosis and tuberculosis are widespread in other parts of the world, so that the conditions could easily return.

Cloning is more commonly used in intensive farming to create animals that are more efficient money-making units. For the proud owners of pedigree beasts you can see the attraction: it offers a way of creating multiple copies of their prize animals, albeit at huge cost. The farm animals that merit the trouble and expense of cloning are the supermodels of the industry, themselves the product of highly sophisticated selective breeding techniques designed to make them ever more meaty and milky.

Selective breeding is all about controlling sexual reproduction, and has been going on since animal rearing began. Only the beefiest cows and the most productive milkers are allowed to mate, so that their superior genes are not mixed with those of flabbier, less impressive creatures. By picking out the animals with the most desirable characteristics, culling the rest, and using judicious inbreeding, it's possible to develop a bloodline that is even better than the original base stock. Naturally, these bovine and porcine Schwarzeneggers are worth a fortune, particularly for their superior ova and semen, but there's never any guarantee that their offspring will be as spectacular as the original – which is where cloning comes in. It offers a way to produce multiple 'twin' copies of top-producing animals.

As the US-based cloning company ViaGen boasts on its website, by having multiple animals that are genetically identical to your top performers you can 'greatly expand their reproductive potential'. There are other benefits too: it's a way to 'keep up with demand for offspring, embryos and semen' from top animals and insure against your prize bull being accidentally injured or killed. You can make sure your herd of animals is more consistently outstanding.

ViaGen, which offers an international cloning service for prime cattle, pigs and horses, makes it sound simple. All they need is a sample of an animal's genes. They provide a biopsy kit and a shipping container, the customer sends off a nick of the animal's flesh, and ViaGen's scientists do the rest. They culture the cells, transfer their DNA into eggs whose genetic material has been suctioned out, and put the resulting embryos into an incubator for a few days. Then they're transferred into a surrogate mother, and after the usual gestation period, the cloned calves, piglets or foals emerge. As soon as they're weaned and given a clean bill of health by a vet the youngsters are shipped off to the customer.[40]

Put like that it seems quite benign, but the truth is that cloning is a way of propagating breeds that have been pushed to their physical limits to produce ever more meat and milk. It threatens to accelerate the use of highly intensive genetics in farm animals, causing serious animal welfare problems. In short, cloning threatens to multiply animals that are genetically programmed to suffer. It is a way of locking in misery.

This is on top of the serious animal-welfare problems associated with the cloning process itself. ViaGen's website doesn't appear to mention it, but for every successfully cloned prize bull many others are likely to have died. From deformed hearts and other organs to being grossly oversized, cloned animals and their surrogate mothers suffer from a range of appalling health problems and deformities that are rarely seen otherwise.[41]

The scientists who created Dolly the Sheep implanted a total of 277 embryos in the process, from which only thirteen pregnancies resulted. Only one of these, Dolly, was a successful live birth. Despite years of research and efforts to perfect the technique, nature appears highly resistant to this contrivance and success rates remain incredibly low, at around one to thirty in every 1,000.[42]

One particularly nasty problem associated with clone pregnancies is hydroallantois (hydrops), a set of typically fatal conditions in which one of the foetal sacs swells with fluid to grotesque proportions. According to data from the US Food & Drug Administration (FDA), hydrops has occurred in 13–40 per cent of clone cow pregnancies; one study reports it in over half of cases. It's a condition that rarely occurs in pregnancies produced through normal artificial insemination or natural selective breeding.[43]

Late miscarriage is another major problem: studies in Europe show that most cloned embryos die during pregnancy.[44] Surviving the pregnancy is only the first hurdle. Cloned newborns are likely to suffer serious health issues, including deformities, lowered immune systems, and problems with their heart, lungs and other major organs. One reason Korean scientists won the race to produce the world's first cloned dog was that they had a ready supply of dogs of all shapes and sizes for experiments. In a culture that eats dogs for dinner and likely has few animal welfare restrictions, scientists had no difficulty sourcing canine guinea pigs and probably don't have to bother keeping a body count.

Establishing the scale of suffering is difficult: much of the information that is readily available is selective and biased, and long-term data is limited. However, when the European Food Safety Authority looked into the issue, it concluded that the health and welfare of 'a significant proportion' of clones was 'adversely affected, often severely and with a fatal outcome'.[45] Meanwhile the European Group on Ethics in Science concluded that the level of suffering and health problems experienced by surrogate dams and animal clones meant that it saw no 'convincing arguments to justify the production of food from clones or their offspring'.[46]

Encouragingly, New Zealand's leading farm-animal research centre, AgResearch, has abandoned its cloning research programme. Closure was blamed on unacceptable death rates. The catalogue of welfare concerns came to light only when reports were issued under the Official Information Act legislation, which gives the public a right to obtain certain previously secret data if it is felt to be in the public interest. The reports showed that animals had suffered from chronic arthritis, pneumonia, lameness and blood poisoning, and that only about 10 per cent of clones survived the trials.[47]

Whatever the research about suffering says, cloning animals for food is something I have always instinctively felt is wrong. The majority of the British public appears to agree. When Compassion in World Farming ran a campaign on the issue, some 2,000 of its supporters lobbied the government minister responsible in the UK. Nearly 3,000 also wrote to their MEPs, sending over 14,500 emails. Europe-wide polls show that nearly two-thirds of people see animal cloning as 'morally wrong'. Over half of those surveyed felt that cloning for food production was unjustifiable. Most said they would not buy meat or milk from cloned animals and eight out of ten said that it should be labelled if it ever became available in the shops.[48]

All the same, it managed to do so. In the UK, milk and beef from the offspring of cloned animals slipped into the food chain in 2010, unannounced and unauthorised. Its discovery caused a public furore. In August that year it emerged that beef from a bull whose mother was cloned had somehow found its way onto dinner tables. The bull was one of a pair produced by a cloned cow created using stem cells from the ear of a prize-winning dairy cow in America. Embryos produced by the clone and a normal bull were frozen and flown to the UK, where they were implanted in surrogates and born on a farm in Shropshire. Two of the animals ended up on a farm in Nairn in Scotland, where they sired ninety-six pedigree calves.

When this was exposed by the UK media, there was a public outcry.[49] David Cameron had been prime minister for just three months, but I was keen to make the most of the public reaction, so at

Compassion in World Farming we organised a stunt involving forty Cameron lookalikes. Actors wearing identical suits, shirts, ties and face masks were let loose in central London. They headed for Downing Street so we could make our point. Passers-by pressed their faces against car windows, stared out of buses and hailed the Cameron army as it marched down Whitehall.

'One Cameron was enough!' a Londoner told me. 'Blinking cruel to let forty of 'em loose at once.' If I'd had a pound for every person who said something similar that day, I'd have gone home quids in.

Outside Downing Street, the lookalikes unfurled a giant white banner condemning cloning as cruel. I was carrying a pile of petitions – thousands of signatures we'd collected in just a few hours. I was ushered through the gates to a security booth and on to the famous black door of Number 10. I banged on it theatrically. A polite but not particularly welcoming official opened it and listened dutifully as I said my piece. He took the petitions. We shook hands, smiled for the cameras, and then I left.

Next stop was the Houses of Parliament. The media took pictures of forty mirror-image prime ministers on Westminster Bridge as Big Ben chimed midday. Then we headed to the London Eye. I remember thinking what a great picture it was going to be – all those clones crammed into a glass pod high above the capital. Sadly, we weren't allowed on. I wondered how many other smartly dressed but totally innocent tourists would be turned away in error that day.

The stunt was fun, but the message was serious. Consumers had – and still have – no way of knowing whether the food on their plate had come from the offspring of cloned animals. We wanted government action, especially as there had been an overwhelming vote in favour of banning food from clones and their offspring by the European Parliament just a month earlier. Sadly, the complicated world of European politics meant it didn't count for anything other than an opinion; the vote was won but nothing changed.

'Frankenstein' food from imported descendants of cloned US animals had all the hallmarks of a winnable issue. I genuinely thought

David Cameron would throw his weight behind us. Has there ever been a more colourful example of dishonesty in the food chain? Yet the prime minister did what governments usually do when they're in a hole: announce an investigation and let it drag on long enough for voters to lose interest. The Food Standards Agency looked into the issue on the basis that the sale of meat and milk from clones is unauthorised under a law called the Novel Foods Regulation. They eventually declared that there is no rule against the sale of meat and milk from the 'progeny' of cloned animals, and that it was, in their opinion, safe to eat.[50] The question left hanging is: how do they really know? And why ride roughshod over public and political opinion? It seemed to me an example of the weak leadership that is allowing our food system to drift into the abyss.

In America, the FDA approved in principle the sale of cloned food as safe for consumers in January 2008, despite widespread protests from consumer and animal welfare groups, environmental organisations, the public, the dairy industry and Congress.[51]

Cloning is just the most extreme example of the scientific wizardry being deployed by companies in the hope of creating farm animals that can produce even more for less. Shielded from the public eye, in laboratories all over the world, men and women in white coats are developing all sorts of weird artificial creatures they hope will be prototypes for future generations of farm animals.

Take broiler chickens. Cram thousands of chickens into a tiny space and you get all sorts of problems, not least the same issue you have if you crowd thousands of humans into a confined space: it gets hot. There are many solutions, ranging from the obvious one of keeping fewer birds in one place, or opening the doors and letting some out, to the more expensive alternative of installing air conditioning.

Boffins may have found a more grotesque solution: featherless chickens. Not only do they stay cooler in tropical climates, but without the inconvenient coats nature provided, they take up less room. And behold, that means you can cram even more into the same space.

There's a further selling point for naked birds: once slaughtered, you don't have to bother to pluck them.[52]

The featherless chickens were created using traditional selective breeding techniques, rather than through cloning or GM. The Israeli geneticist who created the bare-skinned prototype in 2002 believes they could represent the future of mass poultry production in warmer climates. They look just like the raw chickens you buy in the supermarket, only they still have their heads – and unfortunately for them, they're still alive.

They are not the only farming freak show in town. How about cows that have been genetically modified to produce 'human' milk? It sounds like an April Fool trick, but within the next ten years it's conceivable that some kind of hybrid cow/human milk will be available from supermarkets, sold alongside nappies and jars of baby food. At China Agricultural University, scientists have reportedly introduced human genes into 300 dairy cows, enabling them to produce milk with some of the key properties of human breast milk. Apparently it tastes stronger than the usual stuff, but it contains proteins that boost an infant's immune system, just as a mother's milk would. The research team performed various other tricks that gave the bovine milk other human qualities. According to newspaper reports, Professor Ning Li, the scientist who led the team, described the result as 'a possible substitute for human milk'.[53] The mind boggles at how many genetically modified cows would be needed if plans press ahead for commercial production, not to mention the unknown health implications of bringing up a baby on it.

No doubt most scientists involved in cloning and genetically modifying animals genuinely believe their work is for the greater good. Kraemer himself, one of the pioneers, is a wonderful advertisement for what is a highly questionable business, with a kindly air, softly spoken and looking like everybody's grandad. He is clearly passionate about his research and wholeheartedly believes he is making the world a better place. On his office wall is his motto: 'Ask not only what nature can do for you, but also what you can do for nature.'

These days, his research is devoted to preserving endangered species. But are wacky technofixes the right response to the coming food crisis? Or do they simply take factory farming to newer extremes, pushing already overworked animals ever further, to the detriment of their welfare and the quality of food they produce? Thankfully, you don't need a degree in genetics to work that out.

15

China

Mao's mega-farm dream comes true

No exposé of factory farming would be complete without travelling to China. The People's Republic is already a major player in global food production and will undoubtedly have a big say in the shape of things to come. I therefore wanted to see how the most populous country on Earth, the jaw of the fast-growing Asian tiger economies, was getting on with the business of developing its agriculture. After all, half the world's pigs are in China. Steamed, roasted, barbecued or minced in dim sum dishes, the Chinese are big on pork, devouring 34 kg of the stuff per person per year, compared with 25 kg a head in Britain.[1]

Although many pigs are still reared on traditional smallholdings, big players in China's food-production industry are eagerly importing the most intensive pig-rearing techniques they can find in the West, and see the UK and US as role models.

In 2011 the British government signed a multi-million-pound livestock trade agreement with China,[2] involving the live export of thousands of prime breeding pigs. After decades of selective breeding programmes, the British pig industry has production down to a fine art, with the average UK-bred sow producing twenty-two piglets a year, compared with fourteen in China. Chinese farmers want a slice of these supernatural fertility rates and have taken the direct approach of buying British pigs, chartering entire Boeing 747 planes (at a cost of £330,000 per trip) and flying them 9,000 kilometres east.

The *Daily Mail* reported this unusual new trade as if it were an amusing curiosity, making much of the style the animals travelled in. An article highlighted their extensive leg room and freedom to 'relax and stretch out' while in the air, contrasting the conditions with those endured by human passengers in economy seats. A spokesperson for the Yorkshire-based company involved, JSR Genetics, said they are 'targeting China as a major growth area in the coming years', using their 'advanced technology' breeding programmes to develop blood-lines like the company's 'faster-finishing' boar.[3] The piece did not question what conditions the animals were destined for, nor explore what lies behind the new Sino-British porcine relationship – China's rapidly growing appetite for meat. It is this huge surge in demand for meat to feed China's growing middle classes that makes the People's Republic so important to any debate about the future of farming.

Its appetite for large-scale pig farms is actually nothing new. It dates back to the middle of the last century and the Great Leap Forward, Chairman Mao's disastrous campaign to catch up with – even overtake – the Western world's economy in less than fifteen years. Between 1958 and 1961, in an attempt to boost industrial and agricultural produc-tivity, he reorganised his country's vast population into large-scale rural communes. Private farming was banned and those who dared engage in it were persecuted as counter-revolutionaries. The Hong Kong-based historian Frank Dikötter has described how the chair-man's grand plan included 'extravagant schemes for giant piggeries that would bring meat to every table'. One scheme, designed to mark the tenth anniversary of the Chinese Revolution of 1949, included a 'pig city'. According to Dikötter: 'Many hundreds of houses set back from the street were destroyed to make room for the project.'[4]

The Great Leap Forward was catastrophic. Despite the huge invest-ment that went into the project, the overall increase in national production was negligible. The ban on private smallholdings ruined peasant life at the most basic level. Villagers simply could not make a living, because they were no longer able to exploit their own land. The result was huge-scale famine. Dikötter calculates that as many as 45 million people were 'worked, starved or beaten to death'.[5]

Sixty years on, China has become a fearsome superpower with the potential to usurp America as the mightiest nation in the world. Yet it still struggles to feed its people. A 2008 UN study found that while the developmental level in Beijing and Shanghai is comparable to Cyprus and Portugal, some provinces such as Guizhou are more like Namibia or Botswana.[6] Poverty in China remains a largely rural matter. There's a vast and growing gulf between those in urban areas and the rural poor. Around 10 per cent of the nation still lives in poverty: 130 million people.[7]

With such a huge job on its hands, it's easy to see why the government has been seduced into believing large-scale industrial farming is the future. The regime is importing factory-farming models sold enthusiastically by the West. The industrial farming system has many champions and powerful interests who are only too eager to cash in on the potential new business from agricultural trade deals with the East.

So far, there is little evidence that it's feeding the rural hungry. There may be a lot of extra meat, but much of it seems to be feeding cities, not poor country folk. The scale and investment needed means that the rural poor are pushed out of the equation and denied the opportunity to provide food for the cities, perpetuating the poverty trap. Bizarrely, I learned of considerable amounts of Chinese pigmeat being shipped to Japan. It doesn't add up.

In 2011, I set off with a journalist and a cameraman for the eastern side of China, to see some of the biggest pig farms in the world. It was October and we spent the first twenty-four hours in Beijing, where we hooked up with Jeff Zhou, Compassion in World Farming's man on the ground in China. It was not my first time in Beijing – I'd been to a conference in the city three years previously – so the smog that greeted us as we stepped off the plane was all too familiar. I was wearing my binoculars through sheer force of habit – there was little hope of seeing any birdlife through the thick beige blanket.

Compassion in World Farming has been trying to get a foothold in China for a decade. Our first trip out there revealed that there were no words for 'animal welfare' in the Chinese language. In the years that followed, a lot of our efforts met with official bemusement. But something has changed. As we embarked on a fifteen-hour train journey from Beijing to the heart of pig country, Jeff explained why. China is undergoing a food-confidence crisis. For years consumers have simply been told what they want to hear, whether they're buying grapes, pork or fish. Today, thanks to education, access to the Internet and irrepressible evidence, the public can no longer ignore the shady origins of much of China's food. Finally, officials are listening.

In September 2008, a major scandal propelled food safety to the top of the political agenda. It emerged that powdered baby milk had been illegally adulterated with the chemical melamine,[8] apparently to give the appearance of higher protein levels when tested. It was genius for the factories. They could sell watered-down milk and no one would know the difference. Mothers would happily feed it to their babies. But one after the other, children were developing kidney stones and acute kidney failure. Melamine – the chemical also used in plastics as a fire retardant – turned out to be a killer. Thousands of babies were affected and six died. Hundreds underwent treatment for kidney failure. The authorities showed no mercy to the bosses of the baby milk company, executing two of them by firing squad.[9]

Jeff said the tragedy had made people think about food in a new way: 'For the first time they wanted to know where it came from.' Of course it was bad news for all the legitimate milk farmers in China. Millions of people swapped from dairy to soya or insisted on buying imported milk. But perhaps more importantly for the Chinese government, it was a big international embarrassment, reported in the UK and elsewhere.

Not long after that, pig farming came under the spotlight, in what Jeff called the scandal of the 'work-out pig'. Today's Chinese consumers like lean meat. The problem for factory farmers is that pigs in cages don't have much space to exercise. In any case, factory farmers are reluctant to 'waste' expensive feed by allowing animals to use precious energy moving around rather than growing. As a result, their meat is fatty. Chinese pig farmers were keen to find a way of keeping a pig in a cage and making it grow fast without getting fat. They hit on a solution: feeding pigs a steroid called Clenbuterol, which is sometimes used (illegally) by body builders to build muscle without fat. They did it quietly, but two things gave the game away.

First, the pigs grew so big that their skinny legs couldn't support their enormous bodies (an all-too-familiar complaint on intensive chicken farms). Pictures began to circulate on the Internet of pigs that couldn't stand. Second, people started to get ill. Clenbuterol is illegal in most countries, including China, for a good reason: it has serious side effects, including palpitations and stiffening of the heart muscle.[10]

In April 2011 the 'work-out pig' scandal hit the headlines.[11] It was another dangerous breach of public trust in a country walking a tightrope between communism and modernity, echoing an incident five years earlier in which the Chinese newspaper *People's Daily* reported that over 300 people had been poisoned by meat contaminated with Clenbuterol.[12]

The new public concern about how food is produced may explain the surprisingly warm welcome I received from officials when I arrived for this trip. Whereas I used to struggle to get any kind of a hearing from the Chinese government, this time my colleagues and I were treated to an exquisite dinner with special wine. More importantly, we learned the exciting news that finally the government has set up an organisation to promote animal welfare. It's clear that public health and international reputations are at stake in China, and with them internal stability and international trading opportunities. As a CEO campaigning for international welfare standards, suddenly I matter more than I did a decade ago. The question is how long the new attitude of officialdom will take to change anything on the ground.

While most decision-makers are city-based, the animals are in the countryside, albeit often surrounded by concrete. And so we found ourselves bound for Henan Province, the cradle of Chinese civilisation and apparently the centre of China's pig-farming universe. I was not sure what we might actually see. The Muyuan pig farm we hoped to visit had declined our request to be shown around, and a run-in with local security guards did not appeal.

We took a night train from Beijing to Nanyang, a twelve-hour journey through mile upon mile of maize fields. They stretched along the length of the railway track for at least the last hour of our journey. The crop, I was told, was destined for animal feed and biofuel. I was struck by the sheer scale of land devoted to feeding factory farms and cars rather than people.

After a restless night on a hard berth, we piled out of the train wondering if we'd ever see blue sky again. Even 700 kilometres from Beijing, the pollution was thick and deadening. We drove through the busy streets of Nanyang. A scooter pulled up beside us with a small Pekinese dog riding at the owner's feet. Another scooter whizzed by, this time with live chickens dangling by their legs like feathery saddle bags. Their wings were burning on the hot metal and their legs must have been cracking with every jolt of the bike. A few minutes later we saw a black dog caged outside a restaurant. The world bustled by. At last we left the traffic of the city for the open road. We passed through more miles of maize. A couple of storks' nests sat atop telegraph poles, looking like little pyres.

We arrived at our destination, a small village next to a mega-pig farm with 5,000 breeding sows and their attendant progeny. I was eager to hear about life so close to so many industrially reared pigs. After a quick shower and a bite to eat at a local hotel we set off in the hope of finding the pigs. Our first stop was a village just a kilometre downwind from the second of the twenty-one Muyuan pig farms that dominate Henan Province today. This farm is also the company headquarters of one of China's biggest pig producers.[13] Since our request for an official visit had been declined we decided it would be more discreet to abandon the car and get as close as we could on foot.

Before setting off, we researched the company's background. It turned out to be supported by the International Finance Corporation

(IFC), the private lending arm of the World Bank.[14] Indirectly, therefore, it is subsidised by taxpayers around the world. The IFC website offered a bit of information about Muyuan farm company, talking about the firm having a biogas digester, a machine that takes the pig muck and turns it into energy for use on the farm. Not a bad idea, but our calculations indicate that at best this might only provide for about 6 per cent of the total energy required to raise the estimated 450,000 pigs a year that Muyuan used to rear at the time the investment proposal was written.[15] Now they produce a million. The Muyuan company is growing at a phenomenal rate. It has outgrown twenty farm sites and there are reports that IFC taxpayer money is contributing to yet another farm. This twenty-first site will take the twenty-first-century pig one step further from the natural world.

We drove through kilometre after kilometre of maize to a tiny village near the firm's headquarters, then parked and made our way to the farm gates, knowing we were unlikely to go unnoticed for long. We were just a few hundred metres away from thousands of pigs – the smell told us that – but all we could see was vast sheds, so we got close enough to take some photos and then retreated, anxious not to blow our chances so early on the first day in Henan.

Back in the village where we had parked, people seemed happy to talk. A man in his fifties, smoking a cigarette and leaning up against a crumbling village wall, struck up conversation – Jeff translating. He spoke with authority. He gave us his name, but it is changed here to Mr Chan to protect his identity. Life in China has got better, Chan told us, but the farm had done nothing for the locals. It was ruining the road that the community built themselves and destroying their water source. Whereas once they were able to dig individual wells by their homes, now pollution of the ground water makes it undrinkable. They have to fetch water from a tank at the edge of the village.

'We never used to get mosquitoes, but now they're everywhere,' he added. 'We need nets just to open our windows. We used to sleep and eat outside when the weather was hot. That's impossible now – we have to shelter in our houses.'

This particular farm was built on the villagers' communal land. I was told they had been given compensation, but it was a one-off payment, and nearly fifteen years later, as roads, the railway and the farm circled ever closer, the land left available to cultivate had been halved. With so little land there was no work for the locals. 'We tried to prevent the building of the road to the factory,' Chan explained, pointing at the route that cuts across their field. 'But someone hired mafia men to scare us off with knives and sticks.'

Other bits of land have been polluted with pig effluent. When the villagers complained the company paid out compensation and took the land, which was badly poisoned. Despite bearing all these environmental costs, none of the villagers was employed on the farm. 'They don't trust us,' the man claimed. 'They think we'll steal things.' I asked if they benefited from the meat it produced, and he told me they did not. It was destined for city markets, and they could not buy direct from the farm.

As we talked a crowd gathered and children approached to touch the car, which looked flashy and out of place surrounded by stray dogs and wandering ducks. People were smiling, but we knew we could not afford to hang around with our cameras and notebooks. Though we were breaking no laws, we were reluctant to draw attention to ourselves. So we said our goodbyes and headed off in the car, hoping to find out more elsewhere.

Then, about a mile from the village, I noticed something strange about the trees back towards the farm. While most of them were perfectly healthy, in one area the trunks were bare. As there was nobody in sight, we decided to investigate. We scrambled along the edge of a maize field and then up a steep bank, and there we saw the source of the problem: a huge lagoon of putrid watery muck. Poplar trees poked their way out of the slime, but they were drowning in the toxic soup. Already leaves and branches were wilting.

The lagoon was no accidental spillage. Someone had built flimsy mud banks around the edge. We were told the weather had been dry lately, but it was nearly full to the brim. It was painfully obvious that it could overflow at any point, yet it was surrounded by crops intended for consumption. This was the contaminated land the villagers had mentioned and the effluent came straight from the nearby farm. Off-site and out of sight, here was evidence that Muyuan – for all its fancy biogas digesters – wasn't coping with its pig muck.

Ironically, the farm had been accredited by the UN on the basis of its environmental record. The letters 'UN CDM' are emblazoned in red across its towering biogas funnels. Under the international carbon-trading scheme, this very farm is one of the places in the developing world in which international companies can invest to say 'sorry' for polluting too much in the rich world. By supporting Muyuan, under the Clean Development Mechanism (CDM), a Japanese factory can earn the right to keep pumping out fumes at home. Muyuan seemed to me to represent a double whammy: not only was it damaging the environment, but with its proud UN badge, it could be enabling pollution by another filthy friend, thousands of miles away. Our worst environmental suspicions had been confirmed – but we still had not seen a single pig.

While figuring out a game plan, we decided to have a look at a backyard farm. I was keen to see how it works at the other end of the scale, because I am all too aware that small isn't always beautiful.

Jeff easily charmed his way into a smallholding not far from Muyuan, where pigs were being reared. We stepped through a private home into a tiny courtyard divided into concrete pens. Until recently this is what farming across China looked like. The picture is changing fast, but at the time of writing, around 70 per cent of all pigs in China are still kept in informal family farms like the one we were about to see.

Covered with a rickety roof, and drained by nothing more than a sloping floor, the pigpen was as low-tech as it gets. Ten growing pigs, one mother sow and her eleven piglets were rubbing themselves sore against the walls and gnawing at the concrete floor out of sheer bore-dom. Any water they had been given had run out. I was shocked – it was simple animal cruelty.

The only good thing about the farm was that the pigs were fed slops rather than mass-produced chemical-soaked feed, shipped across the world. Pigs can be fantastic at recycling if we let them. But it hardly compensated for the appalling conditions. Nor could the farm give us any clue about how Muyuan is run; its squalor belonged to another age.

Our next stop was a medium-sized farm down the road. We were delighted to hear that it was an eco-farm and curious about what we might find. It came as a huge disappointment. Tens of pregnant sows were kept in sow stalls (also known as gestation crates) – a system banned in the UK on grounds of animal cruelty. The narrow stalls are no bigger than the pig herself and sows are kept incarcerated in them for nearly four months before giving birth. Pregnant pigs housed in this type of accommodation can't turn round in their individual cages, let alone forage or root. For four months they have nothing more to do than stand or lie on a slatted floor ten centimetres above a pit of their own waste. When they are ready to give birth, they are moved to the farrowing crate, an even more restricting metal contraption that will confine them until their piglets are taken away. Their distress is painful to watch, but it is the reality of factory farming, large and small, in many countries. It started in the West and, from what we could see, has very much been sold to the East.

Keen to get out, we said our thank yous to the owners of the 'eco-farm', stripped off our overalls and returned to the car, to run into a nasty surprise. The driver had just been phoned by our hotel, to say that the police were interested in why we'd been filming. Jeff warned that they could be waiting for us when we got back. I had been well aware of the notion that the Communist regime in China still operates an ear in every village, no matter how remote. Now I knew it was true.

In the safety of the car we thrashed out a strategy. We planned to be upfront and honest about the purpose of our visit – to understand pig farming in China better – but to do whatever we could to protect our film footage. Jim, our cameraman, removed the memory card from the camera and hid it. Notebooks were stashed out of sight. We all had visions of a night in a Chinese cell on trumped-up charges of trespassing or spying – which we thought would hardly be unusual.

Drawing up to the hotel, we took a deep breath and did our best to adopt an air of nonchalance, but our nerves were not helped when the reception staff at once removed our passports. We were told the police wanted to check them and verify who we were. I was worried – to be this deep in China and this far into an investigation with no passport was not good. But there was little we could do.

An hour later we met for dinner in the hotel lobby. We left Jeff doing some administration, since he was the only Mandarin speaker, while the hotel looked after us impeccably, ushering us to a private dining room for dinner. We had only just sat down when Jeff burst into the room with a huge grin as if he'd won the lottery. 'Great news!' he exclaimed. 'We've got new friends for dinner! From Muyuan farm!'

Sure enough, just behind him was a middle-aged man who introduced himself as Mr Chen, Muyuan's Environmental Manager. He was accompanied by two young women. 'I've told them we've been trying to contact them for weeks and I persuaded them to join us for dinner!' Jeff beamed. Apparently the Muyuan officials had 'bumped into' him in the hotel lobby, where he'd been glad to invite them to join us.

With smiles and what I hoped were welcoming-looking gestures, I leapt to my feet. Taking Jeff's lead, we pulled up seats for our unexpected guests, and ushered them warmly to the table, hiding our bemusement. Our exaggerated displays of hospitality seemed to defuse the tension. I handed them my business card. Before we knew it, our informal dinner had turned into a full meeting, with beer and much outward joviality.

Behind the smiles however, the pig man and I were sizing each other up. His curiosity about what I was doing with a camera near his farm was as great as my need to know what actually went on behind those concrete walls. What both of us knew, but were too discreet to mention, was that the grapevine in this small town was clearly a match for the Internet. There were eyes and ears everywhere. It was no coincidence that the pig man was in the hotel lobby that night. Our presence had been noted.

I took the opportunity to tell our guests about Compassion's international work, not least with major food companies. I mentioned some of the household names that we work with, like McDonald's. I talked about how we support and encourage improvements in animal welfare and food-quality standards and how these things can benefit business and consumers. Chen smiled. He told us about environmental measures undertaken by his company. He seemed to have come with a message to deliver.

During dinner, he slipped out to make a phone call, presumably to update his colleagues or the police on the status of the foreigners who had been seen taking pictures of the farm. We could see a police car out of the window, and Chen at the hotel entrance, talking on his mobile. The police car switched on its flashing lights, and drove off into the night. Chen returned, seeming more relaxed.

He declined to speak directly about the work of his department, deferring instead to his junior colleagues to explain, but he seemed to know the pig numbers inside out. He told us Muyuan produced a million pigs for sale each year and insisted it dealt effectively with all its waste. Though we had observed shocking evidence to the contrary, with the police so close it was not the time to argue, so we just smiled and let him continue. 'By 2017, Muyuan will rear up to nine million pigs a year in five different regions,' he told us. If true – and there is no reason to disbelieve it – these figures are incredible: Muyuan alone will be producing the same number of pigs as the entire British pig industry.

So now we knew it: this pig company, already vast, has ambitions way beyond its current boundaries. Before too long, what we had witnessed during the day could be replicated, perhaps all over China. Lagoons of filth, mosquitoes, villagers deprived of land, and no doubt desperate pigs crushed into cages built with the help of Western technology and perhaps supported by the UK and US taxpayer through the World Bank and the UN.

By the end of the meal I was exhausted but elated. Without us even getting through the Muyuan door, the pig man had provided a wealth of information. At the end of the meal, Chen not only graciously paid the bill, he also invited us to visit the farm and meet the staff next morning. We all shook hands and our guests left. It had been an extraordinary evening. Apparently, our encouraging smiles and friendly toasts to UK–Chinese cooperation had paid off – we were going to see inside those sheds after all. And our passports were returned.

The following morning, we headed back to the farm, this time as official guests. Chen wasn't there – presumably his job had been done the night before. We were ushered into a sterile meeting room, but no matter how loudly they talked, our hosts could not disguise the horrible screaming of pigs being loaded into lorries nearby.

We spent two hours on the premises, and learned that 70 per cent of Muyuan's pigs are sold domestically to major towns in China, travelling in open-sided trucks for between twenty and thirty hours. The remaining 30 per cent are destined for Japan. Some of the breeding pigs are imported – most recently from Canada – because of the more commercially productive pig breeds available abroad. We were told that the pigs are fed on imported soya, fishmeal from Peru, Chinese wheat and added vitamins, minerals and amino acids. Despite its prolific local production, maize is not on the menu.

Our requests to see the pigs were turned down for reasons of 'infection control' and our offers to don the usual white overalls and walk through anti-infection spray fell on deaf ears. It was clear that Muyuan's management did not want us to see the animals. We used our meeting with the staff to find out as much as we could. Muyuan's lead spokesperson was its Deputy General Manager, Tian Fangping, who was friendly and helpful, drawing diagrams on a whiteboard of the different types of pig accommodation they have. He told us that Muyuan started twenty years ago with twenty-two sows, and since then they have experimented with all types of rearing processes, each time modifying the last to make it more efficient. Predictably, he confirmed that they use the same sow stalls and farrowing crates we saw at the small-holding the day before.

All this work has led to phenomenal success, he told us. In 2010, the company made over £8 million profit and in the first half of 2011 alone they had already hit the £10-million profit mark. I asked if there was anything to limit Muyuan's expansion. He replied that the main constraints were availability of land and the environment. 'For some companies money may be a problem, but thanks to profits and easily available loans, for us finance is not an issue,' Fangping said, explaining that Muyuan gets preferential loans from the Chinese government.

In fact the company is positively rolling in cash. In 2010, it received US$10 million investment from the International Finance Corporation.[16] So it's no surprise that they are building farm number 21 down the road, and that this one will be a dozen times bigger than the average industrial breeding herd in Britain.[17] To our surprise, the Muyuan management offered to take us there for a tour – perhaps because as yet there are no pigs on the premises. They seemed eager to do anything for us, except let us see the animals themselves.

We bundled into a Muyuan van with Fangping and pulled on the Muyuan-branded white wellies. The new farm looked like a post-apocalyptic holiday resort designed by someone with a sick sense of humour. Row upon row of white 'holiday cabins', half built, surrounded by diggers, mud and cement mixers, house thousands of empty steel pigpens. Inside we observed every imaginable bit of technology required to strip the humans out of pig farming. Pipes, tubes, nozzles, fans, wires and of course a slatted floor for pigs to sleep on – like bacon already on a grill pan – so their excrement can be conveniently collected below. It was so automated that a single stockman can 'take care' of 3,000 pigs. It was fashioned in the West; among equipment suppliers, Fangping listed companies from Europe. This is the ultimate factory farm, inhumane and utterly divorced from nature.

'Some small farms don't realise the importance of technology,' the Deputy General Manager mused, gazing at the series of tanks and filters for the gallons of muck this farm will produce. 'Small farms will have to expand or go out of business.'

So there we have it – and this man was in a position to know. Muyuan and its twenty-first-century factory farms are seen as the future for food production in China unless something's done fast. Remembering the appalling conditions I'd seen on some of the smaller farms, I couldn't help thinking that, for China's pigs, it's out of the frying pan, into the fire.

To top it all, with an almost touching naivety, Fangping admitted his meat isn't cheap. This is not food for the hungry masses. 'We're not really reducing the cost of meat for the average consumer because we're catering for the high end and export market,' he told us, rattling off the names of various fancy hotels in China's big cities. Average consumers don't give Muyuan the kind of profits they need.

I had heard enough: I doubt I will ever forget that building site. Stomping across it in my shirt, tie and wellington boots heavy with mud, I thought gloomily about the fifteen-hour train journey ahead to our next destination. At least we were getting out of Muyuan and its pig cities.

It was a relief to move on to Wuxi, a city on the shores of Lake Taihu, China's third-largest freshwater lake, which laps the shores of Jiangsu Province. Tourism is booming. For centuries, the city was the seat of the emperors, and it attracts swathes of Chinese day-trippers and holidaymakers. There are boat trips and guided garden tours as well as 3,000-year-old tombs from the Zhou Dynasty. We were due to visit just before a major Chinese holiday and the hotels and restaurants were taking on extra staff in anticipation. For the first time in days we were due to stay in a really comfortable hotel.

The purpose of our trip was to look at the lake; we had heard it was not as picturesque as it used to be. There were reports that muck was seeping out of some of the pig and poultry farms nearby and into the water, turning it bright green and making it smell of rotten eggs. We wanted to see this direct consequence of intensive agriculture for ourselves.

After careful questioning, our hotel receptionist let slip that in 2007 the water from Taihu Lake – which usually supplies water to millions of residents in Wuxi and beyond – became so polluted it was undrinkable, despite going through the usual urban water-purification processes. When you turned on the tap, she told us, it came out cloudy and smelled foul. The Chinese premier, Wen Jiabao, ordered a clean-up[18] and dashed to the city to drink a glass of Wuxi tap water in front of the media cameras.[19] Since then, despite further efforts, every now and then the water becomes undrinkable.

I set out to the lakeshore from our hotel with my binoculars slung around my neck, hoping to see some birdlife. The lake was misty with the usual Chinese smog and I could just make out the ghostly outline of high-rise flats on the other side. Nonetheless it was a beautiful spot. Before too long I spotted a flash of cobalt blue: a kingfisher. The lake may be polluted, but for the time being the birdlife is wonderful.

For a closer look at the state of the lake, we bundled onto a tourist boat along with hordes of Chinese visitors. From the top deck there was a glorious view of the many-masted fishing vessels, their chopping-board sails fading in and out of focus in the mist. For the first time on the trip our video camera didn't feel out of place. Aboard were hordes of glamorous young couples, old ladies and families, enjoying a trip to so-called Fairy Island, one of many islands on the 2,300-square-kilometre lake. Shaped like a turtle drifting quietly in the water, Fairy Island was covered in lush green trees, buildings and pavilions. Everyone was excited, and many had camcorders, so nobody batted an eyelid when we got our filming equipment out.

Nonetheless, the pollution is a sensitive issue, and we knew we would have to be careful about what we said and who was in earshot. In 2007 a Chinese environmental activist in this region, Wu Lihong, was arrested on charges of fraud and blackmail and jailed for three years,[20] where he alleged he was subjected to physical torture because of his outspoken views.[21] After that brush with the police in Henan Province we were in a cautious mood.

We docked at Fairy Island. Through the trees, we could see Taoist temples and statues, but they were not old. During the Cultural Revolution in the 1960s, temples that had stood on the island for over 700 years were torn down by youths under instruction to destroy all that was old about China. China does nothing by halves. Today's versions are gaudy replicas, and whereas our fellow tourists were keen to see them – and in some cases even kneel before them – we were more interested in looking for muck.

The other passengers left the boat and soon dispersed across the island following neatly scripted signposts that read 'Matchmaker God Temple', 'Enlightenment Bay' or 'Heavenly Street'. With so many heavenly options, it felt a bit mean to be following our own noses to the green stink, but we didn't have to look far. The moment we stepped off the boat we spotted the problem.

Where the water met the land it was thick as paint, tattooing the rocks and tree roots with a luminous halo. It looked like a classic case of algal bloom caused by nitrates from fertilisers and manure getting into the water system. The algae multiply rapidly, thanks to nutrient-rich pollution, before dying and giving way to more of the same. The decaying mass of dead algae strips the water of oxygen, killing fish and other aquatic life. The Chinese government's own figures show that in 2010 nearly ten Olympic-sized swimming pools worth of nitrates poured into these waters. No wonder the lake is dying.[22]

Just a few metres off the tourist trail, we found two old men standing by a motorised pump, jetting green water away from the water's edge. Their job, we discovered, was to prevent the algae from accumulating by the shore. If it is allowed to stagnate, they told us, the stink gets worse and puts off tourists. The tiny pump and two old men seemed a pathetic effort to tackle such an enormous problem, but we quickly discovered that the two pensioners were part of a small army of workers engaged in this absurdly labour-intensive cosmetic surgery. They told us that between May and October, when the problem is at its worst, the government employs a thousand or more people of all ages to keep the muck at bay. 'This is not bad,' one of them told us with a toothy grin, pointing at the green tidemark on the shore. 'In summer we go out on boats with machines to sieve out the algae. It gets pumped into lorries and sent off to a processing centre.'

Of course, China has unlimited human resources to deploy on such clean-up tasks, but the impact is only superficial. The fish are dying. The men told us prices have soared, rising by 20 per cent in the last year alone. The fishermen are forced to tie up their boats most of the year in a government attempt to preserve what is left of the wildlife. Meanwhile intensive farms continue to spew untreated manure into the rivers. The men and their pumps could be no match for it.

In a small way, however, the island also offered some hope. A crowd of young people thronged at the feet of an enormous gold statue of Lao Tse, the grand master of Taoism. Taoism is an important part of Chinese spiritual life and there were a number of smartly dressed youths wearing white uniforms and name badges who had been hired to explain the principles to tourists. We asked them about the green sludge. They would not acknowledge or condemn it – after all they want to promote this spiritual island as a tourist destination – but they were willing to quote Lao Tse on the importance of respect for nature: 'People follow the earth, the earth follows heaven, heaven follows Tao, and Tao follows nature.'

We used our final day in the countryside to investigate the source of the algae on Lake Taihu. Not all of it is from agriculture; the dumping of sewage and industrial waste is also blamed. But I wanted to see for myself whether pig farming was at least part of the problem. After all, there are said to be more than 2,000 intensive livestock farms around the lake.[23] So I set off to find out if effluent from these farms was definitely flowing into the Taihu River basin.

Once again we squeezed into a taxi and set off to find pigs. It was no surprise to find that we were not welcome at the biggest farms. Visitors are generally banned on the ground of disease control – which is not unreasonable – though I suspected there were other things to hide. So we opted for a smaller farm, near a river that feeds into the lake.

We bumped along a track lined with bright green rice fields. Although we were just minutes from the high-rise buildings of Wuxi, the ramshackle farm at the end of the road, where the nearest thing to a tractor was a bicycle wheelbarrow, could have been in deepest rural China. It was run by three families whose eyes widened as we stepped out of the car. We appeared to be the first foreigners they had ever met.

There were at least one hundred pigs on this farm, so although it was small it was a commercial operation, and according to the owners, a viable one. The pigs spanned three generations. We saw tiny piglets nuzzling at their mothers; medium 'growers' getting ready for market; and enormous sows, pregnant and ready to pop.

Sadly, as in all the other pig farms we had seen in China, the animals were living in a totally barren environment, in dark indoor sheds and even on the ground floor of one family's home – right next to their kitchen. It may have been a small-scale operation, but I would still define it as a factory farm, because the animals were being reared intensively, clearly treated as if they were simply another component in an industrial process of turning feed into meat. The sows were stuffed into tiny stalls, lined up as if they were sausages already in a packet. One was so large she could not fit into her crate. Her hind legs were resting on the back of the pig in the neighbouring stall and her own back was pressed hard against the bars. Every time she flinched she gave an involuntary kick to the pig beside her. The suffering was palpable.

I did my best to focus on the task in hand, looking for evidence that farms like this are responsible for fouling up one of China's most beautiful natural lakes. The bright green river at the back of the farmhouse suggested we'd come to the right place. One of the farmers spoke frankly about how things worked. 'The government has said we're not allowed to dispose of the pig waste in the river, but if we give the local officers some benefits they will close their eyes. Nobody bothers checking on us.' He told us that some manure went straight into the river, and the rest was used as fertiliser for crops. When it rains, that too would run off into the river.

As we turned to leave, we noticed one last horror. Next to the kitchen was a pile of old medicine and injection bottles, needles sticking out. They were antibiotics, clearly being administered in an entirely haphazard way in an attempt to prevent the pigs getting sick. 'We have no medical background,' the lady farmer admitted. 'We don't really use a vet, so for all diseases we just use antibiotics.' Surrounded by feed bags, sprays, buckets, medicines and flies, it was clear that this farmer in her jeans and gumboots had no idea about the wider implications of this sort of behaviour. Money is tight and she was willing to do anything to prevent her pigs from succumbing to disease.

Looking around, I admit I felt some sympathy. It was clearly not a wealthy household. The family toilet – which I used in desperation – was proof enough of that. It was a hole in the ground, dug under the house, and sited right next to a pig pen. The family was living cheek by jowl with their pigs. Nonetheless, what they were doing was dangerous. The widespread prophylactic use of antibiotics on factory farms all over the world has devastating public-health implications. It is a practice that is unchecked in China where, the farmer confirmed, there is no need for a licence or vet's prescription to buy antibiotics. They are readily available over the counter. Translate a policy like this into big farms like Muyuan and you are putting global human health at grave risk.

Factory farmers large and small are breeding disease while mortgaging the global medicine cupboard. To them, more pigs in more airless sheds means more money. For everyone else, one consequence is frightening diseases out of control.

By that point the stench of ammonia, the squealing and crashing of piglets fighting in the background and the discussion of medicines was making me feel nauseous. My colleagues were also looking pale. It was time to get out. As I left, I stood for a moment beside the river and nearby pig sheds. Listening to frantic pig squeals, I watched aghast as smelly brown liquid poured from a pipe, down the bank and into the water.

A few hours later we were on the train from Wuxi to Shanghai, heading for our flight back to the UK. We glided through a countryside beaten into industrial submission. For mile upon mile, all we could see were factories, mines, building sites, car parks, cranes and more green-looking rivers and canals. Sitting on the train, I reflected on our outward journey, from Beijing to pig country. Buoyed by the unexpected welcome I'd received in the capital when we set off, I had been full of optimism. Now that I have some insight into the scale of the challenge to stop factory farming taking hold across China, I am daunted.

It is hard to work out what is worse – millions of tiny pig farms that are almost impossible to regulate, many meting out cruelty on a small scale, or massive global climate-changing mega-pig farms, that are equally inhumane and where the profits and jobs go nowhere near the average rural family. It was a stark illustration that it's not necessarily the scale of farming that is at issue, but the nature of the operation: small isn't always beautiful.

In a country the size of China, you have to think big to make any difference. It makes sense to focus on the big operators like Muyuan. If they change for the better, it improves the lives of millions of pigs at once.

There was one ray of hope. During our discussions, Muyuan's representatives did tell me about trials they were undertaking into more humane ways to keep breeding pigs, in more spacious group accommodation instead of narrow crates. I vowed to return to do whatever I could to turn that tiny trial into the norm, just as we've already done in Europe. I remembered those earlier discussions with the new government-backed organisation keen to consider animal welfare as a way of ensuring better food safety.

As we approached Shanghai, a rainbow of neon lights on skyscrapers lit up the city skyline. We left the station feeling excited. The place was vibrant, throbbing with money, life and possibilities. Bang in the city centre, next to the iconic 'Oriental Pearl' building, was a circular pedestrian skyway, like a ring of pavement on stilts. It was lined with red flags ready for a public holiday the following day. Buzzing with night life, it was a great place to while away half an hour on our final evening and watch the world go by.

A young woman with high heels swung her Louis Vuitton bag over her shoulder and strolled slowly around the circuit. An old couple, hand in hand, followed closely behind. Tourists snapped away and touts offered to take our photo. A nearby McDonald's was packed. In a last burst of investigative enthusiasm, I wandered round the circuit to the golden arches. Through the window I could see a young couple, lovingly feeding their toddler chicken and chips.

There's no doubt that many of these people are enjoying better lives than their parents did eking out a living on small farms, but the countryside is bearing the cost of this seductive change. China's rural poor see furnaces and funnels and struggle to find drinking water as their rivers turn to pea soup. Meanwhile, China's elite and growing middle classes increasingly munch on American-style fast food and sip on Starbucks as they watch the city lights twinkle.

As the taxi whisked us to the airport I reflected that if only a fraction of the effort that went into Shanghai's beautiful skyline went into improving animal welfare in China, the world would have millions of happier pigs and healthier people.

Chinese consumers need to start flexing some economic muscle. The food scares have left them understandably nervous. My time in China convinced me that it is not green lakes or sow stalls that will force change here, but consumer demand for healthy food. The Chinese finally can afford to choose.

An unhappy pig is an unhealthy pig, and an unhealthy pig makes unhealthy food.

16

Kings, Commoners and Supermarkets

Where the power lies

Judging by the sign he has erected by the gate to his country seat, the future King of England has a wry sense of humour. Visitors approaching the driveway to Highgrove House in Gloucestershire are greeted by a notice warning them to beware: they are about to enter 'an old fashioned establishment'. It's an unexpected touch, all the stranger because it is next to another sign which adds rather incongruously that his Majesty's residence is a 'GMO Free Zone'.

There were certainly no genetically modified organisms when Prince Charles was a boy, but his evident pride in declaring that his Gloucestershire estate is GM-free suggests he's considerably more up to date than most of his subjects on a very modern issue. He may be old-fashioned about some things, but he has long been ahead of his time about others.

Highgrove House, the family home of Charles and Camilla, Duke and Duchess of Cornwall, lies in the heart of the Cotswolds and is a model of sustainability. There's a bespoke reedbed sewage system to process royal excrement; bottles, cans, newspapers, cardboard and shredded white office paper are all recycled; leftover scraps from breakfasts and banquets are churned into a composting system. The chandeliers run off energy-saving light bulbs, and the staff car parks are lit with solar power.

The nearest place is Tetbury, an old market town with medieval streets lined with gorgeous antique shops and boutiques full of beautiful things. There are bijou cafés serving Earl Grey from delicate china cups on gingham tablecloths, sweet shops with jars of old-fashioned lemon sherbets and bonbons, and bakeries offering organic honey cakes and piles of strawberry meringues that look like little pink and white puffy clouds.

This luscious corner of England has long had royal connections: Gatcombe Park, Princess Anne's country residence, is six miles to the north, while the international polo tournaments at nearby Westonbirt and the horse trials a few miles south of Tetbury at Badminton are a magnet for the rich and famous. Tetbury is also home to the Duke of Beaufort's Hunt, one of the oldest and largest fox-hunting packs in England and still going strong, even though hunting wild animals with packs of dogs has been declared illegal. But while hotels and restaurants in Tetbury rake it in from the well-heeled local clientele and coachloads of tourists, one business is struggling to make ends meet: the organic farm on the Prince of Wales's estate.

In an era when EU and US subsidies incentivise intensive farming, Home Farm at Highgrove illustrates the huge financial challenges faced by those who are repelled by factory farming and want to rear animals and manage land in a gentler, more natural way. The Prince of Wales can of course afford to shoulder losses. However, he is no Marie Antoinette dabbling in this business as a gentle distraction from the burden of royal duty. He is a champion of agricultural sustainability who uses his influence as heir apparent to the throne to press the case to policymakers, the food industry and philanthropists, both privately and on an international stage. He has learned the hard way how challenging it is to make 'doing the right thing' commercially viable.

The Prince decided to convert his farm to a completely organic system in the mid-1980s, hoping to showcase the environmental and commercial benefits. Almost three decades on, his website describes his operation as a successful and viable working farm – 'a flagship for the benefits of an organic and sustainable form of agriculture' – but it has not been easy.[1] The truth is that Home Farm does not always make a profit: some years, it is a struggle just to break even.

The farm is home to 180 dairy cows, 150 suckler cows, 130 breeding ewes that produce around 200–220 lambs a year, and a few rare-breed pigs. It works on a crop-rotation system, a seven-year cycle designed to maximise the richness of the soil. Organic mutton from Home Farm is sent to Calcot Manor, a luxury hotel near Tetbury, and to the Ritz in London. The Prince is enthusiastic about restoring mutton (meat from a two-year-old sheep) to the dinner tables of the nation after speaking to sheep farmers who found they could no longer get a decent price for older ewes. Other products are sold to Duchy Organics, now a partnership with Waitrose super-market.

His Royal Highness may be selling to the luxury market, but he doesn't get fancy prices just because of his name. Like most small farmers he is having to diversify and is currently considering producing cheese. 'If you turn [milk] into cheese, it's worth three times as much. It doesn't cost three times as much to turn into cheese, so it's something we're looking at,' his farm manager says.

Three days a week, the Prince's farm sells vegetables to locals from an old cattle shed. Though he's yet to be seen manning the farm shop himself, behind the scenes he is surprisingly hands-on. I was lucky enough to be invited for a private tour of the farm, joining a small group of charities and business people with an interest in sustainable food. It was a fascinating day out, not least for little insights we gleaned into the Prince – like his passion for hedge laying, a highly specialised and intricate operation involving cutting and weaving branches. But the revelation that his organic farm faces an up-hill battle to remain commercially viable, despite all the cachet and advantages of his name and status, was a sobering reminder of the extent to which the odds are stacked against farmers who refuse to intensify.

The US and European agricultural system has been geared towards intensification since the post-war years. The original motive was laudable, at least in part. Governments wanted to end the years of austerity, ration books and food shortages. The nightmare of German U-boats sinking vital food supplies was still fresh in the public mind. Post-war governments were determined to make national food production more self-sufficient, and quick to pass new legislation that would set the tone for decades to come. The new strategy was focused on increasing production. Vast amounts of public money were used to support and encourage farmers to maximise output, with little thought about longer-term consequences.

The 1947 Agriculture Act was a defining moment in British farming, kick-starting factory farming in the UK.[2] In the US, by then it was already evolving, thanks to Congress passing a subsidy package for farming known as the Farm Bill in 1933. In the UK, farmers were encouraged to make use of the latest chemicals, machines and techniques. Mixed farms with their varied patchworks of crops and animals were abandoned as farmers began specialising in particular crops or species of farm animal. The age-old natural cycle, where crops would be rotated with livestock whose manure would replenish tired soil, disappeared, as artificial fertilisers were used instead. Farming was now an industry, like producing cars or TV sets. Quality was sacrificed for quantity. An agricultural revolution was under way.[3]

Government policy and subsidies were lined up behind the new methods. Agricultural colleges taught the next generation of farmers the way and legions of advisers and salesmen fanned out across the countryside to spread the message to the farmers of the day: either get with intensive production or get out. One of those farmers was Peter Roberts.

Many other farmers took the advice, whether in relation to chickens or other livestock and crops. Roberts was unusual in that he shunned the new way, fearing for animal welfare and the environment. Many were seduced onto the intensive-farming treadmill and a lot fell victim, being forced out of business. Just after the Second

World War, the UK had around half a million farmers. By the 1980s, numbers had fallen by nearly two-thirds.[4] The figures continue to fall today.

Farming was now in the grip of agribusinesses, a raft of ancillary industries spawned to support the 'modern' farmer: tractor and equipment companies, fertiliser and chemical manufacturers, seed, feed and pharmaceutical suppliers. Small farmers who would not, or could not, embrace the new system became hard pressed, often forced to the wall. Between 1947 and 2002, figures from another rapidly industrialising nation, Canada, show that farm revenues nearly doubled, but the money in farmers' pockets – actual net farm incomes – fell by more than half.[5] The suppliers to industrial farming thrived; farmers did not.

In 1964 Ruth Harrison's book *Animal Machines* described how life on the factory farm revolves 'entirely round profits, and animals are assessed purely for their ability to convert food into flesh, or "saleable products"'.[6] There was now some public awareness of what intensive farming meant for animals behind closed doors. It was the trigger that would fire the conscience of perhaps the greatest campaigning champion for farm animal welfare – Roberts, who at that stage, was still on his Hampshire farm.

In founding the charity Compassion in World Farming, with its mission to 'abolish the needless misery of factory farmed animals' and 'establish kindness and compassion', he set the wheels in motion for a decades-long struggle against industrial farming.

At first, it was a real cottage charity, being run with his wife, Anna, from their kitchen table. Up against the might and money of a well-funded system, it was to be the ultimate David and Goliath battle. Changing farming systems for the better is a more and more complex process involving multiple power brokers based in different parts of the world. Though theoretically 'the consumer is king', eradicating cruel, unsustainable and environmentally damaging farming methods involves far more than persuading the public to turn their back on certain products.

When talking to people about how campaigning works, I often draw a 'power pyramid'. To achieve change, campaigns need to pressurise and ultimately persuade each part of the pyramid. At the summit is the person or body that has the ultimate say in the matter in hand. Up until 1992, Britain presented campaigners with a classic power pyramid. At the top was the minister of agriculture, who could propose profound change and push it through government, or conversely, block it. Moving down, you had the permanent secretary or chief civil servant, the 'Sir Humphrey' of the British TV comedy *Yes, Minister*. While ministers would come and go with reshuffles and changes of government, these powerful unelected mandarins would stick around for years, sometimes decades. That could mean entrenched attitudes and a bias towards the status quo. One level down in the pyramid were influential MPs, followed by rank-and-file backbenchers. At the base of the pyramid was a large segment, distant from direct power, but if mobilised in large enough numbers, greatly influential: ordinary people, or consumers – you and me.

The way to gain influence was to mobilise each successive layer of the power pyramid at Westminster. I remember one of Compassion in World Farming's early victories, in 1991. It was a mild January morning and I was walking down Whitehall with the actress Joanna Lumley, a senior MP and a retinue of media hacks. Lumley, one of the UK's best-known actresses, had joined forces with a Conservative backbencher, Sir Richard Body, to demand a ban on the use of chains and restraining collars on pregnant pigs. These instruments of torture, along with narrow gestation crates – sow stalls – were being used to confine expectant sows for months at a time. They were a way of keeping a lot of pigs in a very small space without the animals fighting. By chaining them up, or keeping them in such tiny stalls that they could not even turn around, let alone bite each other, they could be kept in rows like parked cars, without the inevitable aggression. It was bad enough for any pig, but for a heavily pregnant animal, not being able to move was particularly cruel.

We were clutching bundles of papers tied together with red and blue ribbons: petitions. Lumley was also carrying a chain and restraining collar to show people what they looked like. As she held them up to the press, I remember the camera shutters going into overdrive.

Sir Richard had been an MP for an East Anglian farming constituency for over twenty-five years. Farming was in his blood: he'd been a farmer himself, as well as a stock breeder, writer and critic of post-war farming. He was passionate about pigs. He had come second in the annual parliamentary lottery whereby twenty MPs drawn at random can propose a new law. We had urged him to take up this cause for pigs. I was a new campaigner back then and remember sitting in our tiny office above a health-food store in Petersfield, listening to my then boss, Joyce D'Silva, talking nervously to Sir Richard about whether he would stick with it. He had been inundated with requests for bills and was under pressure to choose another issue. Would he hold firm? I heard Joyce shriek with delight – it was game on.

Looking back, it was a fairly clinical campaign. We had a committed champion in Parliament, who happened to be from the governing party of the time. We had a glamorous and high-profile celebrity endorsing the campaign. Plus ultimately we only needed to persuade one person: the agriculture minister. We mobilised concerned voters to write to their MPs to build up a head of steam. Petitions were collected the hard way – with pens and paper on the street – in those pre-Internet days. Press releases and stunts were organised to catch media attention.

The day came for the debate and vote on Sir Richard's bill. We needed at least 100 MPs voting in the House that day. It may not sound many, but the turnout for so-called Private Members' Bills can be very low, with MPs rarely compelled to attend. The mood in the chamber was supportive, though some politicians, including William Hague, then at the beginning of his political career, were worried about Britain going it alone while other European countries continued to use the restraints. Hague was slapped down by a chorus of voices calling for immediate action.

The Tory MP Michael Brown drew on Britain's colonial past to illustrate the importance of taking a principled stand. He declared:

> In the seventeenth and eighteenth centuries, when France and Britain both had empires and both had slavery to maintain those empires, this House unilaterally decided to abolish slavery. Does my honourable friend think that some honourable Members opposed that on the grounds of timing, and said that we should wait until France and other empires had abolished slavery?

Eventually, a vote was called. We were jubilant: 118 votes in favour; only two against. We had won – or so we thought. Sadly that was not entirely the end of it: the draft bill faced two more parliamentary hurdles, and like so many private members' bills, it eventually ran out of time. Its opponents used the highly undemocratic device of 'talking it out', deliberately wasting vital debating time by waffling. We were furious, and determined to find another way to implement what was by now a popular reform. In the end we pulled it off after convincing the agriculture minister himself. The government took up the bill, and it was passed into law.

Since 1993, changing agricultural policy has become a more complicated business. The establishment of the single European market has meant a new strategy, involving influencing power pyramids in key member states throughout the European Union. Influencing the power brokers is a bit like getting to a cherry in the middle of a cake. To get the cherry – Brussels – to vote the way you want, you have to influence as many of the power pyramid 'slices' of the cake as you can. There are now twenty-eight countries in the European Union, and likely to be more soon. Hence, it is much harder to get things done. The plus side is that when reform is achieved, it takes effect in all member states, not just one.

Perhaps the single biggest obstacle to a radical shift away from factory farming in Europe and America is the subsidy system. Agricultural subsidies are hugely powerful players in the food-production game and conspire against producers, like HRH Prince Charles, who eschew intensification. The Common Agricultural Policy (CAP) may make eyes glaze over, but it is central to any debate about factory farming. Ultimately, it underpins the system.

Designed to provide farmers with a reasonable standard of living, consumers with decent-quality food at fair prices, and to preserve rural heritage, the CAP's complex system of protections and incentives has driven the decline in traditional mixed farming and is partly to blame for the great disappearing act of animals from the land. Four out of five of Europe's farm animals are now reared on industrial farms.

Historically, the CAP paid farmers direct subsidies to produce more, benefiting larger arable farmers the most. The system was heavily criticised for producing surpluses – butter and grain mountains, milk and wine lakes – that were frequently dumped on export markets outside the EU, undercutting local producers. In recent years it was reformed to 'decouple' payments from production, instead making payments based on the area of farmland growing goods eligible for subsidy; but even with decoupling and a limit on maximum payments, larger farmers and landowners continue to receive the biggest handouts. It dates back to the beginnings of the Common Market, when France insisted on a system of agricultural subsidies in exchange for agreeing to free trade in industrial goods. Now it is the most expensive and controversial scheme in the EU, costing around £48 billion a year and accounting for almost half of the EU's entire budget. Handouts are supposed to be conditional on meeting environmental or animal welfare standards, but how well this is enforced is open to question.

America has its own subsidy system that props up industrial agriculture, the Farm Bill, a multi-billion-dollar programme of government support. The 2008 package approved nearly £300 billion of spending on agricultural policy over five years.[7] Through this programme, US farmers receive billions of dollars of subsidy.[8] The most heavily subsidised crop in the US is corn (maize), the key feed ingredient of the US 'cheap' meat culture, which accounted for US$77 billion of handouts between 1995 and 2010.[9] The resulting meat may seem good value at the point of sale, but it comes at a high price for the animals reared indoors in horrible conditions, on below-cost cereals and soya beans, thanks to generous taxpayer-funded subsidies.

Through my work at Compassion, I've been campaigning for CAP reform for over two decades. It's a slow process, especially with so many powerful vested interests involved across Europe. Real opportunities for change come along every five years. I remember joining forces with various other charities and organisations, including the RSPB and National Consumer Council, to take the message out to the public. We had someone dressed up as the Grim Reaper handing out money on street corners. Our message was that the CAP was handing out public money to support often damaging farming practices. I'd like to think we've made some progress, but getting Europe's mammoth subsidy system to support a wholesale move away from industrial farming is still a dim and distant goal. Today, just 0.1 per cent of the CAP budget is spent on improving animal welfare – a minuscule amount given the size of the budget and the scale of the problem.

In the battle for a new system, it is possible to achieve more immediate change by winning over retailers rather than regulators. The key players here are supermarkets, fast-food restaurants and the big food-manufacturing companies, because of their huge market share. The big five supermarkets in the UK – Tesco, Sainsbury, Morrisons, ASDA and the Co-op – control roughly 80 per cent of the country's grocery market. Globally, the American-based chain Walmart, which owns ASDA, is number one, France's Carrefour is second, and the UK company Tesco ranks third.[10]

There's a similar picture of big-name domination in the global restaurant market with McDonald's as the world's number one, followed by Yum! Brands, which owns KFC and Pizza Hut. These are followed by the coffee chain Starbucks, Burger King and the sandwich makers Subway.[11]

Such companies have huge influence over the food system. They are a force for good as well as for bad. By working with them, and other influential firms, it is possible to make a radical difference to the entire food and farming chain, from the way farmers rear their animals to the final product. If they decide to make a change, for example to stock only milk from pasture-based cows or cage-free eggs, they can do it far more swiftly and decisively than governments. In the UK, Sainsbury's, the Co-op, Waitrose and Marks & Spencer now stock only free-range eggs. They took a company-wide decision to sell only cage-free, and rolled it out without compromise. Thanks to corporate commitment, the battery-egg-free high street is not far away in the UK.

Contrast that with the EU's attempts to ban battery cages. When Brussels made the decision in 1999, producers were given twelve years to change – more than enough time to get their act together, or so you would think. Yet when the new legislation came into force on New Year's Day 2012, nearly half of the countries were not ready, and were still keeping tens of millions of hens in what were by then illegal cages.

In any case, the new law itself is far from perfect, allowing as it does so-called 'enriched' cages. The new legal variety gives each bird a picture postcard-sized area of extra space. The chickens must now have somewhere to perch and scratch, but they still never see daylight, and are forced to stand on or above a sloping wire floor. It makes egg collection easier, as the eggs roll away once laid, but remains very uncomfortable for the birds' feet. So while the legislation was a huge milestone, the reality is that hen welfare still has a very long way to go in Europe before we can be proud; and it took a very long time to bring about change this way.

When retailers decide what products to stock and how discerning they should be about how the food they sell is produced, consumer views are obviously a powerful lever. Shoppers and diners in most EU countries are becoming increasingly aware of farm animal welfare, and it influences their purchasing decisions. Research and polling suggest that most people believe the issue is important – roughly three-quarters of consumers in the UK and France believe it matters, and the figures are even higher in Hungary and Sweden (83 per cent) and Norway and Italy (84 and 87 per cent respectively).[12] As a result, leading food companies are more and more interested in stocking animal-friendly farm products.

Compassion in World Farming holds an annual awards ceremony to celebrate companies for making animal-friendly policy commitments. I gave my first opening address at our inaugural 'Good Egg Awards' ceremony in the House of Commons in 2007. Since then, we have recognised nearly 500 companies as having made serious commitments to improving the animal welfare standards of the food they sell. Companies like Subway, Starbucks, Sainsbury's, Unilever and McDonald's, have taken pledges to use cage-free eggs in the UK and Europe. ASDA, the Co-op and the ice-cream manufacturer Ben & Jerry's have been recognised for pledges to source milk from cows allowed on grass, instead of locked indoors, during the grazing season. We also give out an award to the overall 'most compassionate' supermarket in the UK, a title that has passed back and forth between Waitrose and Marks & Spencer for over a decade.

From time to time, I'm asked to speak at film screenings. One environmental film company asked me to introduce the US film *Food Inc*, which explores what it calls the 'highly mechanised underbelly' of the industrial farming system in the United States. It's a shocking exposé, but watching it, I could see that it was more than a critique of the system. There was also a message of hope, with examples of companies and producers switching to more humane and sustainable ways of farming. One interviewee described how the sustainable and organic food movement needs to move beyond being David and become Goliath. It needs to be championed by the biggest companies in the world. That's why Compassion develops partnerships with some of the greatest 'Goliaths' in the industry.

It's an approach that involves sticks as well as carrots, but making big companies change their ways doesn't have to mean aggressive confrontation, as our 'Hetty the hen' campaign showed in the 1990s – a protest aimed at the supermarket chain Tesco. It started on an ordinary day, in a supermarket aisle in a sleepy Cornish town. A young man in an in-store security uniform was helping customers with their queries. He was also keeping a watchful eye for light hands: shoplifting was a constant threat, but he knew how to deal with it. What happened next did not feature in his training manual. He turned towards the store entrance and came face to face with someone dressed up as a six-foot hen. The eye-catching costume was of a tatty battery bird, with raw pink skin, flapping arms and sad eyes. This forlorn-looking creature was pushing a shopping trolley round the aisles with a small group of protesters following behind, who in turn were followed by journalists.

Flustered, the security guard stepped in and asked them to leave, but the hen wasn't going anywhere until he'd spoken to the manager. The stunt was part of a nationwide effort to persuade the supermarket to label battery eggs more clearly as 'eggs from caged hens', instead of using weasel words like 'country fresh' or 'farm fresh', which implied something more pleasant.

Earlier that morning, I'd gone head to head on live radio with a Tesco spin doctor who claimed the company's labelling policy was perfectly clear. He was very silver-tongued and probably sounded convincing to listeners. What he didn't know was that Hetty was about to hit roughly thirty Tesco stores at random over a two-week period, complete with local media entourage. We were poking at the soft underbelly of the corporate giant, and it quickly grew clear that they didn't like it.

Within minutes of Hetty's first store appearance, the Tesco spin doctor was on the phone to me. 'Call it off! Call it off!' he appealed. 'We're going to change our label.' From then on, Tesco's battery eggs bore the label 'eggs from caged hens', and the company began stocking more free-range options. Several years later, the EU made it compulsory to label battery eggs with the new, clearer term.

Compassion in World Farming's founder, Peter Roberts, himself learned the power of consumer pressure in the 1980s, in a seminal campaign involving some monks. The Norbertine 'white canons' of Our Lady of England Priory in Storrington, West Sussex, had been generating income for the monastery by rearing veal calves using a 'crating' system. Taken from their mothers at birth, the animals were confined inside 60-centimetre-wide enclosures throughout their short lives. Often chained by the neck to the front of the stall, they could not turn around, stretch their limbs or lie down comfortably. They were reared on milk and nothing else, and by the time they were slaughtered at six months, often they were too weak even to walk to their death. The aim of this shockingly cruel system was to produce very tender white meat by making the calves borderline anaemic.

The monks of Storrington were doing nothing unusual: this was how veal was produced at the time, in the UK, Europe and America. However, Roberts rightly figured that the spectre of such cruelty being inflicted by religious ascetics would capture the public imagination, and decided that Compassion should bring a private prosecution. The monks were charged with nine counts of cruelty under the 1911 Protection of Animals Act and the 1968 Agricultural Act.

Compassion's day in court was a frustrating one. Though the ladies loved Roberts – a Richard Burton lookalike in a tweed jacket who drove to the hearing in a bright yellow Triumph Spitfire convertible – he lost, when the judge ruled that veal crates did not cause 'unnecessary suffering'. Compassion was ordered to pay £12,000 in costs, a huge sum for what was then a tiny charity.[13]

It was only a temporary setback, however. The media loved the story and people could see that the system caused huge suffering. As Roberts would often say: 'Even a damned fool can see it's cruel.' Horrified consumers made their feelings known and the offending meat was left on supermarket shelves. Veal became a dirty word. Not long after the court case, the monks announced they were selling the farm. Not only had it become uneconomic; it had turned into a public-relations disaster.

The then Conservative government began to feel the heat, and Roberts was invited to attend a ministerial meeting on the issue. That same day, the government announced that veal crates would be banned. By then, however, the move was almost academic. Such was the strength of popular disgust that only a handful of veal crate farms were left in the country. The rest had either changed their rearing systems or packed up for good. It was a shining example of consumer power.

The trouble is that consumer power is limited by lack of information, and vested interests work hard to keep people in the dark, shielding them from the often ugly truth about how meat and dairy products are produced. Of course many people prefer not to know, but an increasing number do want to be in a position to make an informed choice. That's why labelling remains such a big issue. While eggs in Europe now have to be labelled according to how they've been produced, there is no such law for meat and milk, allowing retailers and restaurants to use all sorts of ruses to make things sound more appetising.

Take The Ivy, one of the most exclusive eateries in London. Among the regular items on its menu is a dish described as 'corn-fed chicken' – as if 'corn-fed' were a virtue. The Adjournment restaurant in the House of Commons does the same. The truth is that just about all chickens are corn-fed – the vast majority of them on factory farms. So that fancy 'corn-fed' chicken is most likely just factory-reared, by another name.

Logos and symbols on meat designed to imply that it has been produced without needless cruelty can also be misleading. The UK's 'Red Tractor' symbol, for example (officially known as the 'Assured Food Standards' mark), boasts that it stands for 'choosing high animal welfare standards'.[14] In reality, it often guarantees little more than compliance with minimum legislation and government guidelines. In 2002, Compassion in World Farming analysed the scheme against the RSPCA's Freedom Food label and the standards of the leading organic certifier the Soil Association. The Red Tractor scheme's assurances on higher animal welfare were hollow, allowing mothering pigs to be kept in narrow crates, piglets to be mutilated and chickens to be crammed into factory farms. Ten years later on, the study was repeated. Little had changed. Red Tractor again ranked lowest on animal welfare.

The Soil Association logo is the gold standard for animal welfare, though the RSPCA Freedom Food also delivers a genuinely higher welfare choice for consumers shopping for meat, milk and eggs.

The case for better labelling of meat and milk products was under-lined for me when a nine-year-old boy got up in front of an invited audience of MPs and lobbyists to launch a film he'd made called 'How was this animal kept?' Dressed in a blue blazer and wearing a radio microphone, the grandson of the UK government's business secretary, Vince Cable, called for a new law that would see meat and dairy products labelled according to how the animals were reared. It heralded a new campaign, 'Labelling Matters', run by Compassion in World Farming, the RSPCA, the Soil Association and the World Society for the Protection of Animals (WSPA), that calls for mandatory labelling of meat and milk according to method of production in the UK and Europe, in much the same way as we've already achieved for eggs.

The cornerstone of consumer choice is knowing what we're buying. For too long, shoppers have been sold factory-farmed produce under misleading labels. Enough is enough. Labels should be in plain words, not potentially confusing symbols or logos, and there should be guide-lines about prominence, otherwise there will be a temptation to bury the facts in tiny print.

In the end, changing the system matches campaigners against immense vested interests, from the feed manufacturers who convert grain into easily stored and transported 'compound' food, to the equipment manufacturers who make money out of selling farmers the very latest kit; the chemical and fertiliser companies who make fortunes selling pesticides and weedkillers to farms on the intensive treadmill; the pharmaceutical companies and vets who peddle anti-biotics to ward off the diseases inevitable when so many animals are packed into such a small space; and lastly to the farmers themselves.

In between is the farming media, whose publications all too often rely on advertising from chemical, drug and equipment firms, so much so that they often seem to act as cheerleaders for intensification, branding anyone worried about the general direction of travel as 'anti-farmer'.[15] Throw in the constant barrage of advice farmers receive from salesmen keen to sell the latest product, and from government and industry agencies who all too often have got behind the factory-farm model as the only way to go, and it is little wonder that they feel pushed and cajoled, and become so doggedly defensive in the face of criticism. After all, they are only following the path that everyone who ever banged on their door told them was the way ahead. As a result, many invested so heavily in the system that they now feel well and truly stuck.

17

New Ingredients

Rethinking our food

It was my first visit to Hackney City Farm. A dozen hens and ducks stood in a courtyard surrounded by small stables with some sheep, goats and a solitary pig. The surroundings offer a haven amidst the slightly run-down if Bohemian setting of this part of London. Its rustic café was full of character and pleasant for lunch with its menu of buffalo mozzarella, Italian sausage and rabbit ragu. As far as a farm goes, the resemblance ended there; it's little more than a petting zoo.

Tristram Stuart is an author and advocate for reducing food waste. He arrived with fluorescent cycling jacket and helmet in hand and a bee buzzing round his head. I moved to help. 'Don't touch it!' he exclaimed. 'It might be one of mine!' He recently moved from the country to the capital and, missing rural life, consoles himself by keeping bees.

Tristram talked to me about the scale of the problem:

Clearly, we need to feed people, but at present one-third to a half of the world's food is wasted. At the same time, a billion people are hungry and we are extending the agricultural frontier further and further into the world's remaining forests in the quest to grow more and more food. Reducing food waste is one of the simplest

ways of reducing pressure on agricultural land and increasing food
availability globally in a way that involves little or no sacrifice.

Around the middle of the last century, very little food was wasted by
UK households. Now, about a quarter of our food gets tossed in the
bin.[1] In the US, it's even worse: consumers waste about 30 per cent of
their food.[2]

Food is now wasted all along the food supply chain, from farm to
processor to retailer to consumer. In industrialised countries, by far
the majority of this wastage is traceable to shops, catering and households.[3] Supermarkets waste food at their distribution centres, where it
can be trashed before it ever reaches the shelves, or at stores once it
passes its sell-by date. Stuart calculates that the irrigation water used
globally to grow wasted food would be enough for the domestic needs
of 9 billion people – the number expected on the planet by 2050. He
reckons it would be a far better use of the land, oil and water that went
into making the food in the first place to recycle it through pigs and
poultry. 'That's what we domesticated those animals for,' he says
emphatically. With the rise of industrial agriculture, 'we turned the
entire rationale of animal agriculture on its head by feeding them
foods that humans could eat and wasting the food waste that we
should be feeding to them instead.'

His book on the subject, *Waste*, looks at how our throwaway society affects the poor in faraway countries. As a staple food, wheat for
example is traded internationally on the commodity market at a
global price determined by supply and demand. When demand goes
up, so does the price. And with a third of the world's cereal harvest
already being fed to an increasing population of industrially reared
animals, there's little wonder that we are beginning to see food prices
rise worldwide. 'Putting food like this in the bin really is equivalent
to taking it off the world market and out of the mouths of the starving,' Stuart concludes.

The EU disposes of millions of tonnes of valuable food waste while
at the same time importing 40 million tonnes of livestock feed from

South America every year. Stuart argues: 'We need to take account of the economic and environmental costs of continuing with the present system of producing livestock.' Whichever way you look at it, reducing the mountain of food we currently throw away makes good sense. It would go some way to restoring the natural order in the way we produce food too.

In step with common sense, science has a role to play in the menu of the future. I met two 'forward-to-nature' brains at the University of Wageningen in the Netherlands: Willem Brandenburg and René Wijffels. They are exploring the possibility of growing seaweed and algae on a massive scale.

Dr Brandenburg, a fifty-nine-year-old plant scientist, showed me round his glasshouse. It was filled with gurgling, bubbling tanks full of seaweed. He plucked out some slimy strips to show me. 'We only need three hundred and sixty thousand square kilometres of seaweed farming to feed the protein requirements of ten billion people,' he told me. 'That's an area of sea four times the size of Portugal.' Given that 70 per cent of the planet's land surface is covered by ocean, that is a lot of food for not a lot of sea.

Brandenburg is in no doubt about the scale of the challenge we face: 'Over the coming four decades, we'll need to be producing twice as much food with half the inputs; that's why we're looking at plants, plants and plants as the way forward.' It was a pleasant surprise to learn that seaweed is easily digested and compares favourably with meat for protein.

In an interesting combination of tomorrow's technologies, Brandenburg has a vision of floating wind farms connected by a lattice of seaweed farms below the surface. He already has an experimental farm off the Dutch coast of Zeeland and talks enthusiastically about seaweed as the 'engine in doubling plant production', thereby boosting food supplies without taking over more land.

Wijffels, another plant enthusiast, heads up the university's 'Algae Park', a miniature industrial site where racks of glass tubes hum like long fluorescent light bulbs, some filled with green water, some clear.

He is looking into growing algae as a source of protein and biofuel. He believes it could be much more efficient than traditional land-based agriculture and could replace the soya currently being imported into Europe, largely as animal feed. Although his work is in its infancy, Wijffels estimates it could be scalable commercially by 2025.

Some years ago, I looked into free-range farming of fish like salmon, more usually reared in cages. One promising alternative was ocean ranching, where juvenile fish are hatched and reared in captivity before being released into the sea. The liberated fish then live naturally in the wild before returning to their imprinted release point as adults, where they can be caught for harvest. I was delighted then to learn that nearly half the salmon caught commercially in Alaska in 2010 were ocean-ranched.[4] Alaska has really embraced ranching since it banned cage-farming of fish, fearing damage to wild salmon runs from disease and escapees.[5]

Japan has traditionally been big on ocean ranching, and at least ninety species have been released either commercially or experimentally.[6] A wide range of other countries, among them Scotland, Sweden and Iceland, have actively looked at this form of farming the sea.[7]

The way meat is produced 'hasn't changed much' over the last hundred years, according to the Microsoft magnate Bill Gates. That could be about to change: in August 2013 the world's most expensive beefburger was cooked and eaten in London in front of massed ranks of press. The burger, made from *in vitro* – laboratory-produced – meat, was the work of Professor Mark Post of Maastricht University and cost about £200,000 to develop. It was funded by Google's co-founder Sergey Brin, and was made of about 3,000 tiny strips of artificial beef grown from the stem cells of a cow. Speaking after the public tasting, Brin was quoted as saying: 'It's really just a proof of concept right now, from there I am optimistic we can scale by leaps and bounds.'[8]

Bill Gates believes that innovation in meat production has 'tremendous market potential'. As it stands, the basic process remains unchanged, of relying on feeding plants to animals that then return a fraction of the calories and protein they consume in the form of meat, milk and eggs. The outlook is for a near-doubling of global demand for meat by 2050, placing a huge strain on the planet's already over-stretched resources. As Gates puts it, meeting that demand isn't sustainable: 'There's no way to produce enough meat for nine billion people' – the number expected on the planet by mid-century. In an online presentation, Gates describes how food scientists are 'reinventing' meat and eggs, creating alternatives that are 'just as healthful, are produced more sustainably'. It's not about asking everyone to be vegetarian, he explains, but looking at fresh options for producing 'planet-friendly' meat. He sees the future being in the 'perfect fake'.[9] He's in good company. Winston Churchill saw the potential when he said: 'Fifty years hence, we shall escape the absurdity of growing a whole chicken in order to eat the breast or wing by growing these parts separately under a suitable medium.'

'I couldn't tell the difference between Beyond Meat and real chicken,' says Gates enthusiastically in the presentation. He enlists the help of the best-selling author of *The Omnivore's Dilemma*, Michael Pollan, to further explain what they describe as 'three principal motivators' that make reducing meat consumption a good idea: health, environment and animal welfare:

> Health, because we know high consumption of red meat correlates with higher chances of certain cancers; and the environment, because we know that conventional meat production is one of the biggest drivers of climate change, as well as water and pollution; and ethics, since the animal factories that produce most of our meat and milk are brutal places where animals suffer needlessly.[10]

According to New Harvest, an organisation funding research into *in vitro* or 'cultured' meat in the US and Europe, a single cell could, in theory, produce enough meat to feed the global population for a year. New Harvest's Jason Matheny told me what it would taste like when it's fully developed. 'Well, it should taste the same as conventional meat because it's made out of the same stuff . . . we think we can match that same taste and texture by producing meat in culture in a way that's much safer, much more efficient and much healthier for the consumer.'

There is a long way to go before large-scale *in vitro* meat production is realistic – and a mountain to climb to overcome the 'yuck' factor. However, Matheny is confident that it is worth pursuing, not least for the potential health benefits:

> In cultured meat, you can precisely control the amount of fat so you can have more of the healthy fats like omega-3 and less of the unhealthy fats. So we can have hamburgers that actually prevent heart attacks rather than cause them. The yuck factor should really be focused on conventional meat and the way it's produced right now, which is simply unhealthy, unsafe and unsustainable.

18

The Solution

How to avert the coming food crisis

I was in the deep south of America, Georgia, staying at the home of fifth-generation farmers. They were proud of their heritage and upbeat about the future. Will Harris, my host, was an imposing character, a fifty-eight-year-old cowboy whose uniform of choice was a terracotta shirt, worn jeans, lace-up boots and a tattered Stetson. He was warm, determined and ambitious and spoke with a thick southern drawl, smooth as molasses. His family had been raising cattle on their farm, White Oak Pastures, since the American Civil War a century and a half ago.

'Nature abhors a monoculture,' he told me. 'That's why we have a rotational mix of species on this farm.' Harris farms 1,060 hectares of land and a lot of animals: 1,800 cattle, 50,000 chickens for meat, 1,000 laying hens, 800 sheep and various other species. He was big on manners and extremely hospitable. The previous night, he'd drained a bottle of wine into plastic cups as we set off round the farm. Now it was morning, and back to business. He was going to give me a proper tour of the farm in an open-topped jeep. After a quick coffee, we set off. We bumped along a track through pine trees and found ourselves in a sort of swamp, a series of small lakes teeming with wildlife.

'There's plenty of fish, snakes and turtles in there,' Harris said. An osprey flapped lazily overhead. A 2½-metre alligator glided silently across the surface of the still water on one of the lakes. My host told me how he and his siblings loved to camp in this spot when they were little. 'We used to swim in that pond, gators an' all,' he grinned. They breed them tough in Georgia.

The jeep took us on a grass safari across endless meadows. There was nothing monotonous about the scenery, and Harris saw all the greenery through the eyes of his animals, pointing out the different types of grass as if reading through a menu: 'That's crimson clover, the *crème brulée* of grasses . . . that's ryegrass, rocket fuel for cattle . . . and this one's Smutt grass, the one cows and sheep like least . . .' Young black heifers with catlike whiskers snorted softly and ripped up the grass with their tongues.

The farm operated a wholesome type of rotational grazing system. It involved the animals moving round the pasture in succession: big animals (cattle) followed by smaller animals (sheep) followed by poultry, each of them fed and returning manure to the land in their own fashion. It's a way of doing things that is not only healthy for the land but also helps with disease control. Parasites and pathogens often differ between cattle, sheep and chickens. Rotate animals by species, and it's harder for them to build up.

We stopped to admire some chicks with newly sprouted feathers. They were zipping about like clockwork toys. The small huts they were housed in at night were moved every couple of weeks. 'When the cattle defecate, the chickens work it, eating the bugs,' Harris told me. The field we were in held around 10,000 chickens, but you'd never have guessed it – there was not a shed in sight.

As we drove along a dirt track through the meadow, other fruits of the rotational system were revealed. 'Dung beetles are God's gift to pasture!' Harris exclaimed as the jeep shuddered to a halt. We jumped out and before I knew it we were on our hands and knees, inspecting a cowpat. Never let it be said that I don't know how to have fun . . .

The cowpat was peppered with little blooms of red soil pushed up by busy dung beetles playing an important role aerating the soil. Harris dug around in the mess with a paper cup: "They've riddled the soil with holes, even in this hard-assed track! A healthy dung-beetle population is an indicator of a healthy soil.' He gently lifted one out, a little coffee bean on legs: nature's seal of approval on White Oak Pastures.

The farm had not always been like this. Following the Second World War, Harris's father started farming industrially. 'It was all about pounds of beef produced and nothing about the quality,' he told me. He admitted he found the industrial way exciting and that he and his father had been good at it. They fed the cattle grain and used hormone implants to make them grow faster. Antibiotics were mixed in feed and pastures doused with chemicals. But Harris grew disenchanted with the artifice of it all. When he took over the business from his dad in his forties, he dispensed with the props of intensification, bringing the farm full circle.

Now it is a celebrated model of environmental sustainability, animal welfare and good food. His office walls were covered in newspaper cuttings, full of glowing references to the way he runs the farm and the quality of his produce. His clients include the catering giants Sodexo and the retailers Publix and Whole Foods Market. Later that day, a Publix store assistant showed me some White Oak Pastures beef. 'We can't keep it in the store, it sells so fast,' she said.

So what do these happy cattle, living in a land of plenty, looked after by friendly farmers and consumed by contented customers, have to do with feeding a growing population with diminishing natural resources? Certainly, Georgia is a million miles from Malawi or Ethiopia, where a good beef steak is the stuff of daydreams. Yet there are more answers here than there are on high-tech factory farms where cattle are crammed into sheds and plied with antibiotics. As I have argued throughout this book, factory farming is not feeding the world, because the grain-feeding of confined animals uses more food than it produces.[1] It's part of a highly resource-intensive and wasteful food system. By contrast, no ground-up fish meal from Peru is flown across continents to feed the cows on White Oak Pastures; no oil-based fertilisers are poured onto the land. This is farming as nature intended.

Could it be done on a big scale? Harris has no doubt. 'I know I could scale up tenfold,' he said. He already employs eighty people and has plans to expand.

However, when it comes to world food, there's big and there's global. The United Nations estimates that food supply needs to increase by 70–100 per cent by 2050.[2] To achieve this without factory farming we need a common-sense approach, based on three principles: putting people first, reducing food waste, and farming as if tomorrow matters.

HOW PUTTING PEOPLE FIRST HELPS ANIMALS

A third of the world's cereal harvest is fed to farm animals.[3] If it went directly to humans instead, it would feed about 3 billion people.[4] Cereals are a big deal. They're not just for breakfast: worldwide they provide around half of the total calories for humans, in bread, pastries, pasta, tortillas, pies, pizzas – you name it. So how can it make sense to shovel this human staple into factory farms? It's not only cereals: in Argentina we had a glimpse of how 90 per cent of the world's soya production is destined to feed industrially reared animals.[5] As we saw in Peru too, factory farming's seemingly insatiable hunger extends to plundering the oceans to feed farmed fish, pigs and poultry. Given the chance, those confined animals would convert things that people don't or won't eat into something suitable for human consumption. For example cows and sheep will turn grass – often growing on land that can't be used for anything else – into meat and milk. Chickens will search pasture, woodlands and orchards for food, producing meat and laying eggs. Along with pigs, they will recycle food waste with great enthusiasm.

Yet industrial animal rearing has thrown farm animals directly into competition with people for food – and we are not winners in the process. For every 6 kilograms of plant protein such as cereals fed to livestock, only 1 kilogram of animal protein on average is given back in the form of meat or other livestock products for humans.[6] Factory farms are food factories in reverse: they waste it, not make it, and squander valuable cropland in the process.

Reducing the amount of grains fed to farm animals by half would go a long way toward a saner food system. People don't have to choose between eating cereals or meat. Both can be produced far more effectively with the right kind of farming.

The Solution

Rear ruminants on pasture not in sheds. Food from ruminant animals, such as beef, mutton, lamb and milk, should be produced by grazing on mixed, rotational farms, permanent pastures or marginal lands. This converts plant life that humans can't eat into edible food. End the wasteful practice of feeding grain to confined cattle for intensively reared beef or milk.

Feed fish to people, not to livestock. Up to a third of the fish landed in the world is not consumed directly by people. It is used mostly as feed for farmed fish and other livestock.[7] Overfishing and the practice of throwing back dead or dying fish are now well documented. The plundering of our seas to feed confined farmed animals is less well known. Ending the practice would take pressure off our often overexploited seas.

Strong action from governments, consumers and corporations alike, in the form of legislation, subsidy incentives, purchasing policies, research and advice, is needed to achieve these two recommendations.

REDUCING FOOD WASTE

North America and Europe waste up to half their food – enough to satisfy the hunger of the world's billion undernourished people between three and seven times over.[8] It's a staggering statistic. Whether it be from shops, catering companies or tossed into our bins at home, food is wasted all along the food supply chain in industrialised countries.[9] Tristram Stuart's groundbreaking book on the subject, *Waste*, shows graphic images of a whole crop left to rot in the field after being rejected by a supermarket; of potatoes rejected for cosmetic reasons; of masses of 'imperfect' bananas dumped in a ditch. It is not just fruit and veg that gets thrown away; in the UK alone, householders waste the meat equivalent of 50 million chickens, 1.5 million pigs and 100,000 beef cattle every year.

Traditionally, rearing farm animals was a land-based business. They grazed or foraged for food or, in the case of pigs and poultry, ate scraps from the kitchen. The system was based on diversity, maximising resources, working with natural processes and avoiding waste. The animals provided manure and food as part of the natural rhythm of farm life. With the rise of industrial agriculture, Stuart argues, 'we turned the entire rationale of animal agriculture on its head by feeding them foods that humans could eat and wasting the food waste that we should be feeding to them instead.'

Could we be recycling waste food for farm animals? Is this practical on a significant scale? I went to Dagenham, East London's windswept 'waste corridor' and a hotbed of sustainable industries, to look at some relevant recycling projects. I watched as used drink cans leapt off conveyor belts streaming with plastic bottles. Metal paddles stirred silver-grey liquid. A giant washing machine separated plastics from other rubbish. A vast spaghetti junction of pipes, frames and clanking machines stood within a warehouse, reincarnating plastic bottles. Outside, great bales of squashed bottles waited their turn. The company running the place was called Closed Loop. It transforms 875 million discarded bottles a year into something usable. 'Ten years ago, we didn't have the technology to build a plant like this to convert rubbish into resources. That's where food waste is right now,' the company's Nick Cliffe told me.

In the EU, it is currently illegal to feed animal by-products to farm animals. The ban was a panic measure following the 2001 outbreak of

foot-and-mouth disease. As a result, a lot of food waste ends up in
landfill – and in the UK and many other countries, landfill space is
running out. Sites in the densely populated southeast of England
could be full within five years. Already, waste is travelling many miles
simply to be dumped up north. As space diminishes to bury rubbish,
there is a new impetus for more adventurous recycling.

Pure plant-based food waste that has undergone strict processing
can already be fed to animals. Closed Loop is working with a major
frozen-food manufacturer in England that processes waste vegetables
for animal feed. About a million tonnes of industrial by-products like
whey and vegetable leavings are said currently to go to pigs in the
UK.[10] There is the potential to do much more.

There are also some smaller, more local schemes. In the village of
Pince in northwestern France, householders are being offered chickens
in an attempt to reduce the amount of food going to waste.[11] In the
London borough of Tower Hamlets, there's a weekly food-waste
collection. Since it is mixed waste, it goes for composting, which is
better than sending it to landfill, though not as efficient as using it as
animal fodder.[12] The fact that it is only collected weekly is also likely
to deter many people, especially those without much space in their
homes, from taking part.

The Japanese, South Korean and Taiwanese governments are all
ahead of the game. They have grasped the fact that feeding food waste
to livestock is the most efficient way of recycling it and have set up
food-waste collection and recycling centres that ensure the leftovers
are properly sterilised and safe for animals intended for human
consumption.[13] If Britain and the EU made food recycling easier, the
40 million tonnes of livestock feed imported from South America
every year could be reduced, there would be less pressure on landfill
sites, and pigs would enjoy a more varied diet.

Waste is of course also linked to commercially driven food cultures
that encourage people to eat far more meat than they need, with serious
health implications, where profit comes before feeding people properly.

Developing countries have food-waste problems too. However,

their waste is more often caused by a lack of basic technologies and infrastructures than by profligacy. Losses of up to half of a staple crop are all too common simply for want of simple technologies like decent storage facilities, refrigeration and transport.[14] Improving food security for countries like these is as much about improving these basics as it is about growing more.

Worldwide, the UN suggests that about a third of food is wasted through being binned or left to rot.[15] It estimates that 28 per cent of the world's agricultural land is used to produce food that is wasted at an economic cost of about US$750 billion, equivalent to the GDP of Switzerland.[16] If this were reduced by half, it could provide enough food for an extra billion people.[17] Recycling as much of the remaining waste as possible by feeding it to pigs and poultry would add further efficiency savings.

As I set out in previous chapters, avoiding overconsumption of meat deserves serious consideration in developed countries, for our own good.

The Solution

Feed pigs and poultry on food waste and encourage foraging. Pigs and poultry are nature's great foragers and recyclers, the perfect recipients of food waste. The current practice of feeding them cereals and soya squanders vast amounts of food. They should no longer be factory-farmed. Instead, make them integral to mixed farms where they can forage and turn food waste into eggs and meat. Governments must play their part by ensuring there are legislation and policy incentives in place to enable this.

Invest in waste reduction. Governments, civil society and corporations should encourage a reduction in food waste at every level, from farmer to corporation to consumer, through incentives, purchasing policies and the provision of research and advice.

Avoid overeating meat. Rejecting copious junk food in favour of high-quality meat will benefit us and the planet. Research has shown that too much saturated fat from meat and dairy products can be harmful to health and may contribute to obesity, type-2 diabetes and heart disease.[18] Reducing consumption of these saturated animal fats by 30 per cent would lead to about a 15 per cent reduction in heart disease in the UK and Brazil.[19] Consumers, governments, corporations and civil society should work together to promote healthy, sustainable balanced diets that, in Western countries, avoid overconsumption and instead include better-quality meat from animals kept in higher-welfare conditions. This would both benefit human health in the West and reduce pressure on the environment.

FARM AS IF TOMORROW MATTERS

The Chinese philosopher Confucius said: 'For all Man's supposed accomplishments, his continued existence is completely dependent upon six inches of topsoil and the fact that it rains.'

As a result of intensive farming and climate change, that vital topsoil is disappearing – and so, in some places, is any guarantee of rain.

Over the last half-century, many farm animals have disappeared from fields and been confined in sheds, in an agricultural system that has become divorced from the land and separated from the so-called 'nutrient cycle'. The natural cycle in which sun and rain fed grass, which fed animals, whose manure enriched the soil, has been replaced by a new system dependent on fossil fuel-based synthetic fertilisers. Monocultures, heavily reliant on chemical pesticides and artificial fertilisers, are hammering the soil and the environment. The UN has warned that the world's farmland could decline in productivity by a quarter this century.[20] Soil erosion already affects almost a third of the world's cropland,[21] and is widespread in the EU.[22] Meanwhile land is being lost to urbanisation, contaminated by irrigation,[23] and becoming desert as fast if not faster than we're adding to it.[24] There is now also the added pressure from land use for biofuels and the continued growth of industrial livestock.

Much greater emphasis is needed on soil-healthy rotational farming with a mix of crops, pastures and farm animals, reducing reliance on artificial fertilisers, as well as providing better animal welfare. Measures such as crop rotation, green manure, reforestation and taking unsuitable land (such as steep slopes) out of production would help to reverse soil erosion and land degradation.

The Solution

> *Produce food from mixed farms of crops and animals to enhance soil sustainability*. Mixed farms where animals are rotated with soil-enhancing crop rotations should be encouraged. Most pigs and poultry in Europe and the USA are currently confined on factory farms. Restoring the natural link between farm animals and the land needn't require huge amounts of extra space. The UK, for example, rears over 800 million meat chickens a year. Keeping them free-range would need an area around a third of the size of the Isle of Wight – less than one-thousandth of the nation's total farmland.[25] Integrating them within mixed farming systems would benefit animal welfare and soil quality and sustainability.

As a global society, we are wasting as much as half of all the food we produce, by feeding it to farm animals, throwing it away, or letting it rot for want of basic technology. Land is often being driven so hard that we are playing off tomorrow's sustainable harvests against today's short-term gains. With the prospect of 2 billion more people to feed by 2050, our food system needs to be 70–100 per cent more effective. That cannot mean simply doubling farm outputs in a business-as-usual fashion.

Just to double output from our current food system would be like a water company with badly leaking pipes simply laying down a second set of equally leaky pipes. Yes, it would increase the water to people's homes. It would also increase the waste. Far better to repair the pipes.

In the course of writing this book, I have travelled through Europe, North and South America, China and elsewhere. What I have seen and heard has only strengthened my belief that we need an urgent rethink about feeding the world. Of one thing I am certain: industrial farming is not the answer. The veneer of efficiency is no more real than the emperor's new clothes. The system actually wastes food rather than making it.

What I have also discovered is that the means to feed the world's population today and for the foreseeable future are already with us. Globally, enough food is produced to feed around 11 billion people, if only we didn't waste it. Future harvests will need a hefty dose of common sense in their production if we are to feed people properly and fairly. Ending the competition for food between people and farm animals seems a good place to start, along with reducing and recycling food waste and taking animals out of factory sheds and restoring them to the land in fields. With that, we have the recipe for truly sustainable food on an increasingly crowded yet shrinking planet.

Global food production (in calories), with estimated losses, conversion and wastage in the supply chain

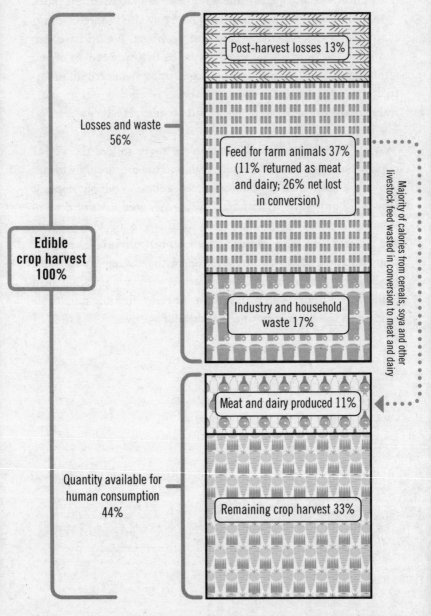

Illustration based on Lundquist, 2008 in C. Nellemann et al., *The Environmental Food Crisis – The Environment's Role in Averting Future Food Crises*. A UNEP rapid response assessment, February 2009.

Wasted: farmed for food but thrown away

Amount of meat wasted worldwide each year in equivalent number of animals

11,600 million chickens wasted

Each icon represents 100 million chickens

270 million pigs wasted

59 million cattle wasted

Each icon represents
10 million pigs or cows

19

Consumer Power

What you can do

Bringing about a better food future is something that everyone can now get involved in. Each of us has three great opportunities a day to help make a kinder, saner food system through the choices we make.

The celebrity chef Hugh Fearnley-Whittingstall is a big champion of great cooking with integrity at his River Cottage on the south coast of England. 'I see the link between good animal welfare and good food all the time,' he told me. 'You see the difference on the plate, you taste the difference, and you feel the difference in your body.'

Like Hugh, I recognise that consumers have real power and believe the way we shop can change farming methods for the better, including animal welfare. Compassionate consumerism is a great way to choose wonderful food and save the world from Farmageddon. I advise: buy foods from the land – reared on farms, not factories; love leftovers, so as to reduce food waste; and choose a balanced diet without eating too much meat.

For food reared on the land, look out for products labelled free-range, pasture-raised, outdoor-reared or organic. Buying meat and milk from ruminants – sheep and cattle – is the most sustainable option, as these convert grass into food.

Free-range and organic pigs and poultry come with the drawback that, for now, they are largely reared on grain and soya. EU laws forbid the feeding of food waste to farm animals. As this inevitably changes, nature's great recyclers can roam the land once again, converting food that people won't eat into food that they will. But for the time being, buying pasture-raised pork, chicken and eggs will provide a better food choice for the consumer and a decent life for the animal, while speeding bigger changes in the future.

Avoid labels that just say things like 'farm fresh', 'country fresh', 'natural' or just 'fresh': it's probably from a factory farm. Labels that just say something like 'corn-fed' should be avoided too. In the UK, the ubiquitous 'Red Tractor' label means that the meat or milk has been produced to British standard and in itself is no guarantee of higher animal welfare.

In the UK, the RSPCA monitors the Freedom Food scheme to assure higher standards of animal welfare. In the Netherlands, look out for the Beter Leven (Better Life) scheme, and in the US, the Animal Welfare Assured (AWA) and Global Animal Partnership (GAP) label. A good choice in Australia is anything with the 'RSPCA approved farming' logo.

Lamb

Sheep usually spend more time in the great outdoors than most other farm animals. So, if in doubt about what to buy to assure grass-fed meat, then products from lamb and sheep are a good choice. But always check the label or ask the seller. Pasture-reared lamb is healthier than intensively reared lamb, having higher levels of the omega-3 essential fatty acids that are linked to reducing many chronic diseases.

Beef

Pasture-reared or grass-fed beef is a good choice, enabling animals to express their natural behaviours and feed in a more natural way. Beef

raised this way has considerably less saturated fat and more vitamin E
and healthy beta-carotene than intensively reared beef.

All too often, cattle are reared intensively, confined indoors on
uncomfortable slatted floors or crowded into barren feedlots and
fed grain or soya. In the US, unless the packaging says grass-fed or
pasture-raised, cattle are likely to be finished on a confined feedlot, which leads to severe welfare problems. So, again, really check
the label.

Dairy

Milk has a wholesome natural image from cows left to gently graze in
fields of green. In Britain, this is still largely the case in the summer-
time. In the US and other countries, dairy cows are being 'zero-grazed':
confined permanently indoors or in feedlot-style pens and never grazing a fresh blade of grass. Sadly, there isn't much on the labels to help
shoppers choose. In the absence of decent pack descriptions, choose
organic, or milk produced under one of the dedicated animal welfare
schemes like Freedom Food.

In the US, look for Animal Welfare Approved standards. If
unavailable, buy Certified Humane or USDA Organic. Or you could
use dairy alternatives such as soy, coconut or almond milk, yogurts
or desserts. Look for labels that say rBGH-free or rBST-free, which
means that the cows were not dosed with genetically engineered
milk-boosting hormones. While these are banned in the EU, they
are still widely used in the US.

Don't forget to check the label on your cheese and yoghurt too!
When you eat out, check whether the milk in your tea or the cheese
in your sandwich is organic. If it doesn't say so on the menu, it probably isn't.

But isn't organic milk the domain of the well-off and haute cuisine?
In Britain, McDonald's restaurants offer only organic milk for their
hot drinks.

Eggs

Many eggs are still produced from caged hens. Barren battery cages are commonly used worldwide, where several hens are crammed into tiny all-wire cages, unable to even stretch their wings for life.

Barren cages are now banned in the EU. However, a proportion of eggs are still produced using so-called 'enriched' cages, which have rudimentary features like perches and a bit more space, but still prevent hens from carrying out proper exercise or other natural behaviours.

In the EU, eggs are the exception in that they have to be labelled by law according to the way they are produced. Look out for free-range and organic; under these terms, the hens will have been given access to the outdoors. Free-range eggs often contain more healthy omega-3 fatty acids and antioxidants than cage eggs, and are a richer source of vitamin E. The general rule is that the more an animal is given access to the outdoors and able to eat grass and mixed foods, the healthier the resulting food.

'Barn' eggs will come from hens kept in big sheds and able to move around but not given access to the outdoors. Avoid eggs labelled 'eggs from caged hens' or 'enriched or colony cage' eggs; these are the rare example of the label telling you they are factory-farmed.

Outside the EU, in the absence of a recognised term denoting outdoor access, then the eggs are probably from battery caged hens. Again, avoid 'fresh' or 'natural'. The term is often confused with free-range, but means nothing of the kind.

In the US, most eggs still come from barren battery cages. Only buy cage-free eggs, those labelled pasture-raised, or eggs that meet animal welfare certification standards like Animal Welfare Approved and Certified Humane.

Remember that additional foods such as mayonnaise, cakes, cookies, pasta and quiches contain egg; unless the ingredients explicitly say 'cage-free' or free-range, they are likely from caged hens.

Chicken meat

By investing a little more in the chicken you choose you can make a big difference to its life. It can also give you a healthier product. Free-range and organic chicken contains up to 50 per cent less fat than its factory-farmed equivalent.

The chickens most commonly found on our supermarket shelves are bred and fed to reach their slaughter weight in around six weeks. Free-range chickens will usually be slaughtered at eight weeks and organic or pasture-raised at around twelve weeks.

Free-range, pasture-raised or organic chickens will have had access to the outdoors during their lives. In the US, look out for chicken certified Animal Welfare Approved (AWA), Global Animal Partnership (GAP), or Certified Humane. Avoid chicken that just says 'fresh' or 'farm-fresh', 'corn-fed', 'vegetarian-fed' or 'natural'. Some of the more outrageous labels found on factory-reared chickens include 'Fresh all natural' and 'cage-free'! Chicken is perhaps the meat most fraught with phony labels.

A good halfway house would be chickens that are reared indoors but to higher welfare standards, for example under the RSPCA Freedom Food scheme in the UK or the Certified Humane scheme in the US.

Turkey meat

As with chicken meat, look for labelling terms like free-range and organic. In the UK, also look for the RSPCA-monitored Freedom Food label. A good compassionate shopping tip is to look for slow-growing traditional turkey breeds like Norfolk Black, Black Wing Bronze and Cambridge Bronze.

Pork, bacon and sausages

Along with chickens, pigs have tended to be the most factory-farmed of animals. They are often kept in crowded indoor pens, bred in confinement crates and fed copious amounts of cereals, soya and other people-type food.

However, there has been a resurgence in recent decades of keeping pigs outdoors, at least for breeding, with some granted much of their lives outside. This is good news, as pigs are naturally inquisitive creatures and need to be able to root around and explore their world. As intelligent as the average dog, they can quickly become bored and suffer if their needs are denied.

In the EU, look out for free-range or organic – the pigs are born and reared in systems with outdoor space where they can roam outdoors. Free-range pigmeat is richer in vitamin E and iron than meat from intensively reared pigs.

'Outdoor reared' is another good choice – the pigs are born in systems with outdoor space and spend around half their life outdoors.

'Outdoor bred' means that the pigs will have been born outside – better for the breeding animals – but then reared indoors.

In the US, avoid pork or bacon from gestation-crate systems – cruel ways of confining pregnant pigs – and look for certifications that do not permit these. These include Animal Welfare Approved, Certified Humane, 5-Step Animal Welfare Rating Program, Organic and American Humane Certified.

As always, avoid general labels that sound good but are meaningless, like 'fresh'.

Farmed fish

This is a tricky area, as many wild fish species are increasingly threatened. Useful advice here is to avoid carnivorous species of farmed fish, such as salmon, trout, cod and halibut, as these are likely raised on feed made from wild fish – Peruvian anchovies and the like.

If sustainably caught, wild salmon and trout is by far the better purchase than their farmed cousins. Wild salmon has up to 60 per cent less fat than farmed.

If you do buy farmed fish, look out for those produced under an accredited scheme like organic or RSPCA Freedom Food.

Buy local

Choosing foods produced closer to home makes good sense. Of course, just being local doesn't mean that the food hasn't come from a factory farm. But buying from a supplier you know or a local farmers' market, or buying local produce from the supermarket, are all good ways to reduce your carbon 'footprint' – less transport needed to get the food to you. It also means you are more likely to strike up a rapport with the farmer, and perhaps even see for yourself how the food is produced. And it reinforces local food communities. It's naturally easier to find out what's going on with your food if it's produced close to home than if it is part of a global supply chain. As the British food writer Joanna Blythman points out, factory farms 'don't go out of their way to welcome us in for "Doors Open" days. So what chance do we have of knowing what's really going on in similar operations thousands of miles away?'[1]

Love leftovers, waste nothing

Avoiding wasting food is perhaps the simplest way to make a major contribution to a better food system. It will help reduce the amount of land, water and oil, not to mention animal suffering, that currently goes into feeding landfill. It cuts down your food bill too.

Avoid overeating meat

There is a growing body of evidence to show that people in the developed world are generally eating too much meat, eggs and dairy for their own health. Cutting down – going meat-free on Mondays for example – is a simple step towards avoiding factory-farmed produce and helping to balance your diet. One quick and easy way to eat well and be light on the planet is to have meat-, egg- and dairy-free meals.

Avoiding Farmageddon is easy. As long as we buy products from animals reared on the land (free-range, organic), favour local producers or retailers that we trust, eat what we buy and thereby reduce food waste, and avoid overeating meat, we can fill our plates in ways that benefit the countryside, our health and animal welfare.

Epilogue

Midsummer, and the garden looks amazing. Our little wildflower meadow has run riot; the poppies, cornflowers and hollyhocks have shot up and are now taller than I am. My wife and I planted them as ground cover for our hens, and to help preserve the butterflies and bees that need a helping hand.

In the cool of the morning, I open the coop and six reddish-brown bundles of feathers scurry out and start pecking and scratching as they busily go about their breakfast. Later, the sun gets the hens sunbathing; lying gently on their side, wings flicking, eyes bright and bulging with delight. They remind me every day that animals are individuals with their own wants and needs.

Black cattle with white faces graze the hill that looms over our village, so close it often sounds like they too are in the garden. Our small 200-year-old cottage nestles in the rural heart of the English South Downs National Park. This community and countryside has survived many changes, not least occupation by the Romans, then as King Alfred the Great's Royal Manor, before the Domesday Book and the iron grasp of William the Conqueror. Hitler's Luftwaffe tried hard to leave its mark, raining thousands of bombs on the area; the sole loss of life was a single unfortunate pig.

Despite the changes, and often because of them, the countryside looks stunning, shaped by history and stewarded by generations of farmers reaping harvests from the land. I listen to the trickle of the chalk stream and the low chatter of village folk drinking in the pub next door and reflect on two years of travel, of exploring the issues behind the brewing storm: the Farmageddon scenario that jeopardises food, the countryside and all of us.

Fifty years ago, Rachel Carson's *Silent Spring* delivered a stark warning of the perils of treating the countryside like just another industrial process. She revealed the realities of the new chemical era on the land, where pitched battles are fought with nature, using pesticide sprays as the weapon of choice. The mother of the modern environmental movement sparked controversy at the highest level – even the US president of the time, John F. Kennedy, was talking about it. She won some important reforms, but overall, her clarion call – that the industrial way was the wrong way – has been ignored.

While writing this book, I have seen what happens when we farm without proper care for nature; the far-reaching consequences of converting varied patchworks of fields into vast, monotonous food factories. From the grievous pollution facing the famous French beaches of Brittany, to the battle to preserve drinking water from toxic algal blooms in China, and the decline of the world-renowned Chesapeake Bay in the US, in each case intensive farming has been implicated. I remember the eerie feeling of looking down from a helicopter on a factory farm rearing three-quarters of a million chickens with not a bird in sight; of driving through a Mexican valley producing a million pigs without seeing a single animal. I've seen the tears of people driven from their land to make way for soya, or suffering ill-effects from the fumes of fishmeal plants, both to feed distant factory farms. I've wept too at the plight of a large-scale intensive dairy farmer, crippled by debt, who shot himself and left five children fatherless.

There is a story behind everything we eat. I've spoken to countless people involved; their personal stories are reflected in this book. Everyday meals like chicken and chips or bacon and eggs or just a glass of milk come with their very own cast of characters and consequences. Startling changes don't have to be accompanied by dramatic scenes. Sometimes, they can be as subtle as disappearing birds, bees and butterflies; as gradual as the erosion of the goodness of food and the quiet ebbing of our quality of life. Sometimes it has been difficult not to feel overwhelmed by the sheer scale of the problem, the Farmageddon scenario that threatens to engulf the countryside and society.

At the same time, I set out to find seeds of hope. I wanted to find better ways than the industrial ascendancy. I discovered they are often all around us; here small acorns in need of nurturing; elsewhere thriving oaks. I felt privileged to see the pasture plains of free-ranging animals in Georgia; uplifted by fields in Argentina dancing defiantly with butterflies in contrast to their lifeless pesticide-soaked GM neighbours. I learned a lot from watching chickens in China roaming woodlands and pigs living inquisitive, active lives on a model farm in Beijing. I've been inspired by extensive farms in Britain and Europe, taking care of their animals, looking after the environment, and producing great food to boot. I found so much more than the few scattered crumbs of comfort I was expecting – and on my own doorstep too.

At our annual village beer festival, it was heartening to listen to a local farmer speaking proudly of his award-winning mixed farm, combining the job of producing food with caring for wildlife. He is not alone. Close by, through leafy lanes and rolling hills, more animals are free-range, and it's a joy to see. Brown and white cattle with tiny calves afoot chomp on grass up to their knees. Pigs scamper amongst scattered huts resembling a holiday camp. Such scenes are far from unique, but not nearly common enough. In the industrialised nations where food factories replace countryside, these oases shine as beacons of hope.

With nearly a third of the planet's land surface devoted to rearing farm animals or growing their feed, perhaps I shouldn't have been surprised at how big the environmental impact can be. I was struck too by the link between how animals are kept and the quality of the food they produce. Generally, the more that animals are reared on the land with natural, varied diets, the healthier and tastier the food. We instinctively know this, which is why terms like 'natural' and 'free-range' are so attractive. It also explains why marketers all too often try to mask factory-farmed food behind labels showing false depictions of green fields, small farmhouses accompanied by comforting terms like 'farm fresh' and 'country fresh'. It speaks of the need for better labelling, to inform people how their food is produced; to remove the blindfold that hinders so many from truly exercising the consumer choice that is their right.

I also explored some of the big questions facing humankind: not least, how to feed the coming world of 9 billion people – 2 billion more than we have as I write. Surely, if we don't industrialise food production, some say, make it even more intensive – perhaps using a greenwash term like 'sustainable intensification' – then we are bound to starve? What I found is that this couldn't be further from the truth; that our planet already produces more than enough food for everyone, now and into the future, if only we didn't waste it. About 11 billion people could be fed on what the world currently produces, many more than today's 7 billion. The problem lies not with producing enough food, but with the extent to which it is wasted, from the simple act of throwing food away in our homes or at the supermarket, to letting it rot in developing countries for want of simple, low-tech assets like decent grain stores. What has become evident to me is the way that one of the biggest causes of food waste is so often overlooked: cereals, soya and fish fed to factory-farmed animals, who then return a fraction of the protein and calories in the form of meat, milk and eggs. If the animals were reared on the land instead, they would swell the global food basket, instead of pilfering from it.

The food system today is like a leaky bucket; it wastes half of what it produces. Simply churning out more without fixing the leaks will cause yet more waste and intensify pressure on already overstretched ecosystems. Of course, higher productivity in some parts of the world would be a good thing, particularly in some developing countries. But as a general strategy, it's likely to be a costly failure. Plugging the leaks seems a much more economic, more sensible way to go. But how much more food could be made available this way? Halving the amount dumped or rotting would free up enough food to feed an extra billion people. A further billion or more people could be fed by halving the amount of cereals destined for industrially reared animals.[1] In this way, enough food could be made available to feed future populations with little or no extra cost to the environment.

The industrial food system is geared towards producing food in volume, regardless of quality, in ways that rely on large amounts of finite resources, including land, soil, oil and water. What will happen when these essential ingredients for the agricultural machine start to run out? 'Man has lost the capacity to foresee and forestall,' said Albert Schweitzer. Avoiding Farmageddon will require us to revive this capacity. Through researching this book, I have seen how profits are often put before feeding people. I have seen compelling evidence of how, if something isn't done soon, warnings of a Farmageddon future could become a reality, one that brings with it a deeply diminished countryside, surging disease, unhealthy food, and growing world hunger. Thankfully, I have also found it doesn't have to be like this.

At a dinner party with representatives of leading businesses, I was asked, given three minutes with the US president or UK prime minister, what I would ask them. I would urge them to support food production that puts animals back on the farm instead of in factories; extensive farming connected to the land, providing more nutritious food in ways that are better for the countryside and animal welfare. Governments could help improve the health of their nation and safeguard future food supplies by building on natural resources: the pasturelands that cover a quarter of farmland worldwide, and two-thirds in Britain, for example. For a generation of consumers shielded from the realities of factory farming, brought up as they are on picture-book images of Old Macdonald and his small farmyard idyll, reinforced by advertising and often misleading labels, the truth often comes as a shock. Putting farm animals back on the farm could be a big vote-winner too, as many people believe that is where they are anyway.

As awareness grows, things are beginning to change. In Europe, some forms of factory farming, like veal crates that prevent calves from ever turning around, and barren hen cages where the birds can never flap their wings, are banned. Animals now have legal recognition as 'sentient beings', capable of feeling pain and suffering – a view long denied in some quarters. Major food companies in Europe – many household names – are increasingly moving to only cage-free eggs, higher-welfare chicken and the like.

During the course of my work as CEO of Compassion in World Farming, I always hoped that getting changes in Europe would lead to similar moves elsewhere. In the United States, where factory farming first emerged, changes are indeed afoot. Kindred spirits in the US, together with concerned citizens, consumers and policymakers, are bringing about reforms to some of the worst examples of factory farming. It was something that Compassion's founder, Peter Roberts, always dreamt of.

I remember vividly the day in late autumn 2006 when I visited Peter's bedside. Compassion's headquarters team had been rejoicing on hearing the news from the USA that Arizona had agreed to ban veal crates. For Peter, the cruel veal crate had been a cause célèbre, and a battle he had won in the UK two decades earlier. Peter was lying in a hospital bed and in a bad way. I hastily rearranged my schedule and rushed through the whitewashed hospital corridors, heart in mouth. He was in a side room surrounded by his wife and three daughters. For several days he had been unconscious and unresponsive. To see him lie so still, with his eyes firmly shut, choked me with emotion. After greeting his distraught family, I leant over, took Peter's hand and, trembling, began to deliver the news: that what he started in the UK had sent ripples across the Atlantic. As I did so, Peter's eyes opened, fixing mine with a listening attention as each sentence rolled into the next. When I stopped talking, his eyes gently closed. Shortly afterwards, he passed away.

Through writing this book, I have become convinced that we can all make a big difference. It has been a privilege to listen to so many people and to tell their stories. I have been left in no doubt about the tremendous power we have as consumers; the difference we can make three times a day with every meal. I have learned how choices we all make can have a real effect, not only on the people, animals and environments behind the food we eat, but on ourselves and our families. Simple measures like eating what we buy instead of throwing some away, and eating less but better meat, can make that difference, and when consumers choose alternatives to industrial factory farming – like free-range, pasture-raised, organic or the like – then supermarkets and policymakers take note. Things begin to change – from Farmageddon to a better future for people, animals and the planet.

Philip Lymbery
Hampshire, England
July 2013

Acknowledgements

When I first joined the animal welfare organisation Compassion in World Farming in the spring of 1990, I had no idea that I would be taking the first step on a journey of exploration that would span three decades. This book feels like the culmination of thoughts, observations and conversations with a vast number of people spanning all of that time. I owe a huge debt of gratitude to the late Peter and Anna Roberts, founders of Compassion in World Farming, for having the faith to employ me when I had little to offer except energy and enthusiasm. I am thankful that Peter saw through my initial failings, not least plummeting through his favoured glass coffee table just days before my trial period was up.

In those early days, I made it my goal to see as many farming systems firsthand as possible and to talk to the people involved, listening to their perspectives and looking for ways to improve things for everyone. This book feels like a natural extension of that early endeavour. I am grateful to all those, too numerous to mention individually, who have hosted me on their farm or business and taken the trouble to explain what they do. I appreciate the often lengthy discussions with colleagues in the animal welfare and environmental community that have informed my thinking and perspective.

When I set out to write this book in early 2011, I had no idea how much would be involved. Many people have given generously of their time and insights along the way, for which I am eternally grateful. I would like to give particular acknowledgement to my co-author and collaborator Isabel Oakeshott, for her hard work, her instinct for storytelling, for teaching me the value of crafting prose with 'colour', rather than just facts, and for enduring some pretty awful sights during our travels.

I owe a huge debt of thanks to our researcher, Jacky Turner, without whose thoroughness and commitment this book would have been much the poorer. Grateful thanks go too to Tina Clark for her endless patience, for reading and commenting on drafts time after time, planning field trips and providing essential support so reliably with unfaltering ease. To Veronica Oakeshott for documenting our trip through China so well, and to Jeff Zhou for seeing us through some difficult moments; thanks to Jim Wickens for his encyclopedic insight into the issues, where to go and who to talk to; to our camera crews and support: Alejandro Reynoso, Brian Kelley and Jim Philpott. Many thanks to Luke Starr for research related to our field trips, to Luke Oakeshott for additional research and to Laurence Stephenson for help in editing pictures and video arising from the trips. Thanks also to Pru Elliott for coming up with the initial graphic ideas that adorn this book.

Grateful thanks to all who kindly commented on drafts – Blake Lee-Harwood, Dil Peeling, Emily Lewis-Brown, Heather Pickett, Phil Brooke, Jeff Zhou, John Robins, Joyce D'Silva, Leah Garcés, Richard Young, Steve McIvor; to Sir David Madden and René Olivieri for spending so much time reading early drafts of the manuscript, providing invaluable advice; to Peter Stevenson for careful attention to detail and for stimulating discussions on how best to feed the world without cruelty to animals.

Special thanks to our editorial team at Bloomsbury, Bill Swainson, Elizabeth Woabank and Steve Cox, and to our literary agent, Robin Jones, for support and encouragement.

A special mention is due to Paul Blanchard for inspiring the book in the first place, for bringing the authors together, and for his boundless enthusiasm for the project. Invaluable help, advice and guidance has been received during our field trips from Alberto Villareal in Argentina; Maria Elena Foronda Farro from Natura, Chimbote; Ron Lane of the Food Animal Initiative; Stefan Austermuhle from Mundo Azul, Peru; Tom Frantz in California.

Grateful thanks to Valerie James, Jeremy Hayward, Sir David Madden, Reverend Professor Michael Reiss, Michel Vandenbosch, Rosemary Marshall, Sarah Petrini and Teddy Bourne for steadfast belief and support throughout.

Finally, huge thanks to my wife, Helen, for her support and understanding, particularly during long periods away from home so soon after our marriage.

Notes

PREFACE

1 R. Harrison, *Animal Machines*, Vincent Stuart, London, 1964.
2 M. Bittman, 'Don't End Agricultural Subsidies, Fix Them', *New York Times*, 1 March 2011, http://opinionator.blogs.nytimes.com/2011/03/01/dont-end-agricultural-subsidies-fix-them/.

INTRODUCTION: OLD MACDONALD

1 G. Dvorsky, 'China's worst self-inflicted environmental disaster', 2012, http://io9.com/5927112/chinas-worst-self+inflicted-disaster the-campaign-to-wipe-out-the-common-sparrow (accessed 8 May 2013).
2 Ibid.
3 Jonathan Leake, 'Farmers to be paid to feed starving birds', *Sunday Times*, 13 May 2012, http://www.thesundaytimes.co.uk/sto/news/uk_news/Environment/article1037693.ece (accessed 8 May 2013).
4 Defra, *Agriculture in the UK 2010*.
5 Ibid.
6 Government Office for Science, *Foresight Project on Global Food and Farming Futures Synthesis Report C1: Trends in food demand and production*, 2011; S. Msangi and M. Rosegrant, *World agriculture in a dynamically-changing environment: IFPRI's long term outlook for food and agriculture under additional demand and constraints*, paper written in support of Expert Meeting on 'How to Feed the World in 2050', Rome, FAO, 2009; H. Steinfeld et al., *Livestock's Long Shadow, environmental issues and options*, FAO, Rome, 2006, Introduction, p. 12.

7   FAO, *State of the World Fisheries and Aquaculture 2010*, UN Food and Agriculture Organization, Rome.

8   Calculated from FAOSTAT online figures for global grain harvest (2009) and food value of cereals. Based on a calorific intake of 2,500 kcalories per person per day.

9   WHO press release, 'World Health Day 2011, Urgent action necessary to safeguard drug treatments', 6 April 2011, http://www.who.int/media-centre/news/releases/2011/whd_20110406/en/index.html.

10 Food Safety Authority Ireland, press release, 'FSAI Survey Finds Horse DNA in Some Beef Burger Products', 15 January 2013, http://www.fsai.ie/news_centre/press_releases/horseDNA15012013.html (accessed 20 June 2013); http://www.guardian.co.uk/world/2013/feb/08/how-horse-meat-scandal-unfolded-timeline.

11 European Commission press release, 'Commission publishes European test results on horse DNA and Phenylbutazone: no food safety issues but tougher penalties to apply in the future to fraudulent labelling', 16 April 2013, http://europa.eu/rapid/press-release_IP-13-331_en.htm (accessed 20 June 2013).

12 BBC, 'Horsemeat in Tesco burgers prompts apology in UK papers', 17 January 2013, http://www.bbc.co.uk/news/uk-21054688 (accessed 20 June 2013).

13 Simon Neville, 'Frozen beefburger sales down 43% since start of horsemeat scandal', 26 February 2013, http://www.guardian.co.uk/uk/2013/feb/26/frozen-beefburger-sales-down-43-horsemeat (accessed 20 June 2013).

14 BBC, 'Horsemeat in Tesco burgers prompts apology in UK papers', 17 January 2013, http://www.bbc.co.uk/news/uk-21054688 (accessed 20 June 2013).

I RUDE AWAKENINGS

1   *Daily Telegraph*, 14 June 2012, Where do milk, eggs and bacon come from? One in three youths don't know, http://www.telegraph.co.uk/foodanddrink/foodanddrinknews/9330894/Where-do-milk-eggs-and-bacon-come-from-One-in-three-youths-dont-know.html (accessed 13 September 2013).

1 CALIFORNIA GIRLS: A VISION OF THE FUTURE?

1   http://www.esri.com/mapmuseum/mapbook_gallery/volume23/agriculture1.html (accessed 13 July 2012).

2 http://www.nass.usda.gov/Statistics_by_State/California/Publications/ California_Ag_Statistics/2010cas-ovw.pdf (accessed 13 July 2012).

3 Calculated from formula; 200 dairy cows produce as much manure as a town of 10,000 people: *Animal waste pollution in America: an emerging national problem, 1997. Environmental risks of livestock and poultry production*. A report by the Minority Staff of the US Senate Committee on Agriculture, Nutrition and Forestry for Senator Tom Harkin.

4 http://www.farmland.org/programs/states/futureisnow/default.asp (accessed 13 July 2012).

5 http://www.sraproject.org/wp-content/uploads/2007/12/dairytalking-points.pdf.

2 HENPECKED: THE TRUTH BEHIND THE LABEL

1 *Guardian*, 'Why are we all keeping hens', 1 August 2011, http://www.guardian.co.uk/lifeandstyle/2011/aug/01/keeping-hens (accessed 2 October 2012); 'Your chickens', article, http://www.yourchickens.co.uk/home/ advertise (accessed 2 October 2012); *Daily Mail*, 'How to . . . keep hens and harvest your own eggs in the comfort of your garden', 15 February 2010, http://www.dailymail.co.uk/femail/article-1251042/How--hens.html (accessed 2 October 2012).

2 Defra, *Outbreak of Highly Pathogenic h5n1 Avian Influenza in Suffolk in January 2007: a Report of the Epidemiological Findings by the National Emergency Epidemiology Group*, 5 April 2007, http://archive.defra.gov. uk/foodfarm/farmanimal/diseases/atoz/ai/documents/epid_findings070405.pdf; Cabinet Office, UK Resilience, 2007, http:// webarchive.nationalarchives.gov.uk/+/http://www.cabinetoffice.gov.uk/ ukresilience/response/recovery_guidance/case_studies/grey1_bmatthews.aspx (accessed 2 October 2012); BBC *News*, 'Bird flu virus is Asian strain', 3 February 2007, http://news.bbc.co.uk/1/hi/uk/6328161. stm (accessed 2 October 2012).

II NATURE

1 http://www.macla.co.uk/nocton/index.php; http://www.allsaintsnocton .org.uk/history.htm; http://www.nocton.org/#today; http://en.wikipedia .org/wiki/Nocton.

2 http://www.youtube.com/watch?v=nnWb9WJ8anU&feature=related; http://www.youtube.com/watch?v=0ht1741iqlM&feature=related.

3 Butterfly Conservation website: http://www.butterfly-conservation.org/

Butterfly/32/Butterfly.html?ButterflyId=15 (accessed 2 October 2012); Bumblebee Conservation Trust leaflet, Farms, crofts and bumbles, http://www.snh.gov.uk/docs/A463311.pdf (accessed 2 October 2012).

3 SILENT SPRING: THE BIRTH OF FARMING'S CHEMICAL AGE

1 Rachel Carson, *Silent Spring*, Penguin, London, 1962 (2000 reprint); Conor Mark Jameson, *Silent Spring Revisited*, Bloomsbury, London 2012.
2 W. J. L. Sladen et al., 'DDT residues in Adélie penguins and a crabeater seal from Antarctica', *Nature*, 210, 14 May 1966, pp. 670–3, http://www.nature.com/nature/journal/v210/n5037/abs/210670a0.html (accessed 10 May 2013).
3 *Baltimore Sun*, 'Geese's movie careers take flight: Scientist Dr. William J. L. Sladen is director of environmental studies at the Airlie Sanctuary in Virginia, home to several of the geese that star in the movie "Fly Away Home"', 6 September 1996, http://articles.baltimoresun.com/1996-09-06/news/1996250099_1_sladen-igor-swans (accessed 10 May 2013).
4 National Wildlife Federation, 'Chesapeake Bay', http://www.nwf.org/wildlife/wild-places/chesapeake-bay.aspx (accessed 7 August 2012).
5 http://www.chesapeakebay.net/issues/issue/agriculture#inline (accessed 7 August 2012).
6 US EPA, Chesapeake Bay Program, 'Health of Freshwater Streams in the Chesapeake Bay Watershed', www.chesapeakebay.net/status_stream-health.aspx?menuitem=50423.
7 Pew Environment Group, *Big Chicken: Pollution and industrial poultry production in America*, Pew, Washington, July 2011.

4 WILDLIFE: THE GREAT DISAPPEARING ACT

1 J. R. Krebs et al., 'The second Silent Spring?', *Nature*, 400, 12 August 1999, pp. 611–12.
2 British Trust for Ornithology (BTO), 'Breeding birds in the wider countryside 2010. Trends in numbers and breeding performance for UK birds', accessed July 2011, http://www.bto.org/about-birds/bird-trends; British Trust for Ornithology (BTO), 'Breeding birds in the wider countryside 2010, Trends in numbers and breeding performance of UK birds. Section 4.2, Latest long term alerts', http://www.bto.org/birdtrends2010/discussion42.shtml.

3 Defra, 'Wild bird populations: farmland birds in England 2009', news release 29 July 2010.

4 BirdLife International, 'Europe-wide monitoring schemes highlight declines in widespread farmland birds', 2008, presented as part of the BirdLife State of the world's birds website, available from http://www.birdlife.org/datazone/sowb/casestudy/62.

5 North American Bird Conservation Initiative, US Committee, *The State of the Birds 2011: report on public lands and waters, United States of America*, 2011.

6 North American Bird Conservation Initiative, US Committee *The State of the Birds 2009, United States of America*, 2009.

7 Defra statistical release, 'Wild bird populations in the UK [to 2009]', 20 January 2011, http://archive.defra.gov.uk/evidence/statistics/environment/wildlife/download/pdf/110120-stats-wild-bird-populations-uk.pdf.

8 Krebs et al., 'The second Silent Spring?'.

9 R. Watson, S. Albon et al., *UK National Ecosystem Assessment: Synthesis of the Key Findings*, 2011, National Ecosystem Assessment project, Defra, NERC, together with agencies in Scotland, Wales and Northern Ireland.

10 New World Encyclopedia, http://www.newworldencyclopedia.org/entry/earthworm (accessed 24 July 2012); Wikipedia, http://en.wikipedia.org/wiki/Earthworm (accessed 24 July 2012).

11 S. Kragten et al., 'Abundance of invertebrate prey for birds on organic and conventional arable farms in the Netherlands', *Bird Conservation International* (2011) 21, pp. 1–11.

12 BirdLife International, 'Grassland birds are declining in North America', 2004, presented as part of the BirdLife State of the world's birds website, www.birdlife.org/datazone/sowb/casestudy/63.

13 See F. Dikötter, *Mao's Great Famine*, Bloomsbury, London, 2010.

14 Bumblebee Conservation Trust, www.bumblebeeconservation.org.uk (accessed July 2011).

15 C. Carvell et al., 'Comparing the efficacy of agri-environment schemes to enhance bumble bee abundance and diversity on arable field margins', *Journal of Applied Ecology* (2007), 44, pp. 29–40.

16 'Colony Collapse Disorder and the Human Bee', 12 August 2008, http://www.articlesbase.com/environment-articles/colony-collapse-disorder-and-the-human-bee-519377.html (accessed 21 May 2013).

17 Ibid.

18 *Daily Mail*, 'Rescuers battle 17 million angry bees after flatbed trailer crashes in fatal U.S. accident', 25 May 2010, http://www.dailymail.

    co.uk/news/article-1281226/Truck-carrying-17million-bees-crashes-Minnesota.html.

19 *Los Angeles Times*, 'Hives for hire', 3 March 2012, http://articles.latimes.com/2012/mar/03/business/la-fi-california-bees-20120304 (accessed 21 May 2013); *Slate* magazine, 'Rent-a-hive', 27 June 2008, http://www.slate.com/articles/news_and_politics/explainer/2008/06/rentahive.html (accessed 21 May 2013).

20 A. Benjamin and B. McCullum, *A World Without Bees*, Guardian Books, London, 2008.

21 Ibid.

22 *Los Angeles Times*, 'Hives for hire', 3 March 2012.

23 A. Manriquez, Apinews, 'China – hand pollination', 17 December 2010, http://www.apinews.com/en/news/item/12780-china-hand-pollination/ (accessed 19 July 2012).

24 *Los Angeles Times*, 'Pesticides suspected in mass die-off of bees', 29 March 2012, http://articles.latimes.com/2012/mar/29/science/la-sci-bees-pesticides-20120330 (accessed 21 May 2013); *Natural News*, 'Confirmed: Common pesticide crashing honeybee populations around the world', 10 April 2012, http://www.naturalnews.com/035518_honey_bees_pesticides_science.html (accessed 21 May 2013).

25 BBC, 'Bee deaths: EU to ban neonicotinoid pesticides', 29 April 2013, http://www.bbc.co.uk/news/world-europe-22335520 (accessed 21 May 2013).

26 Ibid.

27 D. Goulson, University of California website, 'David Goulson: Ecology and Conservation of Bumble Bees', 17 April 2013, http://entomology.ucdavis.edu/News/David_Goulson____Ecology_and_Conservation_of_Bumble_Bees/ (accessed 21 May 2013).

28 Achim Steiner, quoted in 'Bees under bombardment: Report shows multiple factors behind pollinator losses', United Nations Environment Programme, Geneva/Nairobi, 10 March 2011, http://www.unep.org.

29 J. Simpson, 'Chasing butterflies: The Victorian hobby of entomology', 2010, http://suite101.com/article/chasing-butterflies-the-victorian-hobby-of-entomology-a222953 (accessed 27 July 2012).

30 http://www.monarchwatch.org/ (accessed 27 July 2012).

31 http://nationalzoo.si.edu/Animals/Invertebrates/News/monarchmigration.cfm; http://www.sciencelatest.com/2011/12/the astounding monarch butterfly voyage/.

32 *The McGill Tribune*, 2010, 'Follow the butterflies: A monarch migration under threat' (updated 21 September 2010), http://www.mcgilltribune.

com/2.12327/follow-the-butterflies-a-monarch-migration-under-threat-1.1626302 (accessed 27 July 2012).

33 http://www.unityserve.org/butterfly/urquharts.html; http://www.science-latest.com/2011/12/the astounding monarch butterfly voyage/; http://www.ecology.info/monarch butterfly page 3.htm; http://www.monarch butterfly.com/monarch butterflies facts.html; http://www.worldwildlife.org/species/finder/monarchbutterflies/monarchbutterflies.html#; http://www.monarchwatch.org/news/urquhart.htm.

34 Butterfly Conservation, Natural England and FWAG, *Butterflies and Farmland*, http://www.butterfly-conservation.org/uploads/bc0011%20Butterflies%20and%20Farmland(1).pdf, no date, bc0011.

35 UK Butterfly Monitoring Scheme (2010), UK Summary of changes table 2010, http://www.ukbms.org/docs/reports/2010/Summary.

36 http://news.bbc.co.uk/1/hi/sci/tech/3568321.stm (accessed 27 July 2012).

37 http://www.butterfly-conservation.org/article/9/103/large_blue_butter-flies_back_in_britain.html (accessed 27 July 2012).

5 FISH: FARMING TAKES TO THE WATER

1 'Mauritius Aquaculture Masterplan goes green', The FishSite.com, 17 April 2009, http://www.thefishsite.com/fishnews/9633/mauritius-aqua-culture-masterplan-goes-green (accessed 2 March 2012); Department of Environment: Ministry of Environment and national development unit, 2009. EIA guidelines for fish farming in the sea, Mauritius, http://environment.gov.mu.

2 IPS Inter Press News Service Agency, 'Our sea and lagoon are not for sale', 1 August 2007, http://ipsnews.net/africa/nota.asp?idnews=38753 (accessed 2 March 2012).

3 'Fish farms: Emerging threats coming ashore', The FishSite.com., 28 June 2007, http://www.thefishsite.com/fishnews/4615/fish-farms-emerging-threats-coming-ashore (accessed 2 March 2012).

4 P. Coppens, 'The sacred island of the moon', http://www.philipcoppens.com/lochmaree.html (accessed 1 August 2012); http://www.ovguide.com/isle-maree-9202a8c04000641f8000000006ecboao# (accessed 1 August 2012); http://www.ancientsites.com/aw/Post/1264229&authorid=238 (accessed 1 August 2012).

5 S. Millar, 'Last leap for the wild salmon', *Observer*, 11 June 2000, available from http://www.guardian.co.uk/uk/2000/jun/11/stuartmillar.theobserver (accessed 2 August 2012); Wester Ross Fisheries Trust, article, 'Sea trout:

River Ewe and Loch Maree', http://www.wrft.org.uk/fisheries/seatrout.cfm (accessed 2 August 2012).

6 H. Davis, L. Lamb and S. Frost, 'Fishing in the Gairloch area', http:// www.gairloch-fishing.co.uk/maree.htm.

7 Wester Ross Fisheries Trust, article, 'Sea trout: River Ewe and Loch Maree', http://www.wrft.org.uk/fisheries/seatrout.cfm (accessed 2 August 2012).

8 Wester Ross Fisheries Trust Review (2011), available from www.wrft. org.uk/files/WRFT.

9 J. Owen, 'Sea trout loss linked to salmon farm parasite', *National Geographic* News, 22 October 2002, http://news.nationalgeographic. com/news/2002/10/1022_021022_seatroutfish.html (accessed 2 August 2012).

10 A. Mood, *Worse things happen at sea: the welfare of wild-caught fish,* 2010, www.fishcount.org.uk.

11 FAOSTAT online database, http://www.faostat.fao.org.

12 Lenfest Ocean Program (2008), 'Research series: global assessment of aquaculture impacts on wild salmon', Lenfest, www.lenfestocean.org/ publication/global-assessment-aquaculture-impacts-wild-salmon.

13 A. G. J. Tacon and M. Metian, 'Global overview on the use of fish meal and fish oil in industrially compounded aquafeeds: Trends and future prospects', *Aquaculture*, 285 (2008), pp. 146–58.

14 R. L. Naylor et al., 'Feeding aquaculture in an era of finite resources', *PNAS*, 106(36) (2009), pp.15103–10; Tacon and Metian, 'Global overview'.

15 FAO, *State of the World Fisheries and Aquaculture*, 2010.

16 Ibid.; Henk Westhoek et al., *The protein puzzle: the consumption and production of meat, dairy and fish in the European Union*, PBL Netherlands Environmental Assessment Agency, The Hague, 2011.

17 Naylor et al., 'Feeding aquaculture'.

18 Seafish, Annual Review of the status of the feed grade fish stocks used to produce fishmeal and fish oil for the UK market, March 2012, http:// www.seafish.org/media/publications/SeafishAnnualReviewFeed FishStocks_201203.pdf.

19 Naylor et al., 'Feeding aquaculture'.

20 Lenfest Ocean Program (2009), *Summary: Important protein sources for the world's impoverished in competition with aquaculture and animal feed.* Lenfest, September 2009 www. lenfestocean.org/publication/important-protein-source-world%E2%80%99s-impoverished-competition-aquaculture-and-animal-feed.

21 FAO, *State of the World Fisheries.*

NOTES TO PAGES 84–90

22 B. A. Costa-Pierce et al., 'Responsible use of resources for sustainable aquaculture', *Global Conference on Aquaculture 2010*, 22–25 September 2010, Phuket, Thailand, Food and Agriculture Organization of the United Nations (FAO), Rome, Italy, 2011.

23 P. Stevenson, *Closed Waters: The welfare of farmed Atlantic Salmon, Rainbow Trout, Atlantic Cod and Atlantic Halibut*, WSPA/CIWF, London, 2007; P. Lymbery, *In Too Deep: The welfare of intensively farmed fish*, CIWF Trust, Petersfield, 2002.

24 FAO, *State of the World Fisheries*.

25 Atlantic Salmon Trust, 'Salmon farming in Scotland: economic success or ecological failure?', http://www.atlanticsalmontrust.org/concerns/salmon-farming-in-scotland-economic-success-or-ecological-failure.html (accessed 2 August 2012).

26 M. J. Costello, 'How sea lice from salmon farms may cause wild salmonid declines in Europe and North America and be a threat to fishes elsewhere', *Proc. R. Soc. B*, 276 (2009), pp. 3385–94; M. Krkošek et al., 'Epizootics of wild fish induced by farm fish', *PNAS*, 103(42) (2008), pp. 15506–10; M. Krkošek et al., 'Sea lice and salmon population dynamics: effects of exposure time for migratory fish', *Proc. R. Soc. B*, 276 (2009), pp. 2819–28.

27 SEPA, *Regulation and monitoring of marine cage fish farming in Scotland*, Annex H, Scottish Environment Protection Agency, May 2005.

28 The FishSite, 'Sea Lice: a Parasite of Fish and Farmers Alike', 6 February 2009, http://www.thefishsite.com/articles/616/sea-lice-a-parasite-of-fish-and-farmers-alike (accessed 2 August 2012).

29 Krkošek et al., 'Sea lice and salmon population dynamics'.

30 The FishSite, 'Sea Lice'.

31 Atlantic Salmon Trust, 'Salmon farming in Scotland: economic success or ecological failure?', http://www.atlanticsalmontrust.org/concerns/salmon-farming-in-scotland-economic-success-or-ecological-failure.html (accessed 2 August 2012).

32 D. Mackay, 'Perspectives on the Environmental Effects of Aquaculture', presented to the Aquaculture Europe Conference, Norway, August 1999, Scottish Environment Protection Agency.

33 M. Krkošek et al., 'Declining Wild Salmon Populations in Relation to Parasites from Farm Salmon', *Science*, vol. 318 no. 5857 (2007), pp. 1772–5.

34 BC Pacific Salmon Forum, *Final report and recommendations to the Government of British Columbia*, January 2009.

35 B. Harvey, *Sea lice and salmon farms: a second look*, prepared for the BC Pacific Salmon Forum, 2009.

36 A. A. Rosenberg, 'The price of lice', *Nature*, 451 (2008), pp. 23–4.

37 The FishSite, 'Salmon Escape ill-timed as Data Published on Global Incidents', 4 December 2007, http://www.thefishsite.com/fishnews/5834/salmon-escape-illtimed-as-data-published-on-global-incidents (accessed 2 August 2012).

38 FRS Marine Laboratory, *Scottish Fish Farms Annual Production Survey 2000*, Fisheries Research Services Marine Laboratory, Aberdeen, 2001.

39 Orr, 'The Way to Save the Salmon', *Independent*, 30 July 1999.

40 Mackay, 'Perspectives on the Environmental Effects of Aquaculture'.

41 *Orcadian*, 28 September 2000, cited in P. Lymbery, *In Too Deep: The welfare of intensively farmed fish*, CIWF Trust, Petersfield, 2002.

42 SSPO, 'Confirmed breaches of containment, 1st January – 1st Dec 2010', www.scottishsalmon.co.uk/userFiles/769/SSPO_breaches_of_containment_2010.pdf.

43 USDA Nutrient Database, 2011, USDA Agricultural Research Service, National Agricultural Library, Nutrient Database for Standard Reference, Release 24, Nutrient Data Laboratory Home Page: http://ndb.nal.usda.gov/ndb/foods/list (accessed 2 October 2012).

44 R. A. Hites et al., 'Global Assessment of Organic Contaminants in Farmed Salmon', *Science*, vol. 303 no. 5655 (2004), pp. 226–9, http://www.sciencemag.org/content/303/5655/226.short.

45 L. Madsen, J. Arnbjerg and I. Dalsgaard, 'Spinal deformities in triploid all-female rainbow trout (Oncorhynchus mykiss)', *Bull. Eur. Ass. Fish Pathol.*, 20 (5) (2000), pp. 206–8; R. Johnstone, *Production and Performance of Triploid Atlantic Salmon in Scotland*, Marine Laboratory, The Scottish Office Agriculture and Fisheries Department, 1992; S. Willoughby, *Manual of Salmonid Farming*, Fishing News Books, Blackwell Science, Oxford, 1999; A. E. Wall and R. H. Richards, 'Occurrence of cataracts in triploid Atlantic salmon (Salmo salar) on four farms in Scotland', *Veterinary Record*, 131 (1992), pp. 553–7.

46 BBC, 'Headless seals may have been shot', 20 May 2008, http://news.bbc.co.uk/1/hi/scotland/highlands_and_islands/7410701.stm (accessed 2 August 2012).

47 iWatch News, '"Free for all" decimates fish stocks in the South Pacific', http://www.iwatchnews.org/2012/01/25/7900/free-all-decimates-fish-stocks-southern-pacific (accessed 3 August 2012).

48 Institut de Recherche pour le développement, 'Scientists working for responsible fishing in Peru', scientific bulletin no. 349 – May 2010, http://www.en.ird.fr/content/download/17178/146692/.../4/.../

FAS349a-web.pdf; C. E. Paredes, 'Reforming the Peruvian anchoveta sector', Instituto del Peru, July 2010, http://www.ebcd.org/pdf/presentation/164-Paredes.pdf; Y. Evans and S. Tveteras, *Status of Fisheries and Aquaculture Development in Peru: Case Studies of Peruvian Anchovy Fishery, Shrimp Aquaculture, Trout and Scallop Aquaculture,* FAO, Rome, 2011, available from www.fao.org/.

49 Seafish, 'Seafish publishes comprehensive review of feed fish stocks used to produce fishmeal and fish oil for the UK market', 16 April 2012, http://www.seafish.org/about-seafish/news/seafish-publishes-comprehensive-review-of-feed-fish-stocks-used-to-produce-fishmeal-and-fish-oil-for-the-uk-market (accessed 3 August 2012).

50 Seafish, *Fishmeal and Fish Oil Facts and Figures,* 2011, www.seafish.org.

51 http://www.mapsofworld.com/peru/provinces-and-cities/chimbote.html; iWatch News, '"Free for all" decimates fish stocks in the South Pacific'; *New York Times*, 'In Mackerel's Plunder, Hints of Epic Fish Collapse', 25 January 2012, http://www.nytimes.com/2012/01/25/science/earth/in-mackerels-plunder-hints-of-epic-fish-collapse.html?pagewanted=all (accessed 3 August 2012).

52 J. Del Hoyo, A. Elliott and J. Sargatal (eds), *Handbook of the Birds of the World*, vol. 1, Lynx Edicions, Barcelona, 1992.

53 W. M. Mathew, 'Peru and the British guano market 1840–1870', The Economic History Review New Series, vol. 23, no. 1 (April 1970), pp. 112–28, Blackwell Publishing; http://is.gd/F1HbXg (accessed 3 August 2012).

54 I. Newton, *Population Limitation in Birds*, Academic Press, London, 1998.

55 *The Economist*, 'Fishing in Peru: The next anchovy coming to a pizza near you', 5 May 2011, http://www.economist.com/node/18651372 (accessed 3 August 2012).

56 Evans and Tveteras, *Status of Fisheries and Aquaculture Development in Peru*.

57 J. Jacquet et al., 'Conserving wild fish in a sea of market-based efforts', *Oryx*, 44(1) (2009), pp. 45–6.

58 The Goldman Environmental Prize, http://www.goldmanprize.org/2003/southcentralamerica (accessed 3 August 2012).

6 ANIMAL CARE: WHAT HAPPENED TO THE VET?

1 Defra website: http://www.defra.gov.uk/food-farm/animals/welfare/slaughter/premises/ (accessed 13 August 2012).

2 Food Standards Agency website: http://www.food.gov.uk/enforcement/
monitoring/mhservice/workwithindustry/workforce (accessed 13 August
2012).

3 Animal Aid, 'The "humane" slaughter myth', http://www.animalaid.org.
uk/h/n/CAMPAIGNS/slaughter/ (accessed 13 August 2013).

4 Compassion in World Farming, 'Suffering at slaughter exposed by
new film', http://www.ciwf.org.uk/news/factory_farming/suffering_
exposed_by_film.aspx (accessed 13 August 2013).

5 British Veterinary Association, personal communication, 2012.

6 J. Mackey, *Conscious capitalism: Creating a new paradigm for business*,
2007, http://www.wholeplanetfoundation.org/files/uploaded/John_
Mackey-Conscious_Capitalism.pdf (accessed 13 August 2013).

7 http://thinkexist.com/quotation/i-hope-to-make-people-realize-how-
totally/380118.html (accessed 13 August 2012).

III HEALTH

1 http://en.wikipedia.org/wiki/Guy's_Hospital; http://www.bbc.co.uk/his-
tory/historic_figures/bevan_aneurin.shtml.

2 http://www.nhs.uk/NHSEngland/thenhs/about/Pages/overview.aspx.

7 BUGS 'N' DRUGS: THE THREAT TO PUBLIC HEALTH

1 Hansard, 13 May 1953, 1327–43.

2 Chief Medical Officer's Annual Report, 2008, chapter: 'Antimicrobial
resistance: up against the ropes'.

3 Dr Margaret Chan, Director-General World Health Organization
(WHO), speaking on World Health Day, 7 April 2011.

4 *Case Study of a Health Crisis: How Human Health Is under Threat from
Over-use of Antibiotics in Intensive Livestock Farming*, a report for the
Alliance to Save Our Antibiotics, Godalming, 2011.

5 The Joint Committee on the use of Antibiotics in Animal Husbandry
and Veterinary Medicine, chaired by Professor M. M. Swann, was
appointed jointly by Health and Agriculture Ministers in July 1968. Its
report was issued in November 1969.

6 J. Harvey and L. Mason, *The Use and Misuse of Antibiotics in UK Agriculture*, Part 1, *Current Usage*, Soil Association, Bristol, 1968.

7 *Case Study of a Health Crisis*.

8 C. Nathan, 'Antibiotics at the crossroads', *Nature*, 431 (2004), pp. 899–
902; World Health Organization, 2011, 'World Health Day 2011.

Urgent action necessary to safeguard drug treatments', http://www.who. int/mediacentre/news/releases/2011whd_20110406/en/.

9 M. Mellon, C. Benbrook and K. L. Benbrook, 'Hogging It. Estimates of Antimicrobial Abuse in Livestock', Union of Concerned Scientists, 2001; K. M. Shea, 'Antibiotic resistance: what is the impact of agricultural uses of antibiotics on children's health?', *Pediatrics*, 112(1) (2003), pp. 253–8.

10 Committee for Medicinal Products for Veterinary Use, *Reflection Paper on the Use of Fluoroquinolones in Food-producing Animals in the European Union: Development of Resistance and Impact on Human and Animal Health*, 2006, www.emea.europa.eu/pdfs/vet/srwp/18465105en.pdf.

11 Naheeda Portocarero, 'Continued focus on food security and welfare', *World Poultry*, vol. 27, no. 5, online version 23 August 2011; X. Manteca, 'Physiology and disease', in M. C. Appleby et al. (eds), *Long Distance Transport and Welfare of Farm Animals*, CABI, 2008, pp. 69–76.

12 M. McKenna, 'Turning grief into action: Moms and antibiotic misuse', *Wired.com.*, 2011, available from http://www.wired.com/wiredscience/ 2011/05/grief-moms-antibiotics/#more-59108 (accessed 2 August 2011).

13 V. Jones, 'Deadly Bacteria (MRSA) Kill A Baby Boy, Part 1', *Revolution Health*, 2008, available from http://www.revolutionhealth.com/blogs/ valjonesmd/deadly-bacteria-mrsa-15730 (accessed 2 August 2011).

14 Ibid.

15 *Huffington Post*, http://www.huffingtonpost.com/everly-macario-scd-ms-edm (accessed 21 May 2013).

16 IDSA (Infectious Diseases Society of America), March 2006, article: http://www.idsociety.org/Simon_Macario/ (accessed 21 May 2013).

17 http://www.wired.com/wiredscience/2011/05/grief-moms-antibiotics/ #more-59108.

18 European Commission, *Staff working paper of the services of the Commission on antimicrobial resistance*, SANCO/6876/2009r6, 18 November 2009.

19 C. Nunan and R. Young, *MRSA in farm animals and meat: a new threat to human health*, Soil Association, 2007.

20 Ibid.

21 E. de Boer et al., 'Prevalence of methicillin-resistant Staphylococcus aureus in meat', *Int. J Food Microbiology*, 134(1–2) (2009), pp. 52–6.

22 EFSA Panel on Biological Hazards, 'Foodborne antimicrobial resistance as a biological hazard. Scientific Opinion', *EFSA Journal*, 765 (2008), pp. 1–87.

23 L. Garcia-Alvarez et al., 'Meticillin-resistant *Staphylococcus aureus* with a novel *mecA* homologue emerging in human and bovine populations in

the UK and Denmark: a descriptive study', *Lancet Infectious Diseases*, 2011.

24 Defra, *Zoonoses Report: UK 2010*, 2011.

25 EFSA-ECDC, 'The European Union Summary Report on Trends and Sources of Zoonoses, Zoonotic Agents and Food-borne Outbreaks in 2009', *EFSA Journal*, 2011, 9(3), 2090.

26 Defra, *Zoonoses Report: UK 2010*.

27 Chief Medical Officer's Annual Report, 2008, 'Antibiotic resistance'.

28 E. Scallan et al., 'Foodborne illness acquired in the United States—major pathogens', *Emerging Infectious Diseases,* 2011 (Epub ahead of print), DOI: 10.3201/eid1701.P11101.

29 *MeatPoultry* Staff, 'Salmonella Heidelberg infections rise to 107', 12 August 2011, http://www.meatpoultry.com/News.

30 B. Salvage, 'Salmonella Heidelberg infections rise to 119: CDC', Meat-Poultry.com, 15 September 2011, http://www.meatpoultry.com/News/.

31 FAWC, *Report on the welfare of laying hens*, 1997, Section: History, Table 1, http://www.fawc.org.uk/reports/layhens/lhgre007.htm.

32 Defra, *Egg Statistical Notice*, 4 August 2011.

33 Defra, *Zoonoses Report: UK 2010*.

34 C. Snow et al., 'Investigation of risk factors for Salmonella on commercial egg-laying farms in Great Britain, 2004–2005', *Veterinary Record*, 166 (2010), pp. 579–86.

35 Pew Commission on Industrial Farm Animal Production, *Putting Meat on the Table: Industrial Farm Animal Production in America*, 2008; http://www.pewtrusts.org/uploadedFiles/wwwpewtrustsorg/Reports/Industrial_Agriculture/PCIFAP_FINAL.pdf; M. Greger, 'The Human/Animal Interface: Emergence and Resurgence of Zoonotic Infectious Diseases', *Critical Reviews in Microbiology*, 33 (2007) pp. 243–99, DOI: 10.1080/10408410701647594.

36 FAOSTAT, http://fasotat.fao.org.

37 WHO (2010) Cumulative Number of Confirmed Human Cases of Avian Influenza A/(H5N1), reported to WHO, 9 August 2011, www.who.int/csr/disease/avian_influenza/country/cases_table_2011_08_09/en/index.html.

38 M. Du Ry van Beest Holle, 'Human-to-human transmission of avian influenza A/H7N7, The Netherlands, 2003', *Eurosurveillance* 10(12), 1 December 2005, pp. 264–8; http://www.eurosurveillance.org/em/v10n12/1012-222.asp.

39 D. MacKenzie, 'Five easy mutations to make bird flu a lethal pandemic', *New Scientist*, 24 September 2011, p. 14 (online article 21 September).

40 Editorial, 'The risk of an influenza pandemic is fact, not fiction', *New Scientist*, 24 September 2011, p. 3.

41 C. J. L. Murray et al., 'Estimation of potential global pandemic influenza mortality on the basis of vital registry data from the 1918–20 pandemic: a quantitative analysis', *Lancet*, 368 (2006), pp. 2211–18.

42 Oxford Centre for Animal Ethics, news release, 'Bird Flu Will Remain A Threat As Long As Factory Farms Exist', 17 February 2012, http://www.oxfordanimalethics.com/2012/02/news-release-bird-flu-will-remain-a-threat-as-long-as-factory-farms-exist/.

43 GCM website, http://www.granjascarroll.com/ing/ing_historia.php (accessed December 2011).

44 GCM website, http://www.granjascarroll.com/ing/ing_preguntas.php (accessed 27 July 2012).

45 GCM website, http://www.granjascarroll.com/ing/ing_preguntas.php (accessed 3 October 2012).

46 *Guardian*, 'La Gloria, swine flu's ground zero, is left with legacy of anger', 23 April 2010, http://www.guardian.co.uk/world/2010/apr/23/swine-flu-legacy-la-gloria (accessed 27 July 2012).

47 *Washington Post*, 'Mexicans blame industrial hog farms', 10 May 2009, http://www.washingtonpost.com/wp-dyn/content/article/2009/05/09/ (accessed 27 July 2012).

48 S. M. Burns, 'H1N1 Influenza is here', *Journal of Hospital Infection*, 17 July 2009, http://download.thelancet.com/flatcontentassets/H1N1-flu/epidemiology/epidemiology-76.pdf (accessed 27 July 2012).

49 World Health Organization, South East Asia Regional Office (SEARO), 'Message from the Regional Director', November 2009, http://www.searo.who.int/linkfiles/news_letters_nov2010.pdf (accessed 30 May 2012).

50 World Health Organization, 2010, http://www.who.int/csr/don/2010_05_14/en/index.html (accessed 30 May 2012).

51 *Guardian*, 'La Gloria, swine flu's ground zero, is left with legacy of anger'; Encyclopedia Britannica, 'Influenza pandemic (H1N1) of 2009', http://www.britannica.com/EBchecked/topic/1574480/influenza-pandemic-H1N1-of-2009#toc281756 (accessed 27 July 2012).

52 Ibid.

53 GCM website, http://www.granjascarroll.com/ing/ing_preguntas.php (accessed 27 July 2012).

54 Ibid.

55 Ibid.

8 EXPANDING WAISTLINES: FOOD QUALITY TAKES A NOSE-DIVE

1 *Daily Mail*, '58st and a £500,000 bill: But I deserve NHS support says world's fattest man', 14 February 2012, http://www.dailymail.co.uk/news/article-2100052/Worlds-fattest-man-Keith-Martin-lives-London-58-stone.html (accessed 5 October 2012).

2 News-Medical, 'World's "fattest man" needs an army of carers', 15 February 2012, http://www.news-medical.net/news/20120215/Worlde28099s-e2809cfattest-mane2809d-needs-an-army-of-carers.aspx (accessed 15 August 2012); *Daily Mail*, '"I ate 20,000 calories a day, had a 6ft waist ... and last left my house on 9/11": Horrifying life of the blonde-haired little boy who grew into the world's fattest man', 6 May 2012; http://www.dailymail.co.uk/news/article-2140307/Keith-Martin-Horrifying-life-worlds-fattest-man.html#ixzz23b7wAYay; (accessed 15 August 2012); *Daily Mail*, 'Britain's fattest man who weighs 58 stone when he was a regular-sized 15st', 9 April 2012, http://www.dailymail.co.uk/news/article-2126848/Keith-Martin-Britains-fattest-man-weighs-58stone-regular-sized-15st.html#ixzz23b8aJom8; (accessed 15 August 2012); *Daily Telegraph*, 'Fire crews demolish walls to release Britain's fattest teen from house after she posted plight on Facebook', 25 May 2012, http://www.telegraph.co.uk/health/healthnews/9288612/ Fire-crews-demolish-walls-to-release-Britains-fattest-teen-from-house-after-she-posted-plight-on-Facebook.html (accessed 15 August 2012).

3 W. H. Dietz, 'Reversing the tide of obesity', *Lancet*, 378 (2011), pp. 744–6.

4 S. Friel et al., 'Public health benefits of strategies to reduce greenhouse-gas emissions: food and agriculture', *Lancet*, 374, 9706 (2009), pp. 2016–25.

5 G. L. Huber, 'Fats and the prevention of coronary heart disease', http://www.livingheartfoundation.org/fatscoronaryprevention.pdf (accessed 17 August 2012); *Lancet*, 'Plasma Lipid and Lipoprotein Pattern in Greenlandic West-Coast Eskimos', June 1971, http://www.thelancet.com/journals/lancet/article/PIIS0140-6736(71)91658-8/abstract (accessed 17 August 2012).

6 A. P. Simopoulos, 'The importance of the omega-6/omega-3 fatty acid ratio in cardiovascular disease and other chronic diseases', *Experimental Biology and Medicine*, 233 (2008), pp. 674–88.

7 C. A. Daley et al., 'A Literature Review of the Value-Added Nutrients found in Grass-fed Beef Products', June 2005, draft manuscript available at All Things Grass Fed: A cooperative project between California State University, Chico College of Agriculture and University of California

Cooperative Extension, http://www.csuchico.edu/grassfedbeef/; A. P. Simopoulos (2000), 'Human requirement for N-3 polyunsaturated fatty acids', *Poultry Science*, 79(7) (2000), pp. 961–70; A. P. Simopoulos, 'The Importance of the Omega-6/Omega-3 Fatty Acid Ratio in Cardiovascular Disease and Other Chronic Diseases'.

8 H. Pickett, 'Nutritional benefits of higher welfare animal products', 2012, http://www.ciwf.org.uk/includes/documents/cm_docs/2012/n/nutritional_benefits_of_higher_welfare_animal_products_report_june2012.pdf.

9 C. A. Daley et al., 'A review of fatty acid profiles and antioxidant content in grass-fed and grain-fed beef', *Nutrition Journal*, 9:10, 2010, http://www.nutritionj.com/content/9/1/10.

10 J. D. Wood et al., 'Fat deposition, fatty acid composition and meat quality: A review', *Meat Science,* 78 (2008), pp. 343–58.

11 H. Pickett, 'Nutritional benefits of higher welfare animal products'.

12 Ibid.

13 Ibid.

14 Danyel Jennen, *Chicken fatness: from QTL to candidate gene*, PhD thesis, Wageningen University, The Netherlands, 2004, with summary in Dutch, ISBN 90-8504-069-8.

15 Jon Ungoed-Thomas, '"Healthy" chicken piles on the fat', *Times,* 3 April 2005, http://www.timesonline.co.uk/.

16 Richard Young, 'Does organic farming offer a solution?', *The Meat Crisis*, ed. Joyce D'Silva and John Webster, Earthscan, 2010, chapter 5.

17 Z. Hunchar, 'A Hard Story to Swallow: McDonald's Forced to Pay Employee for Weight Gain', *Technorati Blogging*, 2010, available from: http://technorati.com/blogging/article/a-hard-story-to-swallow-mcdonalds/ (accessed 2 September 2011); http://www.neatorama.com/2010/10/30/man-sued-mcdonalds-for-making-him-fat-and-won/ (accessed 2 September 2011); Legal Zoom, 'McDonald's Manager in Brazil Wins Obesity Lawsuit', *Legal Zoom News Sources*, 2010, available from http://www.legalzoom.com/news/politics/international/mcdonalds-manager-brazil-wins (accessed 2 September 2011).

18 Veg Lawyer, 'McDonald's made me fat . . . revisited', *Veg Lawyer's Weblog*, 2007, available from http://veglawyer.wordpress.com/2007/11/23/mcdonalds-made-me-fatrevisited-2/ (accessed 2 September 2011); S. Krum, 'It's fat, fat and more fat', *Guardian*, 27 August 2002, available from http://www.guardian.co.uk/world/2002/aug/27/usa.health (accessed 2 September 2011).

19 Veg Lawyer, 'McDonald's made me fat . . . revisited'.

20 J. Cloud, 'A Food Fight Against McDonald's', *TIME*, 2 December

2002, available from http://www.time.com/time/magazine/article/0,9171,1003804,00.html (accessed 2 September 2011).

21 Ibid.

22 S. English, 'Judge pulls plug on teenagers' claim against McDonald's', *Daily Telegraph*, 6 September 2003, available from http://www.telegraph.co.uk/news/1440693/Judge-pulls-plug-on-teenagers-claim-against-McDonalds.html (accessed 2 September 2011).

23 A. Freeman, 'Fast Food: Oppression Through Poor Nutrition', *California Law Review*, 2007, available from http://www.bfair. net/?p=1054 (accessed 4 September 2011).

24 Dietz, 'Reversing the tide of obesity'.

25 Anthony J. McMichael et al., 'Food, livestock production, energy, climate change, and health', *Lancet*, vol. 370, issue 9594 (2007), pp. 1253–63.

26 Reed Business Media, 'Meat-free drive is impacting consumer markets', Euromonitor, news release 30 August 2011, http://www.foodanddrinkeurope.com/Products-Marketing/Meat-free-drive-is-impacting-consumer-markets-Euromonitor/.

27 World Poultry news online, 'Rabobank predicts sharp decline in meat and poultry production', 30 September 2011, http://www.worldpoultry.net/news/rabobank-predicts-sharp-decline-in-meat-and-poultry-production-9428.html.

28 Meat & Poultry staff, 'Flexitarians will increase in 2012: study', M&P news online, 28 December 2011, http://www.meatpoultry.com/News.

29 World Poultry news online, 'Rabobank predicts'.

IV MUCK

1 *Daily Telegraph*, 'Farmers warn Beckett over "EU manure mountains"', 12 March 2002, http://www.telegraph.co.uk/news/uknews/1387466/Farmers-warn-Beckett-over-EU-manure-mountains.html# (accessed 1 October 2012).

2 Dairy Co, 2011, UK cow numbers, 13 December 2011, http://www.dairyco.org.uk/market-information/farming-data/cow-numbers/uk-cow-numbers/ (accessed 3 October 2012).

3 Natural England, website article 'Capital Grant Scheme', http://www.naturalengland.org.uk/ourwork/farming/csf/cgs/default.aspx (accessed 3 October 2012).

9 HAPPY AS A PIG: TALES OF POLLUTION

1 *Independent*, 'Farmers and greens fight the war of the killer seaweed', 15 August 2011, http://www.independent.co.uk/environment/nature/farmers-and-greens-fight-the-war-of-the-killer-seaweed-2337803.html (accessed 6 August 2012); *The Connexion*, 'Brittany beaches after toxic fumes', 1 September 2011, http://www.connexionfrance.com/50-brittany-beaches-closed-after-toxic-fumes-kill-boar-13715-view-article.html (accessed 6 August 2012); *Daily Mail*, 'Holidaymakers warned of deadly seaweed on Brittany's popular beaches', 28 July 2011, http://www.dailymail.co.uk/travel/article-2019700/Brittany-seaweed-warning-Holidaymakers-told-beware-toxic-fumes-rotting-seaweed.html (accessed 6 August 2012); *Daily Telegraph*, 'Toxic seaweed on French coast sparks health fears', 22 July 2011, http://www.telegraph.co.uk/news/worldnews/europe/france/8655329/Toxic-seaweed-on-French-coast-sparks-health-fears.html (accessed 6 August 2012); *Guardian*, 'Brittany beaches hit by toxic algae', 27 July 2011, http://www.guardian.co.uk/environment/2011/jul/27/brittany-beaches-toxic-algae-boars (accessed 6 August 2012); *The Horse*, 'Horse Dies in Decomposing Seaweed; Toxic Gas Blamed', 6 August 2009, http://www.thehorse.com/ViewArticle.aspx?ID=14674 (accessed 6 August 2012); *Guardian*, 'Lethal algae take over beaches in northern France', 10 August 2009, http://www.guardian.co.uk/world/2009/aug/10/france-brittany-coast-seaweed-algae (accessed 6 August 2012); Science Ray, 'Green algae is fatal to men', 11 September 2011, http://scienceray.com/technology/green-algae-is-fatal-to-men/ (accessed 6 August 2012).

2 S. Heliez et al., 'Risk factors of new Aujeszky's disease virus infection in swine herds in Brittany (France)', *Veterinary Research*, 31 (2000), pp. 146–7, http://www.vetres.org/ (accessed 6 August 2012).

3 *BPEX Weekly*, 'French producers will go bust', 3 December 2010, http://www.bpex.org/bpexWeekly/BW031210.aspx (accessed 12 August 2012).

4 WattAgNet, 'French pig producers are determined to thrive in spite of new welfare, environmental regulations', 10 November 2011, http://www.wattagnet.com/French_pig_producers_are_determined_to_thrive_in_spite_of_new_welfare,_environmental_regulations.html (accessed 6 August 2012).

5 *Daily Telegraph*, 'Toxic seaweed on French coast sparks health fears', 22 July 2011, http://www.telegraph.co.uk/news/worldnews/europe/france/

8655329/Toxic-seaweed-on-French-coast-sparks-health-fears.html
(accessed 6 August 2012).

6 *New York Times*, 'Cultivated environment French farmer pushes green
methods', 28 May 1993, http://www.nytimes.com/1993/05/28/busi-
ness/worldbusiness/28iht-farm.html (accessed 3 October 2012); *Central
Brittany Journal*, website, André Pochon, http://www.thecbj.com/andre-
pochon/ (accessed 3 October 2012).

7 Rodale Institute, extracts from Senate Testimony by Rick Dove, 2002,
http://newfarm.rodaleinstitute.org/depts/pig_page/rick_dove/index.
shtml (accessed 6 August 2012); *Waterkeeper* magazine, Summer 2004,
http://www.waterkeeper.org/ht/a/GetDocumentAction/i/9899
(accessed 6 August 2012); *Waterkeeper* magazine, Fall 2005, http://www.
waterkeeper.org/ht/a/GetDocumentAction/i/9903 (accessed 6 August
2012).

8 North Carolina State University website, http://www.ncat.edu/academ-
ics/schools-colleges1/saes/facilities/farm/swineunit.html (accessed 21
May 2013).

9 North Carolina in the Global Economy website, http://www.soc.duke.
edu/NC_GlobalEconomy/hog/overview.shtml (accessed 21 May 2013).

10 Food and Water Watch website, http://www.factoryfarmmap.org/facts/
(accessed 21 May 2013).

11 Centre for research on globalisation, 'Pork's dirty secret', 4 May 2009,
http://globalresearch.ca/index.php?context=va&aid=13479 (accessed 6
August 2012).

12 J. Trotter, 'Hogwashed', *Waterkeeper* magazine, Summer 2004, http://
www.waterkeeper.org/ht/a/GetDocumentAction/i/9899 (accessed 6
August 2012).

13 J. Tietz, 'Boss Hog', *Rolling Stone*, 14 December 2006, http://regional-
workbench.org/USP2/pdf_files/pigs.pdf (accessed 6 August 2012).

14 H. Steinfeld et al., *Livestock's Long Shadow, environmental issues and
options*, FAO, Rome, 2006.

15 North Carolina Waterkeeper and Riverkeeper Alliance website, 'Hog pollution and our rivers', http://www.riverlaw.us/ (accessed 6 August 2012).

16 US Government Accountability Office, *Concentrated Animal Feeding
Operations: EPA Needs More Information and a Clearly Defined Strategy to
Protect Air and Water Quality*, statement of Anu K. Mittal, Director Natural Resources and Environment, 24 September 2008, highlights of
GAO-08-1177T, a testimony before the Subcommittee on Environment
and Hazardous Materials, Committee on Energy and Commerce, House
of Representatives.

17 S. R. Carpenter and E. M. Bennett, 'Reconsideration of the planetary boundary for phosphorus', *Environmental Research Letters*, 14 February 2011.

18 D. Gurian-Sherman, *CAFOs Uncovered: the untold costs of confined animal feeding operations,* Union of Concerned Scientists, 2008.

19 Ibid.

20 European Commission, 'The EU Nitrates Directive', factsheet, January 2010, http://ec.europa.eu/environment/pubs/factsheets.htm.

10 SOUTHERN DISCOMFORT: THE RISE OF THE INDUSTRIAL CHICKEN

1 Georgians for Pastured Poultry, *Out of Sight, Out of Mind: The Impacts of Chicken Meat Factory Farming in the State of Georgia*, GPP, Decatur, 2012.

2 Red Earth Farm website, http://redearthfarm.weebly.com/ (accessed 7 August 2012).

3 2010 FAOSTAT:http://faostat.fao.org; FAO, *Livestock's Long Shadow: Environmental Issues and Options*, Rome, 2006 – proportion industrially reared.

4 Numbers indicate broiler chickens sold and farms with broiler chicken sales, taken from the USDA Census of Agriculture through 2007, Census information, www.agcensus.usda.gov/Publications/2007/Full_Report/Volume_1,_Chapter_1_US/st99_1_001_001.pdf.

5 Pew Environment Group, *Big Chicken: Pollution and industrial poultry production in America*, Pew, Washington, July 2011.

6 USDA, National Agriculture Statistics Service, 2010, 'Broilers: Inventory by State, US', http://www.nass.usda.gov/Charts_and_Maps/Poultry/brlmap.asp (accessed 1 December 2011); United States Department of Agriculture, *2007 Census of Agriculture – Georgia*, http://www.agcensus.usda.gov/Publications/2007/Full_Report/Volume_1,_Chapter_1_State_Level/Georgia/gav1.pdf (accessed 16 November 2011).

7 FAOSTAT data, 2010, http://faostat.fao.org.

8 The New Georgia Encyclopedia, http://www.georgiaencyclopedia.org/nge/Article.jsp?id=h-1811 (accessed 1 December 2011).

9 Ibid.; 'From Supply Push to Demand Pull: Agribusiness Strategies for Today's Consumers', available from http://www.ers.usda.gov/Amber-Waves/November03/Features/supplypushdemandpull.htm (accessed 2 December 2011).

10 The New Georgia Encyclopedia; Cagle's, Inc., History, http://www.

fundinguniverse.com/company-histories/Cagles-Inc-Company-History.
html (accessed 2 December 2011).

11 *Today in Georgia History*, 'Jesse Jewell', http://www.todayingeorgiahis-
tory.org/content/jesse-jewell (accessed 7 August 2012).

12 The New Georgia Encyclopedia.

13 Ibid.

14 Georgians for Pastured Poultry, *Out of Sight, Out of Mind*.

15 D. L. Cunningham, 'Contract Broiler Production: Questions and
Answers', in Science UoGCoAaE, CAES, 2009; Poultry Workshop,
Public Workshops Exploring Competition in Agriculture, 21 May 2010,
Alabama A&M University, Knight Reception Center, Normal, Alabama,
United States Department of Justice, 2010.

16 Ibid.

17 Cunningham, 'Contract Broiler Production'.

18 Knowles T. G., Kestin S. C., Haslam S. M., Brown S. N., Green L. E., et al.
(2008) 'Leg Disorders in Broiler Chickens: Prevalence, Risk Factors and Prevention'. PLoS ONE 3(2): e1545, doi: 10. 1371/journal. pone. 0001545.
http://www.plosone.org/article/info.

19 Pilgrim's Pride, http://www.turnaround.org/cmaextras/PilgrimsPride.
pdf (accessed 16 December 2011).

20 D. L. Cunningham, 'Cash Flow Estimates for Contract Broiler Production in Georgia: A 30-Year Analysis', The University of Georgia College
of Agricultural and Environmental Sciences, 31 January 2011, http://
www.caes.uga.edu/Publications/pubDetail.cfm?pk_id=7052 (accessed 5
December 2011).

21 Georgians for Pastured Poultry, *Out of Sight, Out of Mind*.

22 Interview conducted by Compassion in World Farming of Southern
Poverty Law Center members who have worked with catchers, 2 November 2011.

23 B. Kiepper, 'Poultry Processing: Measuring True Water Use', University
of Georgia Cooperative Extension, 2011.

24 Bureau of Labor Statistics, Table SNR12, 'Highest incidence rates of total
nonfatal occupational illness cases 2010', Bureau of Labor Statistics US
Department of Labor, October 2011.

25 Georgia Education Agricultural Curriculum Office, 'Broilers: An overview of broiler production in Georgia. Powerpoint presentation', 2006,
http://www.powershow.com/view/108ba-YTFiY/Broilers_An_Over-
view_of_Broiler_Production_in_Georgia_flash_ppt_presentation
(accessed 15 December 2011); 'Injury and Injustice – America's Poultry
Industry', United Food and Commercial Workers International Union,

cited in Southern Poverty Law Center, *Injustice on Our Plates: Immigrant Women in the US Food Industry*, 2010, p. 36; G. Guthey, 'The New Factories in the Fields: Georgia Poultry Workers', *Southern Changes*, vol. 19, no. 3–4, 1997, pp. 23–5.

26 Human Rights Watch, *Blood, Sweat, and Fear: Workers' Rights in US Meat and Poultry Plants*, New York, NY, 2004.

27 Wage and Hour Division, US Department of Labor, Poultry Processing Compliance Survey Fact Sheet, US Department of Labor, 2001.

28 T. Ashdown, 'Poultry Processing', in J. M. Stellman (ed.), *Encyclopaedia of Occupational Health And Safety* III, Geneva, Switzerland, International Labor Organization, 1998.

29 Human Rights Watch, *Blood, Sweat, and Fear*.

30 Ibid.

31 N. Stein and D. Burke, 'Son Of A Chicken Man. As he struggles to remake his family's poultry business into a $24 billion meat behemoth, John Tyson must prove he has more to offer than just the family name', from *Fortune* Magazine, quoted in Human Rights Watch, *Blood, Sweat, and Fear*.

32 *Independent*, 'The true cost of cheap chicken', 4 January 2008, http://www.independent.co.uk/news/uk/home-news/the-true-cost-of-cheap-chicken-768062.html (accessed 9 August 2012).

33 CIWF, 2010, press release, 'UK Consumers Vote for Higher Welfare Chicken and Eggs', 8 April 2010, http://www.ciwf.org.uk/includes/documents/cm_docs/2010/n/nr1009.pdf.

V SHRINKING PLANET

11 LAND: HOW FACTORY FARMS USE MORE, NOT LESS

1 BBC, 'Argentina's forest people suffer neglect', 27 September 2007, http://news.bbc.co.uk/1/hi/programmes/from_our_own_correspondent/7014197.stm (accessed 20 August 2012); University of Pennsylvania website, http://www.sas.upenn.edu/~valeggia/pdf (accessed 20 August 2012); Star of Hope website, http://www.starofhopeusa.org/component/myblog/argentian-some-history-regarding-toba-indians-565.html (accessed 20 August 2012); *Ethnologue*, Languages of Argentina, 2012, http://www.ethnologue.com/show_country.asp?name=AR (accessed 20 August 2012); http://intercontinentalcry.org/peoples/toba-qom/ (accessed 20 August 2012).

2 BBC, 'Argentina's forest people suffer neglect.

3 *Argentina Independent,* 'Qom indigenous leader hit by truck in alleged attack', 10 August 2012, http://www.argentinaindependent.com/tag/formosa/ (accessed 20 August 2012).

4 Calculation by Compassion in World Farming, 2012.

5 Thomas K. Rudel et al., 'Agricultural intensification and changes in cultivated areas, 1970–2005', *PNAS,* 106 (49) (2009), pp. 20675–80.

6 Calculation by Compassion in World Farming, 2011.

7 J. Lundqvist, C. de Fraiture and D. Molden, *Saving Water: From Field to Fork – Curbing Losses and Wastage in the Food Chain,* SIWI Policy Brief, 2008.

8 G. Borgstrom, *The Hungry Planet,* 2nd revised edition, New York, Collier Books, 1972.

9 R. K. Pachauri, 'Global warning! The impact of meat production and consumption on climate change', CIWF Peter Roberts Memorial Lecture, London, 8 September 2008, http://www.ciwf.org.uk/includes/documents/cm_docs/2008/l/1_london_08sept08.pps.

10 OECD-FAO, *Agricultural Outlook 2009–2018: Highlights,* 2009.

11 K. Deininger, D. Byerlee et al., *Rising Global Interest in Farmland: can it yield sustainable and equitable benefits?,* World Bank, 2010.

12 C. Vicente, GRAIN, Buenos Aires, personal communication, 2012.

13 Soybean and Corn Advisor website, http://www.soybeansandcorn.com/Argentina-Crop-Acreage (accessed 21 May 2013).

14 *Chicago Tribune,* 'ANALYSIS – Argentina's soy addiction comes back to bite farmers', 22 April 2013, http://articles.chicagotribune.com/2013-04-22/news/sns-rt-argentina-soy-analysisl2nod6135-20130422_1_soy-yields-corn-yields-pampas-farm-belt (accessed 21 May 2013).

15 Ibid.

16 Harvard Business School, from website, www.losgrobo.com (accessed 20 August 2012); Reuters, 'High yields boost Argentine soy outlook', 3 May 2010, http://www.reuters.com/article/2010/05/03/us-latam-summit-argentina-losgrobo-idUSTRE64259K20100503 (accessed 20 August 2012).

17 L. Cotula, S. Vermeulen, R. Leonard and J. Keeley, *Land Grab or Development Opportunity? Agricultural Investment and International Land Deals in Africa,* IIED/FAO/IFAD, London/Rome, 2009.

18 D. Headley, S. Malaiyandi and F. Shenggen, *Navigating the Perfect Storm: Reflections on the Food, Energy and Financial Crises,* August 2009, IFPRI (Online Resource) available at http://www.ifpri.org/sites/default/files/publications/ifpridp00889.pdf.

19 Deininger, Byerlee et al., *Rising Global Interest in Farmland.*

20 'Global Land Grabbing: Update from the International Conference on Global Land Grabbing', ISS, 2011, http://www.iss.nl/fileadmin/ASSETS/iss/Documents/Conference_programmes/LDPI_conference_summary_May_2011.pdf (accessed 20 August 2012).

21 Chayton Africa website, http://www.chaytonafrica.com/ (accessed 20 August 2012).

22 'Chayton combines good land and secure water assets to grow its Atlas Agricultural operation', *HedgeNews Africa, Journal of the African Alternatives and Hedge Fund Community*, Second Quarter, 2011, vol. 1, 7, http://www.oaklandinstitute.org (accessed 20 August 2012).

23 Crowder, quoted in N. Nyagah, 'Zambia: Rich African farms draw international investors', 2 March 2011, http://allafrica.com/stories/201103021113.html (accessed 20 August 2012).

24 *Guardian*, 'Land deals in Ethiopia bring food self-sufficiency, and prosperity', 4 April 2011, http://www.guardian.co.uk/global-development/poverty-matters/2011/apr/04/ethiopia-land-deals-food-self-sufficiency (accessed 20 August 2012); *Ethiopian Times*, 'Land grab in Africa: demystifying large-scale land investments', 2 April 2012, https://ethiopiantimes.wordpress.com/2012/04/02/land-grab-in-africa-demystifying-large-scale-land-investments/ (accessed 20 August 2012).

25 Karuturi Global website, http://www.karuturi.com/ (accessed 20 August 2012).

26 Karuturi Global website, Welcome to Karuturi Global Limited, http://www.karuturi.com/index.php?option=com_frontpage&Itemid=1 (accessed 3 October 2012).

27 *Guardian*, 'Land deals in Ethiopia'.

28 Bloomberg, 'Ethiopian Government Slashes Karuturi Global Land Concession by Two-Thirds', 4 May 2011, http://www.bloomberg.com/news/2011-05-04/ethiopian-government-slashes-karuturi-global-land-concession-by-two-thirds.html (accessed 20 August 2012).

29 Bloomberg, 'Ethiopian farms lure Bangalore-based Karuturi Global Ltd. As Workers Live in Poverty', 30 December 2009, http://www.bloomberg.com/apps/news?pid=newsarchive&sid=aeuJT_pSE68c (accessed 20 August 2012).

30 Ibid.

31 M. Vermeer and S. Rahmstorf, 'Global sea level linked to global temperature', *Proceedings of the National Academy of Sciences*, 2009; DOI: 10.1073/pnas.0907765106; Potsdam Institute for Climate Impact Research (PIK), 'Sea level could rise from 0.75 to 1.9 meters this century', *ScienceDaily*, 8 December 2009.

32 *Stern Review on the Economics of Climate Change*, HM Treasury and Cabinet Office, 2006, part II, chapters 3 and 4.

33 Ibid.

34 The Rights and Resources Initiative (RRI), *Seeing People through the Trees: Scaling Up Efforts to Advance Rights and Address Poverty, Conflict and Climate Change*, Washington DC, RRI, 2008, http://www.rights-andresources.org/documents/files/doc_737.pdf.

35 *Stern Review on the Economics of Climate Change*, HM Treasury and Cabinet Office, 2006.

36 Soystats, 'World soybean production 2010', 2011, http://www.soystats.com/2011/page_30.htm (accessed 21 August 2012).

37 FAOSTAT: http://faostat.fao.org (accessed 8 September 2012); Soystats, 'Adoption of Bio-tech enhanced soyabean seedstock 1997–2010', 2011, http://www.soystats.com/2011/page_36.htm (accessed 21 August 2012).

38 Soystats, 'World soybean production 2010'.

39 Soystats 2012, http://www.soystats.com/archives/2012/no-frames.htm. (accessed 8 September 2012); USDA Economic Research Service, 'Soybeans and oil crops', 2012, http://www.ers.usda.gov/topics/crops/soybeans-oil-crops/trade.aspx (accessed 21 August 2012); Soystats, 'World soybean meal exports 2010', 2011, shows Argentina accounting for 49% of global exports, ahead of Brazil (23%) and USA (14%), http://www.soystats.com/2011/page_33.htm (accessed 21 August 2012).

40 'What's feeding our food? The environmental and social impacts of the livestock sector', http://www.foe.co.uk/resource/briefings/livestock_impacts.pdf (accessed 8 September 2012).

41 Alternet, 'Feedlot Meat Has Spurred a Soy Boom That Has a Devastating Environmental and Human Cost', 17 March 2011, http://www.alternet.org/story/150277/feedlot_meat_has_spurred_a_soy_boom_that_has_a_devastating_environmental_and_human_cost (accessed 21 May 2013).

42 International Rivers website, 'Paraguay-Paraná Hidrovia', 2012, http://www.internationalrivers.org/zh-hans/node/2348 (accessed 23 August 2012); *The South American Hidrovia Parana – Paraguay*, http://www.chasque.net/rmartine/hidrovia/Envxtrad.html (accessed 23 August 2012).

43 International Rivers website, 'Paraguay-Paraná Hidrovia', http://www.internationalrivers.org/campaigns/paraguay-paran%C3%A1-hidrovia (accessed 21 May 2013).

44 Reuters, 'Eyeing flood waters, Argentine ranchers move cattle', 30 October 2009, http://www.reuters.com/article/2009/10/30/idUSN30241232 (accessed 23 August 2012).

45 IPS, 'ARGENTINA: Countryside No Longer Synonymous with Healthy Living', 4 March 2009, http://www.ipsnews.net/2009/03/argentina-countryside-no-longer-synonymous-with-healthy-living/ (accessed 21 May 2013).

46 J. Richardson, 'Feedlot Meat Has Spurred a Soy Boom That Has a Devastating Environmental and Human Cost', *Axis of Logic*, 26 March 2011, http://axisoflogic.com/artman/publish/Article_62629.shtml (accessed 24 August 2012).

47 Soystats, 'Adoption of biotech-enhanced soybean seedstock 1997–2010', 2011, http://www.soystats.com/2011/Default-frames.htm (accessed 5 October 2012).

48 *Guardian*, 'GM soya "miracle" turns sour in Argentina', 16 April 2004, http://www.guardian.co.uk/science/2004/apr/16/gm.food (accessed 6 September 2012).

49 CAST (Council for Agricultural Science and Technology), Issue Paper 49, *Herbicide-Resistant Weeds Threaten Soil Conservation Gains: Finding a Balance for Soil and Farm Sustainability*, February 2012.

50 Faculty of Medical Sciences, National University of Cordoba, *Report from the 1st National meeting of Physicians in the Crop-Sprayed Towns*, 27 August 2010, http://www.reduas.fcm.unc.edu.ar/statement-from-the-1st-national-meeting-of-physicians-in-the-crop-sprayed-towns/ (accessed 21 May 2013).

51 Ibid.

52 Le Monde/World Crunch, 'Where Soy Is King: In Argentina, Local Health Costs Rise As Agro Booms', 15 August 2011, http://ww.worldcrunch.com/culture-society/where-soy-is-king-in-argentina-local-health-costs-rise-as-agro-booms-/c3s3581/ (accessed 6 August 2012).

53 Monsanto, Corporate Profile, 2012, http://www.monsanto.com/investors/pages/corporate-profile.aspx (accessed 6 August 2012).

12 THICKER THAN WATER: DRAINING RIVERS, LAKES AND OIL WELLS

1 Energy Information Administration, *Analysis of Crude Oil Production in the Arctic National Wildlife Refuge*, May 2008, http://www.eia.gov/oiaf/servicerpt/anwr/pdf/sroiaf(2008)03.pdf (accessed 7 September 2012).

2 CIA Factbook, https://www.cia.gov/library/publications/the-world-factbook/rankorder/2174rank.html (accessed 7 September 2012) (based on US daily consumption of 19.15 million barrels and EU, 13.68 million).

3 *Alaska Journal of Commerce*, 'USGS estimates on Slope shale oil, gas puts Alaska near top', 1 March 2012, http://www.alaskajournal.com/Alaska-Journal-of-Commerce/AJOC-March-4-2012/USGS-estimates-on-Slope-shale-oil-gas-puts-Alaska-near-top/ (accessed 7 September 2012).

4 *Time*, 12 November 2007, The Eco vote, http://www.time.com/time/2007/includes/eco_vote.pdf (accessed 7 September 2012).

5 National Review Online, 'The Campaign Spot', 16 January 2008, http://www.nationalreview.com/campaign-spot/10699/john-mccain-im-raising-hundreds-thousands-day-new-hampshire (accessed 7 September 2012).

6 *New York Times*, 'New and Frozen Frontier Awaits Offshore Oil Drilling', 23 May 2012, http://www.nytimes.com/2012/05/24/science/earth/shell-arctic-ocean-drilling-stands-to-open-new-oil-frontier.html (accessed 7 September 2012).

7 *Daily Telegraph*, 'Total insists Shtokman Russian Arctic gas project not delayed "indefinitely"', 31 August 2012, http://www.telegraph.co.uk/finance/newsbysector/energy/oilandgas/9512809/Total-insists-Shtokman-Russian-Arctic-gas-project-not-delayed-indefinitely.html (accessed 7 September 2012).

8 *Guardian*, 'Arctic oil rush will ruin ecosystem, warns Lloyd's of London', 12 April 2012, http://www.guardian.co.uk/world/2012/apr/12/lloyds-london-warns-risks-arctic-oil-drilling (accessed 7 September 2012).

9 D. Pimentel, *Impacts of Organic Farming on the Efficiency of Energy Use in Agriculture: An Organic Center State of Science Review*, The Organic Center, August 2006. This review cites findings in several other of the Pimentel group papers.

10 World Bank, *World Development Report 2008. Agriculture for Development*, 2008, chapter 2, 'Agriculture's performance, diversity and uncertainties', http://siteresources.worldbank.org.

11 FAO, *The energy and agriculture nexus*, Environment and natural resources working paper 4. Rome, 2000, chapter 2, 'Energy for agriculture', http://www.fao.org/.

12 Pimentel, *Impacts of Organic Farming*.

13 A. A. Bartlett, 'Forgotten Fundamentals of the Energy Crisis', 1978, http://www.npg.org/specialreports/bartlett_section3.htm (accessed 7 September 2012).

14 World Bank, *World Development Report 2008*, chapter 2.

15 National Petroleum Council, *Summary Discussions on Peak Oil*, working document of the NPC Global Oil & Gas Study, Topic Paper #15, July 2007,

http://downloadcenter.connectlive.com/events/npc071807/pdf-down-loads/Study_Topic_Papers/15-STG-Peak-Oil-Discussions.pdf.

16 International Energy Agency, *World Energy Outlook 2010*, presentation to the press, London, 9 November 2010.

17 Oil Depletion Analysis Centre, 'Peak Oil Primer', 24 November 2009, http://www.odac-info.org/.

18 D. Howden, 'World oil supplies are set to run out faster than expected, warn scientists', *Independent*, 14 June 2007, http://www.independent.co.uk/news/science/world-oil-supplies-are-set-to-run-out-faster-than-expected-warn-scientists-453068.html.

19 Shell International, *Shell Energy Scenarios to 2050: an era of volatile transitions*, 2011.

20 A. Coecup, letter to the editor, *The Times*, 30 May 2012.

21 Pimentel, *Impacts of Organic Farming*.

22 Ibid.

23 Soil Association, *Energy efficiency of organic farming: analysis of data from existing Defra studies*, published 31 January 2007.

24 Lloyd's, 'Investment in the Arctic could reach $100bn in ten years', 12 April 2012, http://www.lloyds.com/Lloyds/Press-Centre/Press-Releases/2012/04/Investment-in-the-Arctic-could-reach-USD100bn-in-ten-years.

25 *Guardian*, 'Arctic oil rush will ruin ecosystem'.

26 Fiji Water, official website, http://www.fijiwater.com/ (accessed 10 September 2012).

27 *Daily Telegraph*, 'Fiji Water accused of environmentally misleading claims', 20 June 2011, http://www.telegraph.co.uk/earth/earthnews/8585182Fiji-Water-accused-of-environmentally-misleading-claims.html (accessed 10 September 2012).

28 Ibid.

29 Fiji Water, official website, FAQ: About our water, http://www.fijiwater.com/company/faq/about-fiji-water/ (accessed 12 September 2012).

30 S. Parente and E. Lewis-Brown, *Freshwater Use and Farm Animal Welfare*, CIWF/WSPA, 2012, http://www.ciwf.org.uk/includes/documents/cm_docs/2012/f/freshwater_use_and_farm_animal_welfare_12_page.pdf.

31 World Economic Forum, 'The bubble is close to bursting: A forecast of the main economic and geopolitical water issues likely to arise in the world during the next two decades', draft for discussion at the World Economic Forum Annual Meeting 2009.

32 Based on a bathtub holding 175 litres of water, and published figures showing the total water footprint for a kilo of beef, pork and chicken

396

NOTES TO PAGES 241–244

being 15,500, 5,900 and 4,300 litres consecutively; from Water Footprint Network, http://www.waterfootprint.org/?page=files/Animal-products (accessed 10 September 2012).

33 P. W. Gerbens-Leenes, M. M. Mekonnen and A. Y. Hoekstra, *A Comparative Study on the Water Footprint of Poultry, Pork and Beef in Different Countries and Production Systems*, University of Twente, September 2011.

34 Ibid.

35 C. S. Smith, 'Al Kharj Journal; Milk Flows From Desert At a Unique Saudi Farm', *New York Times*, 31 December 2002, http://www.nytimes.com/2002/12/31/world/al-kharj-journal-milk-flows-from-desert-at-a-unique-saudi-farm.html (accessed 10 September 2012).

36 Ibid.

37 Ibid.

38 2nd UN World Water Development Report, *Water, A Shared Responsibility*, chapter 4, 'The State of the Resource', UNESCO, WMO and IAEA, 2006., http://www.unesco.org/water/wwap/wwdr/wwdr2/pdf/wwdr2_ch_4.pdf.

39 H. Steinfeld et al., *Livestock's Long Shadow: environmental issues and options*, chapter 4, Food and Agriculture Organization of the United Nations, Rome, 2006, http://www.virtualcentre.org/en/library/key_pub/longshad/A0701E00.htm.

40 Y. Wada et al., 'Global depletion of groundwater resources', *Geophysical Research Letters*, vol. 37 (2010), L20402, doi:10.1029/2010GL044571; American Geophysical Union (AGU), 'Groundwater depletion rate accelerating worldwide', AGU Release No. 10–30, 23 September 2010, www.agu.org/news/press/pr_archives/2010/2010-30.shtml.

41 ibid.

42 *Wired* magazine, 'Peak Water: Aquifers and Rivers Are Running Dry. How Three Regions Are Coping', 21 April 2008, http://www.wired.com/science/planetearth/magazine/16-05/ff_peakwater?currentPage=all (accessed 10 September 2012); R. Courtland, 'News briefing, Enough water to go around?', *Nature*, published online 19 March 2008, doi:10.1038/news.2008.678, www.nature.com/news/2008/080319/full/news.2008.678.html.

43 M. Barlow and T. Clarke, *Blue Gold: The Battle against Corporate Theft of the World's Water*, Earthscan, London, 2002.

44 B. Bates et al., *Climate Change and Water*, IPCC Technical paper VI, IPCC, WMO and UNEP, 2008, http://www.ipcc.ch/pdf/technical-papers/climate-change-water-en.pdf.

45 Steinfeld et al., *Livestock's Long Shadow*.

46 BBC, 'Summer "wettest in 100 years" Met Office figures show', 30 August

2012, http://www.bbc.co.uk/news/uk-19427139 (accessed 11 September 2012).

47 *Guardian*, 'Drought tanker ships considered', 17 May 2006, http://www.guardian.co.uk/environment/2006/may/17/water.uknews (accessed 4 October 2012).

48 *Daily Mail*, 'Icebergs considered to help beat the drought', 17 May 2006, http://www.dailymail.co.uk/news/article-386582/Icebergs-considered-help-beat-drought.html (accessed 4 October 2012).

49 BBC, 'Salt water plant opened in London', 2 June 2010, http://www.bbc.co.uk/news/10213835 (accessed 11 September 2012); N. Larkin, 'London Mayor appeals Thames Water desalination plant (Update 2)', Bloomberg, 21 August 2007, http://www.bloomberg.com (accessed 4 October 2012).

50 Thames Water, 'Thames gateway water treatment works', 20 August 2012, http://www.thameswater.co.uk/your-water/9942.htm (accessed 4 October 2012).

51 *Economist*, 'Australia's water shortage: The big dry', 26 April 2007, http://www.economist.com/node/9071007 (accessed 8 May 2013).

52 ABC Riverland, 'Scientists quit flawed Murray-Darling process', 21 May 2011, http://www.abc.net.au/local/stories/2011/05/23/3223924.htm (accessed 8 May 2013).

53 World Economic Forum, 'The bubble is close to bursting'.

54 World Resources Institute, UN Environment Programme, UN Development Programme and the World Bank, *World Resources 1998–99: Environmental change and human health*, 1998, http://www.wri.org/publication/content/8261.

55 American Geophysical Union (AGU), 'Groundwater depletion rate accelerating worldwide'.

56 World Economic Forum, 'The bubble is close to bursting'.

57 A. Mukherji et al., *Revitalizing Asia's irrigation: to sustainably meet tomorrow's food needs*, Colombo, Sri Lanka, International Water Management Institute; Rome, Italy, Food and Agriculture Organization of the United Nations, 2009.

58 IWMI and FAO, 'IWMI-FAO report: Revitalising Asia's irrigation: to sustainably meet tomorrow's food needs', press release, 19 August 2009, http://www.fao.org/nr/water/docs/iwmi-fao-report-revitalizing-asias-irrigation-to-sustainably-meet-tomorrows-food-needs.pdf.

59 J. Liu, H.Yang and H. H. G. Saveniji, 'China's move to high-meat diet hits water security', *Nature*, 454 (2008), p. 397.

60 J. Bruinsma, *The resource outlook to 2050: by how much do land, water and*

crop yields need to increase by 2050?, Expert meeting on 'How to Feed the World in 2050', Rome, FAO, 24–26 June 2009.

61 Parente and Lewis-Brown, *Freshwater Use and Farm Animal Welfare*.

13 HUNDRED-DOLLAR HAMBURGER: THE ILLUSION OF CHEAP FOOD

1 *Independent*, 'Tesco hits a new low with arrival of the £1.99 chicken', 6 February 2008, http://www.independent.co.uk/life-style/food-and-drink/news/tesco-hits-a-new-low-with-arrival-of-the-163199-chicken-778672.html (accessed 12 September 2012).

2 *Guardian*, 'High food prices are here to stay – and here's why', 17 July 2011, http://www.guardian.co.uk/lifeandstyle/2011/jul/17/food-prices-rise-commodities (accessed 12 September 2012).

3 M. Cacciottolo, 'The "hidden hunger" in British families', BBC news, 7 October 2010, http://www.bbc.co.uk/news/magazine-11427207.

4 J. Owen and B. Brady, 'Jobcentres to send poor and hungry to charity food banks', *Independent*, 18 September 2011, http://www.independent.co.uk/news/uk/politics/jobcentres-to-send-poor-and-hungry-to-charity-food-banks-2356578.html.

5 UN Department of Social and Economic Affairs, *World Economic and Social Survey 2011*, United Nations, New York.

6 *Guardian*, 'High food prices are here to stay'.

7 Government Office for Science, *Foresight Project on Global Food and Farming Futures. Synthesis Report C1: Trends in food demand and production*, January 2011; S. Msangi and M. W. Rosegrant, *Agriculture in a dynamically-changing environment: IFPRI's long-term outlook for food and agriculture under additional demand and constraints*, paper written in support of Expert Meeting on 'How to feed the World in 2050', Rome, FAO, 2009, http://www.fao.org/wsfs/forum2050/wsfs-background-documents/wsfs-expert-papers/en/; H. Steinfeld et al., *Livestock's Long Shadow, environmental issues and options*, FAO, Rome, 2006, Introduction, p. 12.

8 Oxfam, *4-a-week: changing food consumption in the UK to benefit people and the planet*, Oxfam GB Briefing Paper, 2009.

9 Foresight, *The Future of Food and Farming: challenges and choices for global sustainability*, Final Project Report, The Government Office for Science, London, 2011.

10 *The Economist*, 'Food and the Arab Spring: Let them eat Baklava; Today's Policies are Recipe for Instability in Middle East', 17 March 2012, http://www.economist.com/node/21550328; UN, 'Soaring cereal tab

continues to afflict poorest countries, UN agency warns', UN News Centre, 11 April 2008, www.un.org/apps/news/story.asp?NewsID=2628 9&Cr=food&Cr1=prices.

11 *New York Times*, 'Bread, the (subsidized) stuff of life in Egypt', 16 January 2008, http://www.nytimes.com/2008/01/16/world/africa (accessed 12 September 2012).

12 *Economist*, 'Food and the Arab Spring: Let them eat Baklava'.

13 Msangi and Rosegrant, *Agriculture in a dynamically-changing environment*.

14 Calculated from FAOSTAT online figures for global grain harvest (2009) and food value of cereals, based on a calorific intake of 2,500 kcalories per person per day.

15 David Pimentel et al., 'Reducing energy inputs in the US food system', *Human Ecology*, 36 (2008), pp. 459–71.

16 Vaclav Smil, *Feeding the world: a Challenge for the Twenty-first Century*, MIT Press, 2000.

17 BBC, 'Hundred-Dollar hamburger?', 14 June 2011, http://www.bbc.co.uk/news/business-13764242 (accessed 12 September 2012).

18 *Daily Mail*, 'The GM genocide: Thousands of Indian farmers are committing suicide after using genetically modified crops', 3 November 2008, http://www.dailymail.co.uk/news/article-1082559/The-GM-genocide-Thousands-Indian-farmers-committing-suicide-using-genetically-modified-crops.html (accessed 12 September 2012).

19 Ibid.

20 Center for Human Rights and Justice, *Every Thirty Minutes: Farmer Suicides, Human Rights, and the Agrarian Crisis in India*, New York, NYU School of Law, 2011.

21 *Independent*, 'Charles: "I blame GM crops for farmers' suicides"', 5 October 2008, http://www.independent.co.uk/environment/green-living/charles-i--blame-gm-crops-for-farmers-suicides-951807.html (accessed 12 September 2012).

22 P. Sainath, 'Farm suicides: a 12-year saga', *The Hindu*, 25 January 2010, http://www.thehindu.com/opinion/columns/sainath/article94324.ece (accessed 12 September 2012).

23 Tata Institute of Social Sciences, *Causes of Farmer Suicides in Maharashtra: An Enquiry*, Final Report Submitted to the Mumbai High Court,15 March 2005, http://mdmu.maharashtra.gov.in/pdf/Farmers_suicide_TISS_report.pdf.

24 J. Mencher, commenting on online ISIS press release, 10 February 2010, http://www.i-sis.org.uk (accessed 12 September 2012).

25 S. Ashley, S. Holden and P. Bazeley, *Livestock in Poverty-Focused Development*, Livestock in Development, Crewkerne, UK, 1999, http://www.theidlgroup.com/documents/IDLRedbook_000.pdf.

26 IFAD, *Rural Poverty Report*, 2011, http://www.ifad.org/rpr2011/index.htm (accessed 12 September 2012).

27 L. R. Brown, *Plan B 4.0: Mobilizing to save civilization*, Earth Policy Institute, W. W. Norton, 2009.

28 Earth Policy Institute, *World grain consumption and stocks, 1960–2009*, supporting dataset for chapter 2 of Brown, *Plan B 4.0,* http://www.earth-policy.org/index.php?/books/pb4/pb4_data.

29 P. J. Gerber, H. Steinfeld, B. Henderson, A. Mottet, C. Opio, J. Dijkman, A. Falucci and G. Tempio, 'Tackling climate change through livestock – a global assessment of emissions and mitigation opportunities', Food and Agriculture Organisation of the United Nations (FAO), Rome, 2013.

30 *Stern Review on the Economics of Climate Change*, HM Treasury and Cabinet Office, 2006, part II, chapters 3 and 4, http://www.hm-treasury.gov.uk/sternreview_index.htm; Joachim von Braun, *The world food situation: new driving forces and required actions*, IFPRI, Washington DC, December 2007, http://www.ifpri.org/pubs/fpr/pr18.pdf; D. S. Battisti and R. L. Naylor, 'Historical Warnings of Future Food Insecurity with Unprecedented Seasonal Heat', *Science*, 323 (2009), pp. 240–4.

31 The Rights and Resources Initiative (RRI), *Seeing People Through The Trees*, RRI, Washington DC, 2008, http://www.rightsandresources.org/documents/files/doc_737.pdf.

32 Cabinet Office, *Food Matters: towards a strategy for the 21st century*, Strategy Unit, July 2008.

VI TOMORROW'S MENU

14 GM: FEEDING PEOPLE OR FACTORY FARMS?

1 I. Potrykus, 'The "Golden Rice" tale', Agbioworld: http://www.agbioworld.org/biotech-info/topics/goldenrice/tale.html (accessed 30 July 2012).

2 *New York Times*, 'Scientist at work: Ingo Potrykus; Golden Rice in a Grenade-Proof Greenhouse', 21 November 2000, http://www.nytimes.com/2000/11/21/science/scientist-at-work-ingo-potrykus-golden-rice-in-a-grenade-proof-greenhouse.html?

3 http://agropedia.iitk.ac.in/?q=content/golden-rice (accessed 30 July 2012).

4 Golden Rice Project, Golden Rice Humanitarian Board website, http:// www.goldenrice.org/index.php (accessed 30 July 2012).

5 Golden Rice Project website, 'Frequently Asked Questions (2)', http:// goldenrice.org/Content3-Why/why3a_FAQ.php#Pseudo-science (accessed 5 October 2012).

6 Greenpeace press release, 'Golden Rice is a technical failure standing in way of real solutions for vitamin A deficiency', 17 March 2005, http:// www.greenpeace.org/international/en/press/releases/golden-rice-is-a-technical-fai/ (accessed 30 July 2012).

7 World Health Organization (WHO) website article 'Micronutrient deficiencies', http://www.who.int/nutrition/topics/vad/en/index.html (accessed 8 May 2013).

8 Golden Rice Project, Golden Rice Humanitarian Board website, http:// www.goldenrice.org/index.php (accessed 30 July 2012).

9 Defra, 'Farm scale evaluations: Managing GM crops with herbicides: effects on farmland wildlife', 2005.

10 M. A. Altieri, 'The Ecological Impacts of Large-Scale Agrofuel Mono-culture Production Systems in the Americas', *Bulletin of Science Technology Society*, June 2009, 29(3), pp. 236–44, http://bst.sagepub. com/content/29/3/236.

11 FAO Food Outlook, May 2012, http://www.fao.org/giews/english/fo/ index.htm.

12 H. Steinfeld et al., *Livestock's Long Shadow: environmental issues and options*, Food and Agriculture Organization of the United Nations, Rome, 2006.

13 National Corn Growers Association (NCGA), *Corn Facts*, 2011 (accessed 22 June 2011).

14 GMO Compass, Field areas 2009, 2010, www.gmocompass.org/eng/ agri_biotechnology/gmo_planting/257.global_gm_planting_2009. html.

15 Ibid.

16 GMO Compass, 'Genetically modified plants: global cultivation area: Soybeans, Maize', 2010, www.gmo-compass.org/eng/agri_biotechnology/ gmo_planting/342.genetically_modified_soybean_global_area_ under_cultivation.htm; www.gmo-compass.org/eng/agri_biotechnology/ gmo_planting/341.genetically_modified_maize_global_area_under_cul-tivation.html.

17 J. Lundqvist, C. de Fraiture and D. Molden, *Saving Water: From Field to*

Fork – Curbing Losses and Wastage in the Food Chain, SIWI Policy Brief, SIWI, 2008, figure 1.

18 D. L. Barlett and J. B. Steel, 'Monsanto's Harvest of Fear', *Vanity Fair*, May 2008, http://www.vanityfair.com/poli'cs/features/2008/05/monsanto200805 (accessed 30 July 2012).

19 Monsanto, news & views, 'Gary Rinehart', http://www.monsanto.com/newsviews/pages/gary-rinehart.aspx (accessed 30 July 2012).

20 'Who We Are', http://www.monsanto.co.uk/.

21 Bloomberg, '"Mounting evidence" of bug-resistant corn seen by EPA', 5 September 2012, http://www.bloomberg.com/news/2012-09-04/-mounting-evidence-of-bug-resistant-corn-seen-by-epa.html (accessed 8 May 2013).

22 *Farm Industry News*, 'In-field resistance to Bt corn rootworm trait documented', 16 August 2011, http://farmindustrynews.com/corn-root worm-traits/field-resistance-bt-corn-rootworm-trait-documented (accessed 5 October 2012).

23 Bloomberg, 'Monsanto Corn Falls to Illinois Bugs as Investigation Widens', 2 September 2011, http://www.bloomberg.com/news/2011-09-02/monsanto-corn-is-showing-illinois-insect-damage-as-investigation-widens.html (accessed 8 May 2013).

24 OECD-FAO, *OECD-FAO Agricultural Outlook 2011–2020*, Summary and highlights, 2011.

25 BBC News online, 'Germany bans Monsanto's GM maize', 14 April 2009, http://news.bbc.co.uk/1/hi/world/europe/7998181.stm.

26 J. Smith, 'An FDA-Created Crisis Circles the Globe', October 2007, available at http://www.newswithviews.com/Smith/jeffrey17.htm (accessed 30 July 2012).

27 Institute for Responsible Technology, 'Genetically Modified Soy Linked to Sterility, Infant Mortality', 2010, http://www.responsibletechnology.org/article-gmo-soy-linked-to-sterility (accessed 30 July 2012).

28 Science Nordic, 'Growing fatter on a GM diet', 17 July 2012, http://sciencenordic.com/growing-fatter-gm-diet (accessed 30 July 2012).

29 My Sanantonio, 'First cloned cat is turning 10', 18 May 2011, http://www.mysanantonio.com/news/article/first-cloned-cat-is-turning-10-1383604.php (accessed 13 September 2012).

30 http://www.chron.com/life/article/First-cloned-cat-turns-10-1383844.php; http://articles.nydailynews.com/2011-08-16/entertainment/29913126_1_duane-kraemer-genetic-savings-and-clone-bioarts; http://en.wikipedia.org/wiki/CC_(cat).

31 Texas A&M University website, http://vetmed.tamu.edu/rsl/faculty/duane-kraemer (accessed 13 September 2012).

32 Yahoo voices, 'Dr. Duane Carl Kraemer: The Transfer Scientist', 2003, http://voices.yahoo.com/dr-duane-carl-kraemer-transfer-scientist-134787. html (accessed 13 September 2012); L. Hawthorne, 'A Project to Clone Companion Animals', *Journal of Applied Animal Welfare Science*, 5(3) (2002), pp. 229–31, http://www.animalsandsociety.net/assets/library/147_ jaws050307.pdf; M. Warner, 'Inside the Very Strange World of Billionaire John Sperling', 29 April 2002, Center for Genetics & Society, http://www. geneticsandsociety.org/article.php?id=108 (accessed 13 September 2012).

33 Yahoo voices, 'Dr. Duane Carl Kraemer'; Hawthorne, 'A Project to Clone Companion Animals'; Warner, 'Inside the Very Strange World'.

34 BioTechnology, 'Genetic Savings and Clone forced to shut down', 2009, http://biotechnology-industries.blogspot.co.uk/2009/01/genetic-sav-ings-and-clone-forced-to.html (accessed 13 September 2012).

35 *National Geographic*, 'First Dog Clone', 28 October 2010, http://news. nationalgeographic.com/news/2005/08/photogalleries/dogclone/ (accessed 13 September 2012).

36 Genome Alberta, 'Californian pit bull lives on in 5 cloned puppies', 5 August 2008, http://genomealberta.ca/connect-with-us/news-releases/ ge3ls08070801.aspx (accessed 13 September 2012); *Guardian*, 'Pet clon-ing service bears five baby Boogers', 5 August 2008, http://www. guardian.co.uk/science/2008/aug/05/genetics.korea (accessed 13 Sep-tember 2012); *Independent*, 'Saved by a pit bull, Californian owner clones five more', 6 August 2008, http://www.independent.co.uk/news/ world/americas/saved-by-a-pit-bull-californian-owner-clones-five-more-886108.html (accessed 13 September 2012).

37 *New York Post*, 'Adorable little abominations of nature', 14 May 2011, http://www.nypost.com (accessed 13 September 2012).

38 L. Goldwert, 'First cloned cat nears 10 but pet replicating business has not boomed due to money, ethical woes', *NY Daily News*, 16 August 2011, http://articles.nydailynews.com/2011-08-16/entertain-ment/29913126_1_duane-kraemer-genetic-savings-and-clone-bioarts.

39 *Southwest Farm Press* staff, 'Disease resistant bull cloned at Texas A&M', 11 January 2001, Southwest Farm Press, http://southwestfarmpress. com/disease-resistant-bull-cloned-texas-am.

40 Viagen website, http://www.viagen.com/ (accessed 13 September 2012).

41 H. Pickett, 'Farm Animal Cloning', Godalming, Compassion in World Farming, 2010, http://www.ciwf.org.uk/includes/documents/cm_docs/ 2010/c/compassion_2010_farm_animal_cloning_report.pdf.

42 University of Utah website, http://learn.genetics.utah.edu/content/tech/ cloning/cloningrisks/ (accessed 13 September 2012).

43 FDA, *Potential Hazards and Risks to Animals Involved in Cloning*, http://www.fda.gov/animalveterinary/safetyhealth/animalcloning/ucm124840.htm (accessed 13 September 2012).

44 P. Loi, L. della Salda, G. Ptak, J. A. Modliński and J. Karasiewicz, 'Peri- and post-natal mortality of somatic cell clones in sheep', *Animal Science Papers and Reports*, 22 (Suppl. 1) (2004), pp. 59–70.

45 EFSA, 'Scientific Opinion of the Scientific Committee on a request from the European Commission on food safety, animal health and welfare and environmental impact of animals derived from cloning by somatic cell nucleus transfer (SCNT) and their offspring and products obtained from those animals', *EFSA Journal*, 767 (2008), pp. 1–49.

46 EGE, Opinion No. 23: 'Ethical Aspects of Animal Cloning for Food Supply', The European Group on Ethics in Science and New Technologies to the European Commission, 16 January 2008.

47 CIWF, 'Cloned animal suffering forces end to AgResearch programme', 23 February 2011, http://www.ciwf.org.uk/includes/documents/cm_docs/2011/n/nr1103.pdf (accessed 13 September 2012).

48 European Commission, 'Europeans' attitudes towards animal cloning', October 2008, http://ec.europa.eu/public_opinion/flash/fl_238_en.pdf.

49 *Daily Mail*, 'Clone beef's been on sale: After clone milk, now food watchdogs launch an investigation into illegal meat sold in British shops', 4 August 2010, http://www.dailymail.co.uk/news/article-1300097/Clone-beefs-sale-After-clone-milk-investigation-launched-illegal-meat.html (accessed 13 September 2012).

50 FSA, 'Cloned meat is safe – hypothetically speaking', 25 November 2010, http://www.food.gov.uk/news-updates/news/2010/nov/acnfcloned (accessed 13 September 2012).

51 FDA website: http://www.fda.gov/AnimalVeterinary/SafetyHealth/AnimalCloning/default.htm (accessed 13 September 2012).

52 The Poultry Site, 'Israeli scientists breed featherless chicken', 1 November 2011, http://www.thepoultrysite.com/poultrynews/24138/israeli-scientists-breed-featherless-chicken (accessed 13 September 2012); *New Scientist*, 'Featherless chicken creates a flap', 21 May 2002, http://www.newscientist.com/article/dn2307-featherless-chicken-creates-a-flap.html (accessed 13 September 2012).

53 *Daily Telegraph*, 'Genetically modified cows produce "human" milk', 2 April 2011, http://www.telegraph.co.uk/earth/agriculture/geneticmodification/8423536/Genetically-modified-cows-produce-human-milk.html# (accessed 13 September 2012).

15 CHINA: MAO'S MEGA-FARM DREAM COMES TRUE

1 Nationmaster.com, http://www.nationmaster.com/graph/foo_por_con_
   per_cap-food-pork-consumption-per-capita (accessed 18 July 2012).

2 Defra press release, 'Vince Cable signs multi-million pound export deal
   to China', 8 November 2010, http://www.defra.gov.uk/news/2010/11/
   08/export-pig-china/ (accessed 18 July 2012).

3 V. Elliott, 'Why British pigs are flying in jumbo jets to China: Beijing
   snaps up our livestock to boost poor-quality herds', *Daily Mail*, 2 October 2011, http://www.dailymail.co.uk/news/article-2044201/China-
   ship-British-pigs-Beijing-boost-poor-quality-herds.html (accessed 18
   July 2012); JSR Genetics, '900 high genetic merit JSR pigs delivered to
   China', 20 July 2012, http://www.jsrgenetics.com/news.php?sid=121
   (accessed 4 October 2012).

4 F. Dikötter, *Mao's Great Famine*, Bloomsbury, London, 2010.

5 Ibid.

6 UNDP, China Human Development Report, 2007–2008, *Access for all:
   Basic public services for 1.3 billion people*, China Translation and Publishing Corporation, Beijing, 2008, http://hdr.undp.org/en/reports/
   national/asiathepacific/china/China_2008_en.pdf (accessed 23 July
   2012).

7 A. Park, 'Still much to be done in fight against poverty', *China Daily*, 4
   August 2009, http://www.chinadaily.com.cn/opinion/2009-04/08/con-
   tent_7657358.htm (accessed 18 July 2012); CIA World Factbook,
   https://www.cia.gov/library/publications/the-world-factbook/fields/
   2046.html (accessed 23 July 2012).

8 BBC News, 'Chinese baby milk scare "severe"', 13 September 2008,
   http://news.bbc.co.uk/1/hi/world/asia-pacific/7614083.stm (accessed
   18 July 2012).

9 Chinese milk scandal, http://en.wikipedia.org/wiki/2008_Chinese_milk_
   scandal (accessed 23 July 2012); SKY News, 'China Milk: Two sentenced to
   death', 22 January 2009, http://news.sky.com/story/663668/china-milk-
   two-sentenced-to-death (accessed 23 July 2012); *Daily Mail*, 'Two men
   sentenced to death for roles in Chinese milk scandal which killed six babies',
   22 January 2009, http://www.dailymail.co.uk/news/article-1126484/Two-
   men-sentenced-death-roles-Chinese-milk-scandal-killed-babies.html
   (accessed 23 July 2012); BBC News, 'Chinese baby milk scare "severe"'.

10 Clenbuterol side effects website, http://www.clenbuterolsideeffects.org/
   (accessed 4 October 2012); Clenbuterol website, http://www.
   lenbuterol.tv/clenbuterol-side-effects/ (accessed 4 October 2012);

Independent, 'Clenbuterol: The new weight-loss wonder drug gripping planet zero', 20 March 2007, http://www.independent.co.uk/life-style/health-and-families/health-news/clenbuterol-the-new-weightloss-wonder-drug-gripping-planet-zero-441059.html (accessed 4 October 2012).

11 *China Daily*, 'Who can guarantee China's pork is safe?', 6 April 2011, http://www.chinadaily.com.cn/china/2011-04/06/content_12281515.htm (accessed 18 July 2012).

12 *People's Daily*, 'Three arrested in pig meat food poisoning of 300 people in Shanghai', 4 November 2006, http://english.peopledaily.com.cn/200611/04/eng20061104_318172.html (accessed 18 July 2012).

13 International Finance Corporation, 'Muyuan Pig. Summary of proposed investment', 2010, http://www.ifc.org/ifcext/spiwebsite1.nsf/ProjectDisplay/SPI_DP29089 (accessed 18 July 2012).

14 Ibid.

15 Calculation based on typical energy use in conventional pig production of 16–17 MJ energy per kg of pigmeat produced (Basset-Mens et al., 2005; Williams et al., 2006); annual production of 450,000 pigs slaughtered at average carcass weight for China of 76.7 kg (FAOSTAT, 2009); C. Basset-Mens and H. M. G. van der Werf, 'Scenario-based environmental assessment of farming systems: the case of pig production in France', *Agriculture, Ecosystems & Environment*, 105 (1–2) (2005), pp. 127–44; A. G. Williams, E. Audsley and D. L. Sandars, *Determining the environmental burdens and resource use in the production of agricultural and horticultural commodities*, Defra Project report ISO25, Bedford, Cranfield University and Defra, 2006.

16 International Finance Corporation (IFC) press release, 'IFC Equity Investment in Muyuan Food Supports Chinese Farming Sector', 18 August 2010, http://www.ifc.org/ifcext/agribusiness.nsf (accessed 18 July 2012).

17 The average commercial British pig farm has 500 sows, http://www.publications.parliament.uk/pa/cm200809/cmselect/cmenvfru/96/96.pdf; Muyuan farm #21 said to be set to house 6,500 sows.

18 CSRchina.net, 'China's economic engine forced to face environmental deficit', http://www.csrchina.net/page-1231.html (accessed 23 July 2012); http://factsanddetails.com/china.php?itemid=391&catid=10&subcatid=66.

19 *China Daily*, 'Wen urges cleanup of algae-stenched lakes', 1 July 2007, http://www.chinadaily.com.cn/china/2007-07/01/content_907145.htm (accessed 4 October 2012).

20 *Economist*, 'China: A lot to be angry about', 1 May 2008, http://www.economist.com/node/11293734 (accessed 23 July 2012).

21 AFP, 'China environmentalist alleges brutal jail treatment', 11 May 2010, http://www.google.com/hostednews/afp/article (accessed 23 July 2012).

22 Article: 'Development of organic agriculture in Taihu Lake region governance of agricultural nonpoint source pollution", http://eng.hi138.com/?i274195_Development_of_organic_agriculture_in_Taihu_Lake_region_governance_of_agricultural_nonpoint_source_pollution (accessed 23 July 2012). An Olympic-sized swimming pool holds 2,500 square metres of water.

23 Ibid.

16 KINGS, COMMONERS AND SUPERMARKETS: WHERE THE POWER
LIES

1 The Prince of Wales website, http://www.princeofwales.gov.uk/personal-profiles/residences/highgrove/homefarm/ (accessed 14 September 2012).

2 BBC website, http://news.bbc.co.uk/onthisday/hi/dates/stories/december/1/newsid_3204000/3204279.stm (accessed 14 September 2012).

3 R. Body, *Farming in the Clouds*, Maurice Temple Smith, London, 1984.

4 4 January 1986. In R. Body, *Our Food, Our Land*, Rider, London, 1991.

5 Statistics Canada, *Agriculture Economic Statistics*, Cat. No. 21-603, and *Canadian Economic Observer*, Cat. No. 11-210. Cited in D. Qualman and F. Tait, *The Farm Crisis, Bigger Farms, and the Myths of 'Competition and Efficiency'*, Canadian Centre for Policy Alternatives, 2004, http://www.policyalternatives.org/documents/National_Office_Pubs /farm_crisis2004.pdf.

6 R. Harrison, *Animal Machines*, Vincent Stuart Ltd, London, 1964.

7 SNAP to health, website, US Farm Bill: Frequently asked questions, http://www.snaptohealth.org/farm-bill-usda/u-s-farm-bill-faq/ (accessed 14 September 2012); Wikipedia, http://en.wikipedia.org/wiki/Food,_Conservation,_and_Energy_Act_of_2008 (accessed 14 September 2012).

8 M. Bittman, 'Don't End Agricultural Subsidies, Fix Them', *New York Times*, 1 March 2011, http://opinionator.blogs.nytimes.com/2011/03/01/dont-end-agricultural-subsidies-fix-them/; J. Steinhauer, 'Farm Subsidies Become Target Amid Spending Cuts', *New York Times*, 6 May 2011, http://www.nytimes.com/2011/05/07/us/politics/07farm.html.

9 Environmental Working Group, National data from EWG farm subsidy database, 2011, http://farm.ewg.org.

10 Planet Retail, 'Global Retail Rankings, 2011', 2012, http://www.planetretail.net/Presentations/GlobalRetailRankings2011-Grocery.pdf (accessed 14 September 2012).

11 Planet Retail, 'Global Retail Rankings 2011; Food Service', 2012, http://www.planetretail.net.

12 U. Kjaernes, M. Miele and J. Roex, *Attitudes of Consumers, Retailers and Producers to Farm Animal Welfare,* Quality Report Number 2, EU 6th Framework Programme, Cardiff University, 2007, Welfare, March 2007, http://www.welfarequality.net/everyone/37097/7/0/22.

13 *Catholic Herald,* 'Monks of Storrington cease veal production', 27 September 1985, http://archive.catholicherald.co.uk/article/27th-september-1985/1/monks-of-storrington-cease-veal-production (accessed 14 September 2012); A. Johnson, *Factory Farming,* Blackwell, Oxford, 1991.

14 Assured Food Standards website, http://www.redtractor.org.uk/Why-Red-Tractor (accessed 14 September 2012).

15 *Farmers Weekly,* 'Farming under fire', 2011, http://www.fwi.co.uk/business/farming-under-fire/ (accessed 14 September 2012).

17 NEW INGREDIENTS: RETHINKING OUR FOOD

1 J. Parfitt, M. Barthel and S. Macnaughton, 'Food waste within food supply chains: quantification and potential for change to 2050', *Phil. Trans. R. Soc. B*, 365, 27 September 2010, pp. 3065–81.

2 C. Nellemann et al., *The Environmental Food Crisis – The Environment's Role in Averting Future Food Crises*, a UNEP rapid response assessment, February 2009, United Nations Environment Programme, GRID-Arendal, www.unep.org/pdf/foodcrisis_lores.pdf.

3 S. Fairlie, *Meat – a Benign Extravagance*, Permanent Publications, 2010, see pp. 46–50.

4 B. White, *Alaska Salmon Fisheries Enhancement Program Report 2010*, Annual Report, Alaska Department of Fish and Game, 2011, http://www.adfg.alaska.gov/FedAidPDFs/FMR11-04.pdf (accessed 27 September 2012).

5 G. P. Knapp, 'Alaska Salmon Ranching: an Economic Review of the Alaska Salmon Hatchery Programme', in B. R. Howell, E. Moksness and T. Svasand (eds), *Stock Enhancement and Sea Ranching*, Fishing News Books, Blackwell Science, Oxford, 1999, pp. 537–56.

6 M. Kaeriyama, 'Hatchery Programmes and Stock Management of Salmonid Populations in Japan', in Howell et al., *Stock Enhancement and Sea Ranching*.

7 S. D. Sedgwick, *Salmon Farming Handbook*, Fishing News Books, Surrey, 1988.

8 ABC News, 'Google Co-founder: The man behind the $300k test-tube burger', 5 August 2013, http://abcnews.go.com/Technology/google-founder-sergey-brin-man-300k-test-tube/story?id=19872215 (accessed 16 August 2013).

9 B. Gates, *The Future of Food*, The Gates Notes, 2013, http://www.the-gatesnotes.com/Features/Future-of-Food (accessed 21 May 2013).

10 Ibid.

18 THE SOLUTION: HOW TO AVERT THE COMING FOOD CRISIS

1 UN Food and Agriculture Organization (FAO), *World Livestock 2011: Livestock in Food Security*, Rome, 2011.

2 J. Bruinsma, *The resource outlook to 2050: By how much do land, water and crop yields need to increase by 2050?*, FAO Expert Meeting on 'How to Feed the World in 2050', FAO, Rome, 24–26 June 2009; United Nations, *World Economic and Social Survey 2011: The great green technological transformation*, United Nations, New York, 2011.

3 Government Office for Science, *Foresight Project on Global Food and Farming Futures Synthesis Report C1: Trends in food demand and production*, 2011; S. Msangi and M. Rosegrant, *World agriculture in a dynamically-changing environment: IFPRI's long term outlook for food and agriculture under additional demand and constraints*, Expert Meeting on 'How to feed the World in 2050', Rome, FAO; H. Steinfeld et al., *Livestock's Long Shadow, environmental issues and options*, FAO, Rome, 2006, Introduction, p. 12.

4 Calculated from FAOSTAT online figures for global grain harvest (2009) and food value of cereals. Based on a calorific intake of 2,500 kcalories per person per day.

5 Steinfeld et al., *Livestock's Long Shadow*, p. 43.

6 David Pimentel et al., 'Reducing energy inputs in the US food system', *Human Ecology*, 36 (2008), pp. 459–71.

7 FAO, *State of the World Fisheries and Aquaculture*, 2010.

8 T. Stuart, *Waste: Uncovering the global food scandal*, Penguin, 2009.

9 S. Fairlie, *Meat – a Benign Extravagance*, Permanent Publications, 2010, see pp. 46–50.

10 Tristram Stuart, personal communication, 2 May 2012.

11 BBC, 'French village Pince to hand out chickens to cut waste', 28 March 2012, http://www.bbc.co.uk/news/world-europe-17540287 (accessed 17 September 2012).

12 Nick Cliffe, Project Manager, Closed Loop, Dagenham, London, personal communication, 2 May 2012.

13 Stuart, personal communication, 2 May 2012.

14 J. Parfitt, M. Barthel and S. Macnaughton, 'Food waste within food supply chains: quantification and potential for change to 2050', *Phil. Trans. R. Soc. B*, 365, 27 September 2010, pp. 3065–81; Institution of Mechanical Engineers, *Population: One planet, too many people?*, 2011.

15 J. Gustavsson, C. Cederberg, U. Sonesson et al., *Global Food Losses and Food Waste: extent, causes and prevention*, FAO, Rome, 2011, www.fao.org/fileadmin/user_upload/ags/publications/GFL_web.pdf.

16 UN FAO, 2013, Food wastage footprint: impact on natural resources, http://www.fao.org/docrep/018/i3347e/i3347e.pdf (accessed 13th September 2013).

17 P. Stevenson, 'Feeding nine billion: How much extra do we need to produce?', 13 June 2013, http://www.eating-better.org/blog/3/Feeding-nine-billion-how-much-extra-food-do-we-need-to-produce.html (accessed 25 July 2013).

18 K. Lock et al., 'Health, agricultural, and economic effects of adoption of healthy diet recommendations', *Lancet*, vol. 376, issue 9753 (2010), pp. 1699–1709.

19 S. Friel et al., 'Public health benefits of strategies to reduce greenhouse-gas emissions: food and agriculture', *Lancet*, vol. 374, issue 9706 (2009), pp. 2016–25.

20 Stuart, *Waste*.

21 Lester R. Brown, *Plan B 4.0: Mobilizing to save civilization*, Earth Policy Institute, W. W. Norton, 2009.

22 EEA, *The European environment – state and outlook 2010: synthesis,* European Environment Agency, Copenhagen, 2010.

23 European Commission, Joint Reseach Centre, http://eusoils.jrc.ec.europa.eu/library/themes/Salinization/ (accessed 17 September 2012).

24 N. V. Fedoroff et al., 'Radically rethinking agriculture for the 21st century', *Science*, 327 (12 February 2010), pp. 833–4.

25 Own calculation

19 CONSUMER POWER: WHAT YOU CAN DO

1 J. Blythman, *What to Eat*, Fourth Estate, London, 2012.

EPILOGUE

1 P. Stevenson, 'Feeding nine billion: How much extra do we need to produce?', 13 June 2013, http://www.eating-better.org/blog/3/Feeding-nine-billion-how-much-extra-food-do-we-need-to-produce.html (accessed 25 July 2013).

Index

A plea from the author, Philip Lymbery,
CEO of leading farm animal welfare organisation
Compassion in World Farming.

To join our campaign today,
please visit
ciwf.org

JEFF FORSHAW
THE QUANTUM
UNIVERSE:
EVERYTHING THAT
CAN HAPPEN
DOES HAPPEN

UTHORS

Brian Cox is a Professor of Particle Physics and Royal Society University Research Fellow at the University of Manchester, and works at the CERN laboratory in Geneva. He is also a popular presenter on TV and radio.

Jeff Forshaw is Professor of Theoretical Physics at the University of Manchester, specializing in the physics of elementary particles. He was awarded the Institute of Physics Maxwell Medal in 1999 for outstanding contributions to theoretical physics.

BRIAN COX & JEFF FORSHAW
THE QUANTUM UNIVERSE: EVERYTHING THAT CAN HAPPEN DOES HAPPEN

PENGUIN BOOKS

PENGUIN BOOKS

Published by the Penguin Group
Penguin Books Ltd, 80 Strand, London WC2R ORL, England
Penguin Group (USA) Inc., 375 Hudson Street, New York, New York 10014, USA
Penguin Group (Canada), 90 Eglinton Avenue East, Suite 700, Toronto, Ontario,
Canada M4P 2Y3 (a division of Pearson Penguin Canada Inc.)
Penguin Ireland, 25 St Stephen's Green, Dublin 2, Ireland (a division of Penguin Books Ltd)
Penguin Group (Australia), 707 Collins Street, Melbourne, Victoria 3008, Australia
(a division of Pearson Australia Group Pty Ltd)
Penguin Books India Pvt Ltd, 11 Community Centre, Panchsheel Park, New Delhi – 110 017, India
Penguin Group (NZ), 67 Apollo Drive, Rosedale, Auckland 0632, New Zealand
(a division of Pearson New Zealand Ltd)
Penguin Books (South Africa) (Pty) Ltd, Block D, Rosebank Office Park,
181 Jan Smuts Avenue, Parktown North, Gauteng 2193, South Africa

Penguin Books Ltd, Registered Offices: 80 Strand, London WC2R ORL, England

www.penguin.com

First published by Allen Lane 2011
Published in Penguin Books 2012
008

Cover Art Direction by Peter Saville

Thanks to Paul Hetherington for recommending the front-cover font:
Lÿon, designed by Radim Pesko and Karl Nawrot

Typeset by Jouve (UK), Milton Keynes
Printed in England by Clays Ltd, St Ives plc

ISBN: 978-0-241-95270-2

www.greenpenguin.co.uk

MIX
Paper from
responsible sources
FSC™ C018179
www.fsc.org

Penguin Books is committed to a sustainable
future for our business, our readers and our planet.
This book is made from Forest Stewardship
Council™ certified paper.

ALWAYS LEARNING

PEARSON

Contents

Acknowledgements

We'd like to thank the many colleagues and friends who helped us 'get things right' and provided a great deal of valuable input and advice. Particular thanks go to Mike Birse, Gordon Connell, Mrinal Dasgupta, David Deutsch, Nick Evans, Scott Kay, Fred Loebinger, Dave McNamara, Peter Millington, Peter Mitchell, Douglas Ross, Mike Seymour, Frank Swallow and Niels Walet.

We owe a great debt of gratitude to our families – to Naomi and Isabel, and to Gia, Mo and George – for their support and encouragement, and for coping so well in the face of our preoccupations.

Finally, we thank our publisher and agents (Sue Rider and Diane Banks) for their patience, encouragement and very capable support. A special thanks is due to our editor, Will Goodlad.

1. *Something Strange Is Afoot*

Quantum. The word is at once evocative, bewildering and fascinating. Depending on your point of view, it is either a testament to the profound success of science or a symbol of the limited scope of human intuition as we struggle with the inescapable strangeness of the subatomic domain. To a physicist, quantum mechanics is one of the three great pillars supporting our understanding of the natural world, the others being Einstein's theories of Special and General Relativity. Einstein's theories deal with the nature of space and time and the force of gravity. Quantum mechanics deals with everything else, and one can argue that it doesn't matter a jot whether it is evocative, bewildering or fascinating; it's simply a physical theory that describes the way things behave. Measured by this pragmatic yardstick, it is quite dazzling in its precision and explanatory power. There is a test of quantum electrodynamics, the oldest and most well understood of the modern quantum theories, which involves measuring the way an electron behaves in the vicinity of a magnet. Theoretical physicists worked hard for years using pens, paper and computers to predict what the experiments should find. Experimenters built and operated delicate experiments to tease out the finer details of Nature. Both camps independently returned precision results, comparable in their accuracy to measuring the distance between Manchester and New York to within a few centimetres. Remarkably, the number returned by the experimenters agreed exquisitely with that computed by the theorists; measurement and calculation were in perfect agreement.

This is impressive, but it is also esoteric, and if mapping the miniature were the only concern of quantum theory, you might be forgiven for wondering what all the fuss is about. Science, of course, has no brief to be useful, but many of the technological and social changes that have revolutionized our lives have arisen out of fundamental research carried out by modern-day explorers whose only motivation is to better understand the world around them. These curiosity-led voyages of discovery across all scientific disciplines have delivered increased life expectancy, intercontinental air travel, modern telecommunications, freedom from the drudgery of subsistence farming and a sweeping, inspiring and humbling vision of our place within an infinite sea of stars. But these are all in a sense spin-offs. We explore because we are curious, not because we wish to develop grand views of reality or better widgets.

Quantum theory is perhaps the prime example of the infinitely esoteric becoming the profoundly useful. Esoteric, because it describes a world in which a particle really can be in several places at once and moves from one place to another by exploring the entire Universe simultaneously. Useful, because understanding the behaviour of the smallest building blocks of the Universe underpins our understanding of everything else. This claim borders on the hubristic, because the world is filled with diverse and complex phenomena. Notwithstanding this complexity, we have discovered that everything is constructed out of a handful of tiny particles that move around according to the rules of quantum theory. The rules are so simple that they can be summarized on the back of an envelope. And the fact that we do not need a whole library of books to explain the essential nature of things is one of the greatest mysteries of all.

It appears that the more we understand about the elemental nature of the world, the simpler it looks. We will, in due course, explain what these basic rules are and how the tiny building blocks conspire to form the world. But, lest we get too dazzled by the underlying simplicity of the Universe, a word of caution is in order: although the basic rules of the game are simple, their consequences are not necessarily easy to calculate. Our everyday experience of the world is dominated by the relationships between vast collections of many trillions of atoms, and to try to derive the behaviour of plants and people from first principles would be folly. Admitting this does not diminish the point – all phenomena really are underpinned by the quantum physics of tiny particles.

Consider the world around you. You are holding a book made of paper, the crushed pulp of a tree.[1] Trees are machines able to take a supply of atoms and molecules, break them down and rearrange them into cooperating colonies composed of many trillions of individual parts. They do this using a molecule known as chlorophyll, composed of over a hundred carbon, hydrogen and oxygen atoms twisted into an intricate shape with a few magnesium and nitrogen atoms bolted on. This assembly of particles is able to capture the light that has travelled the 93 million miles from our star, a nuclear furnace the volume of a million earths, and transfer that energy into the heart of cells, where it is used to build molecules from carbon dioxide and water, giving out life-enriching oxygen as it does so. It's these molecular chains that form the superstructure of trees and all living things, and the paper in your book. You can read the book and understand the words because you have eyes that can convert the scattered light from the pages into electrical impulses that are interpreted by your brain, the most complex structure we know of in the Universe. We have discovered that all these things are nothing more than assemblies of atoms, and that the wide variety of atoms are constructed using only three particles: electrons, protons and neutrons. We have also discovered that the protons and neutrons are themselves made up of smaller entities called quarks, and that is where things stop, as far as we can tell today. Underpinning all of this is quantum theory.

The picture of the Universe we inhabit, as revealed by modern physics, is therefore one of underlying simplicity; elegant phenomena dance away out of sight and the diversity of the macroscopic world emerges. This is perhaps the crowning achievement of modern science; the reduction of the tremendous complexity in the world, human beings included, to a description of the behaviour of just

1. Unless of course you are reading an electronic version of the book, in which case you will need to exercise your imagination.

a handful of tiny subatomic particles and the four forces that act between them. The best descriptions we have of three of the forces, the strong and weak nuclear forces that operate deep within the atomic nucleus and the electromagnetic force that glues atoms and molecules together, are provided by quantum theory. Only gravity, the weakest but perhaps most familiar of the four, does not at present have a satisfactory quantum description.

Quantum theory does, admittedly, have something of a reputation for weirdness, and there have been reams of drivel penned in its name. Cats can be both alive and dead; particles can be in two places at once; Heisenberg says everything is uncertain. These things are all true, but the conclusion so often drawn – that since something strange is afoot in the microworld, we are steeped in mystery – is most definitely not. Extrasensory perception, mystical healing, vibrating bracelets to protect us from radiation and who-knows-what-else are regularly smuggled into the pantheon of the possible under the cover of the word 'quantum'. This is nonsense born from a lack of clarity of thought, wishful thinking, genuine or mischievous misunderstanding, or some unfortunate combination of all of the above. Quantum theory describes the world with precision, using mathematical laws as concrete as anything proposed by Newton or Galileo. That's why we can compute the magnetic response of an electron with such exquisite accuracy. Quantum theory provides a description of Nature that, as we shall discover, has immense predictive and explanatory power, spanning a vast range of phenomena from silicon chips to stars.

Our goal in writing this book is to demystify quantum theory; a theoretical framework that has proved famously confusing, even to its early practitioners. Our approach will be to adopt a modern perspective, with the benefit of a century of hindsight and theoretical developments. To set the scene, however, we would like to begin our journey at the turn of the twentieth century, and survey some of the problems that led physicists to take such a radical departure from what had gone before.

Quantum theory was precipitated, as is often the case in science, by the discovery of natural phenomena that could not be explained by the scientific paradigms of the time. For quantum theory these were many and varied. A cascade of inexplicable results caused excitement and confusion, and catalysed a period of experimental and theoretical innovation that truly deserves to be accorded that most clichéd label: a golden age. The names of the protagonists are etched into the consciousness of every student of physics and dominate undergraduate lecture courses even today: Rutherford, Bohr, Planck, Einstein, Pauli, Heisenberg, Schrödinger, Dirac. There will probably never again be a time in history where so many names become associated with scientific greatness in the pursuit of a single goal; a new theory of the atoms and forces that make up the physical world. In 1924, looking back on the early decades of quantum theory, Ernest Rutherford, the New-Zealand-born physicist who discovered the atomic nucleus in Manchester, wrote: 'The year 1896 . . . marked the beginning of what has been aptly termed the heroic age of Physical Science. Never before in the history of physics has there been witnessed such a period of intense activity when discoveries of fundamental importance have followed one another with such bewildering rapidity.'

But before we travel to nineteenth-century Paris and the birth of quantum theory, what of the word 'quantum' itself? The term entered physics in 1900, through the work of Max Planck. Planck was concerned with finding a theoretical description of the radiation emitted by hot objects – the so-called 'black body radiation' – apparently because he was commissioned to do so by an electric lighting company: the doors to the Universe have occasionally been opened by the prosaic. We will discuss Planck's great insight in more detail later in the book but, for the purposes of this brief introduction, suffice to say he found that he could only explain the properties of black body radiation if he assumed that light is emitted in little packets of energy, which he called 'quanta'. The word itself means 'packets' or 'discrete'. Initially, he thought that this was purely a mathematical trick, but subsequent work in 1905 by Albert Einstein on a phenomenon called the photoelectric effect gave further support to the quantum hypothesis. These results were suggestive, because little packets of energy might be taken to be synonymous with particles.

The idea that light consists of a stream of little bullets had a long and illustrious history dating back to the birth of modern physics and Isaac Newton. But Scottish physicist James Clerk Maxwell appeared to have comprehensively banished any lingering doubts in 1864 in a series of papers that Albert Einstein later described as 'the most profound and the most fruitful that physics has experienced since the time of Newton'. Maxwell showed that light is an electromagnetic wave, surging through space, so the idea of light as a wave had an immaculate and, it seemed, unimpeachable pedigree. Yet, in a series of experiments from 1923 to 1925 conducted at Washington University in Saint Louis, Arthur Compton and his co-workers succeeded in bouncing the quanta of light off electrons. Both behaved rather like billiard balls, providing clear evidence that Planck's theoretical conjecture had a firm grounding in the real world. In 1926, the light quanta were christened 'photons'. The evidence was incontrovertible – light behaves both as a wave and as a particle. That signalled the end for classical physics, and the end of the beginning for quantum theory.

2.   *Being in Two Places at Once*

Ernest Rutherford cited 1896 as the beginning of the quantum revolution because this was the year Henri Becquerel, working in his laboratory in Paris, discovered radioactivity. Becquerel was attempting to use uranium compounds to generate X-rays, discovered just a few months previously by Wilhelm Röntgen in Würzburg. Instead, he found that uranium compounds emit 'les rayons uraniques', which were able to darken photographic plates even when they were wrapped in thick paper that no light could penetrate. The importance of Becquerel's rays was recognized in a review article by the great scientist Henri Poincaré as early as 1897, in which he wrote presciently of the discovery 'one can think today that it will open for us an access to a new world which no one suspected'. The puzzling thing about radioactive decay, which proved to be a hint of things to come, was that nothing seemed to trigger the emission of the rays; they just popped out of substances spontaneously and unpredictably.

In 1900, Rutherford noted the problem: 'all atoms formed at the same time should last for a definite interval. This, however, is contrary to the observed law of transformation, in which the atoms have a life embracing all values from zero to infinity.' This randomness in the behaviour of the microworld came as a shock because, until this point, science was resolutely deterministic. If, at some instant in time, you knew everything it is possible to know about something, then it was believed you could predict with absolute certainty what would happen to it in the future. The breakdown of this kind of predictability is a key feature of quantum theory: it deals with probabilities rather than certainties, not because we lack absolute knowledge, but because some aspects of Nature are, at their very heart, governed by the laws of chance. And so we now understand that it is simply impossible to predict when a particular atom will decay. Radioactive decay was science's first encounter with Nature's dice, and it confused many physicists for a long time.

Clearly, there was something interesting going on inside atoms, although their internal structure was entirely unknown. The key discovery was made by Rutherford in 1911, using a radioactive source to bombard a very thin sheet of gold with a type of radiation known as alpha particles (we now know them to be the nuclei of helium atoms). Rutherford, with his co-workers Hans Geiger and Ernest Marsden, discovered to their immense surprise that around 1 in 8,000 alpha particles did not fly through the gold as expected, but bounced straight back. Rutherford later described the moment in characteristically colourful language: 'It was quite the most incredible event that has ever happened to me in my life. It was almost as incredible as if you fired a 15-inch shell at a piece of tissue paper and it came back and hit you.' By all accounts, Rutherford was an engaging and no-nonsense individual: he once described a self-important official as being 'like a Euclidean point: he has position without magnitude'.

Rutherford calculated that his experimental results could be explained only if the atom consists of a very small nucleus at the centre with electrons orbiting around it. At the time, he probably had in mind a situation similar to the planets orbiting around the Sun. The nucleus contains almost all the mass of the atom, which is why it is capable of stopping his '15-inch shell' alpha particles and bouncing them back. Hydrogen, the simplest element, has a nucleus consisting of a single proton with a radius of around 1.75×10^{-15} m. If you are unfamiliar with this notation, this means 0.00000000000000175 metres, or in words, just under two thousand million millionths of a metre. As far as we can tell today, the single electron is like Rutherford's self-important official, point-like, and it orbits around the hydrogen nucleus at a radius around 100,000 times larger than the nuclear diameter. The nucleus has a positive electric charge and the electron has a negative electric charge, which means there is an attractive force between them analogous to the force of gravity that holds the Earth in orbit around the Sun. This in turn means that atoms are largely empty space. If you imagine a nucleus scaled up to the size of a tennis ball, then the tiny electron would be smaller than a mote of dust orbiting at a distance of a kilometre. These figures are quite surprising because solid matter certainly does not feel very empty.

Rutherford's nuclear atom raised a host of problems for the physicists of the day. It was well known, for instance, that the electron should lose energy as it moves in orbit around the atomic nucleus, because all electrically charged things radiate energy away if they move in curved paths. This is the idea behind the operation of the radio transmitter, inside which electrons are made to jiggle and, as a result, electromagnetic radio waves issue forth. Heinrich Hertz invented the radio transmitter in 1887, and by the time Rutherford discovered the atomic nucleus there was a commercial radio station sending messages across the Atlantic from Ireland to Canada. So there was clearly nothing wrong with the theory of orbiting charges and the emission of radio waves, and that meant confusion for those trying to explain how electrons can stay in orbit around nuclei.

A similarly inexplicable phenomenon was the mystery of the light emitted by atoms when they are heated. As far back as 1853, the Swedish scientist Anders Jonas Ångstrom discharged a spark through a tube of hydrogen gas and analysed the emitted light. One might assume that a glowing gas would produce all the colours of the rainbow; after all, what is the Sun but a glowing ball of gas? Instead, Ångstrom observed that hydrogen emits light of three very distinct colours: red, blue-green and violet, like a rainbow with three pure, narrow arcs. It was soon discovered that each of the chemical elements behaves in this way, emitting a unique barcode of colours. By the time Rutherford's nuclear atom came along, a scientist named Heinrich Gustav Johannes Kayser had compiled a six-volume, 5,000-page reference work entitled *Handbuch der Spectroscopie*, documenting all the shining coloured lines from the known elements. The question, of course, was why? Not only 'why, Professor Kayser?' (he must have been great fun at dinner parties), but also 'why the profusion of coloured lines?' For over sixty years the science of

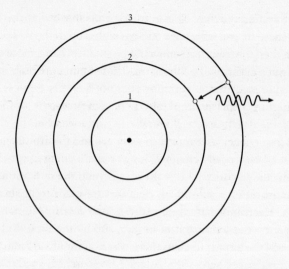

Figure 2.1: Bohr's model of an atom, illustrating the emission of a photon (wavy line) as an electron drops down from one orbit to another (indicated by the arrow).

spectroscopy, as it was known, had been simultaneously an observational triumph and a theoretical wasteland.

In March 1912, fascinated by the problem of atomic structure, Danish physicist Niels Bohr travelled to Manchester to meet with Rutherford. He later remarked that trying to decode the inner workings of the atom from the spectroscopic data had been akin to deriving the foundations of biology from the coloured wing of a butterfly. Rutherford's solar system atom provided the clue Bohr needed, and by 1913 he had published the first quantum theory of atomic structure. The theory certainly had its problems, but it did contain several key insights that triggered the development of modern quantum theory. Bohr concluded that electrons can only take up certain orbits around the nucleus with the lowest-energy orbit lying closest in. He also said that electrons are able to jump between these orbits. They jump out to a higher orbit when they receive energy (from a spark in a tube for example) and, in time, they will fall back down, emitting light in the process. The colour of the light is determined directly by the energy difference between the two orbits. Figure 2.1 illustrates the basic idea; the arrow represents an electron as it jumps from the third energy level down to the second energy level, emitting light (represented by the wavy line) as it does so. In Bohr's model, the electron is only allowed to orbit the proton in one of these special, 'quantized', orbits; spiralling inwards is simply forbidden. In this way Bohr's model allowed him to compute the wavelengths (i.e. colours) of light observed by Ångstrom – they were to be attributed to an electron hopping from the fifth orbit down to the second orbit (the violet light), from the fourth orbit down to the second (the blue-green light) or from the third orbit down to the second (the red light). Bohr's model also correctly predicted that there should be light emitted as a result of electrons hopping down to the first orbit. This light is in the ultra-violet part of the spectrum, which is not visible to the human eye, and so it was not seen by Ångstrom. It had, however, been spotted in 1906 by Harvard physicist Theodore Lyman, and Bohr's model described Lyman's data beautifully.

Although Bohr did not manage to extend his model beyond hydrogen, the ideas he introduced could be applied to other atoms. In particular, if one supposes that the atoms of each element have a unique set of orbits then they will only ever emit light of certain colours. The colours emitted by an atom therefore act like a fingerprint, and astronomers were certainly not slow to exploit the uniqueness of the spectral lines emitted by atoms as a way to determine the chemical composition of the stars.

Bohr's model was a good start, but it was clearly unsatisfactory: just why were electrons forbidden from spiralling inwards when it was known that they should lose energy by emitting electromagnetic waves – an idea so firmly rooted in reality with the advent of radio? And why are the electron orbits quantized in the first place?

And what about the heavier elements beyond hydrogen: how was one to go about understanding their structure?

Half-baked though Bohr's theory may have been, it was a crucial step, and an example of how scientists often make progress. There is no point at all in getting completely stuck in the face of perplexing and often quite baffling evidence. In such cases, scientists often make an ansatz, an educated guess if you like, and then proceed to compute the consequences of the guess. If the guess works, in the sense that the subsequent theory agrees with experiment, then you can go back with some confidence to try to understand your initial guess in more detail. Bohr's ansatz remained successful but unexplained for thirteen years.

We will revisit the history of these early quantum ideas as the book unfolds, but for now we leave a mass of strange results and half-answered questions, because this is what the early founders of quantum theory were faced with. In summary, following Planck, Einstein introduced the idea that light is made up of particles, but Maxwell had shown that light also behaves like waves. Rutherford and Bohr led the way in understanding the structure of atoms, but the way that electrons behave inside atoms was not in accord with any known theory. And the diverse phenomena collectively known as radioactivity, in which atoms spontaneously split apart for no discernible reason, remained a mystery, not least because it introduced a disturbingly random element into physics. There was no doubt about it: something strange was afoot in the subatomic world.

The first step towards a consistent, unified answer is widely credited to the German physicist Werner Heisenberg, and what he did represented nothing less than a completely new approach to the theory of matter and forces. In July of 1925, Heisenberg published a paper throwing out the old hotchpotch of ideas and half-theories, including Bohr's model of the atom, and ushered in an entirely new approach to physics. He began: 'In this paper it will be attempted to secure the foundations for a quantum theoretical mechanics which is exclusively based on relations between quantities which in principle are observable.' This is an important step, because Heisenberg is saying that the underlying mathematics of quantum theory need not correspond to anything with which we are familiar. The job of quantum theory should be to predict directly observable things, such as the colour of the light emitted from hydrogen atoms. It should not be expected to provide some kind of satisfying mental picture for the internal workings of the atom, because this is not necessary and it may not even be possible. In one fell swoop, Heisenberg removed the conceit that the workings of Nature should necessarily accord with common sense. This is not to say that a theory of the subatomic world shouldn't be expected to accord with our everyday experience when it comes to describing the motion of large objects, like tennis balls and aircraft. But we should be prepared to abandon the prejudice that small things behave like smaller versions of big things, if this is what our experimental observations dictate.

There is no doubt that quantum theory is tricky, and absolutely no doubt that Heisenberg's approach is extremely tricky indeed. Nobel Laureate Steven Weinberg, one of the greatest living physicists, wrote of Heisenberg's 1925 paper:

> If the reader is mystified at what Heisenberg was doing, he or she is not alone. I have tried several times to read the paper that Heisenberg wrote on returning from Heligoland, and, although I think I understand quantum mechanics, I have never understood Heisenberg's motivations for the mathematical steps in his paper. Theoretical physicists in their most successful work tend to play one of two roles: they are either sages or magicians ... It is usually not difficult to understand the papers of sage-physicists, but the papers of magician-physicists are often incomprehensible. In that sense, Heisenberg's 1925 paper was pure magic.

Heisenberg's philosophy, though, is not pure magic. It is simple and it lies at the heart of our approach in this book: the job of a theory of Nature is to make predictions for quantities that can be compared to experimental results. We are not mandated to produce a theory that bears any relation to the way we perceive the world at large. Fortunately, although we are adopting Heisenberg's philosophy, we shall be following Richard Feynman's more transparent approach to the quantum world.

We've used the word 'theory' liberally in the last few pages and, before we continue to build quantum theory, it will be useful to take a look at a simpler theory in more detail. A good scientific theory specifies a set of rules that determine what can and cannot happen to some portion of the world. They must allow predictions to be made that can be tested by observation. If the predictions are shown to be false, the theory is wrong and must be replaced. If the predictions are in accord with observation, the theory survives. No theory is 'true' in the sense that it must always be possible to falsify it. As the biologist Thomas Huxley wrote: 'Science is organized common sense where many a beautiful theory was killed by an ugly fact.' Any theory that is not amenable to falsification is not a scientific theory – indeed one might go as far as to say that it has no reliable information content at all. The reliance on falsification is why scientific theories are different from matters of opinion. This scientific meaning of the word 'theory', by the way, is different from its ordinary usage, where it often suggests a degree of speculation. Scientific theories may be speculative if they have not yet been confronted with the evidence, but an established theory is something that is supported by a large body of evidence. Scientists strive to develop theories that encompass as wide a range of phenomena as possible, and physicists in particular tend to get very excited about the prospect of describing everything that can happen in the material world in terms of a small number of rules.

One example of a good theory that has a wide range of applicability is Isaac Newton's theory of gravity, published on 5 July 1687 in his *Philosophiæ Naturalis Principia Mathematica*. It was the first modern scientific theory, and although it has subsequently been shown to be inaccurate in some circumstances, it was so good that it is still used today. Einstein developed a more precise theory of gravity, General Relativity, in 1915.

Newton's description of gravity can be captured in a single mathematical equation:

$$F = G \frac{m_1 m_2}{r^2}$$

This may look simple or complicated, depending on your mathematical background. We do occasionally make use of mathematics as this book unfolds. For those readers who find the maths difficult, our advice is to skip over the equations without worrying too much. We will always try to emphasize the key ideas in a way that does not rely on the maths. The maths is included mainly because it allows us to really explain why things are the way they are. Without it, we should have to resort to the physicist-guru mentality whereby we pluck profundities out of thin air, and neither author would be comfortable with guru status.

Now let us return to Newton's equation. Imagine there is an apple hanging precariously from a branch. The consideration of the force of gravity triggered by a particularly ripe apple bouncing off his head one summer's afternoon was, according to folklore, Newton's route to his theory. Newton said that the apple is subject to the force of gravity, which pulls it towards the ground, and that force is represented in the equation by the symbol F. So, first of all, the equation allows you to calculate the force on the apple if you know what the symbols on the right-hand side of the equals sign mean. The symbol r stands for the distance between the centre of the apple and the centre of the Earth. It's r^2 because Newton discovered that the force depends on the square of the distance between the objects. In non-mathematical language, this means that if you double the distance between the apple and the centre of the Earth, the gravitational force drops by a factor of 4. If you triple the distance, it drops by a factor of 9. And so on. Physicists call this behaviour an inverse square law. The symbols m_1 and m_2 stand for the mass of the apple and the mass of the Earth, and their appearance encodes Newton's recognition that the gravitational force of attraction between two objects depends on the product of their masses. That then begs the question: what is mass? This is an interesting question in itself, and for the deepest answer available today we'll need to wait until we talk about a quantum particle known as the Higgs boson. Roughly speaking, mass is a measure of the amount of 'stuff' in something; the Earth is more massive than the apple. This kind of statement isn't really good enough, though. Fortunately Newton also provided a way of measuring the mass of an object independently of his law of gravitation, and it is encapsulated in the second of his three laws of motion, the ones so beloved of every high school student of physics:

1. Every object remains in a state of rest or uniform motion in a straight line unless it is acted upon by a force;
2. An object of mass m undergoes an acceleration a when acted upon by a force F. In the form of an equation, this reads $F = ma$;
3. To every action there is an equal and opposite reaction.

Newton's three laws provide a framework for describing the motion of things under the influence of a force. The first law describes what happens to an object when no forces act: the object either just sits still or moves in a straight line at constant speed. We shall be looking for an equivalent statement for quantum particles later on, and it's not giving the game away too much to say that quantum particles do not just sit still – they leap around all over the place even when no forces are present. In fact, the very notion of 'force' is absent in the quantum theory, and so Newton's second law is bound for the wastepaper basket too. We do mean that, by the way – Newton's laws are heading for the bin because they have been exposed as only approximately correct. They work well in many instances but fail totally when it comes to describing quantum phenomena. The laws of quantum theory replace Newton's laws and furnish a more accurate description of the world. Newton's physics emerges out of the quantum description, and it is important to realize that the situation is not 'Newton for big things and quantum for small': it is quantum all the way.

Although we aren't really going to be very interested in Newton's third law here, it does deserve a comment or two for the enthusiast. The third law says that forces come in pairs; if I stand up then my feet press into the Earth and the Earth responds by pushing back. This implies that for a 'closed' system the net force acting on it is zero, and this in turn means that the total momentum of the system is conserved. We shall use the concept of momentum throughout this book and, for a single particle, it is defined to be the product of the particle's mass and its speed, which we write $p = mv$. Interestingly, momentum conservation does have some meaning in quantum theory, even though the idea of force does not.

For now though, it is Newton's second law that interests us. $F = ma$ says that if you apply a known force to something and measure its acceleration then the ratio of the force to the acceleration is its mass. This in turn assumes we know how to define force, but that is not so hard. A simple but not very accurate or practical way would be to measure force in terms of the pull exerted by some standard thing; an average tortoise, let us say, walking in a straight line with a harness attaching it to the object being pulled. We could term the average tortoise the 'SI Tortoise' and keep it in a sealed box in the International Bureau of Weights and Measures in Sèvres, France. Two harnessed tortoises would exert twice the force, three would exert three times the force and so on. We could then always talk about any push or pull in terms of the number of average tortoises required to generate it.

Given this system, which is ridiculous enough to be agreed on by any international committee of standards,[1] we can simply pull an object with a tortoise and measure its acceleration, and this will allow us to deduce its mass using Newton's second law. We can then repeat the process for a second object to deduce its mass and then we can put both masses into the law of gravity to determine the force between the masses due to gravity. To put a tortoise-equivalent

1. But not so ridiculous when you consider that an oft-used unit of power, even to this day, is the 'horsepower'.

number on the gravitational force between two masses, though, we would still need to calibrate the whole system to the strength of gravity itself, and this is where the symbol G comes in.

G is a very important number, called 'Newton's gravitational constant', which encodes the strength of the gravitational force. If we doubled G, we would double the force, and this would make the apple accelerate at double the rate towards the ground. It therefore describes one of the fundamental properties of our Universe and we would live in a very different Universe if it took on a different value. It is currently thought that G takes the same value everywhere in the Universe, and that it has remained constant throughout all of time (it appears in Einstein's theory of gravity too, where it is also a constant). There are other universal constants of Nature that we'll meet in this book. In quantum mechanics, the most important is Planck's constant, named after quantum pioneer Max Planck and given the symbol h. We shall also need the speed of light, c, which is not only the speed that light travels in a vacuum but the universal speed limit. 'It is impossible to travel faster than the speed of light and certainly not desirable,' Woody Allen once said, 'as one's hat keeps blowing off.'

Newton's three laws of motion and the law of gravitation are all that is needed to understand motion in the presence of gravity. There are no other hidden rules that we did not state – just these few laws do the trick and allow us, for example, to understand the orbits of the planets in our solar system. Together, they severely restrict the sort of paths that objects are allowed to take when moving under the influence of gravity. It can be proved using only Newton's laws that all of the planets, comets, asteroids and meteors in our solar system are only allowed to move along paths known as conic sections. The simplest of these, and the one that the Earth follows to a very good approximation in its orbit around the Sun, is a circle. More generally, planets and moons move along orbital paths known as ellipses, which are like stretched circles. The other two conic sections are known as the parabola and the hyperbola. A parabola is the path that a cannonball takes when fired from the cannon. The final conic section, the hyperbola, is the path that the most distant object ever constructed by human kind is now following outwards to the stars. Voyager 1 is, at the time of writing, around 17,610,000,000 km from the Earth, and travelling away from the solar system at a speed of 538,000,000 km per year. This most beautiful of engineering achievements was launched in 1977 and is still in contact with the Earth, recording measurements of the solar wind on a tape recorder and transmitting them back with a power of 20 watts. Voyager 1, and her sister ship Voyager 2, are inspiring testaments to the human desire to explore our Universe. Both spacecraft visited Jupiter and Saturn and Voyager 2 went on to visit Uranus and Neptune. They navigated the solar system with precision, using gravity to slingshot them beyond the planets and into interstellar space. Navigators here on Earth used nothing more than Newton's laws to plot their courses between the inner and outer planets and outwards to the stars. Voyager 2 will sail close to Sirius, the brightest star in the skies, in just under 300,000 years. We did all this, and we know all this, because of Newton's theory of gravity and his laws of motion.

Newton's laws provide us with a very intuitive picture of the world. As we have seen, they take the form of equations – mathematical relationships between measurable quantities – that allow us to predict with precision how objects move around. Inherent in the whole framework is the assumption that objects are, at any instant, located somewhere and that, as time passes, objects move smoothly around from place to place. This seems so self-evidently true that it is hardly worth commenting upon, but we need to recognize that this is a prejudice. Can we really be sure that things are definitely here or there, and that they are not actually in two different places at the same time? Of course, your garden shed is not in any noticeable sense sitting in two distinctly different places at once – but how about an electron in an atom? Could that be both 'here' and 'there'? Right now that kind of suggestion sounds crazy, mainly because we can't picture it in our mind's eye, but it will turn out to be the way things actually work. At this stage in our narrative, all we are doing in making this strange-sounding statement is pointing out that Newton's laws are built on intuition, and that is like a house built on sand as far as fundamental physics is concerned.

There is a very simple experiment, first conducted by Clinton Davisson and Lester Germer at Bell Laboratories in the United States and published in 1927, which shows that Newton's intuitive picture of the world is wrong. Although apples, planets and people certainly appear to behave in a 'Newtonian' way, gliding from place to place in a regular and predictable fashion as time unfolds, their experiment showed that the fundamental building blocks of matter do not behave at all like this.

Davisson and Germer's paper begins: 'The intensity of scattering of a homogeneous beam of electrons of adjustable speed incident upon a single crystal of nickel has been measured as a function of direction.' Fortunately, there is a way to appreciate the key content of their findings using a simplified version of their experiment, known as the double-slit experiment. The experiment consists of a source that sends electrons towards a barrier with two small slits (or holes) cut into it. On the other side of the barrier, there is a screen that glows when an electron hits it. It doesn't matter what the source of electrons is, but practically speaking one can imagine a length of hot wire stretched out along the side of the experiment.[2] We've sketched the double-slit experiment in Figure 2.2.

Imagine pointing a camera at the screen and leaving the shutter open to take a long-exposure photograph of the little flashes of light emitted as, one by one, the electrons hit it. A pattern will build up, and the simple question is, what is the pattern? Assuming electrons are simply little particles that behave rather like apples or planets, we might expect the emergent pattern to look something like that shown in Figure 2.2. Some electrons go through the slits, most don't. The ones that make it through might bounce off the

2. Once upon a time, televisions operated using this idea. A stream of electrons generated by a hot wire was gathered, focused into a beam and deflected by a magnetic field on to a screen that glowed when the electrons hit it.

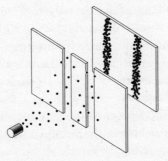

Figure 2.2: An electron-gun source fires electrons towards a pair of slits and, if the electrons behaved like 'regular' particles, we would expect to see hits on the screen that build up a pair of stripes, as illustrated. Remarkably, this is *not* what happens.

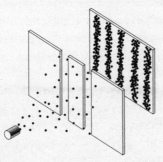

Figure 2.3: In reality the electrons do not hit the screen aligned with the slits. Instead they form a stripy pattern: electron by electron, the stripes build up over time.

edge of the slits a bit, which will spread them out, but the most hits, and therefore the brightest bits of the photograph, will surely appear directly aligned with the two slits.

This isn't what happens. Instead, the picture looks like Figure 2.3. A pattern like this is what Davisson and Germer published in their 1927 paper. Davisson subsequently received the 1937 Nobel Prize for the 'experimental discovery of electron diffraction by crystals'. He shared the prize, not with Germer, but with George Paget Thomson, who saw the same pattern independently in experiments at the University of Aberdeen. The alternating stripes of light and dark are known as an interference pattern, and interference is more usually associated with waves. To understand why, let's imagine doing the double-slit experiment with water waves instead of electrons.

Imagine a water tank with a wall midway down with two slits cut into it. The screen and camera could be replaced with a wave-height detector, and the hot wire with something that makes waves: a plank of wood along the side of the tank attached to a motor that keeps it dipping in and out of the water would do. The waves from the plank will travel across the surface of the water until they reach the wall. When a wave hits the wall, most of it will bounce back, but two small pieces will pass through the slits. These two new waves will spread outwards from the slits towards the wave-height detector. Notice that we used the term 'spread out' here, because the waves don't just carry on in a straight line from the slits. Instead, the slits act as two sources of new waves, each issuing forth in ever increasing semi-circles. Figure 2.4 illustrates what happens.

Figure 2.4. An aerial view of water waves emanating from two points in a tank of water (they are located at the top of the picture). The two circular waves overlap and interfere with each other. The 'spokes' are the regions where the two waves have cancelled each other out and the water there remains undisturbed.

The figure provides a striking visual demonstration of the behaviour of waves in water. There are regions where there are no waves at all, which seem to radiate out from the slits like the spokes of a wheel, whilst other regions are still filled with the peaks and troughs of the waves. The parallels with the pattern seen by Davisson, Germer and Thomson are striking. For the case of electrons hitting the screen, the regions where few electrons are detected correspond to the places in the tank where the water surface remains flat – the spokes you can see radiating outwards in the figure.

In a tank of water it is quite easy to understand how these spokes emerge: it is in the mixing and merging of the waves as they spread out from the slits. Because waves have peaks and troughs, when two waves meet they can either add or subtract. If two waves meet such that the peak of one is aligned with the trough of the other, they will cancel out and there will be no wave at that point. At a different place, the waves might arrive with their peaks in perfect alignment, and here they will add to produce a bigger wave. At each point in the water tank, the distance between it and the two slits will be a little different, which means that at some places the two waves will arrive with peaks together, at others with peaks and troughs aligned and, in most places, with some combination of these two extremes. The result will be an alternating pattern; an interference pattern.

In contrast to water waves, the experimentally observed fact that electrons also produce an interference pattern is very difficult to understand. According to Newton and common sense, the electrons emerge from the source, travel in straight lines towards the slits (because there are no forces acting on them – remember Newton's first law), pass through with perhaps a slight deflection if they glance off the edge of the slit, and continue in a straight line until they hit the screen. But this would not result in an interference pattern – it would give the pair of stripes as shown in Figure 2.2. Now we could suppose that there is some clever mechanism whereby the electrons exert a force on each other so as to deflect themselves from straight lines as they stream through the slits. But this can be ruled out because we can set the experiment up such that we send just one electron at a time from source to screen. You would have to wait, but, slowly and surely, as the electrons hit the screen one after the other, the stripy pattern would build up. This is very surprising because the stripy pattern is absolutely characteristic of waves interfering with each other, yet it emerges one electron at a time – dot by dot. It's a good mental exercise to try to imagine how it could be that, particle by particle, an interference pattern builds up as we fire tiny bullet-like particles at a pair of slits in a screen. It's a good exercise because it's futile, and a few hours of brain racking should convince you that a stripy pattern is inconceivable. Whatever those particles are that hit the screen, they are not behaving like 'regular' particles. It is as if the electrons are in some sense 'interfering with themselves'. The challenge for us is to come up with a theory that can explain what that means.

There is an interesting historical coda to this story, which provides a glimpse into the intellectual challenge raised by the double-slit experiment. George Paget Thomson was the son of J. J. Thomson, who himself received a Nobel Prize for his discovery of the electron in 1899. J. J. Thomson showed that the electron is a particle, with a particular electric charge and a particular mass; a tiny, point-like grain of matter. His son received the Nobel Prize forty years later for showing that the electron doesn't behave as his father might have expected. Thomson senior was not wrong; the electron does have a well-defined mass and electric charge, and every time we see one it appears as a little point of matter. It just doesn't seem to behave *exactly* like a regular particle, as Davisson, Germer and Thomson junior discovered. Importantly, though, it doesn't behave *exactly* like a wave either because the pattern is not built up as a result of some smooth deposition of energy; rather it is built out of many tiny dots. We always detect Thomson senior's single, point-like electrons.

Perhaps you can already see the need to engage with Heisenberg's way of thinking. The things we observe are particles, so we had better construct a theory of particles. Our theory must also be able to predict the appearance of the interference pattern that builds up as the electrons, one after another, pass through the slits and hit the screen. The details of how the electrons travel from source to slits to screen are not something we observe, and therefore need not be in accord with anything we experience in daily life. Indeed, the electron's 'journey' need not even be something we can talk about at all. All we have to do is find a theory capable of predicting that the electrons hit the screen in the pattern observed in the double-slit experiment. This is what we will do in the next chapter.

Lest we lapse into thinking that this is merely a fascinating piece of micro-physics that has little relevance to the world at large, we should say that the quantum theory of particles we develop to explain the double-slit experiment will also turn out to be capable of explaining the stability of atoms, the coloured light emitted from the chemical elements, radioactive decay, and indeed all of the great puzzles that perplexed scientists at the turn of the twentieth century. The fact that our framework describes the way electrons behave when locked away inside matter will also allow us to understand the workings of quite possibly the most important invention of the twentieth century: the transistor.

In the very final chapter of this book, we will meet a striking application of quantum theory that is one of the great demonstrations of the power of scientific reasoning. The more outlandish predictions of quantum theory usually manifest themselves in the behaviour of small things. But, because large things are made of small things, there are certain circumstances in which quantum physics is required to explain the observed properties of some of the most massive objects in the Universe – the stars. Our Sun is fighting a constant battle with gravity. This ball of gas a third of a million times more massive than our planet has a gravitational force at its surface that is almost twenty-eight times that at the Earth, which provides a powerful incentive for it to collapse in on itself. The collapse is prevented by the outward pressure generated by nuclear fusion reactions deep within the solar core as 600 million tonnes of hydrogen are converted into helium every second. Vast though our star is, burning fuel at such a ferocious rate must ultimately have consequences, and one day the Sun's fuel source will run out. The outward pressure will then cease and the force of gravity will reassert its grip unopposed. It would seem that nothing in Nature could stop a catastrophic collapse.

In reality, quantum physics steps in and saves the day. Stars that have been rescued by quantum effects in this way are known as white dwarves, and such will be the final fate of our Sun. At the end of this book we will employ our understanding of quantum mechanics to determine the maximum mass of a white dwarf star. This was first calculated, in 1930, by the Indian astrophysicist Subrahmanyan Chandrasekhar, and it turns out to be approximately 1.4 times the mass of our Sun. Quite wonderfully, that number can be computed using only the mass of a proton and the values of the three constants of Nature we have already met: Newton's gravitational constant, the speed of light, and Planck's constant.

The development of the quantum theory itself and the measurement of these four numbers could conceivably have been achieved without ever looking at the stars. It is possible to imagine a particularly agoraphobic civilization confined to deep caves below the surface of their home planet. They would have no concept of a sky, but they could have developed quantum theory. Just for fun, they may even decide to calculate the maximum mass of a giant sphere of gas. Imagine that, one day, an intrepid explorer chooses to venture above ground for the first time and gaze in awe at the spectacle above: a sky full of lights; a galaxy of a hundred billion suns arcing from horizon to horizon. The explorer would find, just as we have found from our vantage point here on Earth, that out there amongst the many fading remnants of dying stars there is not a single one with a mass exceeding the Chandrasekhar limit.

3.   What Is a Particle?

Our approach to quantum theory was pioneered by Richard Feynman, the Nobel Prize-winning, bongo-playing New Yorker described by his friend and collaborator Freeman Dyson as 'half genius, half buffoon'. Dyson later changed his opinion: Feynman could be more accurately described as 'all genius, all buffoon'. We will follow his approach in our book because it is fun, and probably the simplest route to understanding our Quantum Universe.

As well as being responsible for the simplest formulation of quantum mechanics, Richard Feynman was also a great teacher, able to transfer his deep understanding of physics to the page or lecture theatre with unmatched clarity and a minimum of fuss. His style was contemptuous of those who might seek to make physics more complicated than it need be. Even so, at the beginning of his classic undergraduate textbook series *The Feynman Lectures on Physics*, he felt the need to be perfectly honest about the counterintuitive nature of the quantum theory. Subatomic particles, Feynman wrote, 'do not behave like waves, they do not behave like particles, they do not behave like clouds, or billiard balls, or weights on springs, or like anything that you have ever seen'. Let's get on with building a model for exactly how they do behave.

As our starting point we will assume that the elemental building blocks of Nature are particles. This has been confirmed not only by the double-slit experiment, where the electrons always arrive at specific places on the screen, but by a whole host of other experiments. Indeed 'particle physics' is not called that for nothing. The question we need to address is: how do particles move around? Of course, the simplest assumption would be that they move in nice straight lines, or curved lines when acted upon by forces, as dictated by Newton. This cannot be correct though, because any explanation of the double-slit experiment requires that the electrons 'interfere with themselves' when they pass through the slits, and to do that they must in some sense be spread out. This therefore is the challenge: build a theory of point-like particles such that those same particles are also spread out. This is not as impossible as it sounds: we can do it if we let any single particle be *in many places at once*. Of course, that may still sound impossible, but the proposition that a particle should be in many places at once is actually a rather clear statement, even if it sounds silly. From now on, we'll refer to these counterintuitive, spread-out-yet-point-like particles as quantum particles.

With this 'a particle can be in more than one place at once' proposal, we are moving away from our everyday experience and into uncharted territory. One of the major obstacles to developing an understanding of quantum physics is the confusion this kind of thinking can engender. To avoid confusion, we should follow Heisenberg and learn to feel comfortable with views of the world that run counter to tangible experience. Feeling 'uncomfortable' can be mistaken for 'confusion', and very often students of quantum physics continue to attempt to understand what is happening in everyday terms. It is the resistance to new ideas that actually leads to confusion, not the inherent difficulty of the ideas themselves, because the real world simply doesn't behave in an everyday way. We must therefore keep an open mind and not be distressed by all the weirdness. Shakespeare had it right when Hamlet says, 'And therefore as a stranger give it welcome. There are more things in heaven and earth, Horatio, Than are dreamt of in your philosophy.'

A good way to begin is to think carefully about the double-slit experiment for water waves. Our aim will be to work out just what it is about waves that causes the interference pattern. We should then make sure that our theory of quantum particles is capable of encapsulating this behaviour, so that we can have a chance of explaining the double-slit experiment for electrons.

There are two reasons why waves journeying through two slits can interfere with themselves. The first is that the wave travels through *both of the slits at once*, creating two new waves that head off and mix together. It's obvious that a wave can do this. We have no problem visualizing one long, ocean wave rolling to the shore and crashing on to a beach. It is a wall of water; an extended, travelling thing. We are therefore going to need to decide how to make our quantum particle 'an extended, travelling thing'. The second reason is that the two new waves heading out from the slits are able either to add or to subtract from each other when they mix. This ability for two waves to interfere is clearly crucial in explaining the interference pattern. The extreme case is when the peak of one wave coincides with the trough of another, in which case they completely cancel each other out. So we are also going to need to allow our quantum particle to interfere somehow with itself.

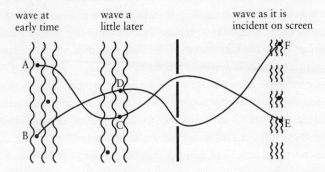

Figure 3.1. How the wave describing an electron moves from source to screen and how it should be interpreted as representing all of the ways that the electron travels. The paths A to C to E and B to D to F illustrate just two of the infinity of possible paths the single electron does take.

The double-slit experiment connects the behaviour of electrons and the behaviour of waves, so let us see how far we can push the connection. Take a look at Figure 3.1 and, for the time being, ignore the lines joining A to E and B to F and concentrate on the waves. The figure could then describe a water tank, with the wavy lines representing, from left to right, how a water wave rolls its way across the tank. Imagine taking a photograph of the tank just after a plank of wood has splashed in on the left-hand side to make a wave. The snapshot would reveal a newly formed wave that extends from top to bottom in the picture. All the water in the rest of the tank would be calm. A second snapshot taken a little later reveals that the water wave has moved towards the slits, leaving flat water behind it. Later still, the water wave passes through the pair of slits and generates the stripy interference pattern illustrated by the wavy lines on the far right.

Now let us reread that last paragraph but replace 'water wave' with 'electron wave', whatever that may mean. An electron wave, suitably interpreted, has the potential to explain the stripy pattern we want to understand as it rolls through the experiment like a water wave. But we do need to explain why the electron pattern is made up of tiny dots as the electrons hit the screen one by one. At first sight that seems in conflict with the idea of a smooth wave, but it is not. The clever bit is to realize that we can offer an explanation if we interpret the electron wave not as a real material disturbance (as is the case with a water wave), but rather as something that simply informs us where the electron is likely to be found. Notice we said 'the' electron because the wave is to describe the behaviour of a single electron – that way we have a chance of explaining how those dots emerge. This is an electron wave, and not a wave of electrons: we must never fall into the trap of thinking otherwise. If we imagine a snapshot of the wave at some instant in time, then we want to interpret it such that where the wave is largest the electron is most likely to be found, and where the wave is smallest the electron is least likely to be found. When the wave finally reaches the screen, a little spot appears and informs us of the location of the electron. The sole job of the electron wave is to allow us to compute the odds that the electron hits the screen at some particular place. If we do not worry about what the electron wave actually 'is', then everything is straightforward because once we know the wave then we can say where the electron is likely to be. The fun comes next, when we try to understand what this proposal for an electron wave implies for the electron's journey from slit to screen.

Before we do this, it might be worth reading the above paragraph again because it is very important. It's not supposed to be obvious and it is certainly not intuitive. The 'electron wave' proposal has all the necessary properties to explain the appearance of the experimentally observed interference pattern, but it is something of a guess as to how things might work out. As good physicists we should work out the consequences and see if they correspond to Nature.

Returning to Figure 3.1, we have proposed that at each instant in time the electron is described by a wave, just as in the case of water waves. At an early time, the electron wave is to the left of the slits. This means that the electron is in some sense located somewhere within the wave. At a later time, the wave will advance towards the slits just as the water wave did, and the electron will now be somewhere in the new wave. We are saying that the electron 'could be first at A and then at C', or it 'could be first at B and then at D', or it 'could be at A and then at D', and so on. Hold that thought for a minute, and think about an even later time, after the wave has passed through the slits and reached the screen. The electron could now be found at E or perhaps at F. The curves that we have drawn on the diagram represent two possible paths that the electron could have taken from the source, through the slits and onto the screen. It could have gone from A to C to E, and it could have gone from B to D to F. These are just two out of an infinite number of possible paths that the electron could have taken.

The crucial point is that it makes no sense to say that 'the electron could have ventured along each of these routes, but really it went along only one of them'. To say that the electron really ventured along one particular path would be to give ourselves no more of a chance of explaining the interference pattern than if we had blocked up one of the slits in the water wave experiment. We need to allow the wave to go through both slits in order to get an interference pattern, and this means that we must allow all the possible paths for the electron to travel from source to screen. Put another way, when we said that the electron is 'somewhere within the wave' we really meant to say that it is simultaneously everywhere in the wave! This is how we must think because if we suppose the electron is actually located at some specific point, then the wave is no longer spread out and we lose the water wave analogy. As a result, we cannot explain the interference pattern.

Again, it might be worth rereading the above piece of reasoning because it motivates much of what follows. There is no sleight of hand: what we are saying is that we need to describe a spread-out wave that is also a point-like electron, and one possible way to achieve this is to say that the electron sweeps from source to screen following all possible paths at once.

This suggests that we should interpret an electron wave as describing a single electron that travels from source to screen by an infinity of different routes. In other words, the correct answer to the question 'how did that electron get to the screen' is 'it travelled by an infinity of possible routes, some of which went through the upper slit and some of which went though the lower one'. Clearly the 'it' that is the electron is not an ordinary, everyday particle. This is what it means to be a quantum particle.

Having decided to seek a description of an electron that mimics in many ways the behaviour of waves, we need to develop a more precise way to talk about waves. We shall begin with a description of what is happening in a water tank when two waves meet, mix and interfere with each other. To do this, we must find a convenient way of representing the positions of the peaks and troughs of each wave. In the technical jargon, these are known as phases. Colloquially things are described as 'in phase' if they reinforce one another in some way, or 'out of phase' if they cancel each other out. The word is also used to describe the Moon: over the course of around twenty-eight days, the Moon passes from new to full and back again in a continuous waxing and waning cycle. The etymology of the word 'phase' stems from the Greek *phasis*, which means the appearance and disappearance of an astronomical phenomenon, and the regular appearance and disappearance of the bright lunar surface

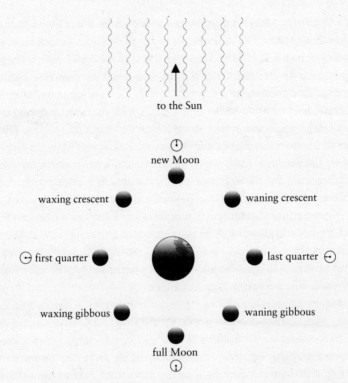

Figure 3.2. The phases of the Moon.

seems to have led to its twentieth-century usage, particularly in science, as a description of something cyclical. And this is a clue as to how we might find a pictorial representation of the positions of the peaks and troughs of water waves.

Have a look at Figure 3.2. One way to represent a phase is as a clock face with a single hand rotating around. This gives us the freedom to represent visually a full 360 degrees worth of possibilities: the clock hand can point to 12 o'clock, 3 o'clock, 9 o'clock and all points in between. In the case of the Moon, you could imagine a new Moon represented by a clock hand pointing to 12 o'clock, a waxing crescent at 1:30, the first quarter at 3, the waxing gibbous at 4:30, the full Moon at 6 and so on. What we are doing here is using something abstract to describe something concrete; a clock face to describe the phases of the Moon. In this way we could draw a clock with its hand pointing to 12 o'clock and you'd immediately know that the clock represented a new Moon. And even though we haven't actually said it, you'd know that a clock with the hand pointing to 5 o'clock would mean that we are approaching a full Moon. The use of abstract pictures or symbols to represent real things is absolutely fundamental in physics – this is essentially what physicists use mathematics for. The power of the approach comes when the abstract pictures can be manipulated using simple rules to make firm predictions about the real world. As we'll see in a moment, the clock faces will allow us to do just this because they are able to keep track of the relative positions of the peaks and troughs of waves. This in turn will allow us to calculate whether they will cancel or reinforce one another when they meet.

Figure 3.3 shows a sketch of two water waves at an instant in time. Let's represent the peaks of the waves by clocks reading 12 o'clock and the troughs by clocks reading 6 o'clock. We can also represent places on the waves intermediate between peaks and troughs with clocks reading intermediate times, just as we did for the phases of the Moon between new and full. The distance between the successive peaks and troughs of the wave is an important number; it is known as the wavelength of the wave.

The two waves in Figure 3.3 are out of phase with each other, which means that the peaks of the top wave are aligned with the troughs of the bottom wave, and vice versa. As a result it is pretty clear that they will entirely cancel each other out when we add them together. This is illustrated at the bottom of the figure, where the 'wave' is flat-lining. In terms of clocks, all of the 12 o'clock clocks for the top wave, representing its peaks, are aligned with the 6 o'clock clocks for the bottom wave, representing its troughs. In fact, everywhere you look, the clocks for the top wave are pointing in the opposite direction to the clocks for the bottom wave.

Using clocks to describe waves does, at this stage, seem like we

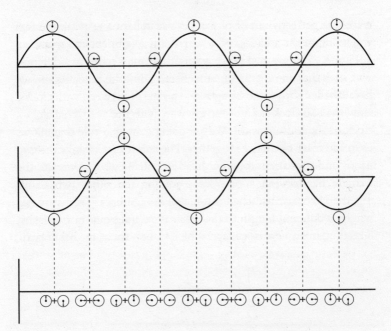

Figure 3.3. Two waves arranged such that they cancel out completely. The top wave is out of phase with the second wave, i.e. peaks align with troughs. When the two waves are added together they cancel out to produce nothing – as illustrated at the bottom where the 'wave' is flat-lining.

are over-complicating matters. Surely if we want to add together two water waves, then all we need to do is add the heights of each of the waves and we don't need clocks at all. This is certainly true for water waves, but we are not being perverse and we have introduced the clocks for a very good reason. We will discover soon enough that the extra flexibility they allow is absolutely necessary when we come to use them to describe quantum particles.

With this in mind, we shall now spend a little time inventing a precise rule for adding clocks. In the case of Figure 3.3, the rule must result in all the clocks 'cancelling out', leaving nothing behind: 12 o'clock cancels 6 o'clock, 3 o'clock cancels 9 o'clock and so on. This perfect cancellation is, of course, for the special case when the waves are perfectly out of phase. Let's search for a general rule that will work for the addition of waves of any alignment and shape.

Figure 3.4 shows two more waves, this time aligned in a different way, such that one is only slightly offset against the other. Again, we have labelled the peaks, troughs and points in between with clocks. Now, the 12 o'clock clock of the top wave is aligned with the 3 o'clock clock of the bottom wave. We are going to state a rule that allows us to add these two clocks together. The rule is that we take the two hands and stick them together head to tail. We then complete the triangle by drawing a new hand joining the other two hands together. We have sketched this recipe in Figure 3.5. The new hand will be a different length to the other two, and point in a different direction; it is a new clock face, which is the sum of the other two.

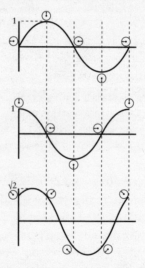

Figure 3.4. Two waves offset relative to each other. The top and middle waves add together to produce the bottom wave.

We can be more precise now and use simple trigonometry to calculate the effect of adding together any specific pair of clocks. In Figure 3.5 we are adding together the 12 o'clock and 3 o'clock clocks. Let's suppose that the original clock hands are of length 1 cm (corresponding to water waves of peak height equal to 1 cm). When we place the hands head-to-tail we have a right-angled triangle with two sides each of length 1 cm. The new clock hand will be the length of the third side of the triangle: the hypotenuse. Pythagoras' Theorem tells us that the square of the hypotenuse is equal to the sum of the squares of the other two sides: $h^2 = x^2 + y^2$. Putting the numbers in, $h^2 = 1^2 + 1^2 = 2$. So the length of the new clock hand h is the square root of 2, which is approximately 1.414 cm. In what direction will the new hand point? For this we need to know the angle in our triangle, labelled θ in the figure. For the particular example of two hands of equal length, one pointing to 12 o'clock and one to 3 o'clock, you can probably work it out without knowing any trigonometry at all. The hypotenuse obviously points at an angle of 45 degrees, so the new 'time' is half way between 12 o'clock and 3 o'clock, which is half past one. This example is a special case, of course. We chose the clocks so that the hands were at right angles and of the same length to make the mathematics easy. But it is obviously possible to work out the length of the hand and time resulting from the addition of any pair of clock faces.

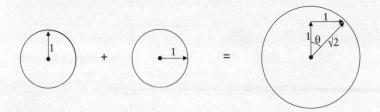

Figure 3.5. The rule for adding clocks.

Now look again at Figure 3.4. At every point along the new wave, we can compute the wave height by adding the clocks together using the recipe we just outlined and asking how much of the new clock hand points in the 12 o'clock direction. When the clock points to 12 o'clock this is obvious – the height of the wave is simply the length of the clock hand. Similarly at 6 o'clock, it's obvious because the wave has a trough with a depth equal to the length of the hand.

It's also pretty obvious when the clock reads 3 o'clock (or 9 o'clock) because then the wave height is zero, since the clock hand is at right angles to the 12 o'clock direction. To compute the wave height described by any particular clock we should multiply the length of the hand, h, by the cosine of the angle the hand makes with the 12 o'clock direction. For example, the angle that a 3 o'clock makes with 12 o'clock is 90 degrees and the cosine of 90 degrees is zero, which means the wave height is zero. Similarly, a time of half-past-one corresponds to an angle of 45 degrees with the 12 o'clock direction and the cosine of 45 degrees is approximately 0.707, so the height of the wave is 0.707 times the length of the hand (notice that 0.707 is $1/\sqrt{2}$). If your trigonometry is not up to those last few sentences then you can safely ignore the details. It's the principle that matters, which is that, given the length of a clock hand and its direction you can go ahead and calculate the wave height – and even if you don't understand trigonometry you could make a good stab at it by carefully drawing the clock hands and projecting on to the 12 o'clock direction using a ruler. (We would like to make it very clear to any students reading this book that we do not recommend this course of action: sines and cosines are useful things to understand.)

That's the rule for adding clocks, and it works a treat, as illustrated in the bottom of the three pictures in Figure 3.4, where we have repeatedly applied the rule for various points along the waves.

In this description of water waves, all that ever matters is the projection of the 'time' in the 12 o'clock direction, corresponding to

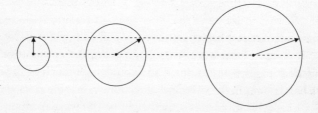

Figure 3.6. Three different clocks all with the same projection in the 12 o'clock direction.

just one number: the wave height. That is why the use of clocks is not really necessary when it comes to describing water waves. Take a look at the three clocks in Figure 3.6: they all correspond to the same wave height and so they provide equivalent ways of representing the same height of water. But clearly they are different clocks and, as we shall see, these differences do matter when we come to use them to describe quantum particles because, for them, the length of the clock hand (or equivalently the size of the clock) has a very important interpretation.

At some points in this book and at this point especially, things are abstract. To keep ourselves from succumbing to dizzying confusion, we should remember the bigger picture. The experimental results of Davisson, Germer and Thomson, and their similarity with the behaviour of water waves, have inspired us to make an ansatz: we should represent a particle by a wave, and the wave itself can be represented by lots of clocks. We imagine that the electron wave propagates 'like a water wave', but we haven't explained how that works in any detail. But then we never said how the water wave propagates either. All that matters for the moment is that we recognize the analogy with water waves, and the notion that the electron is described at any instant by a wave that propagates and interferes like water waves do. In the next chapter we will do better than this and be more precise about how an electron actually moves around as time unfolds. In doing that we will be led to a host of treasures, including Heisenberg's famous Uncertainty Principle.

Before we move on to that, we want to spend a little time talking about the clocks that we are proposing to represent the electron wave. We emphasize that these clocks are not real in any sense, and their hour hand has absolutely nothing to do with what time of day it is. This idea of using an array of little clocks to describe a real physical phenomenon is not so bizarre a concept as it may seem. Physicists use similar techniques to describe many things in Nature, and we have already seen how they can be used to describe water waves.

Another example of this type of abstraction is the description of the temperature in a room, which can be represented using an array of numbers. The numbers do not exist as physical objects any more than our clocks do. Instead, the set of numbers and their association with points in the room is simply a convenient way of representing the temperature. Physicists call this mathematical structure a field. The temperature field is simply an array of numbers, one for every point. In the case of a quantum particle, the field is more complicated because it requires a clock face at each point rather than a single number. This field is usually called the wavefunction of the particle. The fact that we need an array of clocks for the wavefunction, whilst a single number would suffice for the temperature field or for water waves, is an important difference. In physics jargon, the clocks are there because the wavefunction is a 'complex' field, whilst the temperature or water wave heights are both 'real' fields. We shall not need any of this language, because we can work with the clock faces.[1]

We should not worry that we have no direct way to sense a wavefunction, in contrast to a temperature field. The fact that it is not something we can touch, smell or see directly is irrelevant. Indeed, we would not get very far in physics if we decided to restrict our description of the Universe to things we can directly sense.

In our discussion of the double-slit experiment for electrons, we said that the electron wave is largest where the electron is most likely to be. This interpretation allowed us to appreciate how the stripy interference pattern can be built up dot by dot as the electrons arrive. But this is not a precise enough statement for our purposes now. We want to know what the probability is to find an electron at a particular point – we want to put a number on it. This is where the clocks become necessary, because the probability that we want is not simply the wave height. The correct thing to do is to

---

[1]. For those who are familiar with mathematics, just exchange the words as follows: 'clock' for 'complex number', 'size of the clock' for 'modulus of the complex number' and 'the direction of the hour-hand' for 'the phase'. The rule for adding clocks is nothing more than the rule for adding complex numbers.

interpret the *square* of the length of the clock hand as the probability to find the particle at the site of the clock. This is why we need the extra flexibility that the clocks give us over simple numbers. That interpretation is not meant to be at all obvious, and we cannot offer any good explanation for why it is correct. In the end, we know that it is correct because it leads to predictions that agree with experimental data. This interpretation of the wavefunction was one of the thorny issues facing the early pioneers of quantum theory.

The wavefunction (that is our cluster of clocks) was introduced into quantum theory in a series of papers published in 1926 by the Austrian physicist Erwin Schrödinger. His paper of 21 June contains an equation that should be etched into the mind of every undergraduate physics student. It is known, naturally enough, as the Schrödinger equation:

$$i\hbar \frac{\partial}{\partial t}\Psi = \hat{H}\Psi$$

The Greek symbol Ψ (pronounced 'psi') represents the wavefunction, and the Schrödinger equation describes how it changes as time passes. The details of the equation are irrelevant for our purposes because we are not going to follow the Schrödinger approach in this book. What is interesting, though, is that, although Schrödinger wrote down the correct equation for the wavefunction, he initially got the interpretation wrong. It was Max Born, one of the oldest of the physicists working on the quantum theory in 1926, who, at the grand old age of forty-three, gave the correct interpretation in a paper submitted just four days after Schrödinger's. We mention his age because quantum theory during the mid 1920s gained the nickname 'Knabenphysik' – 'boy physics' – because so many of the key protagonists were young. In 1925 Heisenberg was twenty-three, Wolfgang Pauli, whose famous Exclusion Principle we shall meet later on, was twenty-two, as was Paul Dirac, the British physicist who first wrote down the correct equation describing the electron. It is often claimed that their youth freed them from the old ways of thinking and allowed them fully to embrace the radical new picture of the world represented by quantum theory. Schrödinger, at thirty-eight, was an old man in this company and it is true that he was never completely at ease with the theory he played such a key role in developing.

Born's radical interpretation of the wavefunction, for which he received the Nobel Prize for physics in 1954, was that the square of the length of the clock hand at a particular point represents the probability of finding a particle there. For example, if the hour-hand on the clock located at some place has a length of 0.1 then squaring this gives 0.01. This means that the probability to find the particle at this place is 0.01, i.e. one in a hundred. You might ask why Born didn't just square the clocks up in the first place, so that in the last example the clock hand would itself have a length of 0.01. That will not work, because to account for interference we are going to want to add clocks together and adding 0.01 to 0.01 say (which gives 0.02) is not the same as adding 0.1 to 0.1 and then squaring (which gives 0.04).

We can illustrate this key idea in quantum theory with another example. Imagine doing something to a particle such that it is described by a specific array of clocks. Also imagine we have a device that can measure the location of particles. A simple-to-imagine-but-not-so-simple-to-build device might be a little box that we can rapidly erect around any region of space. If the theory says that the chance of finding a particle at some point is 0.01 (because the clock hand at that point has length 0.1), then when we erect the box around that point we have a one in a hundred chance of finding the particle inside the box afterwards. This means that it is unlikely that we'll find anything in the box. However, if we are able to reset the experiment by setting everything up such that the particle is once again described by the same initial set of clocks, then we could redo the experiment as many times as we wish. Now, for every 100 times we look in the little box we should, on average, discover that there is a particle inside it once – it will be empty the remaining ninety-nine times.

The interpretation of the squared length of the clock hand as the probability to find a particle at a particular place is not particularly difficult to grasp, but it does seem as if we (or to be more precise, Max Born) plucked it out of the blue. And indeed, from a historical perspective, it proved very difficult for some great scientists, Einstein and Schrödinger among them, to accept. Looking back on the summer of 1926, fifty years later, Dirac wrote: 'This problem of getting the interpretation proved to be rather more difficult than just working out the equations.' Despite this difficulty, it is noteworthy that by the end of 1926 the spectrum of light emitted from the hydrogen atom, one of the great puzzles of nineteenth-century physics, had already been computed using both Heisenberg's and Schrödinger's equations (Dirac eventually proved that their two approaches were in all cases entirely equivalent).

Einstein famously expressed his objection to the probabilistic nature of quantum mechanics in a letter to Born in December 1926. 'The theory says a lot but does not really bring us any closer to the secret of the "old one". I, at any rate, am convinced that *He* is not playing at dice.' The issue was that, until then, it had been assumed that physics was completely deterministic. Of course, the idea of probability is not exclusive to quantum theory. It is regularly used in a variety of situations, from gambling on horse races to the science of thermodynamics, upon which whole swathes of Victorian engineering rested. But the reason for this is a lack of knowledge about the part of the world in question, rather than something fundamental. Think about tossing a coin – the archetypal game of chance. We are all familiar with probability in this context. If we toss the coin 100 times, we expect, on average, that fifty times it will land heads and fifty times tails. Pre-quantum theory, we were obliged to say that, if we knew everything there is to know about the coin – the precise way we tossed it into the air, the pull of gravity, the details of little air currents that swish through the room, the temperature of the air, etc. – then we could, *in principle*, work out whether the coin would land heads or tails. The emergence of probabilities in this context is therefore a reflection of our lack of knowledge about the system, rather than something intrinsic to the system itself.

The probabilities in quantum theory are not like this at all; they are fundamental. It is not the case that we can only predict the probability of a particle being in one place or another because we are ignorant. We can't, *even in principle*, predict what the position of a particle will be. What we can predict, with absolute precision, is the probability that a particle will be found in a particular place if we look for it. More than that, we can predict with absolute precision how this probability changes with time. Born expressed this beautifully in 1926: 'The motion of particles follows probability laws but the probability itself propagates according to the law of causality.' This is exactly what Schrödinger's equation does: it is an equation that allows us to calculate exactly what the wavefunction will look like in the future, given what it looks like in the past. In that sense, it is analogous to Newton's laws. The difference is that, whilst Newton's laws allow us to calculate the position and speed of particles at any particular time in the future, quantum mechanics allows us to calculate only the probability that they will be found at a particular place.

This loss of predictive power was what bothered Einstein and many of his colleagues. With the benefit of over eighty years of hindsight and a great deal of hard work, the debate now seems somewhat redundant, and it is easy to dismiss it with the statement that Born, Heisenberg, Pauli, Dirac and others were correct and Einstein, Schrödinger and the old guard were wrong. But it was certainly possible back then to believe that quantum theory was incomplete in some way, and that the probabilities appear, just as in thermodynamics or coin tossing, because there is some information about the particles that we are missing. Today that idea gains little purchase – theoretical and experimental progress indicate that Nature really does use random numbers, and the loss of certainty in predicting the positions of particles is an intrinsic property of the physical world: probabilities are the best we can do.

4. *Everything That Can Happen Does Happen*

We've now set up a framework within which we can explore quantum theory in detail. The key ideas are very simple in their technical content, but tricky in the way they challenge us to confront our prejudices about the world. We have said that a particle is to be represented by lots of little clocks dotted around and that the length of the clock hand at a particular place (squared) represents the probability that the particle will be found at that place. The clocks are not the main point – they are a mathematical device we'll use to keep track of the odds on finding a particle somewhere. We also gave a rule for adding clocks together, which is necessary to describe the phenomenon of interference. We now need to tie up the final loose end, and look for the rule that tells us how the clocks change from one moment to the next. This rule will be the replacement of Newton's first law, in the sense that it will allow us to predict what a particle will do if we leave it alone. Let's begin at the beginning and imagine placing a single particle at a point.

Figure 4.1. The single clock representing a particle that is definitely located at a particular point in space.

We know how to represent a particle at a point, and this is shown in Figure 4.1. There will be a single clock at that point, with a hand of length 1 (because 1 squared is 1 and that means the probability to find the particle there is equal to 1, i.e. 100 per cent). Let's suppose that the clock reads 12 o'clock, although this choice is completely arbitrary. As far as the probability is concerned, the clock hand can point in any direction, but we have to choose something to start with, so 12 o'clock will do. The question we want to answer is the following: what is the chance that the particle will be located somewhere else at a later time? In other words, how many clocks do we have to draw, and where do we have to place them, at the next moment? To Isaac Newton, this would have been a very dull question; if we place a particle somewhere and do nothing to it, then it's not going to go anywhere. But Nature says, quite categorically, that this is simply wrong. In fact, Newton could not be more wrong.

Here is the correct answer: the particle *can be anywhere else in the Universe at the later time*. That means we have to draw an infinite number of clocks, one at every conceivable point in space. That sentence is worth reading lots of times. Probably we need to say more.

Allowing the particle to be anywhere at all is equivalent to assuming nothing about the motion of the particle. This is the most unbiased thing we can do, and that does have a certain ascetic appeal to it,[1] although admittedly it does seem to violate the laws of common sense, and perhaps the laws of physics as well.

A clock is a representation of something definite – the likelihood that a particle will be found at the position of the clock. If we know that a particle is at one particular place at a particular time, we represent it by a single clock at that point. The proposal is that if we start with a particle sitting at a definite position at time zero, then at 'time zero plus a little bit' we should draw a vast, indeed infinite, array of new clocks, filling the entire Universe. This admits the possibility that the particle hops off to *anywhere and everywhere* else in an instant. Our particle will simultaneously be both a nanometre away and also a billion light years away in the heart of a star in a distant galaxy. This sounds, to use our native northern vernacular, daft. But let's be very clear: the theory must be capable of explaining the double-slit experiment and, just as a wave spreads out if we dip a toe into still water, so an electron initially located somewhere

1. Or aesthetic appeal, depending on your point of view.

must spread out as time passes. What we need to establish is exactly how it spreads.

Unlike a water wave, we are proposing that the electron wave spreads out to fill the Universe in an instant. Technically speaking, we'd say that the rule for particle propagation is different from the rule for water wave propagation, although both propagate according to a 'wave equation'. The equation for water waves is different from the equation for particle waves (which is the famous Schrödinger equation we mentioned in the last chapter), but both encode wavy physics. The differences are in the details of how things propagate from place to place. Incidentally, if you know a little about Einstein's theory of relativity you might be getting nervous when we speak of a particle hopping across the Universe in an instant, because that would seem to correspond to something travelling faster than the speed of light. Actually, the idea that a particle can be here and, an instant later, somewhere else very far away is not in itself in contradiction with Einstein's theories, because the real statement is that *information* cannot travel faster than the speed of light, and it turns out that quantum theory remains constrained by that. As we shall learn, the dynamics corresponding to a particle leaping across the Universe are the very opposite of information transfer, because we cannot know where the particle will leap to beforehand. It seems we are building a theory on complete anarchy, and you might naturally be concerned that Nature surely cannot behave like this. But, as we shall see as the book unfolds, the order we see in the everyday world really does emerge out of this fantastically absurd behaviour.

If you are having trouble swallowing this anarchic proposal – that we have to fill the entire Universe with little clocks in order to describe the behaviour of a single subatomic particle from one moment to the next – then you are in good company. Lifting the veil on quantum theory and attempting to interpret its inner workings is baffling to everyone. Niels Bohr famously wrote that 'Those who are not shocked when they first come across quantum mechanics cannot possibly have understood it', and Richard Feynman introduced volume III of *The Feynman Lectures on Physics* with the words: 'I think I can safely say that nobody understands quantum mechanics.' Fortunately, following the rules is far simpler than trying to visualize what they actually mean. The ability to follow through the consequences of a particular set of assumptions carefully, without getting too hung up on the philosophical implications, is one of the most important skills a physicist learns. This is absolutely in the spirit of Heisenberg: let us set out our initial assumptions and compute their consequences. If we arrive at a set of predictions that agree with observations of the world around us, then we should accept the theory as good.

Many problems are far too difficult to solve in a single mental leap, and deep understanding rarely emerges in 'eureka' moments. The trick is to make sure that you understand each little step and after a sufficient number of steps the bigger picture should emerge. Either that or we realize we have been barking up the wrong tree and have to start over from scratch. The little steps we've outlined so far are not difficult in themselves, but the idea that we have decided to take a single clock and turn it into an infinity of clocks is certainly a tricky concept, especially if you try to imagine drawing them all. Eternity is a very long time, to paraphrase Woody Allen, especially near the end. Our advice is not to panic or give up and, in any case, the infinity bit is a detail. Our next task is to establish the rule that tells us what all those clocks should actually look like at some time after we laid down the particle.

The rule we are after is the essential rule of quantum theory, although we will need to add a second rule when we come to consider the possibility that the Universe contains more than just one particle. But first things first: for now, let's focus on a single particle alone in the Universe – no one can accuse us of rushing into things. At one instant in time, we'll suppose we know exactly where it is, and it's therefore represented by a single, solitary clock. Our specific task is to identify the rule that will tell us what each and every one of the new clocks, scattered around the Universe, should look like at any time in the future.

We'll first state the rule without any justification. We will come back to discuss just why the rule looks like it does in a few paragraphs, but for now we should treat it as one of the rules in a game. Here's the rule: at a time t in the future, a clock a distance x from the original clock has its hand wound in an anti-clockwise direction by an amount proportional to x^2; the amount of winding is also proportional to the mass of the particle m and inversely proportional to the time t. In symbols, this means we are to wind the clock hand anti-clockwise by an amount proportional to mx^2/t. In words, it means that there is more winding for a more massive particle, more winding the further away the clock is from the original, and less winding for a bigger step forward in time. This is an algorithm – a recipe if you like – that tells us exactly what to do to work out what a given arrangement of clocks will look like at some point in the future. At every point in the universe, we draw a new clock with its hand wound around by an amount given by our rule. This accounts for our assertion that the particle can, and indeed does, hop from its initial position to each and every other point in the Universe, spawning new clocks in the process.

To simplify matters we have imagined just one initial clock, but of course at some instant in time there might already be many clocks, representing the fact that the particle is not at some definite location. How are we to figure out what to do with a whole cluster of clocks? The answer is that we are to do what we did for one clock, and repeat that for each and every one of the clocks in the cluster. Figure 4.2 illustrates this idea. The initial set of clocks are represented by the little circles, and the arrows indicate that the particle hops from the site of every initial clock to the point X, 'depositing' a new clock in the process. Of course, this delivers one new clock to X for every initial clock, and we must add all these clocks together in order to construct the final, definitive clock at X. The size of this final clock's hand gives us the chance of finding the particle at X at the later time.

It is not so strange that we should be adding clocks together when several arrive at the same point. Each clock corresponds to a different way that the particle could have reached X. This addition of the clocks is understandable if we think back to the double-slit experiment; we are simply trying to rephrase the wave description in terms of clocks. We can imagine two initial clocks, one at each slit. Each of these two clocks will deliver a clock to a particular point on the screen at some later time, and we must add these two clocks together in order to obtain the interference pattern.[2] In summary therefore, the rule to calculate what the clock looks like at any point is to transport all the initial clocks to that point, one by one, and then add them together using the addition rule we encountered in the previous chapter.

Since we developed this language in order to describe the propa-

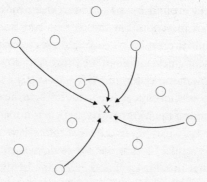

Figure 4.2. Clock hopping. The open circles indicate the locations of the particle at some instant in time; we are to associate a clock with each point. To compute the probability to find the particle at X we are to allow the particle to hop there from all of the original locations. A few such hops are indicated by the arrows. The shape of the lines does not have any meaning and it certainly does not mean that the particle travels along some trajectory from the site of a clock to X.

2. If you are having trouble with that last sentence try replacing the word 'clock' with 'wave'.

gation of waves, we can also think about more familiar waves in these terms. The whole idea, in fact, goes back a long way. Dutch physicist Christiaan Huygens famously described the propagation of light waves like this as far back as 1690. He did not speak about imaginary clocks, but rather he emphasized that we should regard each point on a light wave as a source of secondary waves (just as each clock spawns many secondary clocks). These secondary waves then combine to produce a new resultant wave. The process repeats itself so that each point in the new wave also acts as a source of further waves, which again combine, and in this way a wave advances.

We can now return to something that may quite legitimately have been bothering you. Why on earth did we choose the quantity mx^2/t to determine the amount of winding of the clock hand? This quantity has a name: it is known as the *action*, and it has a long and venerable history in physics. Nobody really understands why Nature makes use of it in such a fundamental way, which means that nobody can really explain why those clocks get wound round by the amount they do. Which somewhat begs the question: how did anyone realize it was so important in the first place? The action was first introduced by the German philosopher and mathematician Gottfried Leibniz in an unpublished work written in 1669, although he did not find a way to use it to make calculations. It was reintroduced by the French scientist Pierre-Louis Moreau de Maupertuis in 1744, and subsequently used to formulate a new and powerful principle of Nature by his friend, the mathematician Leonard Euler. Imagine a ball flying through the air. Euler found that the ball travels on a path such that the action computed between any two points on the path is always the smallest that it can be. For the case of a ball, the action is related to the *difference* between the kinetic and potential energies of the ball.[3] This is known as 'the principle of least action',

3. The kinetic energy is equal to $mv^2/2$ and the potential energy is mgh when the ball is a height h above the ground. g is the rate at which all objects accelerate in

and it can be used to provide an alternative to Newton's laws of motion. At first sight it's a rather odd principle, because in order to fly in a way that minimizes the action, the ball would seem to have to know where it is going before it gets there. How else could it fly through the air such that, when everything is done, the quantity called the action is minimized? Phrased in this way, the principle of least action sounds teleological – that is to say things appear to happen in order to achieve a pre-specified outcome. Teleological ideas generally have a rather bad reputation in science, and it's easy to see why. In biology, a teleological explanation for the emergence of complex creatures would be tantamount to an argument for the existence of a designer, whereas Darwin's theory of evolution by natural selection provides a simpler explanation that fits the available data beautifully. There is no teleological component to Darwin's theory – random mutations produce variations in organisms, and external pressures from the environment and other living things determine which of these variations are passed on to the next generation. This process alone can account for the complexity we see in life on Earth today. In other words, there is no need for a grand plan and no gradual ascent of life towards some sort of perfection. Instead, the evolution of life is a random walk, generated by the imperfect copying of genes in a constantly shifting external environment. The Nobel-Prize-winning French biologist Jacques Monod went so far as to define a cornerstone of modern biology as 'the systematic or axiomatic denial that scientific knowledge can be obtained on the basis of theories that involve, explicitly or not, a teleological principle'.

As far as physics is concerned, there is no debate as to whether or not the least action principle actually works, for it allows calculations to be performed that correctly describe Nature and it is a cornerstone of physics. It can be argued that the least action principle is not teleological at all, but the debate is in any case neutralized

---

the vicinity of the Earth. The action is their difference integrated between the times associated with the two points on the path.

once we have a grasp of Feynman's approach to quantum mechanics. The ball flying through the air 'knows' which path to choose because it actually, secretly, explores every possible path.

How was it discovered that the rule for winding the clocks should have anything to do with this quantity called the action? From a historical perspective, Dirac was the first to search for a formulation of quantum theory that involved the action, but rather eccentrically he chose to publish his research in a Soviet journal, to show his support for Soviet science. The paper, entitled 'The Lagrangian in Quantum Mechanics', was published in 1933 and languished in obscurity for many years. In the spring of 1941, the young Richard Feynman had been thinking about how to develop a new approach to quantum theory using the Lagrangian formulation of classical mechanics (which is the formulation derived from the principle of least action). He met Herbert Jehle, a visiting physicist from Europe, at a beer party in Princeton one evening, and, as physicists tend to do when they've had a few drinks, they began discussing research ideas. Jehle remembered Dirac's obscure paper, and the following day they found it in the Princeton Library. Feynman immediately started calculating using Dirac's formalism and, in the course of an afternoon with Jehle looking on, he found that he could derive the Schrödinger equation from an action principle. This was a major step forward, although Feynman initially assumed that Dirac must have done the same because it was such an easy thing to do; easy, that is, if you are Richard Feynman. Feynman eventually asked Dirac whether he'd known that, with a few additional mathematical steps, his 1933 paper could be used in this way. Feynman later recalled that Dirac, lying on the grass in Princeton after giving a rather lacklustre lecture, simply replied, 'No, I didn't know. That's interesting.' Dirac was one of the greatest physicists of all time, but a man of few words. Eugene Wigner, himself one of the greats, commented that 'Feynman is a second Dirac, only this time human.'

To recap: we have stated a rule that allows us to write down the whole array of clocks representing the state of a particle at some instant in time. It's a bit of a strange rule – fill the Universe with an infinite number of clocks, all turned relative to each other by an amount that depends on a rather odd but historically important quantity called the action. If two or more clocks land at the same point, add them up. The rule is built on the premise that we must allow a particle the freedom to jump from any particular place in the Universe to absolutely anywhere else in an infinitesimally small moment. We said at the outset that these outlandish ideas must ultimately be tested against Nature to see whether anything sensible emerges. To make a start on that, let's see how something very concrete, one of the cornerstones of quantum theory, emerges from this apparent anarchy: Heisenberg's Uncertainty Principle.

Heisenberg's Uncertainty Principle

Heisenberg's Uncertainty Principle is one of the most misunderstood parts of quantum theory, a doorway through which all sorts of charlatans and purveyors of tripe[4] can force their philosophical musings. He presented it in 1927 in a paper entitled 'Über den anschaulichen Inhalt der quantentheoretischen Kinematik und Mechanik', which is very difficult to translate into English. The difficult word is *anschaulich*, which means something like 'physical' or 'intuitive'. Heisenberg seems to have been motivated by his intense annoyance that Schrödinger's more intuitive version of quantum theory was more widely accepted than his own, even though both formalisms led to the same results. In the spring of 1926, Schrödinger was convinced that his equation for the wavefunction provided a physical picture of what was going on inside atoms. He thought that his wavefunction was a thing you could visualize, and was related to the distribution of electric charge inside the atom. This turned out to be incorrect, but at least it made physicists feel good during the

---

4. Wikipedia describes 'tripe' as 'a type of edible offal from the stomachs of various farm animals', but it is colloquially used to mean 'nonsense'. Either definition is appropriate here.

first six months of 1926: until Born introduced his probabilistic interpretation.

Heisenberg, on the other hand, had built his theory around abstract mathematics, which predicted the outcomes of experiments extremely successfully but was not amenable to a clear physical interpretation. Heisenberg expressed his irritation to Pauli in a letter on 8 June 1926, just weeks before Born threw his metaphorical spanner into Schrödinger's intuitive approach. 'The more I think about the physical part of Schrödinger's theory, the more disgusting I find it. What Schrödinger writes about the *Anschaulichkeit* of his theory . . . I consider *Mist.*' The translation of the German word *mist* is 'rubbish' or 'bullshit' . . . or 'tripe'.

What Heisenberg decided to do was to explore what an 'intuitive picture', or *Anschaulichkeit*, of a physical theory should mean. What, he asked himself, does quantum theory have to say about the familiar properties of particles such as position? In the spirit of his original theory, he proposed that a particle's position is a meaningful thing to talk about only if you also specify how you measure it. So you can't ask where an electron actually is inside a hydrogen atom without describing exactly how you'd go about finding out that information. This might sound like semantics, but it most definitely is not. Heisenberg appreciated that the very act of measuring something introduces a disturbance, and that as a result there is a limit on how well we can 'know' an electron. Specifically, in his original paper, Heisenberg was able to estimate what the relationship is between how accurately we can simultaneously measure the position and the momentum of a particle. In his famous Uncertainty Principle, he stated that if Δx is the uncertainty in our knowledge of the position of a particle (the Greek letter Δ is pronounced 'delta', so Δx is pronounced 'delta x') and Δp is the corresponding uncertainty in the momentum, then

$$\Delta x \Delta p \sim h$$

where h is Planck's constant and the '$\sim$' symbol means 'is similar in size to'. In words, the product of the uncertainty in the position of a particle and the uncertainty in its momentum will be roughly equal to Planck's constant. This means that the more accurately we identify the location of a particle, the less well we can know its momentum, and vice versa. Heisenberg came to this conclusion by contemplating the scattering of photons off electrons. The photons are the means by which we 'see' the electron, just as we see everyday objects by scattering photons off them and collecting them in our eyes. Ordinarily, the light that bounces off an object disturbs the object imperceptibly, but that is not to deny our fundamental inability to absolutely isolate the act of measurement from the thing one is measuring. One might worry that it could be possible to beat the limitations of the Uncertainty Principle by devising a suitably ingenious experiment. We are about to show that this is not the case and the Uncertainty Principle is absolutely fundamental, because we are going to derive it using only our theory of clocks.

Deriving Heisenberg's Uncertainty Principle from the Theory of Clocks

Rather than starting with a particle at a single point, let us instead think about a situation where we know roughly where the particle is, but we don't know exactly where it is. If we know that a particle is somewhere in a small region of space then we should represent it by a cluster of clocks filling that region. At each point within the region there will be a clock, and that clock will represent the probability that the particle will be found at that point. If we square up the lengths of all the clock hands at every point and add them together, we will get 1, i.e. the probability to find the particle *somewhere* in the region is 100 per cent.

In a moment we are going to use our quantum rules to perform a serious calculation, but first we should come clean and say that we have failed to mention an important addendum to the clock-winding rule. We didn't want to introduce it earlier because it is a technical detail, but we won't get the correct answers when it comes to calculating actual probabilities if we ignore it. It relates to what we said at the end of the previous paragraph.

If we begin with a single clock, then the hand must be of length 1, because the particle must be found at the location of the clock with a probability of 100 per cent. Our quantum rule then says that, in order to describe the particle at some later time, we should transport this clock to all points in the Universe, corresponding to the particle leaping from its initial location. Clearly we cannot leave all of the clock hands with a length of 1, because then our probability interpretation falls down. Imagine, for example, that the particle is described by four clocks, corresponding to its being at four different locations. If each one has a size of 1 then the probability that the particle is located at any one of the four positions would be 400 per cent and this is obviously nonsense. To fix this problem we must shrink the clocks in addition to winding them anti-clockwise. This 'shrink rule' states that after all of the new clocks have been spawned, every clock should be shrunk by the square root of the total number of clocks.[5] For four clocks, that would mean that each hand must be shrunk by $\sqrt{4}$, which means that each of the four final clocks will have a hand of length $\frac{1}{2}$. There is then a $(\frac{1}{2})^2 = 25$ per cent chance that the particle will be found at the site of any one of the four clocks. In this simple way we can ensure that the probability that the particle is found somewhere will always total 100 per cent. Of course, there may be an infinite number of possible locations, in which case the clocks would have zero size, which may sound alarming, but the maths can handle it. For our purposes, we shall always imagine that there are a finite number of clocks, and in any case we will never actually need to know how much a clock shrinks.

Let's get back to thinking about a Universe containing a single particle whose location is not precisely known. You can treat the

5. Shrinking all clocks by the same amount is strictly only true provided that we are ignoring the effects of Einstein's Special Theory of Relativity. Otherwise, some of the clocks get shrunk more than others. We shan't need to worry about this.

next section as a little mathematical puzzle – it may be tricky to follow the first time through, and it may be worth rereading, but if you are able to follow what is going on then you'll understand how the Uncertainty Principle emerges. For simplicity, we've assumed that the particle moves in one dimension, which means it is located somewhere on a line. The more realistic three-dimensional case is not fundamentally different – it's just harder to draw. In Figure 4.3 we've sketched this situation, representing the particle by a line of three clocks. We should imagine that there are many more than this – one at every possible point that the particle could be – but this would be very hard to draw. Clock 3 sits at the left side of the initial clock cluster and clock 1 is at the right side. To reiterate, this represents a situation in which we know that the particle starts out somewhere between clocks 1 and 3. Newton would say that the particle stays between clocks 1 and 3 if we do nothing to it, but what does the quantum rule say? This is where the fun starts – we are going to play with the clock rules to answer this question.

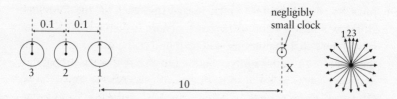

Figure 4.3. A line of three clocks all reading the same time: this describes a particle initially located in the region of the clocks. We are interested in figuring out what the chances are of finding the particle at the point X at some later time.

Let's allow time to tick forward and work out what happens to this line of clocks. We'll start off by thinking about one particular point a large distance away from the initial cluster, marked X in the figure. We'll be more quantitative about what a 'large distance' means later on, but for now it simply means that we need to do a lot of clock winding.

Applying the rules of the game, we should take each clock in the initial cluster and transport it to point X, winding the hand around and shrinking it accordingly. Physically, this corresponds to the particle hopping from that point in the cluster to point X. There will be many clocks arriving at X, one from each initial clock in the line, and we should add them all up. When all this is done, the square of the length of the resulting clock hand at X will give the probability that we will find the particle at X.

Now let's see how this all pans out and put some numbers in. Let's say that the point X is a distance of '10 units' away from clock 1, and that the initial cluster is '0.2 units' wide. Answering the obvious question: 'How far is 10 units?' is where Planck's constant enters our story, but for now we shall deftly side-step that issue and simply specify that 1 unit of distance corresponds to 1 complete (twelve-hour) wind of the clock. This means that the point X is approximately $10^2 = 100$ complete windings away from the initial cluster (remember the winding rule). We shall also assume that the clocks in the initial cluster started out of equal size, and that they all point to 12 o'clock. Assuming they are of equal size is simply the assumption that the particle is equally likely to be anywhere in between points 1 and 3 in the figure and the significance of them all reading the same time will emerge in due course.

To transport a clock from point 1 to point X, we have to rotate the clock hand anti-clockwise 100 complete times, as per our rule. Now let's move across to point 3, which is a further 0.2 units away, and transport that clock to X. This clock has to travel 10.2 units, so we have to wind its hand back a little more than before, i.e. by 10.2^2, which is very close to 104, complete winds.

We now have two clocks landing at X, corresponding to the particle hopping from 1 to X and from 3 to X, and we must add them together in order to start the task of computing the final clock. Because they both got wound around by very close to a whole number of winds, they will both end up pointing roughly to 12 o'clock, and they will add up to form a clock with a bigger hand also pointing to 12 o'clock. Notice that it is only the final direction of the clock hands that matters. We do not need to keep track of how often they wind around. So far so good, but we haven't finished because there are many other little clocks in between the right- and left-hand edges of the cluster.

And so our attention now turns to the clock midway between the two edges, i.e. at point 2. That clock is 10.1 units away from X, which means that we have to wind it 10.1^2 times. This is very close to 102 complete rotations – again a whole number of winds. We need to add this clock to the others at X and, as before, this will make the hand at X even longer. Continuing, there is also a point midway between points 1 and 2 and the clock hopping from there will get 101 complete rotations, which will add to the size of the final hand again. But here is the important point. If we now go midway again between these two points, we get to a clock that will be wound 100.5 rotations when it reaches X. This corresponds to a clock with a hand pointing to 6 o'clock, and when we add this clock we will *reduce* the length of the clock hand at X. A little thought should convince you that, although the points labelled 1, 2 and 3 each produce clocks at X reading 12 o'clock, and although the points midway between 1, 2 and 3 also produce clocks that read 12 o'clock, the points that are $\frac{1}{4}$ and $\frac{3}{4}$ of the way between points 1 and 3 and points 2 and 3 each generate clocks that point to 6 o'clock. In total that is five clocks pointing up and four clocks pointing down. When we add all these clocks together, we'll get a resultant clock at X that has a tiny hand because nearly all of the clocks will cancel each other out.

This 'cancellation of clocks' obviously extends to the realistic case where we consider every possible point lying in the region between points 1 and 3. For example, the point that lies $\frac{1}{8}$ of the way along from point 1 contributes a clock reading 9 o'clock, whilst the point lying $\frac{3}{8}$ of the way reads 3 o'clock – again the two cancel each other out. The net effect is that the clocks corresponding to all of the ways that the particle could have travelled from somewhere in the cluster to point X cancel each other out. This cancellation is illustrated on the far right of the figure. The arrows indicate the clock hands arriving at X from various points in the initial cluster.

The net effect of adding all these arrows together is that they all cancel each other out. This is the crucial 'take home' message.

To reiterate, we have just shown that, provided the original cluster of clocks is large enough and that point X is far enough away, then for every clock that arrives at X pointing to 12 o'clock, there will be another that arrives pointing to 6 o'clock to cancel it out. For every clock that arrives pointing to 3 o'clock, there will be another that arrives pointing to 9 o'clock to cancel it out, and so on. This wholesale cancellation means that there is effectively no chance at all of finding the particle at X. This really is very encouraging and interesting, because it looks rather like a description of a particle that isn't moving. Although we started out with the ridiculous-sounding proposal that a particle can go from being at a single point in space to anywhere else in the Universe a short time later, we have now discovered that this is not the case if we start out with a cluster of clocks. For a cluster, because of the way all the clocks interfere with each other, the particle has effectively no chance of being far away from its initial position. This conclusion has come about as a result of an 'orgy of quantum interference', in the words of Oxford professor James Binney.

For the orgy of quantum interference and corresponding cancellation of clocks to happen, point X needs to be far enough away from the initial cluster so that the clocks can rotate around many times. Why? Because if point X is too close then the clock hands won't necessarily have the chance to go around at least once, which means they will *not* cancel each other out so effectively. Imagine, for example, that the distance from the clock at point 1 to point X is 0.3 instead of 10. Now the clock at the front of the cluster gets a smaller wind than before, corresponding to $0.3^2 = 0.09$ of a turn, which means it is pointing just past 1 o'clock. Likewise, the clock from point 3, at the back of the cluster, now gets wound by $0.5^2 = 0.25$ of a turn, which means it reads 3 o'clock. Consequently, all of the clocks arriving at X point somewhere between 1 o'clock and 3 o'clock, which means they do not cancel each other out but instead add up to one big clock pointing to approximately 2 o'clock.

All of this amounts to saying that there will be a reasonable chance of finding the particle at points close to, but outside of, the original cluster. By 'close to', we mean that there isn't sufficient winding to get the clock hands around at least once. This is starting to have a whiff of the Uncertainty Principle about it, but it is still a little vague, so let's explore exactly what we mean by a 'large enough' initial cluster and a point 'far enough away'.

Our initial ansatz, following Dirac and Feynman, was that the amount the hands wind around when a particle of mass m hops a distance x in a time t is proportional to the action, i.e. the amount of winding is proportional to mx^2/t. Saying it is 'proportional to' isn't good enough if we want to calculate real numbers. We need to know precisely what the amount of winding is equal to. In Chapter 2 we discussed Newton's law of gravitation, and in order to make quantitative predictions we introduced Newton's gravitational constant, which determines the strength of the gravitational force. With the addition of Newton's constant, numbers can be put into the equation and real things can be calculated, such as the orbital period of the Moon or the path taken by the Voyager 2 spacecraft on its journey across the solar system. We now need something similar for quantum mechanics – a constant of Nature that 'sets the scale' and allows us to take the action and produce a precise statement about the amount by which we should wind clocks as we move them a specified distance away from their initial position in a particular time. That constant is Planck's constant.

A Brief History of Planck's Constant

In a flight of imaginative genius during the evening of 7 October 1900, Max Planck managed to explain the way that hot objects radiate energy. Throughout the second half of the nineteenth century, the exact relationship between the distribution of the wavelengths of light emitted by hot objects and their temperature was one of the great puzzles in physics. Every hot object emits light and, as the temperature is increased, the character of the light changes. We are familiar with light in the visible region, corresponding to the colours of the rainbow, but light can also occur with wavelengths that are either too long or too short to be seen by the human eye. Light with a longer wavelength than red light is called 'infra-red' and it can be seen using night-vision goggles. Still longer wavelengths correspond to radio waves. Likewise, light with a wavelength just shorter than blue is called ultra-violet, and the shortest wavelength light is generically referred to as 'gamma radiation'. An unlit lump of coal at room temperature will emit light in the infra-red part of the spectrum. But if we throw it on to a burning fire, it will begin to glow red. This is because, as the temperature of the coal rises, the average wavelength of the radiation it emits decreases, eventually entering the range that our eyes can see. The rule is that the hotter the object, the shorter the wavelength of the light it emits. As the precision of the experimental measurements improved in the nineteenth century, it became clear that nobody had the correct mathematical formula to describe this observation. This problem is often termed the 'black body problem', because physicists refer to idealized objects that perfectly absorb and then re-emit radiation as 'black bodies'. The problem was a serious one, because it revealed an inability to understand the character of light emitted by anything and everything.

Planck had been thinking hard about this and related matters in the fields of thermodynamics and electromagnetism for many years before he was appointed Professor of Theoretical Physics in Berlin. The post had been offered to both Boltzmann and Hertz before Planck was approached, but both declined. This proved to be fortuitous, because Berlin was the centre of the experimental investigations into black body radiation, and Planck's immersion at the heart of the experimental work proved key to his subsequent theoretical tour de force. Physicists often work best when they are able to have wide-ranging and unplanned conversations with colleagues.

We know the date and time of Planck's revelation so well because he and his family had spent the afternoon of Sunday 7 October 1900 with his colleague Heinrich Rubens. Over lunch, they discussed the failure of the theoretical models of the day to explain the details of black body radiation. By the evening, Planck had scribbled a formula on to a postcard and sent it to Rubens. It turned out to be the correct formula, but it was very strange indeed. Planck later described it as 'an act of desperation', having tried everything else he could think of. It is genuinely unclear how Planck came up with his formula. In his superb biography of Albert Einstein, *Subtle is the Lord* ..., Abraham Pais writes: 'His reasoning was mad, but his madness has that divine quality that only the greatest transitional figures can bring to science.' Planck's proposal was both inexplicable and revolutionary. He found that he could explain the black body spectrum, but only if he assumed that the energy of the emitted light was made up of a large number of smaller 'packets' of energy. In other words the total energy is quantized in units of a new fundamental constant of Nature, which Planck called 'the quantum of action'. Today, we call it Planck's constant.

What Planck's formula actually implies, although he didn't appreciate it at the time, is that light is *always* emitted and absorbed in packets, or quanta. In modern notation, those packets have energy $E = hc/\lambda$, where λ is the wavelength of the light (pronounced 'lambda'), c is the speed of light and h is Planck's constant. The role of Planck's constant in this equation is as the conversion factor between the wavelength of light and the energy of its associated quantum. The realization that the quantization of the energy of emitted light, as identified by Planck, arises because the light itself is made up of particles was proposed, tentatively at first, by Albert Einstein. He made the proposition during his great burst of creativity in 1905 – the annus mirabilis which also produced the Special Theory of Relativity and the most famous equation in scientific history, $E = mc^2$. Einstein received the 1921 Nobel Prize for physics (which due to a rather arcane piece of Nobelian bureaucracy he received in 1922) for this work on the photoelectric effect, and not for his better-known theories of relativity. Einstein proposed that light can be regarded as a stream of particles (he did not at that time use the word 'photons') and he correctly recognized that the energy of each photon is inversely proportional to its wavelength. This conjecture by Einstein is the origin of one of the most famous paradoxes in quantum theory – that particles behave as waves, and vice versa.

Planck removed the first bricks from the foundations of Maxwell's picture of light by showing that the energy of the light emitted from a hot object can only be described if it is emitted in quanta. It was Einstein who pulled out the bricks that brought down the whole edifice of classical physics. His interpretation of the photoelectric effect demanded not only that light is emitted in little packets, but that it also interacts with matter in the form of localized packets. In other words, light really does behave as a stream of particles.

The idea that light is made from particles – that is to say that 'the electromagnetic field is quantized' – was deeply controversial and not accepted for decades after Einstein first proposed it. The reluctance of Einstein's peers to embrace the idea of the photon can be seen in the proposal, co-written by Planck himself, for Einstein's membership of the prestigious Prussian Academy in 1913, a full eight years after Einstein's introduction of the photon:

> In sum, one can say that there is hardly one among the great problems in which modern physics is so rich to which Einstein has not made a remarkable contribution. That he may sometimes have missed the target in his speculations, as, for example, in his hypothesis of light quanta, cannot really be held too much against him, for it is not possible to introduce really new ideas even in the most exact sciences without sometimes taking a risk.

In other words, nobody really believed that photons were real. The widely held belief was that Planck was on safe ground because his proposal was more to do with the properties of matter – the little oscillators that emitted the light – rather than the light itself. It was simply too strange to believe that Maxwell's beautiful wave equations needed replacing with a theory of particles.

We mention this history partly to reassure you of the genuine difficulties that must be faced in accepting quantum theory. It is impossible to visualize a thing, such as an electron or a photon, that behaves a little bit like a particle, a little bit like a wave, and a little bit like neither. Einstein remained concerned about these issues for the rest of his life. In 1951, just four years before his death, he wrote: 'All these fifty years of pondering have not brought me any closer to answering the question, what are light quanta?'

Sixty years later, what is unarguable is that the theory we are in the process of developing using our arrays of little clocks describes, with unerring precision, the results of every experiment that has ever been devised to test it.

Back to Heisenberg's Uncertainty Principle

This, then, is the history behind the introduction of Planck's constant. But for our purposes, the most important thing to notice is that Planck's constant is a unit of 'action', which is to say that it is the same type of quantity as the thing which tells us how far to wind the clocks. Its modern value is $6.6260695729 \times 10^{-34}$ kg m^2/s, which is very tiny by everyday standards. This will turn out to be the reason why we don't notice its all-pervasive effects in everyday life.

Recall that we wrote of the action corresponding to a particle hopping from one place to another as the mass of the particle multiplied by the distance of the hop squared divided by the time interval over which the hop occurs. This is measured in kg m^2/s, as is Planck's constant, and so if we simply divide the action by Planck's constant, we'll cancel all the units out and end up with a pure number. According to Feynman, this pure number is the amount we should wind the clock associated with a particle hopping from one place to another. For example, if the number is 1, that means 1 full wind and if it's $\frac{1}{2}$, it means $\frac{1}{2}$ a wind, and so on. In symbols, the precise amount by which we should turn the clock hand to account for the possibility that a particle hops a distance x in a time t is $mx^2/(2ht)$.

Notice that a factor $\frac{1}{2}$ has appeared in the formula. You can either take that as being what is needed to agree with experiment or you can note that this arises from the definition of the action.[6] Either is fine. Now that we know the value of Planck's constant, we can really quantify the amount of winding and address the point we deferred a little earlier. Namely, what does jumping a distance of '10' actually mean?

Let's see what our theory has to say about something small by everyday standards: a grain of sand. The theory of quantum mechanics we've developed suggests that if we place the grain down somewhere then at a later time it *could* be anywhere in the Universe. But this is obviously not what happens to real grains of sand. We have already glimpsed a way out of this potential problem because if there is sufficient interference between the clocks, corresponding to the sand grain hopping from a variety of initial locations, then they will all cancel out to leave the grain sitting still. The first question we need to answer is: how many times will the clocks get wound if we transport a particle with the mass of a grain of sand a distance of, say, 0.001 millimetres, in a time of one second? We wouldn't be able to see such a tiny distance with our eyes, but it is still quite large on the scale of atoms. You can do the calculation quite easily yourself by substituting the numbers into Feynman's winding rule.[7] The answer is something like a trillion complete winds of the clock. Imagine how much interference that

6. For a particle of mass m that hops a distance x in a time t, the action is $\frac{1}{2}m(x/t)^2 t$ if the particle travels in a straight line at constant speed. But this does not mean the quantum particle travels from place to place in straight lines. The clock-winding rule is obtained by associating a clock with each possible path the particle can take between two points and it is an accident that, after summing over all these paths, the result is equal to this simple result. For example, the clock-winding rule is not this simple if we include corrections to ensure consistency with Einstein's Theory of Special Relativity.

7. A sand grain typically has a mass around 1 microgram, which is a billionth of a kilogram.

allows for. The upshot is that the sand grain stays where it is and there is almost no probability that it will jump a discernible distance, even though we really have to consider the possibility that it secretly hopped everywhere in the Universe in order to reach that conclusion.

This is a very important result. If you had put the numbers in for yourself then you'd already have a feel for why this is the case; it's the smallness of Planck's constant. Written out in full, it has a value 0.000000000000000000000000000000000066260695729 kg m²/s. Dividing pretty much any everyday number by that will result in a lot of clock winding and a lot of interference, with the result that the exotic journeys of our sand grain across the Universe all cancel each other out, and we perceive this voyager through infinite space as a boring little speck of dust sitting motionless on a beach.

Our particular interest of course is in those circumstances where clocks do not cancel each other out, and, as we have seen, this occurs if the clocks do not turn by more than a single wind. In that case, the orgy of interference will not happen. Let's see what this means quantitatively.

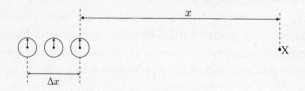

Figure 4.4. The same as Figure 4.3 except that we are now not committing to a specific value of the size of the clock cluster or the distance to the point X.

We are going to return to the clock cluster, which we've redrawn in Figure 4.4, but we'll be more abstract in our analysis this time instead of committing to definite numbers. We will suppose that the cluster has a size equal to Δx, and the distance of the closest point in the cluster to point X is x. In this case, the cluster size Δx refers to the uncertainty in our knowledge of the initial position of the particle; it started out somewhere in a region of size Δx. Starting with point 1, the point in the cluster closest to point X, we should wind the clock corresponding to a hop from this point to X by an amount

$$W_1 = \frac{mx^2}{2ht}$$

Now let's go to the farthest point, point 3. When we transport the clock from this point to X, it will be wound around by a greater amount, i.e.

$$W_3 = \frac{m(x + \Delta x)^2}{2ht}$$

We can now be precise and state the condition for the clocks propagated from all points in the cluster *not* to cancel out at X: there should be less than one full wind of difference between the clocks from points 1 and 3, i.e.

$$W_3 - W_1 < \text{one wind}$$

Writing this out in full, we have

$$\frac{m(x + \Delta x)^2}{2ht} - \frac{mx^2}{2ht} < 1$$

We're now going to consider the specific case for which the cluster size, Δx, is much smaller than the distance x. This means we are asking for the prospects that our particle will make a leap far outside of its initial domain. In this case, the condition for no clock cancellation, derived directly from the previous equation, is

$$\frac{mx\Delta x}{ht} < 1$$

If you know a little maths, you'll be able to get this by multiplying out the bracketed term and neglecting all the terms that involve $(\Delta x)^2$. This is a valid thing to do because we've said that Δx is very small compared to x, and a small quantity squared is a very small quantity.

This equation is the condition for there to be no cancellation of the clocks at point X. We know that if the clocks don't cancel out at a particular point, then there is a good chance that we will find the particle there. So we have discovered that if the particle is initially located within a cluster of size Δx, then at a time t later there is a good chance to find it a long distance x away from the cluster if the above equation is satisfied. Furthermore, this distance increases with time, because we are dividing by the time t in our formula. In other words, as more time passes, the chances of finding the particle further away from its initial position increases. This is beginning to look suspiciously like a particle that is moving. Notice also that the chance of finding the particle a long way away also increases as Δx gets smaller – i.e. as the uncertainty in the initial position of the particle gets smaller. In other words, the more accurately we pin down the particle, the faster it moves away from its initial position. This now looks a lot like Heisenberg's Uncertainty Principle.

To make final contact, let us rearrange the equation a little bit. Notice that for a particle to make its way from anywhere in the cluster to point X in time t, it must leap a distance x. If you actually measured the particle at X then you would naturally conclude that the particle had travelled at a speed equal to x/t. Also, remember that the mass multiplied by the speed of a particle is its momentum, so the quantity mx/t is the measured momentum of the particle. We can now go ahead and simplify our equation some more, and write

$$\frac{p\Delta x}{h} < 1$$

where p is the momentum. This equation can be rearranged to read

$$p\Delta x < h$$

and this really is important enough to merit more discussion, because it looks very much like Heisenberg's Uncertainty Principle.

This is the end of the maths for the time being, and if you haven't followed it too carefully you should be able to pick the thread up from here.

If we start out with a particle localized within a blob of size Δx,

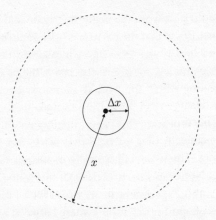

Figure 4.5. A small cluster grows with time, corresponding to a particle that is initially localized becoming delocalized as time advances.

we have just discovered that, after some time has passed, it could be found anywhere in a larger blob of size x. The situation is illustrated in Figure 4.5. To be precise, this means that if we had looked for the particle initially, then the chances are that we would have found it somewhere inside the inner blob. If we didn't measure it but instead waited a while, then there would be a good chance of finding it later on anywhere within the larger blob. This means that the particle could have moved from a position within the small initial blob to a position within the larger one. It doesn't have to have moved, and there is still a probability that it will be within the smaller region Δx. But it is quite possible that a measurement will reveal that the particle has moved as far out as the edge of the bigger blob.[8] If this extreme case were realized in a measurement then we would conclude that the particle is moving with a momentum given by the equation we just derived (and if you have not followed the maths then you will just have to take this on trust), i.e. $p = h/\Delta x$.

8. There is a chance that the particle travels even farther than the 'extreme' case marked out by the large blob in the figure but, as we have shown, the clocks tend to cancel out for such scenarios.

Now, we could start from the beginning again and set everything up exactly as before, so that the particle is once again initially located in the smaller blob of size Δx. Upon measuring the particle, we would probably find it somewhere else inside the larger blob, other than the extreme edge, and would therefore conclude that its momentum is smaller than the extreme value.

If we imagine repeating this experiment again and again, measuring the momentum of a particle that starts out inside a small cluster of size Δx, then we will typically measure a range of values of p anywhere between zero and the extreme value $h/\Delta x$. Saying that 'if you do this experiment many times then I predict you will measure the momentum to be somewhere between zero and $h/\Delta x$' means that 'the momentum of the particle is uncertain by an amount $h/\Delta x$'. Just as for the case of the uncertainty in position, physicists assign the symbol Δp to this uncertainty, and write $\Delta p \Delta x \sim h$. The '$\sim$' sign indicates that the product of the uncertainties in position and momentum is roughly equal to Planck's constant – it might be a little bigger or it might be a little smaller. With a little more care in the mathematics we could get this equation exactly right. The result would depend upon the details of the initial clock cluster, but it is not worth the extra effort to spend time doing that because what we have done is sufficient to capture the key ideas.

The statement that the uncertainty in a particle's position multiplied by the uncertainty in its momentum is (approximately) equal to Planck's constant is perhaps the most familiar form of Heisenberg's Uncertainty Principle. It is telling us that, starting from the knowledge that the particle is located within some region at some initial time, a measurement of the particle's position at some time later will reveal that the particle is moving with a momentum whose value cannot be predicted more accurately than 'somewhere between zero and $h/\Delta x$'. In other words, if we start out by confining a particle to be in a smaller and smaller region, then it has a tendency to want to jump further and further away from that region. This is so important, it is worth restating a third time: the more precisely you know the position of a particle at some instant, the less well you know how fast it is moving and therefore where it will be sometime later.

This is exactly Heisenberg's statement of the Uncertainty Principle. It lies at the heart of quantum theory, but we should be quite clear that it is not in itself a vague statement. It is a statement about our inability to track particles around with precision, and there is no more scope for quantum magic here than there is for Newtonian magic. What we have done in the last few pages is to derive Heisenberg's Uncertainty Principle from the fundamental rules of quantum physics as embodied in the rules for winding, shrinking and adding clocks. Indeed, its origin lies in our proposition that a particle can be anywhere in the Universe an instant after we measure its position. Our initial wild proposal that the particle can be anywhere and everywhere in the Universe has been tamed by the orgy of quantum interference, and the Uncertainty Principle is in a sense all that remains of the original anarchy.

There is something very important that we should say about how to interpret the Uncertainty Principle before we move on. We must not make the mistake of thinking that the particle is actually at some single specific place and that the spread in initial clocks really reflects some limitation in our understanding. If we thought that then we would not have been able to compute the Uncertainty Principle correctly, because we would not admit that we must take clocks from every possible point inside the initial cluster, transport them in turn to a distant point X and then add them all up. It was the act of doing this that gave us our result, i.e. we had to suppose that the particle arrives at X via a superposition of many possible routes. We will make use of Heisenberg's principle in some real-world examples later on. For now, it is satisfying that we have derived one of the key results of quantum theory using nothing more than some simple manipulations with imaginary clocks.

Let's stick a few numbers into the equations to get a better feel for things. How long will we have to wait for there to be a reasonable probability that a sand grain will hop outside a matchbox? Let's assume that the matchbox has sides of length 3 cm and that the sand grain weighs 1 microgram. Recall that the condition for there to be a reasonable probability of the sand grain hopping a given distance is given by

$$\frac{mx\Delta x}{ht} < 1$$

where Δx is the size of the matchbox. Let's calculate what t should be if we want the sand grain to jump a distance $x = 4$ cm, which would comfortably exceed the size of the matchbox. Doing a very simple bit of algebra, we find that

$$t > \frac{mx\Delta x}{h}$$

and sticking the numbers in tells us that t must be greater than approximately 10^{21} seconds. That is around 6×10^{13} years, which is over a thousand times the current age of the Universe. So it probably won't happen. Quantum mechanics is weird, but not weird enough to allow a grain of sand to hop unaided out of a matchbox.

To conclude this chapter, and launch ourselves into the next one, we will make one final observation. Our derivation of the Uncertainty Principle was based upon the configuration of clocks illustrated in Figure 4.4. In particular, we set up the initial cluster of clocks so that they all had hands of the same size and were all reading the same time. This specific arrangement corresponds to a particle initially at rest within a certain region of space – a sand grain in a matchbox, for example. Although we discovered that the particle will most likely not remain at rest, we also discovered that for large objects – and a grain of sand is very large indeed in quantum terms – this motion is completely undetectable. So there is some motion in our theory, but it is motion that is imperceptible for big enough objects. Obviously we are missing something rather important, because big things do actually move around, and remember that quantum theory is a theory of all things big and small. We must now address this problem: how can we explain motion?

5. *Movement as an Illusion*

In the previous chapter we derived Heisenberg's Uncertainty Principle by considering a particular initial arrangement of clocks – a small cluster of them, each with hands of the same size and pointing in the same direction. We discovered that this represents a particle that is approximately stationary, although the quantum rules imply that it jiggles around a little. We shall now set up a different initial configuration; we want to describe a particle in motion. In Figure 5.1, we've drawn a new configuration of clocks. Again it is a cluster of clocks, corresponding to a particle that is initially located in the vicinity of the clocks. The clock at position 1 reads 12 o'clock, as before, but the other clocks in the cluster are now all wound forwards by different amounts. We've drawn five clocks this time simply because it will help make the reasoning more transparent, although as before we are to imagine clocks in between the ones we have drawn – one for each point in space in the cluster. Let's apply the quantum rule as before and move these clocks to point X, a long way outside the cluster, to once again describe the many ways that the particle can hop from the cluster to X.

In a procedure that we hope is becoming more routine, let's take the clock from point 1 and propagate it to point X, winding it around as we go. It will wind around by an amount

$$W_1 = \frac{mx^2}{2ht}$$

Now let's take the clock from point 2 and propagate it to point X. It's a little bit further away, let's say a distance d further, so it will wind a bit more

$$W_2 = \frac{m(x + d)^2}{2ht}$$

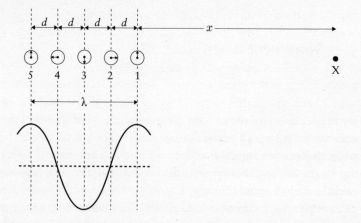

Figure 5.1: The initial cluster (illustrated by the clocks marked 1 to 5) is made up of clocks that all read different times – they are all shifted by three hours relative to their neighbours. The lower part of the figure illustrates how the time on the clocks varies through the cluster.

This is exactly what we did in the previous chapter, but perhaps you can already see that something different will happen for this new initial configuration of clocks. We set things up such that clock 2 was initially wound *forwards* by three hours relative to clock 1 – from 12 o'clock to 3 o'clock. But in carrying clock 2 to point X, we have to wind it *backwards* by a little more than clock 1, corresponding to the extra distance d that it has to travel. If we arrange things so that the initial forward wind of clock 2 is exactly the same as the extra backward wind it gets when travelling to X, then it will arrive at X *showing exactly the same time* as clock 1. This will mean that, far from cancelling out, it will add to clock 1 to make a larger clock, which in turn means that there will be a high probability that the particle will be found at X. This is a completely different situation from the orgy of quantum interference that occurred when we began with all the clocks reading the same time. Let's now consider clock 3, which we have wound forwards six hours relative to clock 1. This clock has to travel an extra distance $2d$ to make it to point X and again, because of the offset in time, this clock will arrive pointing to

12 o'clock. If we set all the offsets in the same manner, then this will happen right across the cluster and all of the clocks will add together constructively at X.

This means that there will be a high probability that the particle will be found at the point X at some later time. Clearly point X is special because it is that particular point where all the clocks from the cluster conspire to read the same time. But point X is not the only special point – all points to the left of X for a distance equal to the length of the original cluster also share the same property that the clocks add together constructively. To see this, notice that we could take clock 2 and transport it to a point a distance d to the left of X. This would correspond to moving it a distance x, which is exactly the same distance that we moved clock 1 when we moved it to X. We could then transport clock 3 to this new point through a distance $x + d$, which is exactly the same distance that we previously moved clock 2. These two clocks should therefore read the same time when they arrive and add together. We can keep on doing this for all the clocks in the cluster, but only until we reach a distance to the left of X equal to the original cluster size. Outside of this special region, the clocks largely cancel out because they are no longer protected from the usual orgy of quantum interference.[1] The interpretation is clear: the cluster of clocks moves, as illustrated in Figure 5.2.

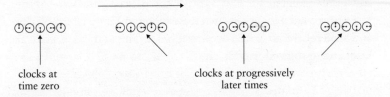

clocks at
time zero

clocks at progressively
later times

Figure 5.2. The cluster of clocks moves at constant speed to the right. This is because the original cluster had its clocks wound relative to each other as described in the text.

1. You might like to check this explicitly for yourself.

This is a fascinating result. By setting up the initial cluster using offset clocks rather than clocks all pointing in the same direction, we have arrived at the description of a moving particle. Intriguingly, we can also make a very important connection between the offset clocks and the behaviour of waves.

Remember that we were motivated to introduce the clocks back in Chapter 2 in order to explain the wave-like behaviour of particles in the double-slit experiment. Look back at Figure 3.3 on page 35, where we sketched an arrangement of clocks that describes a wave. It is just like the arrangement of the clocks in our moving cluster. We've sketched the corresponding wave below the cluster in Figure 5.1 using exactly the same methodology as before: 12 o'clock represents the peak of the wave, 6 o'clock represents the trough and 3 o'clock and 9 o'clock represent the places where the wave height is zero.

As we might have anticipated, it appears that the representation of a moving particle has something to do with a wave. The wave has a wavelength, and this corresponds to the distance between clocks showing identical times in the cluster. We've also drawn this on the figure, and labelled it λ.

We can now work out how far the point X should be away from the cluster in order for adjacent clocks to add constructively. This will lead us to another very important result in quantum mechanics, and make the connection between quantum particles and waves much clearer. Time for a bit more mathematics.

First, we need to write down the extra amount by which clock 2 is wound relative to clock 1 because it has further to travel to point X. Using the results on page 75, this is

$$W_2 - W_1 = \frac{m(x+d)^2 - mx^2}{2ht} \simeq \frac{mxd}{ht}$$

Again, you may be able to work this out for yourself by multiplying out the brackets and throwing away the d^2 bits because d, the distance between the clocks, is very small compared to x, the distance to point X a long way away from the original cluster.

It is also straightforward to write down the criterion for the clocks to read the same time; we want the extra amount of winding due to the propagation of clock 2 to be exactly cancelled by the extra forward wind we gave it initially. For the example shown in Figure 5.1, the extra wind for clock 2 is $\frac{1}{4}$, because we've wound the clock forward by a quarter of a turn. Similarly, clock 3 has a wind of $\frac{1}{2}$, because we've wound it around $\frac{1}{2}$ a turn. In symbols, we can express the fraction of one full wind between two clocks quite generally as d/λ, where d is the distance between the clocks and λ is the wavelength. If you can't quite see this, just think of the case for which the distance between two clocks is equal to the wavelength. Then $d = \lambda$, and therefore $d/\lambda = 1$, which is one full wind, and both clocks will read the same time.

Bringing this all together, we can say that for two adjacent clocks to read the same time at point X we require the extra amount of wind we put into the initial clock to be equal to the extra amount of wind due to the difference in propagation distance:

$$\frac{mxd}{ht} = \frac{d}{\lambda}$$

We can simplify this, as we've done before, by noticing that mx/t is the momentum of the particle, p. So with a little bit of rearrangement, we get

$$p = \frac{h}{\lambda}$$

This result is important enough to warrant a name, and it is called the de Broglie equation because it was first proposed in September 1923 by the French physicist Louis de Broglie. It is important because it associates a wavelength with a particle of a known momentum. In other words it expresses an intimate link between a property usually associated with particles – momentum – and a property usually associated with waves – wavelength. In this way, the wave-particle duality of quantum mechanics has emerged from our manipulations with clocks.

The de Broglie equation constituted a huge conceptual leap. In his original paper, he wrote that a 'fictitious associated wave' should be assigned to all particles, including electrons, and that a stream of electrons passing through a slit 'should show diffraction phenomena'.[2] In 1923, this was theoretical speculation, because Davisson and Germer did not observe an interference pattern using beams of electrons until 1927. Einstein made a similar proposal to de Broglie's, using different reasoning, at around the same time, and these two theoretical results were the catalyst for Schrödinger to develop his wave mechanics. In the last paper before he introduced his eponymous equation, Schrödinger wrote: 'That means nothing else but taking seriously the de Broglie–Einstein wave theory of moving particles.'

We can gain a little more insight into the de Broglie equation by looking at what happens if we decrease the wavelength, which would correspond to increasing the amount of winding between adjacent clocks. In other words, we will reduce the distance between clocks reading the same time. This means that we would then have to increase the distance x to compensate for the decrease in λ. In other words, point X needs to be further away in order for the extra winding to be 'undone'. That corresponds to a faster-moving particle: smaller wavelength corresponds to larger momentum, which is exactly what the de Broglie equation says. It is a lovely result that we have managed to 'derive' ordinary motion (because the cluster of clocks moves smoothly in time) starting from a static array of clocks.

Wave Packets

We would now like to return to an important issue that we skipped over earlier in the chapter. We said that the initial cluster moves in its entirety to the vicinity of point X, but only roughly maintains its

2. 'Diffraction' is a word used to describe a particular type of interference, and it is characteristic of waves.

original configuration. What did we mean by that rather imprecise statement? The answer provides a link back to the Heisenberg Uncertainty Principle, and delivers further insight.

We have been describing what happens to a cluster of clocks, which represents a particle that can be found somewhere within a small region of space. That's the region spanned by our five clocks in Figure 5.1. A cluster like this is referred to as a wave packet. But we have already seen that confining a particle to some region in space has consequences. We cannot prevent a localized particle from getting a Heisenberg kick (i.e. its momentum is uncertain because it is localized), and as time passes this will lead to the particle 'leaking out' of the region within which it was initially located. This effect was present for the case where the clocks all read the same time and it is present in the case of the moving cluster too. It will tend to spread the wave packet out as it travels, just as a stationary particle spreads out over time.

If we wait long enough, the wave packet corresponding to the moving cluster of clocks will have totally disintegrated and we'll lose any ability to predict where the particle actually is. This will obviously have implications for any attempt we might make to measure the speed of our particle. Let's see how this works out.

A good way to measure a particle's speed is to make two measurements of its position at two different times. We can then deduce the speed by dividing the distance the particle travelled by the time between the two measurements. Given what we've just said, however, this looks like a dangerous thing to do because if we make a measurement of the position of a particle too precisely then we are in danger of squeezing its wave packet, and that will change its subsequent motion. If we don't want to give the particle a significant Heisenberg kick (i.e. a significant momentum because we make Δx too small) then we must make sure that our position measurement is sufficiently vague. Vague is, of course, a vague term, so let's make it less so. If we use a particle detection device that is capable of detecting particles to an accuracy of 1 micrometre and our wave packet has a width of 1 nanometre, then the detector won't have much impact on the particle at all. An experimenter reading out the detector might be very happy with a resolution of 1 micron but, from the electron's perspective, all the detector did was report back to the experimenter that the particle is in some huge box, a thousand times bigger than the actual wave packet. In this case, the Heisenberg kick induced by the measurement process will be very small compared to that induced by the finite size of the wave packet itself. That's what we mean by 'sufficiently vague'.

We've sketched the situation in Figure 5.3 and have labelled the initial width of the wave packet d and the resolution of our detector Δ. We've also drawn the wave packet at a later time; it's a little broader and has a width d', which is bigger than d. The peak of the wave packet has travelled a distance L over some time interval t at a speed v. Apologies if that particular flourish of formality reminds you of your long-forgotten school days sitting behind a stained and eroded wooden bench listening to a science teacher's voice fading into the half-light of a late winter's afternoon as you slide into an inappropriate nap. We are covering ourselves in chalk dust for good reason, and it is our hope that the conclusion of this section will jolt you back to consciousness more effectively than the flying board dusters of your youth.

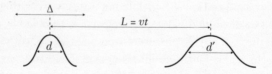

Figure 5.3. A wave packet at two different times. The packet moves to the right and spreads out as time advances. The packet moves because the clocks that constitute it are wound around relative to each other (de Broglie) and it spreads out because of the Uncertainty Principle. The shape of the packet is not very important but, for completeness, we should say that where the packet is large the clocks are large, and where it is small the clocks are small.

Back in the metaphorical science lab, with renewed vigour, we are trying to measure the speed v of the wave packet by making two measurements of its position at two different times. This will give us the distance L that the wave packet has travelled in a time t. But our detector has a resolution Δ, so we won't be able to pin down L exactly. In symbols, we can say that the measured speed is

$$v = \frac{L \pm \Delta}{t}$$

where the combined plus or minus sign is there simply to remind us that, if we actually make the two position measurements, we will generally not always get L but instead 'L plus a bit' or 'L minus a bit', where the 'bit' is due to the fact we agreed not to make a very accurate measurement of the particle's position. It is important to bear in mind that L is not something we can actually measure: we always measure a value somewhere in the range $L \pm \Delta$. Remember also that we need Δ to be much larger than the size of the wave packet otherwise we will squeeze the particle and that will disrupt it.

Let's rewrite the last equation very slightly so that we can better see what's going on:

$$v = \frac{L}{t} \pm \frac{\Delta}{t}$$

It seems that if we take t to be very large then we will get a measurement of the speed $v = L/t$ with a very tiny spread, because we can choose to wait around for a very long time, making t as large as we like and consequently Δ/t as small as we like whilst still keeping Δ comfortably large. This looks like we have a nice way to make an arbitrarily precise measurement of the particle's speed without disturbing it at all; just wait for a huge amount of time between the first and the second measurements. This makes perfect intuitive sense. Imagine you are measuring the speed of a car driving along a road. If you measure how far it has travelled in one minute, you will tend to get a much more precise measurement of its speed than if you measure how far it travelled in one second. Have we dodged Heisenberg?

Of course not – we have forgotten to take something into

account. The particle is described by a wave packet that spreads out as time passes. Given enough time, the spreading out will completely wash out the wave packet and that means the particle could be anywhere. This will increase the range of values we get in our measurement of L and spoil our ability to make an arbitrarily accurate measurement of its speed.

For a particle described by a wave packet, we are ultimately still bound by the Uncertainty Principle. Because the particle is initially confined in a region of size d, Heisenberg informs us that the particle's momentum is correspondingly blurred out by an amount equal to h/d.

There is therefore only one way we can build a configuration of clocks to represent a particle that travels with a definite momentum – we must make d, the size of the wave packet, very large. And the larger we make it, the smaller the uncertainty in its momentum will be. The lesson is clear: a particle of well-known momentum is described by a large cluster of clocks.[3] To be precise, a particle of absolutely definite momentum will be described by an infinitely long cluster of clocks, which means an infinitely long wave packet.

We have just argued that a finite-size wave packet does not correspond to a particle with a definite momentum. This means that if we measured the momentum of very many particles, all described by exactly the same initial wave packet, then we would not get the same answer each time. Instead we would get a spread of answers and it does not matter how brilliant we are at experimental physics, that spread cannot be made smaller than h/d.

We can therefore say that a wave packet describes a particle that is travelling with a range of momenta. But the de Broglie equation implies that we can just substitute the word 'wavelengths' for 'momenta' in the last sentence, because a particle's momentum is

3. Of course if d is very large then one might wonder how we can even measure the momentum. That concern is sidestepped by ensuring that no matter how big d is, L is much bigger than it.

associated with a wave of definite wavelength. This in turn means that a wave packet must be made up of many different wavelengths. Likewise, if a particle is described by a wave with a definite wavelength then that wave must necessarily be infinitely long. It sounds like we are being pushed to conclude that a small wave packet is made up of many infinitely long waves of different wavelengths. We are indeed being pushed down this route, and what we are describing is very familiar to mathematicians, physicists and engineers alike. This is an area of mathematics known as Fourier analysis, named after the French mathematical physicist Joseph Fourier.

Fourier was a colourful man. Amongst his many notable achievements, he was Napoleon's governor of Lower Egypt and the discoverer of the greenhouse effect. He apparently enjoyed wrapping himself up in blankets, which led to his untimely demise one day in 1830 when, tightly wrapped, he fell down his own stairs. His key paper on Fourier analysis addressed the subject of heat transfer in solids and was published in 1807, although the basic idea can be traced back much earlier.

Fourier showed that any wave at all, of arbitrarily complex shape and extent, can be synthesized by adding together a number of sine waves of different wavelengths. The point is best illustrated through pictures. In Figure 5.4 the dotted curve is made by adding together the first two sine waves in the lower graphs. You can almost do the addition in your head – the two waves are both at maximum height in the centre, and so they add together there, whilst they tend to cancel each other out at the ends. The dashed curve is what happens if we add together all four of the waves illustrated in the lower graphs – now the peak in the centre is becoming more pronounced. Finally, the solid curve shows what happens when we add together the first ten waves, i.e. the four shown plus six more of progressively decreasing wavelength. The more waves we add in to the mix, the more detail we can achieve in the final wave. The wave packet in the upper graph could describe a localized particle, rather like the wave packet illustrated in Figure 5.3. In this way it really is possible

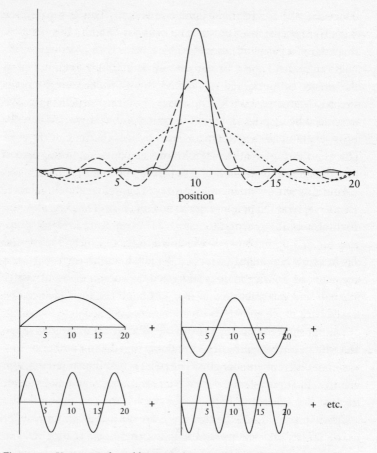

Figure 5.4. Upper graph: Adding together several sine waves to synthesize a sharply peaked wave packet. The dotted curve contains fewer waves than the dashed one, which in turn contains fewer than the solid one. Lower graphs: The first four waves used to build up the wave packets in the upper graph.

to synthesize a wave of any shape at all – it is all achieved by adding together simple sine waves.

The de Broglie equation informs us that each of the waves in the lower graphs of Figure 5.4 corresponds to a particle with a definite momentum, and the momentum increases as the wavelength decreases. We are beginning to see why it is that if a particle is described by a localized cluster of clocks then it must necessarily be made up of a range of momenta.

To be more explicit, let's suppose that a particle is described by the cluster of clocks represented by the solid curve in the upper graph in Figure 5.4.[4] We have just learnt that this particle can also be described by a series of much longer clusters of clocks: the first wave in the lower graphs plus the second wave in the lower graphs, plus the third wave in the lower graphs, and so on. In this way of thinking, there are several clocks at each point (one from each long cluster), which we should add together to produce the single clock cluster represented in upper graph of Figure 5.4. The choice of how to think about the particle is really 'up to you'. You can think of it as being described by one clock at each point, in which case the size of the clock immediately lets you know where the particle is likely to be found, i.e. in the vicinity of the peak in the upper graph of Figure 5.4. Alternatively, you can think of it as being described by a number of clocks at each point, one for each possible value of the momentum of the particle. In this way we are reminding ourselves that the particle localized in a small region does not have a definite momentum. The impossibility of building a compact wave packet from a single wavelength is an evident feature of Fourier's mathematics.

This way of thinking provides us with a new perspective on Heisenberg's Uncertainty Principle. It says that we cannot describe a particle in terms of a localized cluster of clocks using clocks corresponding to waves of a single wavelength. Instead, to get the clocks to cancel outside the region of the cluster, we must necessarily mix in different wavelengths and hence different momenta. So, the price we pay for localizing the particle to some region in space is to admit we do not know what its momentum is. Moreover, the more we restrict the particle, the more waves we need to add in and

4. Recall that when we draw pictures of waves, they are really a convenient way of picturing what the projections of the clock hands in the 12 o'clock direction are.

the less well we know its momentum. This is exactly the content of the Uncertainty Principle, and it is very satisfying to have found a different way of reaching the same conclusion.[5]

To close this chapter we want to spend a little more time with Fourier. There is a very powerful way of picturing quantum theory that is intimately linked to the ideas we have just been discussing. The important point is that any quantum particle, whatever it is doing, is described by a wavefunction. As we've presented it so far, the wavefunction is simply the array of little clocks, one for each point in space, and the size of the clock determines the probability that the particle will be found at that point. This way of representing a particle is called the 'position space wavefunction' because it deals directly with the possible positions that a particle can have. There are, however, many ways of representing the wavefunction mathematically, and the little clocks in space version is only one of them. We touched on this when we said it is possible to think of the particle as also being represented by a sum over sine waves. If you ponder this point for a moment, you should realize that specifying the complete list of sine waves actually provides a complete description of the particle (because by adding together these waves we can obtain the clocks associated with the position space wavefunction). In other words, if we specify exactly which sine waves are needed to build a wave packet, and exactly how much of each sine wave we need to add in to get the shape just right, then we will have a different but entirely equivalent description of the wave packet. The neat thing is that any sine wave can itself be described by a single imaginary clock: the size of the clock encodes the maximum height of the wave and the phase of the wave at some point can be represented by the time that the clock reads. This means that we can choose to represent a particle not by clocks in space but by an alternative list of clocks, one for each possible value of the particle's momentum. This description is just as economical as the 'clocks in space' descrip-

5. This way of arriving at the Uncertainty Principle did, however, rely on the de Broglie equation in order to link the wavelength of a clock wave to its momentum.

tion, and instead of making explicit where the particle is likely to be found we are instead making explicit what values of momentum the particle is likely to have. This alternative array of clocks is known as the momentum space wavefunction and it contains exactly the same information as the position space wavefunction.[6]

This might sound very abstract, but you may well use technology based on Fourier's ideas every day, because the decomposition of a wave into its component sine waves is the foundation of audio and video compression technology. Think about the sound waves that make up your favourite tune. This complicated wave can, as we have just learnt, be broken down into a series of numbers that give the relative contributions of each of a large number of pure sine waves to the sound. It turns out that, although you may need a vast number of individual sine waves to reproduce the original sound wave exactly, you can in fact throw a lot of them away without compromising the perceived audio quality at all. In particular, the sine waves that contribute to sound waves that humans can't hear are not kept. This vastly reduces the amount of data needed to store an audio file – hence your mp3 player doesn't need to be too large.

We might also ask what possible use could this different and even more abstract version of the wavefunction be? Well, think of a particle represented, in position space, by a single clock. This describes a particle located at a certain place in the Universe; the single point where the clock sits. Now think of a particle represented by a single clock, but this time in momentum space. This represents a particle with a single, definite momentum. Describing such a particle using the position space wavefunction would, in contrast, require an infinite number of equally sized clocks, because according to the Uncertainty Principle, a particle with a definite momentum can be found anywhere. As a result, it is sometimes simpler to perform calculations directly in terms of the momentum space wavefunction.

6. In the jargon, the momentum space wavefunctions that correspond to particles with definite momentum are known as momentum eigenstates, after the German word *eigen*, meaning 'characteristic'.

In this chapter, we have learnt that the description of a particle in terms of clocks is capable of capturing what we ordinarily call 'movement'. We have learnt that our perception that objects move smoothly from point to point is, from the perspective of quantum theory, an illusion. It is closer to the truth to suppose that particles move from A to B via all possible paths. Only when we add together all of the possibilities does motion as we perceive it emerge. We have also seen explicitly how the clock description manages to encode the physics of waves, even though we only ever deal with point-like particles. It is time now to really exploit the similarity with the physics of waves as we tackle the important question: how does quantum theory explain the structure of atoms?

6.   The Music of the Atoms

The interior of an atom is a strange place. If you could stand on a proton and gaze outwards into inter-atomic space, you would see only void. The electrons would still be imperceptibly tiny even if they approached close enough for you to touch them, which they very rarely would. The proton is around 10^{-15} m in diameter, 0.000000000000001 metres, and is a quantum colossus compared to the electrons. If you stand on your proton at the edge of England on the White Cliffs of Dover, the fuzzy edge of the atom lies somewhere amongst the farms of northern France. Atoms are vast and empty, which means the full-size you is vast and empty too. Hydrogen is the simplest atom, comprising a single proton and a single electron. The electron, vanishingly small as far as we can tell, might seem to have a limitless arena within which to roam, but this is not true. It is bound to its proton, trapped by their mutual electromagnetic attraction, and it is the size and shape of this generous prison that gives rise to the characteristic barcode rainbow of light meticulously documented in the *Handbuch der Spectroscopie* by our old friend and dinner-party guest Professor Kayser.

We are now in a position to apply the knowledge we have accumulated so far to the question that so puzzled Rutherford, Bohr and others in the early decades of the twentieth century: what exactly is going on inside an atom? The problem, if you recall, was that Rutherford discovered that the atom is in some ways like a miniature solar system, with a dense nucleus Sun at the centre and electrons as planets sweeping around in distant orbits. Rutherford knew that this model couldn't be right, because electrons in orbit around a nucleus should continually emit light. The result should be catastrophic for the atom, because if the electron continually emits light then it must lose energy and spiral inwards on an inevitable collision course with the proton. This, of course, doesn't happen. Atoms tend to be stable things, so what is wrong with this picture?

This chapter marks an important stage in the book, because it is the first time that our theory is to be used to explain real-world phenomena. All our hard work to this point has been concerned with getting the essential formalism worked out so that we have a way to think about a quantum particle. Heisenberg's Uncertainty Principle and the de Broglie equation represent the pinnacle of our achievements, but in the main we have been modest, thinking about a universe containing just one particle. It is now time to show how quantum theory impacts on the everyday world in which we live. The structure of atoms is a very real and tangible thing. You are made of atoms: their structure is your structure, and their stability is your stability. It would not be unduly hyperbolic to say that understanding the structure of atoms is one of the necessary conditions for understanding our Universe as a whole.

Inside a hydrogen atom, the electron is trapped in a region surrounding the proton. We are going to start by imagining that the electron is trapped in some sort of box, which is not very far from the truth. Specifically, we'll investigate to what extent the physics of an electron trapped inside a tiny box captures the salient features of a real atom. We are going to proceed by exploiting what we learnt in the previous chapter about the wave-like properties of quantum particles, because, when it comes to describing atoms, the wave picture really simplifies things and we can make a good deal of progress without having to worry about shrinking, winding and adding clocks. Always bear in mind, though, that the waves are a convenient shorthand for what is going on 'under the bonnet'.

Because the framework we've developed for quantum particles is extremely similar to that used in the description of water waves, sound waves or the waves on a guitar string, we'll think first about how these more familiar material waves behave when they are confined in some way.

Generally speaking, waves are complicated things. Imagine jumping into a swimming pool full of water. The water will slosh around

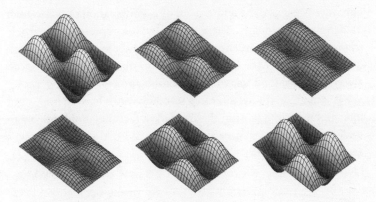

Figure 6.1. Six successive snapshots of a standing wave in a tank of water. The time advances from the top left to the bottom right.

all over the place, and it would seem to be futile to try to describe what is going on in any simple fashion. Underlying the complexity, however, there is hidden simplicity. The key point is that the water in a swimming pool is confined, which means that all the waves are trapped inside the pool. This gives rise to a phenomenon known as 'standing waves'. The standing waves are hidden away in the mess when we disturb the pool by jumping into it, but there is a way to make the water move so that it oscillates in the regular, repeating patterns of the standing waves. Figure 6.1 shows how the water surface looks when it is undergoing one such oscillation. The peaks and troughs rise and fall, but most importantly they rise and fall in exactly the same place. There are other standing waves too, including one where the water in the middle of the tank rises and falls rhythmically. We do not usually see these special waves because they are hard to produce, but the key point is that any disturbance of the water at all – even the one we caused by our inelegant dive and subsequent thrashing around – can be expressed as some combination or other of the different standing waves. We've seen this type of behaviour before; it is a direct generalization of Fourier's ideas that we encountered in the last chapter. There, we saw that any wave packet can be built up out of a combination of waves each of definite wavelength. These special waves, representing particle states of definite momentum, are sine waves. In the case of confined water waves, the idea generalizes so that any disturbance can always be described using some combination of standing waves. We'll see later in this chapter that standing waves have an important interpretation in quantum theory, and in fact they hold the key to understanding the structure of atoms. With this in mind, let's explore them in a little more detail.

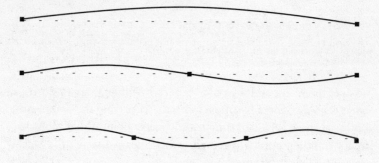

Figure 6.2. The three longest wavelength waves that can fit on a guitar string. The longest wavelength (at the top) corresponds to the lowest harmonic (fundamental) and the others correspond to the higher harmonics (overtones).

Figure 6.2 shows another example of standing waves in Nature: three of the possible standing waves on a guitar string. On plucking a guitar string, the note we hear is usually dominated by the standing wave with the largest wavelength – the first of the three waves shown in the figure. This is known in both physics and music as the 'lowest harmonic' or 'fundamental'. Other wavelengths are usually present too, and they are known as overtones or higher harmonics. The other waves in the figure are the two longest-wavelength overtones. The guitar is a nice example because it's simple enough to see why a guitar string can only vibrate at these special wavelengths. It is because it is held fixed at both ends – by the guitar bridge at one end and your finger pressing against a fret at the other. This means that the string cannot move at these two points, and this determines the allowed wavelengths. If you play the guitar, you'll know this physics instinctively; as you move your fingers up the fret board towards the bridge, you decrease the length of the string and therefore force it to vibrate with shorter and shorter wavelengths, corresponding to higher-pitched notes.

The lowest harmonic is the wave that has only two stationary points, or 'nodes'; it moves everywhere except at the two fixed ends. As you can see from the figure, this note has a wavelength of twice the length of the string. The next smallest wavelength is equal to the length of the string, because we can fit another node in the centre. Next, we can get a wave with wavelength equal to $\frac{2}{3}$ times the length of the string, and so on.

In general, just as in the case of the water confined in a swimming pool, the string will vibrate in some combination of the different possible standing waves, depending on how it is plucked. The actual shape of the string can always be obtained by adding together the standing waves corresponding to each of the harmonics present. The harmonics and their relative sizes give the sound its characteristic tone. Different guitars will have different distributions of harmonics and therefore sound different, but a middle C (a pure harmonic) on one guitar is always the same as a middle C on another. For the guitar, the shape of the standing waves is very simple: they are pure sine waves whose wavelengths are fixed by the length of the string. For the swimming pool, the standing waves are more complicated, as shown in Figure 6.1, but the idea is exactly the same.

You may be wondering why these special waves are called 'standing waves'. It is because the waves do not change their shape. If we take two snapshots of a guitar string vibrating in a standing wave, then the two pictures will only differ in the overall size of the wave. The peaks will always be in the same place, and the nodes will always be in the same place because they are fixed by the ends of the string or, in the case of the swimming pool, by the sides of the pool. Mathematically, we could say that the waves in the two snapshots differ only by an overall multiplicative factor. This factor varies periodically with time, and expresses the rhythmical vibration of the string. The same is true for the swimming pool in Figure 6.1, where each snapshot is related to the others by an overall multiplicative factor. For example, the last snapshot can be obtained from the first by multiplying the wave height at every point by minus one.

In summary, waves that are confined in some way can always be expressed in terms of standing waves (waves that do not change their shape) and, as we have said, there are very good reasons for devoting so much time to understanding them. At the top of the list is the fact that standing waves are *quantized*. This is very clear for the standing waves on a guitar string: the fundamental has a wavelength of twice the length of the string, and the next longest allowed wavelength is equal to the length of the string. There is no standing wave with a wavelength in between these two and so we can say that the allowed wavelengths on a guitar string are quantized.

Standing waves therefore make manifest the fact that something gets quantized when we trap waves. In the case of a guitar string, it is clearly the wavelength. For the case of an electron inside a box, the quantum waves corresponding to the electron will also be trapped, and by analogy we should expect that only certain standing waves will be present in the box, and therefore that something will be quantized. Other waves simply cannot exist, just as a guitar string doesn't play all the notes in an octave at the same time no matter how it is plucked. And just as for the sound of a guitar, the general state of the electron will be described by a blend of standing waves. These quantum standing waves are starting to look very interesting, and, encouraged by this, let's start our analysis proper.

To make progress, we must be specific about the shape of the box inside which we place our electron. To keep things simple, we'll suppose that the electron is free to hop around inside a region of size L, but that it is totally forbidden from wandering outside this region. We do not need to say how we intend to forbid the electron from wandering – but if this is supposed to be a simplified model of an atom then we should imagine that the force exerted by the positively charged nucleus is responsible for its confinement. In the jargon, this is known as a 'square well potential'. We've sketched the situation in Figure 6.3, and the reason for the name should be obvious.

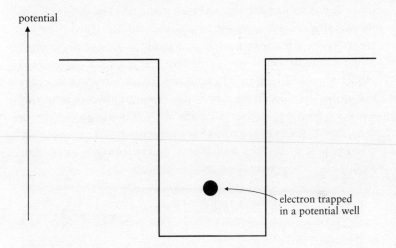

Figure 6.3. An electron trapped in a square well potential.

The idea of confining a particle in a potential is a very important one that we'll use again, so it will be useful to make sure we understand exactly what it means. How do we actually trap particles? That is quite a sophisticated question; to get to the bottom of it we'll need to learn about how particles interact with other particles, which we will do in Chapter 10. Nevertheless, we can make progress provided we don't ask too many questions.

The ability 'not to ask too many questions' is a necessary skill in physics because we have to draw the line somewhere in order to answer any questions at all; no system of objects is perfectly isolated. It seems reasonable that if we want to understand how a microwave oven works, we don't need to worry about any traffic passing by outside. The traffic will have a tiny influence on the operation of the oven. It will induce vibrations in the air and ground which will shake the oven a little bit. There may also be stray magnetic fields that influence the internal electronics of the oven, no matter how well they are shielded. It is possible to make mistakes in ignoring things because there might be some crucial detail that we miss. If this is the case, we'll simply get the wrong answer and have to reconsider our assumptions. This is very important, and goes to the heart of the success of science; all assumptions are ultimately validated or negated by experiment. Nature is the arbiter, not human intuition. Our strategy here is to ignore the details of the mechanism that traps the electron and model it by something called a potential. The word 'potential' really just means 'an effect on the particle due to some physics or other that I will not bother to explain in detail'. We will bother to describe in detail how particles interact later on, but for now we'll talk in the language of potentials. If this sounds a bit cavalier, let us give an example to illustrate how potentials are used in physics.

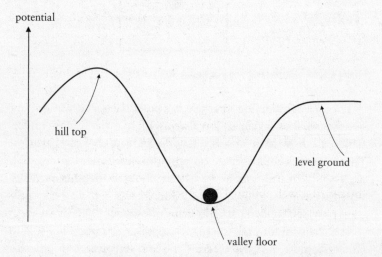

Figure 6.4. A ball sitting on a valley floor. The height of the ground above sea-level is directly proportional to the potential that the particle experiences when it rolls around.

Figure 6.4 illustrates a ball trapped in a valley. If we give the ball a kick then it can roll up the valley, but only so far, and then it will roll back down again. This is an excellent example of a particle trapped by a potential. In this case, the Earth's gravitational field generates the potential and a steep hill makes a steep potential. It should be clear that we could calculate the details of how a ball rolls around in a valley without knowing the precise details of how the valley floor interacts with the ball – for this we'd have to know about the theory of quantum electrodynamics. If it turned out that the details of the inter-atomic interactions between the atoms in the ball and the atoms in the valley floor affected the motion of the ball too much, then the predictions we make would be wrong. In fact, the inter-atomic interactions are important because they give rise to friction, but we can also model this without getting into Feynman diagrams. But we digress.

This example is very tangible because we can literally see the shape of the potential[1]. However, the idea is more general and works for potentials other than those created by gravity and valleys. An example is the electron trapped in a square well. Unlike the case of the ball in a valley, the height of the walls is not the actual height of anything; rather it represents how fast the electron needs to be moving before it can escape from the well. For the case of a valley, this would be analogous to rolling the ball so fast that it climbed up the walls and out of the valley. If the electron is moving slowly enough then the actual height of the potential won't matter much, and we can safely assume that the electron is confined to the interior of the well.

Let us now focus on the electron trapped inside a box described by a square well potential. Since it cannot escape from the box, the quantum waves must fall to zero at the edges of the box. The three possible quantum waves with the largest wavelengths are then

1. The fact that the gravitational potential exactly maps the terrain is because, in the vicinity of the Earth's surface, the gravitational potential is proportional to the height above the ground.

entirely analogous to the guitar-string waves illustrated in Figure 6.2: the longest possible wavelength is twice the size of the box, $2L$; the next longest wavelength is equal to the size of the box, L; and the next again has a wavelength of $2L/3$. Generally, we can fit electron waves with wavelength $2L/n$ in the box where $n = 1, 2, 3, 4$, etc.

Specifically for the square box, therefore, the electron waves are precisely the same shape as the waves on a guitar string; they are sine waves with a very particular set of allowed wavelengths. Now we can go ahead and invoke the de Broglie equation from the last chapter to relate the wavelength of these sine waves to the momentum of the electron via $p = h/\lambda$. In which case, the standing waves describe an electron that is only allowed to have certain momenta, given by the formula $p = nh/(2L)$, where all we did here was to insert the allowed wavelengths into the de Broglie equation.

And so it is that we have demonstrated that the momentum of our electron is quantized in a square well. This is a big deal. However, we do need to take care. The potential in Figure 6.3 is a special case, and for other potentials the standing waves are not generally sine waves. Figure 6.5 shows a photograph of the standing waves on a drum. The drum skin is sprinkled with sand, which collects at the nodes of the standing wave. Because the boundary enclosing the vibrating drum skin is circular, rather than square, the standing waves are no longer sine waves.[2] This means that, as soon as we move to the more realistic case of an electron trapped by a proton, its standing waves will likewise not be sine waves. In turn this means that the link between wavelength and momentum is lost. How, then, are we to interpret these standing waves? What is it that is generally quantized for trapped particles, if it isn't their momentum?

We can get the answer by noticing that in the square well potential, if the electron's momentum is quantized, then so too is its energy. That is a simple observation and appears to contain no important new information, since energy and momentum are simply related to each other. Specifically, the energy $E = p^2/2m$,

2. They are in fact described by Bessel functions.

Figure 6.5. A vibrating drum covered in sand. The sand collects at the nodes of the standing waves.

where p is the momentum of the trapped electron and m is its mass.[3] This is not such a pointless observation as it might appear, because, for potentials that are not as simple as the square well, each standing wave *always* corresponds to a particle of definite energy.

The important difference between energy and momentum emerges because $E = p^2/2m$ is only true when the potential is flat in the region where the particle can exist, allowing the particle to move freely, like a marble on a table top or, more to the point, an electron in a square well. More generally, the particle's energy will not be equal to $E = p^2/2m$; rather it will be the sum of the energy due to its motion and its potential energy. This breaks the simple link between the particle's energy and its momentum.

---

3. This is obtained using the fact that the energy is equal to $\frac{1}{2}mv^2$ and $p = mv$. These equations do get modified by Special Relativity but the effect is small for an electron inside a hydrogen atom.

We can illustrate this point by thinking again about the ball in a valley, shown in Figure 6.4. If we start with the ball resting happily on the valley floor, then nothing happens.[4] To make it roll up the side of the valley, we'd have to give it a kick, which is equivalent to saying that we need to add some energy to it. The instant after we kick the ball, all of its energy will be in the form of kinetic energy. As it climbs the side of the valley, the ball will slow down until, at some height above the valley floor, it will come to a halt before rolling back down again and up the other side. At the moment it stops, high up the valley side, it has no kinetic energy, but the energy hasn't just magically vanished. Instead, all of the kinetic energy has been changed into potential energy, equal to mgh, where g is the acceleration due to gravity at the Earth's surface and h is the height of the ball above the valley floor. As the ball starts to roll back down into the valley, this stored potential energy is gradually converted back into kinetic energy as the ball speeds up again. So as the ball rolls from one side of the valley to the other, the total energy remains constant, but it periodically switches between kinetic and potential. Clearly, the ball's momentum is constantly changing, but its energy remains constant (we have pretended that there is no friction to slow the ball down. If we did include it then the total energy would still be constant but only after including the energy dissipated via friction).

We are now going to explore the link between standing waves and particles of definite energy in a different way, without appealing to the special case of the square well. We'll do this using those little quantum clocks.

First, notice that, if an electron is described by a standing wave at some instant in time, then it will be described by the same standing wave at some later time. By 'the same', we mean that the shape of the wave is unchanged, as was the case for the standing water wave in Figure 6.1. We don't, of course, mean that the wave does not change

---

4. This is a big ball and we don't need to worry about any quantum jiggling. But, if the thought crossed your mind, it is a good sign: your intuition is becoming quantized.

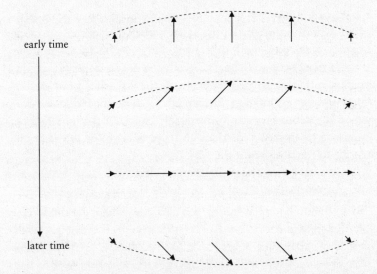

early time

later time

Figure 6.6: Four snapshots of a standing wave at successively later times. The arrows represent the clock hands and the dotted line is the projection onto the '12 o'clock' direction. The clocks all turn around in unison.

at all; the water height does change, but crucially the positions of the peaks and nodes do not. This allows us to figure out what the quantum clock description of a standing wave must look like, and it is illustrated in Figure 6.6 for the case of the fundamental standing wave. The clock sizes along the wave reflect the position of the peaks and nodes, and the clock hands sweep around together at the same rate. We hope you can see why we've drawn this particular pattern of clocks. The nodes must always be nodes, the peaks must always be peaks and they must always stay in the same place. This means that the clocks sitting in the vicinity of the nodes must *always* be very small, and the clocks representing the peaks must *always* have the longest hands. The only freedom we have, therefore, is to allow the clocks to sit where we put them and rotate in sync.

If we were following the methodology of the earlier chapters, we would now start from the configuration of clocks shown in the top row of Figure 6.6 and use the shrinking and turning rules to generate the bottom three rows at later times. This exercise in clock hopping is a hop too far for this book, but it can be done, and there is a nice twist because to do it correctly it is necessary to include the possibility that the particle 'bounces off the walls of the box' before hopping to its destination. Incidentally, because the clocks are bigger in the centre, we can immediately conclude that an electron described by this array of clocks is more likely to be found in the middle of the box than at the edges.

So, we have found that the trapped electron is described by an array of clocks that all whizz around at the same rate. Physicists don't usually talk like this, and musicians certainly don't; they both say that standing waves are waves of definite frequency.[5] High-frequency waves correspond to clocks that whizz around faster than the clocks of low-frequency waves. You can see this because, if a clock whizzes around faster, then the time it takes a peak to turn into a trough and then rise back again (represented by a single rotation of the clock hand) decreases. In terms of water waves, the high-frequency standing waves move up and down faster than the low-frequency ones. In music, a middle C is said to have a frequency of 262 Hz, which means that, on a guitar, the string vibrates up and down 262 times every second. The A above middle C has a frequency of 440 Hz, so it vibrates more rapidly (this is the agreed tuning standard for most orchestras and musical instruments across the world). As we've noted, however, it is only for pure sine waves that these waves of definite frequency also have definite wavelength. Generally speaking, *frequency* is the fundamental quantity that describes standing waves, and this sentence is probably a pun.

The million-dollar question, then, is 'What does it mean to speak of an electron of a certain frequency?' We remind you that these electron states are interesting to us because they are quantized and because an electron in one such state remains in that state for all

5. Actually, musicians probably don't say this either, and drummers definitely don't, because 'frequency' is a word with more than two syllables.

time (unless something enters the region of the potential and gives the electron a whack).

That last sentence is the big clue we need to establish the significance of 'frequency'. We encountered the law of energy conservation earlier in the chapter, and it is one of the few non-negotiable laws of physics. Energy conservation dictates that if an electron inside a hydrogen atom (or a square well) has a particular energy, then that energy *cannot* change until 'something happens'. In other words, an electron cannot spontaneously change its energy without a reason. This might sound uninteresting, but contrast this with the case of an electron that is known to be located at a point. As we know very well, the electron will leap off across the Universe in an instant, spawning an infinity of clocks. But the standing wave clock pattern is different. It keeps its shape, with all the clocks happily rotating away for ever unless something disturbs them. The unchanging nature of standing waves therefore makes them a clear candidate to describe an electron of definite energy.

Once we make the step of associating the frequency of a standing wave with the energy of a particle then we can exploit our knowledge of guitar strings to infer that higher frequencies must correspond to higher energies. That is because high frequency implies short wavelength (since short strings vibrate faster) and, from what we know of the special case of the square well potential, we can anticipate that a shorter wavelength corresponds to a higher-energy particle via de Broglie. The important conclusion, therefore, and all that really needs to be remembered for what follows, is that *standing waves describe particles of definite energy and the higher the energy the faster the clocks whizz round*.

In summary, we have deduced that when an electron is confined by a potential, its energy is quantized. In the physics jargon, we say that a trapped electron can only exist in certain 'energy levels'. The lowest energy the electron can have corresponds to its being described by the 'fundamental' standing wave alone,[6] and this energy

6. i.e. $n = 1$ in the case of the square well potential.

level is usually referred to as the 'ground state'. The energy levels corresponding to standing waves with higher frequencies are referred to as 'excited states'.

Let us imagine an electron of a particular energy, trapped in a square well potential. We say that it is 'sitting in a particular energy level' and its quantum wave will be associated with a single value of n (see page 100). The language 'sitting in a particular energy level' reflects the fact that the electron doesn't, in the absence of any external influence, do anything. More generally, the electron could be described by many standing waves at once, just as the sound of a guitar will be made up of many harmonics at once. This means that the electron will not in general have a unique energy.

Crucially, a measurement of the electron's energy must always reveal a value equal to that associated with one of the contributing standing waves. In order to compute the probability of finding the electron with a particular energy, we should take the clocks associated with the specific contribution to the total wavefunction coming from the corresponding standing wave, square them all up and add them all together. The resulting number tells us the probability that the electron is in this particular energy state. The sum of all such probabilities (one for each contributing standing wave) must add up to one, which reflects the fact that we will always find that the particle has an energy that corresponds to a specific standing wave.

Let's be very clear: an electron can have several different energies at the same time, and this is just as weird a statement as saying that it has a variety of positions. Of course, by this stage in the book this ought not to be such a shock, but it is shocking to our everyday sensibilities. Notice that there is a crucial difference between a trapped quantum particle and the standing waves in a swimming pool or on a guitar string. In the case of the waves on a guitar string, the idea that they are quantized is not at all weird, because the actual wave describing the vibrating string is simultaneously composed of many different standing waves, and all those waves physically contribute to the total energy of the wave. Because they can be mixed together in any way, the actual energy of the vibrating string can take on any value at all. For an electron trapped inside an atom, however, the relative contribution of each standing wave describes the probability that the electron will be found with that particular energy. The crucial difference arises because water waves are waves of water molecules but electron waves are most certainly not waves of electrons.

These deliberations have shown us that the energy of an electron inside an atom is quantized. This means that the electron is simply unable to possess any energy intermediate between certain allowed values. This is just like saying that a car can travel at 10 miles per hour or 40 miles per hour, but at no other speeds in between. Immediately, this fantastically bizarre conclusion offers us an explanation for why atoms do not continuously radiate light as the electron spirals into the nucleus. It is because there is no way for the electron to constantly shed energy, bit by bit. Instead, the only way it can shed any energy is to lose a whole chunk in one go.

We can also relate what we have just learnt to the observed properties of atoms, and in particular we can explain the unique colours of light they emit. Figure 6.7 shows the visible light emitted from the simplest atom, hydrogen. The light is composed of five distinct colours, a bright-red line corresponding to light with a wavelength of 656 nanometres, a light-blue line of wavelength 486 nanometres, and three other violet lines which fade away into the ultraviolet end of the spectrum. This series of coloured lines is known as the

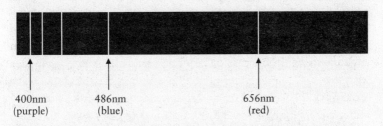

Figure 6.7. The Balmer series for hydrogen: this is what happens when light from hydrogen gas is passed through a spectroscope.

Balmer series, after the Swiss mathematical physicist Johann Balmer, who wrote down a formula able to describe them in 1885. Balmer had no idea why his formula worked, because quantum theory was yet to be discovered – he simply expressed the regularity behind the pattern in a simple mathematical formula. But we can do better, and it is all to do with the allowed quantum waves that fit inside the hydrogen atom.

We know that light can be thought of as a stream of photons, each of energy $E = hc/\lambda$, where λ is the wavelength of the light.[7] The observation that atoms only emit certain colours of light therefore means that they only emit photons of very specific energies. We have also learnt that an electron 'trapped in an atom' can only possess certain very specific energies. It is a small step now to explain the long-standing mystery of the coloured light emitted from atoms: the different colours correspond to the emission of photons when electrons 'drop down' from one allowed energy level to another. This idea implies that the observed photon energies should always correspond to differences between a pair of allowed electron energies. This way of describing the physics nicely illustrates the value of expressing the state of the electron in terms of its allowed energies. If we had instead chosen to talk about the allowed values of the electron's momentum then the quantum nature would not be so apparent and we would not so easily conclude that the atom can only emit and absorb radiation at specific wavelengths.

The particle-in-a-box model of an atom is not accurate enough to allow us to compute the electron energies in a real atom, which is necessary to check this idea. But accurate calculations can be done if we model more accurately the potential in the vicinity of the proton that traps the electron. It is enough to say that these calculations confirm, without any shadow of doubt, that this really is the origin of those enigmatic spectral lines.

7. Incidentally, if you know that $E = cp$ for massless particles, which is a consequence of Einstein's Theory of Special Relativity, then $E = hc/\lambda$ follows immediately by making using of the de Broglie equation.

You may have noticed that we have not explained why it is that the electron loses energy by emitting a photon. For the purposes of this chapter, we do not need an explanation. But something must induce the electron to leave the sanctity of its standing wave, and that 'something' is the topic of Chapter 10. For now, we are simply saying that 'in order to explain the observed patterns of light emitted by atoms it is necessary to suppose that the light is emitted when an electron drops down from one energy level to another level of lower energy'. The allowed energy levels are determined by the shape of the confining box and they vary from atom to atom because different atoms present a different environment within which their electrons are confined.

Up until now, we have made a good fist of explaining things using a very simple picture of an atom, but it isn't really good enough to pretend that electrons move around freely inside some confining box. They are moving around in the vicinity of a bunch of protons and other electrons, and to really understand atoms we must now think about how to describe this environment more accurately.

The Atomic Box

Armed with the notion of a potential, we can be more accurate in our description of atoms. Let's start with the simplest of all atoms, a hydrogen atom. A hydrogen atom is made up of just two particles: one electron and one proton. The proton is nearly 2,000 times heavier than the electron, so we can assume that it is not doing much and just sits there, creating a potential within which the electron is trapped.

The proton has a positive electric charge and the electron has an equal and opposite negative charge. As an aside, the reason why the electric charges of the proton and the electron are *exactly* equal and opposite is one of the great mysteries of physics. There is probably a very good reason, associated with some underlying theory of subatomic particles that we have yet to discover, but, as we write this book, nobody knows.

What we do know is that, because opposite charges attract, the proton is going to tug the electron towards it and, as far as pre-quantum physics is concerned, it could pull the electron inwards to arbitrarily small distances. How small would depend on the precise nature of a proton; is it a hard ball or a nebulous cloud of something? This question is irrelevant because, as we have seen, there is a minimum energy level that the electron can be in, determined (roughly speaking) by the longest wavelength quantum wave that will fit inside the potential generated by the proton. We've sketched the potential created by the proton in Figure 6.8. The deep 'hole' functions like the square well potential we met earlier except that

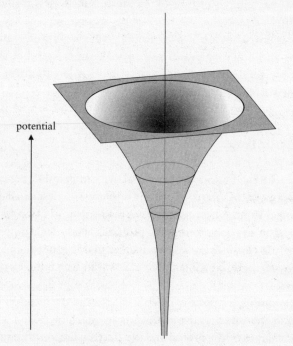

potential

Figure 6.8. The Coulomb potential well around a proton. The well is deepest where the proton is located.

the shape is not as simple. It is known as the 'Coulomb potential', because it is determined by the law describing the interaction between two electric charges, first written down by Charles-Augustin de Coulomb in 1783. The challenge is the same, however: we must find out what quantum waves can fit inside the potential, and these will determine the allowed energy levels of the hydrogen atom.

Being blunt, we might say that the way to do this is to 'solve Schrödinger's wave equation for the Coulomb potential well', which is one way to implement the clock-hopping rules. The details are technical, even for something as simple as a hydrogen atom, but fortunately we do not really learn much more than we have appreciated already. For that reason, we shall jump straight to the answer, and Figure 6.9 shows some of the resulting standing waves for an electron in a hydrogen atom. What is shown is a map of the probability to find the electron somewhere. The bright regions are where the electron is most likely to be. The real hydrogen atom is, of course, three-dimensional, and these pictures correspond to slices through the centre of the atom. The figure on the top left is the ground state wavefunction, and it tells us that the electron is, in this case, typically to be found around 1×10^{-10} m from the proton. The energies of the standing waves increase from the top left to the bottom right. The scale also changes by a factor of eight from the top left to the bottom right – in fact the bright region covering most of the top-left picture is approximately the same size as the small bright spots in the centre of the two pictures on the right. This means the electron is likely to be farther away from the proton when it is in the higher energy levels (and hence that it is more weakly bound to it). It is clear that these waves are not sine waves, which means they do not correspond to states of definite momentum. But, as we have been at pains to emphasize, they do correspond to states of definite energy.

The distinctive shape of the standing waves is due to the shape of the well and some features are worth discussing in a little more detail. The most obvious feature of the well around a proton is that it is spherically symmetric. This means that it looks the same no

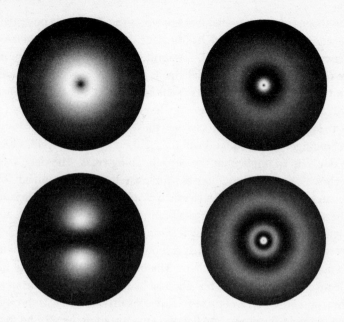

Figure 6.9. Four of the lowest energy quantum waves describing the electron in a hydrogen atom. The light regions are where the electron is most likely to be found and the proton is in the centre. The top-right and bottom-left pictures are zoomed out by a factor of 4 relative to the first and the bottom-right picture is zoomed out by a factor of 8 relative to the first. The first picture is around 3×10^{-10} m across.

matter which angle you view it from. To picture this, think of a basketball with no markings on it: it's a perfect sphere and it will look exactly the same no matter how you rotate it around. Perhaps we might dare to think of an electron inside a hydrogen atom as if it were trapped inside a tiny basketball? This is certainly more plausible than saying the electron is trapped in a square well and, remarkably, there is a similarity. Figure 6.10 shows, on the left, two of the lowest-energy standing sound waves that can be produced within a basketball. Again we have taken a slice through the ball, and the air pressure within the ball varies from black to white as the pressure increases. On the right are two possible electron standing

Figure 6.10: Two of the simplest standing sound waves inside a basketball (left) compared to the corresponding electron waves in a hydrogen atom (right). They are very similar. The top picture for hydrogen is a close-up of the central region in the bottom left picture in Figure 6.9.

waves in a hydrogen atom. The pictures are not identical, but they are very similar. So, it is not entirely stupid to imagine that the electron within a hydrogen atom is being trapped within something akin to a tiny basketball. This picture really serves to illustrate the wavelike behaviour of quantum particles, and it hopefully takes some of the mystery out of things: understanding the electron in a hydrogen atom is not more complicated than understanding how the air vibrates inside a basketball.

Before we leave the hydrogen atom, we would like to say a little more about the potential created by the proton and how it is that the electron can leap from a higher energy level to a lower one with the emission of a photon. We avoided any discussion of how the

proton and the electron communicate with each other, quite legitimately, by introducing the idea of a potential. This simplification
allowed us to understand the quantization of energy for trapped
particles. But if we want a serious understanding of what's going
on, we should try to explain the underlying mechanism for trapping
particles. In the case of a particle moving in an actual box, we might
imagine some impenetrable wall that is presumably made up of
atoms, and the particle is prevented from passing through the wall
by interacting with the atoms within it. A proper understanding
of 'impenetrability' comes from understanding how the particles
interact with each other. Likewise, we said that the proton in a hydrogen atom 'produces a potential' in which the electron moves, and we
said that the potential traps the electron in a manner analogous to the
way a particle is trapped in a box. That too ducks the deeper issue,
because clearly the electron interacts with the proton and it is that
interaction which dictates how the electron is confined.

In Chapter 10 we'll see that we need to supplement the quantum
rules we've articulated so far with some new rules dealing with
particle interactions. At the moment, we have very simple rules:
particles hop around, carrying imaginary clocks which wind back
by clearly specified amounts depending on the size of the hop. All
hops are allowed, and so a particle can hop from A to B via an infinity of different routes. Each route delivers its own quantum clock to
B and we must add up the clocks to determine a single resultant
clock. That clock then tells us the chance of actually finding the
particle at B. Adding interactions into the game turns out to be
surprisingly simple. We supplement the hopping rules with a new
rule, stating that a particle can emit or absorb another particle. If
there was one particle before the interaction, then there can be two
particles afterwards; if there were two particles before the interaction, then there can be one particle afterwards. Of course, if we are
going to work out the maths then we need to be more precise about
which particles can fuse together or split apart, and we need to say
what happens to the clock that each particle carries when it interacts. This is the subject of Chapter 10, but the implications for atoms

should be clear. If there is a rule saying that an electron can interact by emitting a photon, then we have the possibility that the electron in a hydrogen atom can spit out a photon, lose energy and drop down to a lower energy level. It could also absorb a photon, gain energy and leap up to a higher energy level.

The existence of spectral lines indicates that this is what is happening, and this process is ordinarily heavily biased one way. In particular, the electron can spit out a photon and lose energy at any time, but the only way it can gain energy and jump up to a higher energy level is if there is a photon (or some other source of energy) available to collide with it. In a gas of hydrogen, such photons are typically few and far between, and an atom in an excited state is much more likely to emit a photon than absorb one. The net effect is that hydrogen atoms tend to de-excite, by which we mean that emission wins over absorption and, given time, the atom will make its way down to the $n = 1$ ground state. This is not always the case, because it is possible to arrange to continually excite atoms by feeding them energy in a controlled way. This is the basis of a technology that has become ubiquitous: the laser. The basic idea of a laser is to pump energy into atoms, excite them, and collect the photons that are produced when the electrons drop down in energy. Those photons are very useful for reading data with high precision from the surface of a CD or DVD: quantum mechanics affects our lives in myriad ways.

In this chapter, we have succeeded in explaining the origin of spectral lines using the simple idea of quantized energy levels. It would seem we have a way of thinking about atoms that works. But something is not quite right. We are missing one final piece of the jigsaw, without which we have no chance of explaining the structure of atoms heavier than hydrogen. More prosaically, we will also be unable to explain why we don't fall through the floor, and that is problematic for our best theory of Nature. The insight we are looking for comes from the work of Austrian physicist Wolfgang Pauli.

7. The Universe in a Pin-head (and Why We Don't Fall Through the Floor)

That we do not fall through the floor is something of a mystery. To say the floor is 'solid' is not very helpful, not least because Rutherford discovered that atoms are almost entirely empty space. The situation is made even more puzzling because, as far as we can tell, the fundamental particles of Nature are of no size at all.

Dealing with particles 'of no size' sounds problematic, and perhaps impossible. But nothing we said in the previous chapters presupposed or required that particles have any physical extent. The notion of truly point-like objects need not be wrong, even if it flies in the face of common sense – if indeed the reader has any common sense left at this stage of a book on quantum theory. It is, of course, entirely possible that a future experiment, perhaps even the Large Hadron Collider, will reveal that electrons and quarks are not infinitesimal points, but for now this is not mandated by experiment and there is no place for 'size' in the fundamental equations of particle physics. That's not to say that point particles don't have their problems – the idea of a finite charge compressed into an infinitely small volume is a thorny one – but so far the theoretical pitfalls have been circumvented. Perhaps the outstanding problem in fundamental physics, the development of a quantum theory of gravity, hints at finite extent, but the evidence is just not there to force physicists to abandon the idea of elementary particles. To be emphatic: point-like particles are really of no size and to ask 'What happens if I split an electron in half?' makes no sense at all – there is no meaning to the idea of 'half an electron'.

A pleasing bonus of working with elementary fragments of matter that have no size at all is that we don't have any trouble with the idea that the entire visible Universe was once compressed into a volume the size of a grapefruit, or even a pin-head. Mind-boggling though that may seem – it's hard enough to imagine compressing a mountain to the size of a pea, never mind a star, a galaxy, or the 350 billion large galaxies in the observable Universe – there is absolutely no reason why this shouldn't be possible. Indeed, present-day theories of the origins of structure in the Universe deal directly with its properties when it was in such an astronomically dense state. Such theories, whilst outlandish, have a good deal of observational evidence in their favour. In the final chapter we will meet objects with densities, if not at the 'Universe in a pin-head' scale, then certainly in 'mountain in a pea' territory: white dwarves are objects with the mass of a star squashed to the size of the Earth, and neutron stars have similar masses condensed into perfect, city-sized spheres. These objects are not science fiction; astronomers have observed them and made high-precision measurements of them, and quantum theory will allow us to calculate their properties and compare them with the observational data. As a first step on the road to understanding white dwarves and neutron stars, we will need to address the more prosaic question with which we began this chapter: if the floor is largely empty space, why do we not fall through it?

This question has a long and venerable history, and the answer was not established until surprisingly recently, in 1967, in a paper by Freeman Dyson and Andrew Lenard. They embarked on the quest because a colleague had offered a bottle of vintage champagne to anyone who could prove that matter shouldn't simply collapse in on itself. Dyson referred to the proof as extraordinarily complicated, difficult and opaque, but what they showed was that matter can only be stable if electrons obey something called the Pauli Exclusion Principle, one of the most fascinating facets of our quantum universe.

We shall begin with some numerology. We saw in the last chapter that the structure of the simplest atom, hydrogen, can be understood by searching for the allowed quantum waves that fit inside the proton's potential well. This allowed us to understand, at least qualitatively, the distinctive spectrum of the light emitted from hydrogen atoms. If we had had the time, we could have calculated the energy levels in a hydrogen atom. Every undergraduate physics student performs this calculation at some stage in their studies and it works beautifully, agreeing with the experimental data. As far as the last chapter was concerned, the 'particle in a box' simplification was good enough because it contains all the key points that we wanted to highlight. However, there is a feature of the full calculation that we shall need, which comes about because the real hydrogen atom is extended in three dimensions. For our particle in a box example, we only considered one dimension and obtained a series of energy levels labelled by a single number that we called n. The lowest energy level was labelled $n = 1$, the next $n = 2$ and so on. When the calculation is extended to the full three-dimensional case it turns out, perhaps unsurprisingly, that three numbers are needed to characterize all of the allowed energy levels. These are traditionally labelled n, l and m, and they are referred to as quantum numbers (in this chapter, m is not to be confused with the mass of the particle). The quantum number n is the counterpart of the number n for a particle in a box. It takes on integer values ($n = 1$, 2, 3, etc.) and the particle energies tend to increase as n increases. The possible values of l and m turn out to be linked to n; l must be smaller than n and it can be zero, e.g. if $n = 3$ then l can be 0, 1 or 2. m can take on any value ranging from minus l to plus l in integer steps. So if $l = 2$ then m can be equal to -2, -1, 0, 1 or 2. We are not going to explain where those numbers come from, because it won't add anything to our understanding. Suffice to say that the four waves in Figure 6.9 have $(n,l) = (1,0)$, $(2,0)$, $(2,1)$ and $(3,0)$ respectively (all have $m = 0$).[1]

As we have said, the quantum number n is the main number

---

1. Technically, as we mentioned in the previous chapter, because the potential well around the proton is spherically symmetric rather than a square box, the solution to the Schrödinger equation must be proportional to a spherical harmonic. The associated angular dependence gives rise to the l and m quantum numbers. The radial dependence of the solution gives rise to the principal quantum number n.

controlling the values of the allowed energies of the electrons. There is also a small dependence of the allowed energies upon the value of l but it only shows up in very precise measurements of the emitted light. Bohr didn't consider it when he first calculated the energies of the spectral lines of hydrogen, and his original formula was expressed entirely in terms of n. There is absolutely no dependence of the electron energy upon m unless we put the hydrogen atom inside a magnetic field (in fact m is known as the 'magnetic quantum number'), but this certainly doesn't mean that it isn't important. To see why, let's get on with our bit of numerology.

If $n = 1$ then how many different energy levels are there? Applying the rules we stated above, l and m can both only be 0 if $n = 1$, and so there is just the one energy level.

Now let's do it for $n = 2$: l can take on two values, 0 and 1. If $l = 1$, then m can be equal to -1, 0 or $+1$, which is 3 more energy levels, making 4 in total.

For $n = 3$, l can be 0, 1 or 2. For $l = 2$, m can be equal to -2, -1, 0, $+1$, or $+2$, giving 5 levels. So in total, there are $1 + 3 + 5 = 9$ levels for $n = 3$. And so on.

Remember those numbers for the first three values of n: 1, 4 and 9. Now take a look at Figure 7.1, which shows the first four rows of the periodic table of the chemical elements, and count how many elements there are in each row. Divide that number by 2, and you'll get 1, 4, 4 and 9. The significance of all this will soon be revealed.

Credit for arranging the chemical elements in this way is usually

Group	1	2	3	4	5	6	7	8	9	10	11	12	13	14	15	16	17	18
1	1 H																	2 He
2	3 Li	4 Be											5 B	6 C	7 N	8 O	9 F	10 Ne
3	11 Na	12 Mg											13 Al	14 Si	15 P	16 S	17 Cl	18 Ar
4	19 K	20 Ca	21 Sc	22 Ti	23 V	24 Cr	25 Mn	26 Fe	27 Co	28 Ni	29 Cu	30 Zn	31 Ga	32 Ge	33 As	34 Se	35 Br	36 Kr

Figure 7.1. The first four rows of the periodic table.

given to the Russian chemist Dmitri Mendeleev, who presented it to the Russian Chemical Society on 6 March 1869, which was a good few years before anyone had worked out how to count the allowed energy levels in a hydrogen atom. Mendeleev arranged the elements in order of their atomic weights, which in modern language corresponds to the number of protons and neutrons inside the atomic nucleus, although of course he didn't know that at the time either. The ordering of the elements actually corresponds to the number of protons inside the nucleus (the number of neutrons is irrelevant) but for the lighter elements this makes no difference, which is why Mendeleev got it right. He chose to arrange the elements in rows and columns because he noticed that certain elements had very similar chemical properties, even though they had different atomic weights; the vertical columns group together such elements – helium, neon, argon and krypton on the far right of the table are all unreactive gases. Mendeleev didn't just get the pattern right, he also predicted the existence of new elements to fill gaps in his table: elements 31 and 32 (gallium and germanium) were discovered in 1875 and 1886. These discoveries confirmed that Mendeleev had uncovered something deep about the structure of atoms, but nobody knew what.

What is striking is that there are two elements in row one, eight in rows two and three and eighteen in row four, and those numbers are exactly twice the numbers we just worked out by counting the allowed energy levels in hydrogen. Why is this?

As we have already mentioned, the elements in the periodic table are ordered from left to right in a row by the number of protons in the nucleus, which is the same as the number of electrons they contain. Remember that all atoms are electrically neutral – the positive electric charges of the protons are exactly balanced by the negative charges of the electrons. There is clearly something interesting going on that relates the chemical properties of the elements to the allowed energies that the electrons can have when they orbit around a nucleus.

We can imagine building up heavier atoms from lighter ones by adding protons, neutrons and electrons one at a time, bearing in mind that whenever we add an extra proton into the nucleus we should add an extra electron into one of the energy levels. The exercise in numerology will generate the pattern we see in the periodic table if we simply assert that each energy level can contain two and only two electrons. Let's see how this works.

Hydrogen has only one electron, so that would slot into the $n = 1$ level. Helium has two electrons, which would both fit into the $n = 1$ level. Now the $n = 1$ level is full up. We must add a third electron to make lithium, but it will have to go into the $n = 2$ level. The next seven electrons, corresponding to the next seven elements (beryllium, boron, carbon, nitrogen, oxygen, fluorine and neon), can also sit in a level with $n = 2$ because that has four slots available, corresponding to $l = 0$ and $l = 1$, $m = -1, 0$ and $+1$. In that way we can account for all of the elements up to neon. With neon, the $n = 2$ levels are all full and we must move to $n = 3$, starting with sodium. The next eight electrons, one by one, start to fill up the $n = 3$ levels; first the electrons go into $l = 0$, and then into $l = 1$. That accounts for all the elements in the third row, up to argon. The fourth row of the table can be explained if we assume that it contains all of the remaining $n = 3$ electrons (i.e. the ten electrons with $l = 2$) and the $n = 4$ electrons with $l = 0$ and 1 (which makes eight electrons), making the magic number of eighteen electrons in total. We've sketched how the electrons fill up the energy levels for the heaviest element in our table, krypton (which has thirty-six electrons) in Figure 7.2.

To elevate all of what we just said to science rather than numerology we have some explaining to do. Firstly, we need to explain why the chemical properties are similar for elements in the same vertical column. What is clear from our scheme is that the first element in each of the first three rows starts off the process of filling levels with increasing values of n. Specifically, hydrogen starts things off with a single electron in the otherwise empty $n = 1$ level, lithium starts off the second row with a single electron in the $n = 2$ level and sodium starts the third row with a single electron in the otherwise empty $n = 3$ level. The third row is a little odd because the $n = 3$ level can

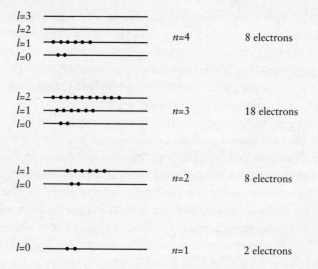

Figure 7.2. Filling the energy levels of krypton. The dots represent electrons and the horizontal lines represent the energy levels, labelled by the quantum numbers n, l and m. We have grouped together levels with different values of m but the same values of n and l.

hold eighteen electrons and there are not eighteen elements in the third row. We can guess at what is happening though – the first eight electrons fill up the $n = 3$ levels with $l = 0$ and $l = 1$, and then (for some reason) we should switch to the fourth row. The fourth row now contains the remaining ten electrons from the $n = 3$ levels with $l = 2$ and the eight electrons from the $n = 4$ levels with $l = 0$ and $l = 1$. The fact that the rows are not entirely correlated with the value of n indicates that the link between the chemistry and the energy-level counting is not as simple as we have been making out. However, it is now known that potassium and calcium, the first two elements in the fourth row, do have electrons in the $n = 4$, $l = 0$ level and that the next ten elements (from scandium to zinc) have their electrons in the belated $n = 3$, $l = 2$ levels.

To understand why the filling up of the $n = 3$ and $l = 2$ levels is deferred until after calcium requires an explanation of why the $n = 4$, $l = 0$ levels, which contain the electrons in potassium and calcium, is of lower energy than the $n = 3$, $l = 2$ levels. Remember, the 'ground state' of an atom will be characterized by the lowest-energy configuration of the electrons, because any excited state can always lower its energy by the emission of a photon. So when we have been saying that 'this atom contains these electrons sitting in those energy levels' we are telling you the lowest energy configuration of the electrons. Of course, we have not made any attempt to actually compute the energy levels, so we aren't really in a position to rank them in order of energy. In fact it is a very difficult business to calculate the allowed electron energies in atoms with more than two electrons, and even the two-electron case (helium) is not so easy. The simple idea that the levels are ranked in order of increasing n comes from the much easier calculation for the hydrogen atom, where it is true that the $n = 1$ level has the lowest energy followed by the $n = 2$ levels, then come the $n = 3$ levels and so on.

The obvious implication of what we just said is that the elements on the far right of the periodic table correspond to atoms in which a set of levels has just been completely filled. In particular, for helium the $n = 1$ level is full, whilst for neon the $n = 2$ level is full, and for argon the $n = 3$ level is fully populated, at least for $l = 0$ and $l = 1$. We can develop these ideas a little further and understand some important ideas in chemistry. Fortunately we aren't writing a chemistry textbook, so we can be brief and, at the risk of dismissing an entire subject in a single paragraph, here we go.

The key observation is that atoms can stick together by sharing electrons – we will meet this idea in the next chapter when we explore how a pair of hydrogen atoms can bind to make a hydrogen molecule. The general rule is that elements 'like' to have all their energy levels neatly filled up. In the case of helium, neon, argon and krypton, the levels are already completely full, and so they are 'happy' on their own – they don't 'bother' reacting with anything. For the other elements, they can 'try' to fill their levels by sharing electrons with other elements. Hydrogen, for example, needs one extra electron to fill its $n = 1$ level. It can achieve this by sharing an electron with another hydrogen atom. In so doing, it forms a hydrogen molecule, with chemical symbol H_2. This is the common form in which hydrogen gas exists. Carbon has four electrons out of a possible eight in its $n = 2$, $l = 0$ and $l = 1$ levels, and would 'like' another four if possible to fill them up. It can achieve this by binding together with four hydrogen atoms to form CH_4, the gas known as methane. It can also do it by binding with two oxygen atoms, which themselves need two electrons to complete their $n = 2$ set. This leads to CO_2 – carbon dioxide. Oxygen could also complete its set by binding with two hydrogen atoms to make H_2O – water. And so on. This is the basis of chemistry: it is energetically favourable for atoms to fill their energy levels with electrons, even if that is achieved by sharing with a neighbour. Their 'desire' to do this, which ultimately stems from the principle that things tend to their lowest energy state, is what drives the formation of everything from water to DNA. In a world abundant in hydrogen, oxygen and carbon we now understand why carbon dioxide, water and methane are so common.

This is very encouraging, but we have a final piece of the jigsaw to explain: why is it that only two electrons can occupy each available energy level? This is a statement of the Pauli Exclusion Principle, and it is clearly necessary if everything we have been discussing is to hang together. Without it, the electrons would crowd together in the lowest possible energy level around every nucleus, and there would be no chemistry, which is worse than it sounds, because there would be no molecules and therefore no life in the Universe.

The idea that two and only two electrons can occupy each energy level does seem quite arbitrary, and historically nobody had any idea why it should be the case when the idea was first proposed. The initial breakthrough was made by Edmund Stoner, the son of a professional cricketer (who took eight wickets against South Africa in 1907, for those who read their *Wisden Cricketers' Almanack*) and a former student of Rutherford's who later ran the physics department at the University of Leeds. In October 1924, Stoner proposed that there should be two electrons allowed in each (n, l, m) energy level. Pauli developed Stoner's proposal and in 1925 he published a rule that Dirac named after him a year later. The Exclusion Principle, as first proposed by Pauli, states that no two electrons in an atom can share the same quantum numbers. The problem he faced was that it appeared that two electrons *could* share each set of n, l and m values. Pauli got round the problem by simply introducing a new quantum number. This was an ansatz; he didn't know what it represented, but it had to take on one of only two values. Pauli wrote that, 'We cannot give a more precise reason for this rule.' Further insight came in 1925, in a paper by George Uhlenbeck and Samuel Goudsmit. Motivated by precise measurements of atomic spectra, they identified Pauli's extra quantum number with a real, physical property of the electron known as 'spin'.

The basic idea of spin is quite simple, and dates back to 1903, well before quantum theory. Just a few years after its discovery, German physicist Max Abraham proposed that the electron was a tiny, spinning electrically charged sphere. If this were true, then electrons would be affected by magnetic fields, depending on the orientation of the field relative to their spin axis. In their 1925 paper, which was published three years after Abraham's death, Uhlenbeck and Goudsmit noted that the spinning ball model couldn't work because, in order to explain the observed data, the electron would have to be spinning faster than the speed of light. But the spirit of the idea was correct – the electron does possess a property called spin, and it does affect its behaviour in a magnetic field. Its true origin, however, is a direct and rather subtle consequence of Einstein's Theory of Special Relativity that was only properly appreciated when Paul Dirac wrote down an equation describing the quantum behaviour of the electron in 1928. For our purposes, we shall need only acknowledge that electrons do come in two types, which we refer to as 'spin up' and 'spin down', and the two are distinguished by having opposite values of their angular momentum, i.e. it is like they are spinning in opposite directions. It's a pity that Abraham died just a few years before the true nature of electron spin was discovered, because he never gave up his conviction that the electron was a little sphere. In his obituary in 1923, Max Born and Max Von Laue wrote: 'He was an honourable opponent who fought with honest weapons and who did not cover up a defeat by lamentation and nonfactual arguments . . . He loved his absolute ether, his field equations, his rigid electron, just as a youth loves his first flame, whose memory no later experience can extinguish.' If only all of one's opponents were like Abraham.

Our goal in the remainder of this chapter is to explain why it is that electrons behave in the strange way articulated by the Exclusion Principle. As ever, we shall make good use of those quantum clocks.

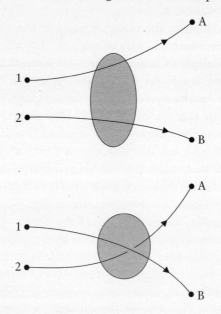

Figure 7.3. Two electrons scattering.

We can attack the question by thinking about what happens when two electrons 'bounce' off each other. Figure 7.3 illustrates a particular scenario where two electrons, labelled '1' and '2', start out somewhere and end up somewhere else. We have labelled the final locations A and B. The shaded blobs are there to remind us that we have not yet thought about just what happens when two electrons interact with each other (the details are irrelevant for the purposes of this discussion). All we need to imagine is that electron 1 hops from its starting place and ends up at the point labelled A. Likewise, electron 2 ends up at the point labelled B. This is what is illustrated in the top of the two pictures in the figure. In fact, the argument we are about to present works fine even if we ignore the possibility that the electrons might interact. In that case, electron 1 hops to A oblivious to the meanderings of electron 2 and the probability of finding electron 1 at A and electron 2 at B would be simply a product of two independent probabilities.

For example, suppose the probability of electron 1 hopping to point A is 45% and the probability of electron 2 hopping to point B is 20%. The probability of finding electron 1 at A *and* electron 2 at B is $0.45 \times 0.2 = 0.09 = 9\%$. All we are doing here is using the logic that says that the chances of tossing a coin and getting 'tails' *and* rolling a dice and getting a 'six' at the same time is one-half multiplied by one-sixth, which is equal to $\frac{1}{12}$ (i.e. just over 8%).[2]

As the figure illustrates, there is a second way that the two electrons can end up at A and B. It is possible for electron 1 to hop to B whilst electron 2 ends up at A. Suppose that the chance of finding electron 1 at B is 5% and the chance of finding electron 2 at A is 20%. Then the probability of finding electron 1 at B *and* electron 2 at A is $0.05 \times 0.2 = 0.01 = 1\%$.

We therefore have two ways of getting our two electrons to A and B – one with a probability of 9% and one with a probability of 1%. The probability of getting one electron at A and one at B, if we don't care which is which, should therefore be $9\% + 1\% = 10\%$. Simple; but wrong.

The error is in supposing that it is possible to say which electron arrives at A and which one arrives at B. What if the electrons are

2. We will learn in Chapter 10 that accounting for the possibility that the two electrons interact with each other means we need to calculate the probability to find electron 1 at A and electron 2 at B 'all at once' because it does not reduce to a multiplication of two independent probabilities. But that is a detail as far as this chapter is concerned.

identical to each other in every way? This might sound like an irrelevant question, but it isn't. Incidentally, the suggestion that quantum particles might be strictly identical was first made in relation to Planck's black body radiation law. A little-known physicist called Ladislas Natanson had pointed out, as far back as 1911, that Planck's law was incompatible with the assumption that photons could be treated as identifiable particles. In other words, if you could tag a photon and track its movements, then you wouldn't get Planck's law.

If electrons 1 and 2 are absolutely identical then we must describe the scattering process as follows: initially there are two electrons, and a little later there are still two electrons located in different places. As we've learnt, quantum particles do not travel along well-defined trajectories, and this means that there really is no way of tracking them, even in principle. It therefore makes no sense to claim electron 1 appeared at A and electron 2 at B. We simply can't tell, and it is therefore meaningless to label them. This is what it means for two particles to be 'identical' in quantum theory. Where does this line of reasoning take us?

Look at the figure again. For this particular process, the two probabilities we associated with the two diagrams (9% and 1%) are not wrong. They are, however, not the whole story. We know that quantum particles are described by clocks, so we should associate a clock with electron 1 arriving at A with a size equal to the square root of 45%. Likewise there is a clock associated with electron 2 arriving at B and it has a size equal to the square root of 20%.

Now comes a new quantum rule – it says that we are to associate a single clock with the process as a whole, i.e. there is a clock whose size squared is equal to the probability to find electron 1 at A and electron 2 at B. In other words, there is a single clock associated with the upper picture in Figure 7.3. We can see that this clock must have a size equal to the square root of 9%, because that is the probability for the process to happen. But what time does it read? Answering this question is the domain of Chapter 10 and it involves the idea of clock multiplication. As far as this chapter is concerned, we don't need to know the time, we only need the important new rule that we have just stated, but which is worth repeating because it is a very general statement in quantum theory: we should associate a *single* clock with each possible way that an *entire process* can happen. The clock we associate with finding a single particle at a single location is the simplest illustration of this rule, and we have managed to get this far in the book with it. But it is a special case, and as soon as we start to think about more than one particle we need to extend the rule.

This means that there is a clock of size equal to 0.3 associated with the upper picture in the figure. Likewise, there is a second clock of size equal to 0.1 (because 0.1 squared is 0.01 = 1%) associated with the lower picture in the figure. We therefore have two clocks and we want a way to use them to determine the probability to find an electron at A and another at B. If the two electrons were distinguishable then the answer would be simple – we would just add together the probabilities (and not the clocks) associated with each possibility. We would then obtain the answer of 10%.

But if there is absolutely no way of telling which of the diagrams actually happened, which is the case if the electrons are indistinguishable from each other, then following the logic we've developed for a single particle as it hops from place to place, we should seek to combine the clocks. What we are after is a generalization of the rule which states that, for one particle, we should add together the clocks associated with all of the different ways that the particle can reach a particular point in order to determine the probability to find the particle at that point. For a system of many identical particles, we should combine together all the clocks associated with all of the different ways that the particles can reach a set of locations in order to determine the probability that particles will be found at those locations. This is important enough to merit reading a few times – it should be clear that this new law for combining clocks is a direct generalization of the rule we have been using for a single particle. You may have noticed that we have been very careful with our wording, however. We did not say that the clocks should necessarily be *added* together – we said that they should be *combined* together. There is a good reason for our caution.

The obvious thing to do would be to add the clocks together. But before leaping in we should ask whether there is a good reason why this is correct. This is a nice example of not taking things for granted in physics – exploring our assumptions often leads to new insights, as it will do in this instance. Let's take a step back, and think of the most general thing we could imagine. This would be to allow for the possibility of giving one of the clocks a turn or a shrink (or expansion) before we add them. Let's explore this possibility in more detail.

What we are doing is saying, 'I have two clocks and I want to combine them to make a single clock, so that I can use that to tell me what the probability is for the two electrons to be found at A and B. How should I combine them?' We are not pre-empting the answer, because we want to understand if adding clocks together really is the rule we should use. It turns out that we do not have much freedom at all, and simply adding clocks is, intriguingly, one of only *two* possibilities.

To streamline the discussion, let's refer to the clock corresponding to particle 1 hopping to A *and* particle 2 hopping to B as clock 1. This is the clock associated with the upper picture in Figure 7.3. Clock 2 corresponds to the other option, where particle 1 hops to B instead. Here is an important realization: if we give clock 1 a turn before adding it to clock 2, then the final probability we calculate must be the same as if we choose to give clock 2 the same turn before adding it to clock 1.

To see this, notice that swapping the labels A and B around in our diagrams clearly cannot change anything. It is just a different way of describing the same process. But swapping A and B around swaps the diagrams in Figure 7.3 around too. This means that if we decide to wind clock 1 (corresponding to the upper picture) before adding it to clock 2, then this must correspond precisely to winding the clock 2 before adding it to clock 1, after we've swapped labels. This piece of logic is crucial, so it's worth hammering home. Because we have assumed that there is no way of telling the difference between the two particles, then we are allowed to swap the labels around.

This implies that a turn on clock 1 *must* give the same answer as when we apply the same turn to clock 2, because there is no way of telling the clocks apart.

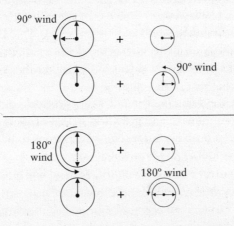

Figure 7.4. The upper part of the figure illustrates that adding clocks 1 and 2 together after winding clock 1 by 90 degrees is not the same as adding them together after winding clock 2 by 90 degrees. The lower part illustrates the interesting possibility that we could wind one of the clocks by 180 degrees before adding.

This is not a benign observation – it has a very important consequence, because there are only *two* possible ways of playing around with the winding and shrinking of clocks before adding them together that will deliver a final clock with the property that it does not depend upon which of the original clocks gets the treatment.

This is illustrated in Figure 7.4. The top half of the figure illustrates that, if we wind clock 1 by 90 degrees and add it to clock 2 then the resultant clock is not of the same size as the resultant we would get if we instead wound clock 2 by 90 degrees and add it to clock 1. We can see this because, if we first wind clock 1, the new hand, represented by the dotted arrow, points in the opposite direction to clock 2's hand, and therefore partly cancels it out. Winding clock 2 instead leaves its hand pointing in the same direction as clock 1's, and now the hands will add together to form a larger hand.

It should be clear that 90 degrees is not special, and that other angles will also give resultant clocks that depend upon which of clocks 1 and 2 we decided to wind.

The obvious exception is a clock wind of zero degrees, because winding clock 1 by zero degrees before adding to clock 2 is obviously exactly the same as winding clock 2 by zero degrees before adding to clock 1. This means that adding clocks together without any wind is a viable possibility. Similarly, winding both clocks by the same amount would work, but that really is just the same as the 'no winding' situation and corresponds simply to redefining what we call '12 o'clock'. This is tantamount to saying that we are always free to wind every clock around by some amount, as long as we do that to every clock. This will never impact on the probabilities we are trying to compute.

The lower part of Figure 7.4 illustrates that there is, perhaps surprisingly, one other way we can combine the clocks: we could turn one of them through 180 degrees before adding them together. This does not produce exactly the same clock in the two cases but it does produce the same size of clock, and that means it leads to the same probability to find one electron at A and a second at B.

A similar line of reasoning rules out the possibility of shrinking or expanding one of the clocks before adding, because if we shrink clock 1 by some fraction before adding to clock 2 then this will not usually be the same as shrinking clock 2 by that same amount before adding it to clock 1, and there are no exceptions to that rule.

So, we have an interesting conclusion to draw. Even though we started out by allowing ourselves complete freedom, we have discovered that, because there is no way of telling the particles apart, there are in fact only two ways we can combine the clocks: we can either add them or we can add them after first winding one or the other by 180 degrees. The truly delightful thing is that Nature exploits both possibilities.

For electrons, we have to incorporate the extra twist before adding the clocks. For particles like photons, or Higgs bosons, we have to add clocks without the twist. And so it is that Nature's particles come in two types: those which need the twist are called fermions and those without the twist are called bosons. What determines whether a particular particle is a fermion or a boson? It is the spin.

The spin is, as the name suggests, a measure of the angular momentum of a particle and it is a matter of fact that fermions always have a spin equal to some half-integer value[3] while bosons always have integer spin. We say that the electron has spin-half, the photon has spin-one and the Higgs boson has spin-zero. We have been avoiding dealing with the details of spin in this book, because it is a technical detail most of the time. However, we did need the result that electrons can come in two types, corresponding to the two possible values of their angular momentum (spin up and spin down), when we were discussing the periodic table. This is an example of a general rule that says particles of spin s generally come in $2s + 1$ types, e.g. spin $\frac{1}{2}$ particles (like electrons) come in two types, spin 1 particles come in three types and spin 0 particles come in one type. The relationship between the angular momentum of a particle and the way we are to combine clocks is known as the spin-statistics theorem, and it emerges when quantum theory is formulated so that it is consistent with Einstein's Theory of Special Relativity. More specifically, it is a direct result of making sure that the law of cause and effect is not violated. Unfortunately, deriving the spin-statistics theorem is beyond the level of this book – actually it is beyond the level of many books. In *The Feynman Lectures on Physics*, Richard Feynman has this to say:

> We apologise for the fact that we cannot give you an elementary explanation. An explanation has been worked out by Pauli from complicated arguments of quantum field theory and relativity. He has shown that the two must necessarily go together, but we have not been able to find a way of reproducing his arguments at an elementary level. It appears to be one of the few places in physics where there is a rule which can be stated very simply, but for which no one has found a simple and easy explanation.

3. In units of Planck's constant divided by 2π.

Bearing in mind that Richard Feynman wrote this in a university-level textbook, we must hold up our hands and concur. But the rule is simple, and you must take our word for it that it can be proved: for fermions, you have to give a twist, and for bosons you don't. It turns out that the twist is the reason for the Exclusion Principle, and therefore for the structure of atoms; and, after all our hard work, this is now something that we can explain very simply.

Imagine moving points A and B in Figure 7.3 closer and closer together. When they are very close together, clock 1 and clock 2 must be of nearly the same size and read nearly the same time. When A and B are right on top of each other then the clocks must be identical. That should be obvious, because clock 1 corresponds to particle 1 ending up at point A and clock 2 is, in this special case, representing exactly the same thing because points A and B are on top of each other. Nevertheless, we do still have two clocks, and we must still add them together. But here is the catch: for fermions, we must give one of the clocks a twist, winding it first by 180 degrees. This means that the clocks will always read exactly 'opposite' times when A and B are in the same place – if one reads 12 o'clock then the other will read 6 o'clock – so adding them together always gives a resultant clock of zero size. That is a fascinating result, because it means that there is always zero chance of finding the two electrons at the same place: the laws of quantum physics are causing them to avoid each other. The closer they get to each other, the smaller the resultant clock, and the less likely that is to happen. This is one way to articulate Pauli's famous principle: electrons avoid each other.

Originally, we set out to demonstrate that no two identical electrons can be in the same energy level in a hydrogen atom. We have not quite shown this to be true yet, but the notion that electrons avoid each other clearly has implications for atoms and for why we do not fall through the floor. Now we can see that not only do the electrons in the atoms in our shoes push against the electrons in the floor because like-charges repel; they also push against each other because they naturally avoid each other, according to the Pauli Exclusion Principle. It turns out that, as Dyson and Lenard proved, it is the electron avoidance that really keeps us from falling through the floor, and it also forces the electrons to occupy the different energy levels inside atoms, giving them a structure, and ultimately leading to the variety of chemical elements we see in Nature. This is clearly a piece of physics with very significant consequences for everyday life. In the final chapter of this book, we will show how Pauli's principle also plays a crucial role in preventing some stars from collapsing under the influence of their own gravity.

To finish, we should explain how it is that, if no two electrons can be at the same place at the same time, then it also follows that no two electrons in an atom can have the same quantum numbers, which means that they cannot have the same energy and spin. If we consider two electrons of the same spin, then we want to show that they cannot be in the same energy level. If they were in the same energy level then necessarily each electron would be described by exactly the same array of clocks distributed through space (corresponding to the relevant standing wave). For each pair of points in space – let's denote them X and Y – there are then two clocks. Clock 1 corresponds to 'electron 1 at X' and 'electron 2 at Y', whilst clock 2 corresponds to 'electron 1 at Y' and 'electron 2 at X'. We know from our previous deliberations that these clocks should be added together after winding one of them by 6 hours in order to deduce the probability to find one electron at X and a second at Y. But if the two electrons have the same energies, then clocks 1 and 2 must be identical to each other before the crucial extra wind. After the wind, they will read 'opposite' times and, as before, add together to make a clock of no size. That happens for any particular locations X and Y, and so there is absolutely zero chance of ever finding a pair of electrons in the same standing wave configuration, and therefore with the same energy. That, ultimately, is responsible for the stability of the atoms in your body.

8.   *Interconnected*

So far we have been paying close attention to the quantum physics of isolated particles and atoms. We have learnt that electrons sit inside atoms in states of definite energy, known as stationary states, although the atom may be in a superposition of different such states. We have also learnt that it is possible for an electron to make a transition from one energy state to another with the concurrent emission of a photon. The emission of photons in this way makes tangible the energy states in an atom; we see the characteristic colours of atomic transitions everywhere. Our physical experience, though, is of vast assemblies of atoms stuck together in clumps, and for that reason alone it is time to start pondering what happens when we stick atoms together.

The contemplation of atomic clusters is going to lead us along a road that will take in chemical bonding, the differences between conductors and insulators and, eventually, to semiconductors. These interesting materials have properties that can be exploited to build tiny devices capable of carrying out operations in basic logic. They are known as transistors, and by stringing many millions of them together we can build microchips. As we shall see, the theory of transistors is deeply quantum. It is difficult to see how they could have been invented and exploited without quantum theory, and difficult to imagine the modern world without them. They are a prime example of serendipity in science; the curiosity-led exploration of Nature that we've spent so much time describing in all its counterintuitive detail, eventually led to a revolution in our everyday lives. The dangers in trying to classify and control scientific research is beautifully summarized in the words of William Shockley, one of the inventors of the transistor and head of the Solid State Physics Group at Bell Telephone Laboratories:[1]

1. This is an excerpt from his 1956 Nobel Prize-winner's speech.

I would like to express some viewpoints about words often used to classify types of research in physics; for example, pure, applied, unrestricted, fundamental, basic, academic, industrial, practical, etc. It seems to me that all too frequently some of these words are used in a derogatory sense, on the one hand to belittle the practical objectives of producing something useful and, on the other hand, to brush off the possible long-range value of explorations into new areas where a useful outcome cannot be foreseen. Frequently, I have been asked if an experiment I have planned is pure or applied research; to me it is more important to know if the experiment will yield new and probably enduring knowledge about nature. If it is likely to yield such knowledge, it is, in my opinion, good fundamental research; and this is much more important than whether the motivation is purely esthetic satisfaction on the part of the experimenter on the one hand or the improvement of the stability of a high-power transistor on the other. It will take both types to confer the greatest benefit on mankind.

Since that comes from the inventor of perhaps the most useful device since the invention of the wheel, policy-makers and managers throughout the world would do well to pay attention. Quantum theory changed the world, and whatever new theories emerge from the cutting-edge physics being done today, they will almost certainly change our lives again.

As ever, we'll start at the beginning and extend our study of a universe containing just one particle to a universe of two. Imagine, in particular, a simple universe containing two isolated hydrogen atoms; two electrons bound in orbit around two protons that are very far apart. In a few pages we are going to start bringing the two atoms closer together to see what happens, but for now we are to suppose that they are very distant from each other.

The Pauli Exclusion Principle says that the two electrons cannot be in the same quantum state, because electrons are indistinguishable fermions. You might at first be tempted to say that, if the atoms are far apart, then the two electrons must be in very different quantum states and there is not much more to be said on the matter. But things are vastly more interesting than that. Imagine putting electron number 1 in atom number 1 and electron number 2 in atom number 2. After waiting a while it doesn't make sense to say that 'electron number 1 is still in atom number 1'. It might be in atom number 2 now because there is always the chance that the electron did a quantum hop. Remember, everything that can happen does happen, and electrons are free to roam the Universe from one instant to the next. In the language of little clocks, even if we started out with clocks describing one of the electrons clustered only in the vicinity of one of the protons, we would be forced to introduce clocks in the vicinity of the other proton at the next instant. And even if the orgy of quantum interference meant that the clocks near the other proton are very tiny, they would not be of zero size, and there would always be a finite probability that the electron could be there. The way to think more clearly about the implications of the Exclusion Principle is to stop thinking in terms of two isolated atoms and think instead of the system as a whole: we have two protons and two electrons and our task is to understand how they organize themselves. Let us simplify the situation by neglecting the electromagnetic interaction between the two electrons – this won't be a bad approximation if the protons are far apart, and it doesn't affect our argument in any important way.

What do we know about the allowed energies for the electrons in the two atoms? We don't need to do a calculation to get a rough idea; we can use what we know already. For protons that are far apart (imagine they are many miles apart), the lowest allowed energies for the electrons must surely correspond to the situation where they are bound to the protons to make two isolated hydrogen atoms. In this case, we might be tempted to conclude that the lowest energy state for the entire two-proton, two-electron system would correspond to two hydrogen atoms sitting in their lowest energy states, ignoring each other completely. But although this sounds right, it cannot be right. We must think of the system as a whole, and just like an isolated hydrogen atom, this four-particle system must have its own unique spectrum of allowed electron energies. And because of the Pauli principle, the electrons cannot both be in exactly the same energy level around each proton, blissfully ignorant of the existence each other.[2]

It seems that we must conclude that the pair of identical electrons in two distant hydrogen atoms cannot have the same energy but we have also said that we expect the electrons to be in the lowest energy level corresponding to an idealized, perfectly isolated hydrogen atom. Both those things cannot be true and a little thought indicates that the way out of the problem is for there to be not one but *two* energy levels for each level in an idealized, isolated hydrogen atom. That way we can accommodate the two electrons without violating the Exclusion Principle. The difference in the two energies must be very small indeed for atoms that are far apart, so that we can pretend the atoms are oblivious to each other. But really, they are not oblivious, because of the tendril-like reaches of the Pauli principle: if one of the two electrons is in one energy state then the other must be in the second, different energy state and this intimate link between the two atoms persists regardless of how far apart they are.

This logic extends to more than two atoms – if there are twenty-four hydrogen atoms scattered far apart across the Universe, then for every energy state in a single-atom universe there are now twenty-four energy states, all taking on almost but not quite the same values. When an electron in one of the atoms settles into a particular state it does so in full 'knowledge' of the states of each of the other twenty-three electrons, regardless of their distance away. And so, every electron in the Universe knows about the state of every other electron. We need not stop there – protons and neutrons are fermions too, and so every proton knows about every other proton and every neutron knows about every other neutron. There is an intimacy between the particles that make up our Universe that extends

2. For the sake of this discussion we are ignoring the electron's spin. What we have said still applies if we imagine that it refers to two electrons of the same spin.

across the entire Universe. It is ephemeral in the sense that for particles that are far apart the different energies are so close to each other as to make no discernible difference to our daily lives.

This is one of the weirdest-sounding conclusions we've been led to so far in the book. Saying that every atom in the Universe is connected to every other atom might seem like an orifice through which all sorts of holistic drivel can seep. But there is nothing here that we haven't met before. Think about the square well potential we thought about in Chapter 6. The width of the well determines the allowed spectrum of energy levels, and as the size of the well is changed, the energy level spectrum changes. The same is true here in that the shape of the well inside which our electrons are sitting, and therefore the energy levels they are allowed to occupy, is determined by the positions of the protons. If there are two protons, the energy spectrum is determined by the position of both of them. And if there are 10^{80} protons forming a universe, then the position of every one of them affects the shape of the well within which 10^{80} electrons are sitting. There is only ever one set of energy levels and when anything changes (e.g. an electron changes from one energy level to another) then everything else must instantaneously adjust itself such that no two fermions are ever in the same energy level.

The idea that the electrons 'know' about each other instantaneously sounds like it has the potential to violate Einstein's Theory of Relativity. Perhaps we can build some sort of signalling apparatus that exploits this instantaneous communication to transmit information at faster-than-light speeds. This apparently paradoxical feature of quantum theory was first appreciated in 1935, by Einstein in collaboration with Boris Podolsky and Nathan Rosen; Einstein called it 'spooky action at a distance' and did not like it. It took some time before people realized that, despite its spookiness, it is impossible to exploit these long-range correlations to transfer information faster than the speed of light and that means the law of cause and effect can rest safe.

This decadent multiplicity of energy levels is not just an esoteric device to evade the constraints of the Exclusion Principle. In fact, it is anything but esoteric because this is the physics behind chemical bonding. It is also the key idea in explaining why some materials conduct electricity whilst others do not and, without it, we would not understand how a transistor works. To begin our journey to the transistor, we are going to go back to the simplified 'atom' we met in Chapter 6, when we trapped an electron inside a potential well. To be sure, this simple model didn't allow us to compute the correct spectrum of energies in a hydrogen atom, but it did teach us about the behaviour of a single atom and it will serve us well here too. We are going to use two square wells joined together to make a toy model of two adjacent hydrogen atoms. We'll think first about the case where there is a single electron moving in the potential created by two protons. The upper picture in Figure 8.1 illustrates how we'll do it. The potential is flat except where it dips down to make two wells, which mimic the influence of the two protons in their ability to trap electrons. The step in the middle helps keep the electron trapped either on the left or on the right, provided it is high enough. In the technical parlance, we say that the electron is moving in a double-well potential.

Our first challenge is to use this toy model to understand what happens when we bring two hydrogen atoms together – we will see that when they get close enough they bind together, to make a molecule. After that, we shall contemplate more than two atoms and that will allow us to appreciate what happens inside solid matter.

If the wells are very deep, we can use the results from Chapter 6 to determine what the lowest-lying energy states should correspond to. For a single electron in a single square well, the lowest energy state is described by a sine wave of wavelength equal to twice the size of the box. The next-to-lowest energy state is a sine wave whose wavelength is equal to the size of the box, and so on. If we put an electron into one side of a double-well, and if the well is deep enough, then the allowed energies must be close to those for an electron trapped in a single deep well, and its wavefunction should

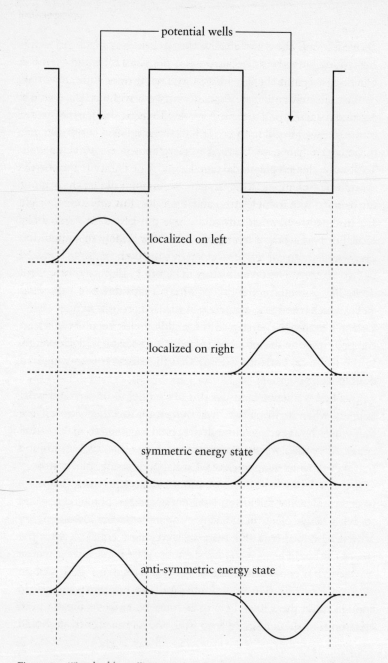

Figure 8.1. The double-well potential at the top and, below it, four interesting wavefunctions describing an electron in the potential. Only the bottom two correspond to an electron of definite energy.

therefore look quite like a sine wave. It is to the small differences between a perfectly isolated hydrogen atom and a hydrogen atom in a distantly separated pair to which we now turn our attention.

We can safely anticipate that the top two wavefunctions drawn in Figure 8.1 correspond to those for a single electron when it is located either in the left well or the right well (remember we use 'well' and 'atom' interchangeably). The waves are approximately sine waves, with a wavelength equal to twice the width of the well. Because the wavefunctions are identical in shape we might say that they should correspond to particles with equal energies. But this can't be right because, as we have already said, there must be a tiny probability that, no matter how deep the wells or how widely separated they are, the electron can hop from one to the other. We have hinted at this by sketching the sine waves as 'leaking' slightly through the walls of the well, representing the fact that there is a very small probability to find non-zero clocks in the adjacent well.

The fact that the electron always has a finite probability of leaping from one well to the other means that the top two wavefunctions in Figure 8.1 cannot possibly correspond to an electron of definite energy, because we know from Chapter 6 that such an electron is described by a standing wave whose shape does not change with time or, equivalently, a bunch of clocks whose sizes never change with time. If, as time advances, new clocks are spawned in the originally empty well then the shape of the wavefunction will most certainly change with time. What then, does a state of definite energy look like for a double well? The answer is that the states must be more democratic, and express an equal preference to find the electron in either well. This is the only way to make a standing wave and stop the wavefunction sloshing back and forth from one well to the other.

The lower two wavefunctions we've sketched in Figure 8.1 have this property. These are what the lowest-lying energy states actually look like. They are the only possible stationary states we can build that look like the 'single-well' wavefunctions in each individual well, and also describe an electron that is equally likely to be found in either well. They are in fact the *two* energy states that we deduced must be present if we are to put two electrons into orbit around two distant protons to make two almost identical hydrogen atoms in a way consistent with the Pauli principle. If one electron is described by one of these two wavefunctions, then the other electron must be described by the other – this is what the Pauli principle demands.[3] For deep enough wells, or if the atoms are far enough apart, the two energies will be almost equal, and almost equal to the lowest energy of a particle trapped in a single isolated well. We should not worry that one of the wavefunctions looks partly upside-down – remember it is only the size of a clock that matters when determining the probability to find the particle at some place. In other words, we could invert all the wavefunctions we've drawn in this book and not change the physical content of anything at all. The 'partly upside-down' wavefunction (labelled 'anti-symmetric energy state' in the figure) therefore still describes an equal superposition of an electron trapped in the left-hand well and an electron trapped in the right-hand well. Crucially though, the symmetric and anti-symmetric wavefunctions are not *exactly* the same (they could not be, otherwise Pauli would be upset). To see this, we need to look at the behaviour of these two lowest-energy wavefunctions in the region *between* the wells.

One wavefunction is symmetric about the centre of the two wells, and the other is anti-symmetric (they are labelled as such in the figure). By 'symmetric' we mean that the wave on the left is the mirror image of the wave on the right. For the 'antisymmetric' wave, the wave on the left is the mirror image of the wave on the right only after it has been turned upside down. The terminology is not very important, but what does matter is that the two waves are different in the region between the two wells. It is this small difference that means that they describe states with very slightly different energies. In fact, the symmetric wave is the one with the lower energy. So turning one of the waves upside down does in fact mat-

3. Recall we have in mind two identical electrons, i.e. they have equal spin.

ter, but only a tiny amount if the wells are deep enough or far enough apart.

It can certainly be confusing to think in terms of particles with definite energy because, as we have just seen, they are described by wavefunctions that are of equal size in either well. This genuinely does mean that there is an equal chance of finding the electron in either well when we look for it, even if the wells are separated by an entire Universe.

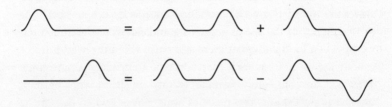

Figure 8.2. Upper: an electron localized in the left well can be understood as the *sum* of the two lowest energy states. Lower: likewise, an electron located in the right well can be understood as the *difference* between the two lowest energy states.

How should we picture what is going on if we actually place an electron into one well and a second electron into the other well? We said before that we expect the initially empty well to fill with clocks in order to represent the fact that the particle can hop from one side to the other. We even hinted at the answer when we said that the wavefunction 'sloshes' back and forth. To see how this works out, we need to notice that we can express a state localized on one of the protons as the sum of the two lowest-energy wavefunctions. We've illustrated this in Figure 8.2 but what does it mean? If the electron is sat in a particular well at some time, then this implies that it does not actually have a unique energy. Specifically, a measurement of its energy will return a value equal to one of the two possible energies corresponding to the two states of definite energy that build up the wavefunction. The electron is therefore in two energy states at once. We hope that, by this stage in the book, this is not a novel concept.

But here is the interesting thing. Because these two states are not of exactly the same energy, their clocks rotate at slightly different rates (as discussed on page 105). This has the effect that a particle initially described by a wavefunction localized around one proton will, after a long enough time, be described by a wavefunction which is peaked around the other proton. We don't intend to go into details, but suffice to say that this phenomenon is quite analogous to the way that two sound waves of almost the same frequency add together to produce a resultant wave that is at first loud (the two waves are in phase) and then, some time later, quiet (as the two waves are out of phase). This phenomenon is known as 'beats'. As the frequency of the waves gets closer and closer, so the time interval between loud and quiet increases until, when the waves are of exactly the same frequency, they combine to produce a pure tone. This will be completely familiar to any musician who, perhaps unknowingly, exploits this piece of wave physics when they make use of a tuning fork. The story runs in exactly the same way for the second electron sat in the second well. It too tends to migrate from one well to the other in a fashion that exactly mirrors the behaviour of the first electron. Although we might start with one electron in one well and a second electron in the other, after waiting long enough the electrons will swap positions.

We are now going to exploit what we have just learnt. The really interesting physics happens when we start to move the atoms closer together. In our model, moving the atoms together corresponds to reducing the width of the barrier separating the two wells. As the barrier gets thinner, the wavefunctions begin to merge together and the electron is increasingly likely to be found in the region between the two protons. Figure 8.3 illustrates what the four lowest-energy wavefunctions look like when the barrier is thin. It is interesting that the lowest-energy wavefunction is starting to look like the lowest-energy sine wave we would get if we had a single electron in a single, wide well, i.e. the two peaks merge together to produce a single peak (with a dimple in it). Meanwhile, the second-lowest-energy wavefunction looks rather like the sine wave corresponding to the next-to-lowest energy for a single, wide well. This is what we should expect, because, as the barrier between the wells gets thinner, its effect diminishes and, eventually, when it has no thickness at all, it has no effect at all and so our electron should behave exactly as if it is in a single well.

Having looked at what happens at the two extremes – the wells widely separated and the wells close together – we can complete the picture by considering how the allowed electron energies vary as we decrease the distance between the wells. We've sketched the results for the lowest four energy levels in Figure 8.4. Each of the four lines represents one of the four lowest energy levels, and we've sketched the corresponding wavefunctions next to them. The right-hand edge of the picture shows the wavefunctions when the wells are widely separated (see also Figure 8.1). As we expect, the difference between the energy levels of the electrons in each well are virtually indistinguishable. As the wells get closer together, however, the energy levels begin to separate (compare the wavefunctions on the left with those in Figure 8.3). Interestingly, the energy level corresponding to the anti-symmetric wavefunction increases, whilst that corresponding to the symmetric wavefunction decreases.

This has a profound consequence for a real system of two protons and two electrons – that is, two hydrogen atoms. Remember that in reality two electrons can actually fit into the same energy level because they can have opposite spins. This means that they can both fit into the lowest (symmetric) energy level and, crucially, this level decreases in energy as the atoms get closer together. This means that it is energetically favourable for two distant atoms to move closer together. And this is what actually happens in Nature:[4] the symmetric wavefunction describes a system in which the electrons are shared more evenly between the two protons than one might anticipate from the 'far apart' wavefunction, and because this 'sharing' configuration is of lower energy, the atoms are drawn towards each other. This attraction is eventually halted because the two protons are positively charged and as such they repel each other

4. Providing the protons are not moving too rapidly relative to each other.

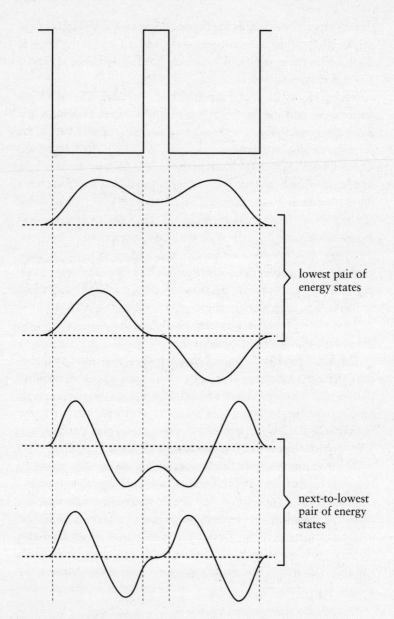

Figure 8.3: Like Figure 8.1 except that the wells are closer together. The 'leakage' into the region in between the wells increases. Unlike Figure 8.1, we also show the wavefunctions corresponding to the pair of next-to-lowest energies.

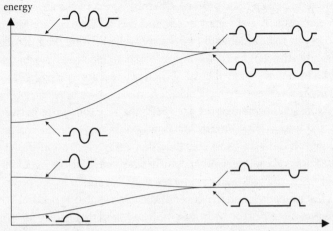

energy

distance between atoms

Figure 8.4: The variation of the allowed electron energies as we change the distance between the wells.

(there is also repulsion due to the fact that the electrons have equal charges), but this repulsion only beats the inter-atomic attraction at distances smaller than around 0.1 nanometres (at room temperature). The result is that a pair of hydrogen atoms at rest will eventually nestle together. This pair of nestled hydrogen atoms has a name: it is a hydrogen molecule.

This preference for two atoms to stick together as a result of sharing their electrons between them is known as a covalent bond. If you look back at the top wavefunction in Figure 8.3, then this is roughly what the covalent bond in a hydrogen molecule looks like. Remember that the height of the wave corresponds to the probability that an electron will be found at that point.[5] There is a peak above each well, i.e. around each proton, which informs us that each electron is still most likely to be in the vicinity of one or other of the protons. But there is also a significant chance that the electrons will

5. This is true for standing waves, where the clock size and the projection onto the 12 o'clock direction are proportional to each other.

spend time between the protons. Chemists speak of the atoms 'sharing' electrons in a covalent bond, and this is what we are seeing, even in our toy model with two square wells. Beyond the hydrogen molecule, this tendency for atoms to share electrons is what we invoked when we were discussing chemical reactions on pages 123–4.

That is a very satisfying conclusion to reach. We have learnt that, for hydrogen atoms that are far apart, the tiny difference between the two lowest-lying energy states was only of academic interest, although it did lead us to conclude that every electron in the Universe knows about every other, which is certainly fascinating. On the other hand, the two states get increasingly separated as the protons get closer together, and the lower of the two eventually becomes the state that describes the hydrogen molecule, and that is very far from being of mere academic interest, because covalent bonding is the reason that you are not a bunch of atoms sloshing around in a featureless blob.

Now we can keep pulling on this intellectual thread and start to think about what happens when we bring more than two atoms together. Three is bigger than two, so let's start there and consider a triple-well potential, as illustrated in Figure 8.5. As ever, we are to imagine that each well is at the site of an atom. There should be three lowest energy states, but looking at the figure you might be tempted to think that there are now four energy states for every state of the single well. The four states we have in mind are illustrated in the figure and they correspond to wavefunctions that are variously symmetric or anti-symmetric about the centre of the two potential barriers.[6] This counting must be wrong, because if it were correct then one could put four identical fermions into these four states and the Pauli principle would be violated. To get the Pauli principle to work out we need just three energy states and this, of course, is

6. You might think there are four more wavefunctions, corresponding to the ones we have drawn turned upside down, but, as we have said, these are equivalent to the ones drawn.

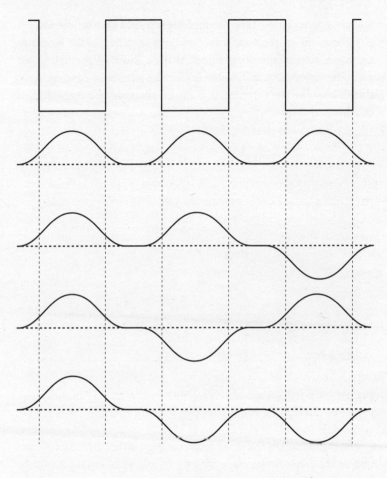

Figure 8.5. The triple well, which is our model for three atoms in a row, and the possible lowest-energy wavefunctions. At the bottom we illustrate how the bottom of the four waves can be obtained from the other three.

what happens. To see this, we need merely spot that we can always write any one of the four wavefunctions sketched in the figure as a combination of the other three. At the bottom of the figure, we have illustrated how that works out in one particular case; we have shown how the last wavefunction can be obtained by a combination of adding and subtracting the other three.

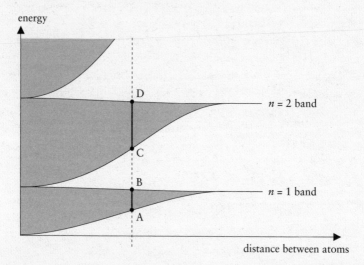

Figure 8.6. The energy bands in a chunk of solid matter and how they vary with the distance between the atoms.

Having identified the three lowest energy states for a particle sitting in the triple-well potential, we can ask what Figure 8.4 looks like in this case, and it should come as no surprise at all to find that it looks rather similar, except that what was a pair of allowed energy states becomes a triplet of allowed states.

Enough of three atoms – we shall now swiftly move our attention to a chain of many. This is going to be particularly interesting because it contains the key ideas that will allow us to explain a lot about what is happening inside solid matter. If there are N wells (to model a chain of N atoms) then for each energy in the single well there will now be N energies. If N is something like 10^{23}, which is typical of the number of atoms in a small chunk of solid material, that is an awful lot of splitting. The result is that Figure 8.4 now looks something like Figure 8.6. The vertical dotted line illustrates that, for atoms that are separated by the corresponding distance, the electrons can only have certain allowed energies. That should be no big surprise (if it is, then you'd better start reading the book again from the beginning), but what is interesting is that the allowed energies come in 'bands'. The energies from A to B are allowed, but no other energies are allowed until we get to C, whence energies from C to D are allowed, and so on. The fact that there are many atoms in the chain means that there are very many allowed energies crammed into each band. So many in fact, that for a typical solid we can just as well suppose that the allowed energies form a smooth continuum in each band. This feature of our toy model is preserved in real solid matter – the electrons there really do have energies that come grouped in bands like this, and that has important implications for what kind of solid we are talking about. In particular, these bands explain why some materials (metals) conduct electricity whilst others (insulators) do not.

How so? Let's begin by considering a chain of atoms (as ever modelled by a chain of potential wells), but now suppose that each atom has several electrons bound to it. This, of course, is the norm – only hydrogen has just the one electron bound to a single proton – and so we are moving from a discussion of a chain of hydrogen atoms to the more interesting case of a chain of heavier atoms. We should also remember that electrons can come in two types; spin up and spin down, and the Pauli principle informs us that we can drop no more than two electrons into each allowed energy level. It follows that for a chain of atoms each containing just one electron per atom (i.e. hydrogen) the $n = 1$ energy band is half-filled. This is illustrated in Figure 8.7, where we have sketched the energy levels for a chain of 5 atoms. This means that each band contains 5 distinct allowed energies. These 5 energy states can accommodate a maximum of 10 electrons, but we only have 5 to worry about so, in the lowest energy configuration, the chain of atoms contains the

5 electrons occupying the bottom half of the $n = 1$ energy band. If we had 100 atoms in the chain then the $n = 1$ band could contain 200 electrons, but for hydrogen, we only have 100 electrons to deal with and so once again the $n = 1$ band is half filled when the chain of atoms is in its lowest energy configuration. Figure 8.7 also shows what happens in the case that there are 2 electrons for every atom (helium) or 3 electrons per atom (lithium). In the case of helium, the lowest-energy configuration corresponds to a filled $n = 1$ band, whilst for lithium the $n = 1$ band is filled and the $n = 2$ band is half filled. It should be pretty clear that this pattern of filled or half-filled continues such that atoms with an even number of electrons always lead to filled bands whilst atoms with an odd number of electrons always lead to half-filled bands. Whether a band is full or not is, as we shall very soon discover, the reason why some materials are conductors whilst others are insulators.

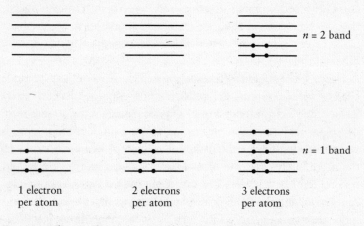

Figure 8.7. The way electrons occupy the lowest available energy states in a chain of five atoms when each atom contains one, two or three electrons. The black dots denote the electrons.

Let's now imagine connecting the ends of our atomic chain to the terminals of a battery. We know from experience that if the atoms form a metal then an electric current will flow. But what does that actually mean, and how does it emerge from our story so far?

The precise action of the battery on the atoms within the wire is, fortunately, something we don't really need to understand. All we need to know is that connecting up the battery provides a source of energy that is able to kick an electron a little, and that kick is always in the same direction. A good question to ask is exactly how a battery does that. To say 'it is because it induces an electric field within the wire and electric fields push electrons' is not entirely satisfying, but it will have to satisfy us as far as this book is concerned. Ultimately, we could appeal to the laws of quantum electrodynamics and try to work the whole thing out in terms of electrons interacting with photons. But we would add absolutely nothing to the current discussion by doing this, so in the interests of brevity, we won't.

Imagine an electron sitting in one of those states of definite energy. We will start by assuming that the action of the battery can only provide very tiny kicks to the electron. If the electron is sat in a low energy state, with many other electrons above it on the energy ladder (we have Figure 8.7 in mind when using this language), it will be unable to receive the energy kick from the battery. It is blocked, because the energy states above it are already filled. For example, the battery might be capable of kicking the electron up to an energy state a few rungs higher, but if all the accessible rungs are already occupied then our target electron must pass up on the opportunity to absorb the energy because there is simply nowhere for it to go. Remember, the Exclusion Principle prevents it from joining the other electrons if the available places are taken. The electron will be forced to behave as if there is no battery connected at all. The situation is different for those electrons with the highest energies. They are lying close to the top of the heap and can potentially absorb a tiny kick from the battery and move into a higher energy state – but only if they are not sitting at the very top of an already full band. Referring back to Figure 8.7, we see that the highest-energy electrons will be able to absorb energy from the battery if the atoms in the chain contain an odd number of electrons. If they contain an even number, then the topmost electrons still cannot go anywhere because there is a big gap in their energy ladder, and they will only overcome this if they are given a large enough kick.

This implies that if the atoms in a particular solid contain an even number of electrons, those electrons may well behave as if the battery had never been connected. A current simply can't flow because there is no way for its electrons to absorb energy. This is a description of an insulator. The only way out of this conclusion is if the gap between the top of the highest filled band and the bottom of the next empty band is sufficiently small – we shall have more to say about that very soon. Conversely, if the atoms contain an odd number of electrons then the topmost electrons are always free to absorb a kick from the battery. As a result they hop up into a higher energy level and, because the kick is always in the same direction, the net effect is to induce a flow of these mobile electrons, which we recognize as an electrical current. Very simplistically, therefore, we might conclude that, if a solid is made up from atoms containing odd numbers of electrons, then they are destined to be conductors of electricity.

Happily, the real world is not that simple. Diamond, a crystalline solid made up entirely of carbon atoms which have six electrons, is an insulator. Graphite, on the other hand, which is also pure carbon, is a conductor. In fact, the odd/even electron rule hardly ever works out in practice, but that is because our 'wells in a line' model of a solid is far too rudimentary. What is absolutely true, though, is that good conductors of electricity are characterized by the fact that the highest-energy electrons have the headroom to leap into higher energy states, whilst insulators are insulators because their topmost electrons are blocked from accessing the higher energy states by a gap in their ladder of allowed energies.

There is a further twist to this tale, and it is a twist that matters when we come to explaining how the current flows in a semiconductor in the next chapter. Let us imagine an electron, free to roam around an unfilled band of a perfect crystal. We say a crystal because we mean to imply that the chemical bonds (possibly covalent) have acted so as to arrange the atoms in a regular pattern. Our one-dimensional model of a solid corresponds to a crystal if all of the wells are equidistant and of the same size. Connect a battery, and an electron will merrily hop up from one level to the next as the applied electric field gently nudges it. As a result, the electric current will steadily increase as the electrons absorb more energy and move faster and faster. To anyone who knows anything about electricity, this should sound rather odd, because there is no sign of 'Ohm's Law', which states that the current (I) should be fixed by the size of the applied voltage (V) according to $V = I \times R$, where R represents the resistance in the wire. Ohm's Law emerges because as the electrons hop their way up the energy ladder they can also lose energy and drop all the way down again – this will only happen if the atomic lattice is not perfect, either because there are impurities within the lattice (i.e. rogue atoms that are different from the majority) or if the atoms are jiggling around significantly, which is what is guaranteed to happen at any non-zero temperature. As a result, the electrons spend most of their time playing a microscopic game of snakes and ladders as they climb up the energy ladder only to fall down again as a result of their interactions with the less than perfect atomic lattice. The average effect is to produce a 'typical' electron energy and that leads to a fixed current. This typical electron energy determines how fast the electrons flow down the wire and that is what we mean by a current of electricity. The resistance of the wire is to be seen as a measure of how imperfect the atomic lattice is through which the electrons are moving.

But that is not the twist. Even without Ohm's Law, the current doesn't just keep increasing. When electrons reach the top of a band, they behave very oddly indeed, and the net effect of this behaviour is to decrease the current and eventually reverse it. This is very odd: even though the electric field is kicking the electrons in one direction, they end up travelling in the opposite direction when they near the top of a band. The explanation of this weird effect is beyond the scope of this book, so we shall just say that the role of the positively charged atomic cores is the key, and they act to push the electrons so that they reverse direction.

Now, as advertised, we will explore what happens when a would-be insulator behaves like a conductor because the gap between the last filled band and the next, empty, band is 'sufficiently small'. At this stage it is worth introducing some jargon. The last (i.e. highest-energy) band of energies that is completely filled with electrons is referred to as the 'valence band', and the next band up (either empty or half-filled in our analysis) is referred to as the 'conduction band'. If the valence and conduction bands actually overlap (and that is a real possibility), then there is no gap at all and a would-be insulator instead behaves as a conductor. What if there is a gap but the gap is 'sufficiently small'? We have indicated that the electrons can receive energy from a battery, so we might suppose that, if the battery is powerful, then it could deliver a mighty enough kick to project an electron sitting near to the top of the valence band up into the conduction band. That is possible, but this is not where our interest lies because typical batteries can't generate a big enough kick. To put some numbers on it, the electric field within a solid is typically of the order of a few volts per metre, and we would need fields of a few volts per nanometre (i.e. a billion times stronger) in order to provide a kick capable of making an electron jump the electron volt[7] or so in energy needed to leap from the valence band to the conduction band in a typical insulator. Much more interesting is the kick that an electron can receive from the atoms that make up the solid. They are not rigidly sitting in the same place, but rather they are jiggling around a little bit – the hotter the solid the more they jiggle and a jiggling atom can deliver far more energy to an electron than a practical battery; enough to make it leap a few electron volts in energy. At room temperature, it is actually very rare to hit an electron that

---

7. The electron volt is a very convenient unit of energy for discussing electrons in atoms and is widely used in nuclear and particle physics. It is the energy an electron would acquire if it were accelerated through a potential difference of 1 volt. That definition is not important, all that matters is that it is a way of quantifying energy. To get a feel for the size, the energy required to completely liberate an electron from the ground state of a hydrogen atom is 13.6 electron volts.

hard, because at 20°C the typical thermal energies are around $\frac{1}{40}$ of an electron volt. But this is only an average, and there are a very large number of atoms in a solid, so it does occasionally happen. When it does, electrons can leap from their valence band prison into the conduction band, where they may then absorb the tiny kicks from a battery and in so doing initiate a flow of electricity.

Materials in which, at room temperature, a sufficient number of electrons can be lifted up from the valence to conduction band in this way have their own special name: they are called semiconductors. At room temperature they can carry a current of electricity, but as they are cooled down, and their atoms jiggle less, so their ability to conduct electricity diminishes and they turn back into insulators. Silicon and germanium are the two classic examples of semiconductor materials and, because of their dual nature, they can be used to great advantage. Indeed, it is no exaggeration to say that the technological application of semiconductor materials revolutionized the world.

9. *The Modern World*

In 1947, the world's first transistor was built. Today, every year the world manufactures over 10,000,000,000,000,000,000, which is one hundred times more than the sum total of all the grains of rice consumed every year by the world's seven billion residents. The world's first transistor computer was built in Manchester in 1953, and had ninety-two of them. Today, you can buy over a hundred thousand transistors for the cost of a single grain of rice and there are around a billion of them in your mobile phone. In this chapter, we are going to describe how a transistor works, surely the most important application of quantum theory.

As we saw in the previous chapter, a conductor is a conductor because some of the electrons are sitting in the conduction band. As a result, they are quite mobile and can 'flow down' the wire when a battery is connected. The analogy with flowing water is a good one; the battery is causing current to flow. We can even use the 'potential' concept to capture this idea, because the battery creates a potential within which the conduction electrons move, and the potential is in a sense, 'downhill'. So an electron in the conduction band of a material 'rolls' down the potential created by the battery, gaining energy as it goes. This is another way to think about the tiny kicks we talked about in the last chapter – instead of a battery inducing tiny kicks that accelerate the electron along the wire, we are invoking a classical analogy akin to water flowing down a hill. This is a good way to think about the conduction of electricity by electrons, and it is the way we will be thinking throughout the rest of this chapter.

In a semiconductor material like silicon, something very interesting happens because the current is not only carried by electrons in the conduction band. The electrons in the valence band contribute to

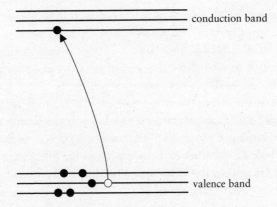

Figure 9.1. An electron-hole pair in a semiconductor.

the current too. To see that, take a look at Figure 9.1. The arrow shows an electron, originally sitting inert in the valence band, absorbing some energy and being lifted up into the conduction band. Certainly the elevated electron is now much more mobile, but something else is mobile too – there is now a hole left in the valence band, and that hole provides some wriggle room for the otherwise inert valence band electrons. As we have seen, connecting a battery to this semiconductor will cause the conduction band electron to hop up in energy, thereby inducing an electric current. What happens to that hole? The electric field created by the battery can cause an electron from some lower energy state in the valence band to hop into the vacant hole. The hole is filled in, but now there is a hole 'deeper' down in the valence band. When electrons in the valence band hop into the vacant hole, the hole moves around.

Rather than bother keeping track of the motion of all the electrons in the almost-full valence band, we can instead decide to keep track of where the hole is and forget about the electrons. That book-keeping convenience is the norm for those working on the physics of semiconductors, and it will make our life simpler to think in that way too.

An applied electric field induces the conduction band electrons to flow, creating a current, and we should like to know what it does to the holes in the valence band. We know that the valence band electrons are not free to move, because they are almost completely trapped by the Pauli principle but they will shuffle along under the influence of the electric field and the hole moves along with them. This might sound counterintuitive, and if you are having trouble with the idea that if electrons in the valence band shuffle to the left then the hole also shuffles to the left, perhaps the following analogy will help. Imagine a line of people all standing in a queue 1 metre apart, except that somewhere in the middle of the line a single person is missing. The people are analogous to electrons and the missing person is the hole. Now imagine that all the people stride 1 metre forwards so that they end up where the person in front of them was standing. Obviously the gap in the line jumps 1 metre forwards too, and so it is with the holes. One could also imagine water flowing down a pipe – a small bubble in the water will move in the same direction as the water, and this 'missing water' is analogous to a hole in the valence band.

But, as if that wasn't enough to be going on with, there is an important added complication; we now need to invoke the piece of physics that we introduced in the 'twist' at the end of the last chapter. If you recall, we said that electrons moving near to the top of a filled band are accelerated by an electric field in the *opposite* direction to electrons moving near to the bottom of a band. This means that the holes, which are near the top of the valence band, move in the opposite direction to the electrons, which are near the bottom of the conduction band.

The bottom line is that we can picture a flow of electrons in one direction and a corresponding flow of holes in the other direction. A hole can be thought of as carrying an electric charge that is exactly opposite to the charge of an electron. To see this, remember that the material through which our electrons and holes flow is, on average, electrically neutral. In any ordinary region there is no net charge, because the charge due to the electrons cancels the positive charge carried by the atomic cores. But if we make an electron–hole pair by exciting an electron out of the valence band and into the conduction band (as we have been discussing), then there is a free electron roaming around, which constitutes an excess of negative charge relative to the average conditions in that region of the material. Likewise, the hole is a place where there is no electron and so it corresponds to a region where there is a net excess of positive charge. The electric current is defined to be the rate at which positive charges flow,[1] and so electrons contribute negatively to the current and the holes contribute positively, if they are flowing in the same direction. If, as is the case in our semiconductor, the electrons and holes flow in opposite directions, then the two add together to produce a larger net flow of charge and hence a larger current.

Whilst all this is a little intricate, the net effect is very straightforward: we are to imagine a current of electricity through a semiconductor material as being representative of the flow of charge, and this flow can be made up of conduction band electrons moving in one direction and valence band holes moving in the opposite direction. This is to be contrasted with the flow of current in a conductor – in that case, the current is dominated by the flow of a large number of electrons in the conduction band, and the extra current coming from electron–hole pair production is negligible.

To understand the utility of semiconductor materials is to appreciate that the current flowing in a semiconductor is not like an uncontrollable flood of electrons down a wire, as it is in a conductor. Instead, it is a much more delicate combination of electron and hole currents and, with a little clever engineering, that delicate combination can be exploited to produce tiny devices that are capable of exquisitely controlling the flow of current through a circuit.

What follows is an inspiring example of applied physics and engineering. The idea is to deliberately contaminate a piece of pure

1. This definition is purely a matter of convention and a historical curiosity. We could just as well define the current to flow in the direction that the conduction band electrons move.

silicon or germanium so as to induce some new available energy levels for the electrons. These new levels will allow us to control the flow of electrons and holes through our semiconductor just as we might control the flow of water through a network of pipes using valves. Of course, anyone can control the flow of electricity through a wire – just pull the plug. But that is not what we are talking about – rather we are talking about making tiny switches that allow the current to be controlled dynamically within a circuit. Tiny switches are the building blocks of logic gates, and logic gates are the building blocks of microprocessors. So how does that all work out?

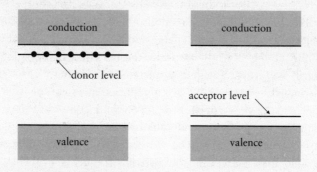

Figure 9.2. The new energy levels induced in a n-type semiconductor (on the left) and a p-type semiconductor (on the right).

The left-hand part of Figure 9.2 illustrates what happens if a piece of silicon is contaminated with phosphorous. The degree of contamination can be controlled with precision and this is very important. Suppose that every now and then within a crystal of pure silicon an atom is removed and replaced with a phosphorous atom. The phosphorous atom snuggles neatly into the spot vacated by the silicon atom, the only difference being that phosphorous has one *more* electron than silicon. That extra electron is very weakly bound to its host atom, but it is not entirely free and so it occupies an energy level lying just below the conduction band. At low temperatures the conduction band is empty, and the extra electrons donated by the phosphorous atoms reside in the donor level marked in the figure.

At room temperature, electron–hole pair creation in the silicon is very rare, and only about one electron in every trillion gets enough energy from the thermal vibrations of the lattice to jump out of the valence band and into the conduction band. In contrast, because the donor electron in phosphorous is so weakly bound to its host, it is very likely that it will make the small hop from the donor level into the conduction band. So at room temperature, for levels of doping greater than one phosphorous atom for every trillion silicon atoms, the conduction band will be dominated by the presence of the electrons donated by the phosphorous atoms. This means it is possible to control very precisely the number of mobile electrons that are available to conduct electricity, simply by varying the degree of phosphorous contamination. Because it is electrons roaming in the conduction band that are free to carry the current, we say that this type of contaminated silicon is 'n-type' ('n' for 'negatively charged').

The right-hand part of Figure 9.2 shows what happens if instead we contaminate the silicon with atoms of aluminium. Again, the aluminium atoms are sprinkled sparingly around among the silicon atoms, and again they snuggle nicely into the spaces where silicon atoms would otherwise be. The difference comes because aluminium has one *fewer* electron than silicon. This introduces holes into the otherwise pure crystal, just as phosphorous added electrons. These holes are located in the vicinity of the aluminium atoms, and they can be filled in by electrons hopping out of the valence band of neighbouring silicon atoms. The 'hole-filled' acceptor level is illustrated in the figure, and it sits just above the valence band because it is easy for a valence electron in the silicon to hop into the hole made by the aluminium atom. In this case, we can naturally regard the electric current as being propagated by the holes, and for that reason this kind of contaminated silicon is known as 'p-type' ('p' for 'positively charged'). As before, at room temperature, the level of aluminium contamination does not need to be much more than one part per trillion before the current due to the motion of the holes from the aluminium is dominant.

So far we have simply said that it is possible to make a lump of

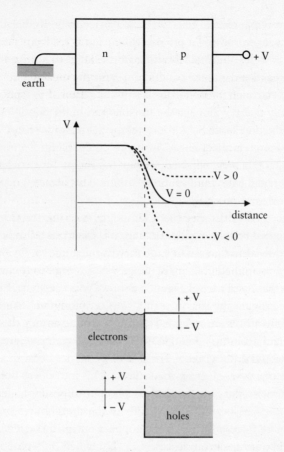

Figure 9.3. A junction formed by joining together a piece of n-type and a piece of p-type silicon.

silicon which is able to transmit a current, either by allowing electrons donated by phosphorous atoms to sail along in the conduction band or by allowing holes donated by aluminium atoms to sail along in the valence band. What is the big deal?

Figure 9.3 illustrates that we are on to something because it shows what happens if we join together two pieces of silicon; one n-type and the other p-type. Initially, the n-type region is awash with electrons from the phosphorous and the p-type region is awash with holes from the aluminium. As a result, electrons from the n-type region drift over into the p-type region, and holes from the p-type region drift over into the n-type region. There is nothing mysterious about this; the electrons and holes simply meander across the junction between the two materials just as a drop of ink spreads out in a bath of water. But as the electrons and holes drift in opposite directions, they leave behind regions of net positive charge (in the n-type region) and net negative charge (in the p-type region). This build up of charge opposes further migration by the 'like sign charges repel' rule, until eventually there is a balance, and no further net migration occurs.

The second of the three pictures in Figure 9.3 illustrates how we might think of this using the language of potentials. What is shown is how the electric potential varies across the junction. Deep in the n-type region, the effect of the junction is unimportant, and since the junction has settled into a state of equilibrium, no current flows. That means the potential is constant inside this region. Before moving on we should once again be clear what the potential is doing for us: it is simply telling us what forces act on the electrons and holes. If the potential is flat, then, just as a ball sitting on flat ground will not roll, an electron will not move.

If the potential dips down then we might suppose that an electron placed in the vicinity of the falling potential will 'roll downhill'. Inconveniently, convention has it the other way and a downhill potential means 'uphill' for an electron, i.e. electrons will flow uphill. In other words, a falling potential acts as a barrier to an electron, and this is what we've drawn in the figure. There is a force pushing the electron away from the p-type region as a result of the build up of negative charge that has occurred by earlier electron migration. This force is what prevents any further net migration of electrons from the n-type to the p-type silicon. Using downhill potentials to represent an uphill journey for an electron is actually not as silly as it seems, because things now make sense from the point of view of the holes, i.e. holes naturally flow downhill. So now we can also see that the way we drew the potential (i.e. going from the high ground on the left to low ground on the right) also correctly accounts for the fact that holes are prevented from escaping from the p-type region by the step in the potential.

The third picture in the figure illustrates the flowing water analogy. The electrons on the left are ready and willing to flow down the wire but they are prevented from doing so by a barrier. Likewise the holes in the p-type region are stranded on the wrong side of the barrier; the water barrier and the step in the potential are just two different ways of speaking about the same thing. This is how things are if we simply stick together an n-type piece of silicon and a p-type piece. Actually, the act of sticking them together takes more care than we are suggesting – the two cannot simply be glued together, because then the junction will not allow the electrons and holes to flow freely from one region to the other.

Interesting things start to happen if we now connect this 'pn junction' up to a battery, which allows us to raise or lower the potential barrier between the n-type and p-type regions. If we lower the potential of the p-type region then we steepen the step and make it even harder for the electrons and holes to flow across the junction. But raising the potential of the p-type region (or lowering the potential of the n-type region) is just like lowering the dam that was holding back the water. Immediately, electrons will flood from n-type to p-type and holes will flood in the opposite direction. In this way a pn-junction can be used as a diode – it can allow a current to flow, but only in one direction. Diodes are, however, not where our ultimate interest resides.

Figure 9.4 is a sketch of the device that changed the world – the transistor. It shows what happens if we make a sandwich, with a layer of p-type silicon in between two layers of n-type silicon. Our explanation of a diode will serve us well here, because the ideas are basically the same. Electrons drift from the n-type regions into the p-type region and holes drift the other way until this diffusion is eventually halted by the potential steps at the junctions between the layers. In isolation, it is as if there are two reservoirs of electrons held apart by a barrier, and a single reservoir of holes that sits brim-full in between.

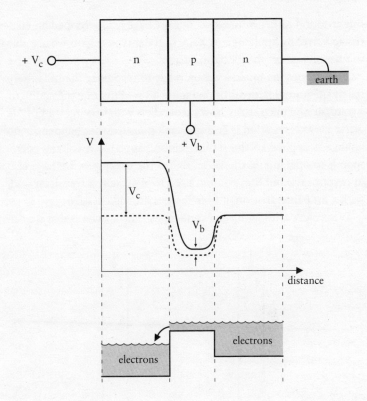

Figure 9.4: A transistor.

The interesting action occurs when we apply voltages to the n-type region on one side and the p-type region in the middle. Applying positive voltages causes the plateau on the left to rise (by an amount V_c) and likewise the plateau in the p-type region (by an amount V_b). We've indicated this by the solid line in the middle diagram in the figure. This way of arranging the potentials has a dramatic effect, because it creates a waterfall of electrons as they flood over the lowered central barrier and into the n-type region on the left (remember, electrons flow 'uphill'). Providing that V_c is larger than V_b, the flow of electrons is one-way and the electrons on the left remain unable to flow across the p-type region. This all might sound rather innocuous, but we have just described an electronic valve. By applying a voltage to the p-type region we are able to turn on and off the electron current.

Now comes the finale – we are ready to recognize the full potential of the humble transistor. In Figure 9.5 we illustrate the action of a transistor by once again drawing parallels with flowing water. The 'valve closed' situation is entirely analogous to what happens if no voltage is applied to the p-type region. Applying a voltage corresponds to opening up the valve. Below the two pipes, we have also drawn the symbol that is often used to represent a transistor and, with a little imagination, it even looks a little like a valve.

What can we do with valves and pipes? The answer is that we can

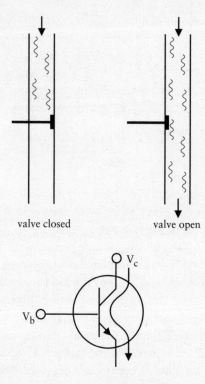

valve closed   valve open

Figure 9.5. The 'water in a pipe' analogy with a transistor.

build a computer and if those pipes and valves can be made small enough then we can make a serious computer. Figure 9.6 illustrates conceptually how we can use a pipe with two valves to construct something called a 'logic gate'. The pipe on the left has both valves open and as a result water flows out of the bottom. The pipe in the middle and the pipe on the right both have one valve closed and obviously no water can then flow out of the bottom. We have not bothered to show the fourth possibility, when both valves are closed. If we were to represent the flow of water out of the bottom of our pipes by the digit '1' and the absence of flow by the digit '0', and if we assign the digit '1' to an open valve and the digit '0' to a closed valve, then we can summarize the action of the four pipes (three drawn and one not) by the equations '1 AND 1 = 1', '1 AND 0 = 0', '0 AND 1 = 0' and '0 AND 0 = 0'. The word 'AND' is here a logical operation and it is being used in a technical way – the system of pipe and valves we just described is called an 'AND gate'. The gate takes two inputs (the state of the two valves) and returns a single output (whether water flows or not) and the only way to get a '1' out is to feed a '1' and a '1' in. We hope it is clear how we can use a pair

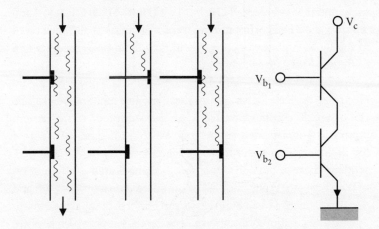

Figure 9.6. An 'AND' gate built using a water pipe and two valves (*left*) or a pair of transistors (*right*). The latter is much better suited to building computers.

of transistors connected in series to built an AND gate – the circuit diagram is illustrated in the figure. We see that only if both transistors are turned on (i.e. by applying positive voltages to the p-type regions, V_{b_1} and V_{b_2}) is it possible for a current to flow, which is just what is needed to implement an AND gate.

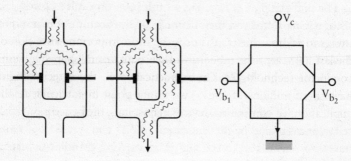

Figure 9.7. An 'OR' gate built using water pipes and two valves (*left*) or a pair of transistors (*right*).

Figure 9.7 illustrates a different logic gate. This time water will flow out of the bottom if either valve is open and only if both are closed will it not flow. This is called an 'OR' gate and, using the same notation as before, '1 OR 1 = 1', '1 OR 0 = 1', '0 OR 1 = 1' and '0 OR 0 = 0'. The corresponding transistor circuit is also illustrated in the figure and now a current will flow in all cases except when both transistors are switched off.

Logic gates like these are the secret behind the power of digital electronic devices. Starting from these modest building blocks one can assemble combinations of logic gates in order to implement arbitrarily sophisticated algorithms. We can imagine specifying a list of inputs into some logical circuits (a series of '0's and '1's), sending these inputs through some sophisticated configuration of transistors that spits out a list of outputs (again a series of '0's and '1's). In that way we can build circuits to perform complicated mathematical calculations, or to make decisions based on which keys are pressed on a keyboard, and feed that information to a unit which then displays the corresponding characters on a screen, or to trigger an alarm if an intruder breaks into a house, or to send a stream of text characters down a fibre optic cable (encoded as a series of binary digits) to the other side of the world, or . . . in fact, anything you can think of, because virtually every electrical device you possess is crammed full of transistors.

The potential is limitless, and we have already exploited the transistor to change the world enormously. It is probably not overstating things to say that the transistor is the most important invention of the last 100 years – the modern world is built on and shaped by semiconductor technologies. On a practical level, these technologies have saved millions of lives – we might point in particular to the applications of computing devices in hospitals, the benefits of rapid, reliable and global communication systems and the uses of computers in scientific research and in controlling complex industrial processes.

William B. Shockley, John Bardeen and Walter H. Brattain were awarded the Nobel Prize in Physics in 1956 'for their researches on semiconductors and their discovery of the transistor effect'. There has probably never been a Nobel Prize awarded for work that directly touches so many people's lives.

10. *Interaction*

In the opening chapters we set up the framework to explain how tiny particles move around. They hop around, exploring the vastness of space without any prejudice, metaphorically carrying their tiny clocks with them as they go. When we add together the multitude of clocks corresponding to the different ways that a particle can arrive at some particular point in space, we obtain one definitive clock whose size informs us of the chance of finding the particle 'there'. From this wild, anarchic display of quantum leaping emerges the more familiar properties of everyday objects. In a sense, every electron, every proton and every neutron in your body is constantly exploring the Universe at large, and only when the sum total of all those explorations is computed do we arrive at a world in which the atoms in your body, fortunately, tend to stay in a reasonably stable arrangement – at least for a century or so. What we have not yet addressed in any detail is the nature of the interactions between particles. We have managed to make a lot of progress without being specific about how particles actually talk to each other, in particular by exploiting the concept of a potential. But what is a potential? If the world is made up solely of particles, then surely we should be able to replace the vague notion that particles move 'in the potential' created by other particles, and speak instead about how the particles move and interact with each other.

The modern approach to fundamental physics, known as quantum field theory, does just this by supplementing the rules for how particles hop around with a new set of rules that explain how those particles interact with each other. These rules turn out to be no more complicated than the rules we've met so far, and it is one of the wonders of modern science that, despite the intricate complexity of the natural world, there are not many of them. 'The eternal mystery of the world is its comprehensibility,' Albert Einstein wrote, and 'the fact that it is comprehensible is a miracle.'

Let's start by articulating the rules of the first quantum field theory to be discovered – quantum electrodynamics, or QED. The origins of the theory can be traced all the way back to the 1920s, when Dirac in particular had an initial burst of success in quantizing Maxwell's electromagnetic field. We've already met the quantum of the electromagnetic field many times in this book – it is the photon – but there were many problems associated with the new theory that were apparent but remained unsolved throughout the 1920s and 1930s. How exactly does an electron emit a photon when it moves between the energy levels in an atom, for example? And, for that matter, what happens to a photon when it is absorbed by an electron, allowing the electron to jump to a higher energy level in the first place? Photons can obviously be created and destroyed in atomic processes, and the means by which this happens is not addressed in the 'old-fashioned' quantum theory that we have met so far in this book.

In the history of science, there are a handful of legendary gatherings of scientists – meetings that certainly appear to have changed the course of science. They probably didn't, in the sense that the participants had usually been working on problems for years, but the Shelter Island Conference of June 1947, held at the tip of Long Island, New York, has a better claim than most for catalysing something special. The participant list alone is worth reciting, because it is short and yet a role-call of the greats of twentieth-century American physics. In alphabetical order: Hans Bethe, David Bohm, Gregory Breit, Karl Darrow, Herman Feshbach, Richard Feynman, Hendrik Kramers, Willis Lamb, Duncan MacInnes, Robert Marshak, John von Neumann, Arnold Nordsieck, J. Robert Oppenheimer, Abraham Pais, Linus Pauling, Isidor Rabi, Bruno Rossi, Julian Schwinger, Robert Serber, Edward Teller, George Uhlenbeck, John Hasbrouck van Vleck, Victor Weisskopf and John Archibald Wheeler. The reader has met several of these names in this book already, and any student of physics probably has heard of most of them. The American writer Dave

Barry once wrote: 'If you had to identify, in one word, the reason why the human race has not achieved, and never will achieve, its full potential, that word would be meetings.' This is doubtless true, but Shelter Island was an exception. The meeting began with a presentation of what has become known as the Lamb shift. Willis Lamb, using high-precision microwave techniques developed during the Second World War, found that the hydrogen spectrum was not, in fact, perfectly described by old-fashioned quantum theory. There was a minute shift in the observed energy levels that could not be accounted for using the theory we have developed so far in this book. It is a tiny effect, but it was a wonderful challenge to the assembled theorists.

We shall leave Shelter Island there, poised after Lamb's talk, and turn to the theory that emerged in the months and years that followed. In doing so we will uncover the origin of the Lamb shift, but, to whet your appetite, here is a cryptic statement of the answer: the proton and electron are not alone inside the hydrogen atom.

QED is the theory that explains how electrically charged particles, like electrons, interact with each other and with particles of light (photons). It is single-handedly capable of explaining all natural phenomena with the exception of gravity and nuclear phenomena. We'll turn our attention to nuclear phenomena later on, and in doing so explain why the atomic nucleus can hold together even though it is a bunch of positively charged protons and zero charge neutrons which would fly apart in an electro-repulsive instant without some sub-nuclear goings-on. Pretty much everything else – certainly everything you see and feel around you – is explained at the deepest known level by QED. Matter, light, electricity and magnetism – it is all QED.

Let's begin by exploring a system we have already met many times throughout the book: a world containing one single electron. The little circles in the 'clock hopping' figure on page 50 illustrate the various possible locations of the electron at some instant in time. To deduce the likelihood of finding it at some point X at a later time, our quantum rules say that we are to allow the electron to hop to X from every possible starting point. Each hop delivers a clock to X, we add up these clocks and then we are done.

We're going to do something now that might look a little over-complicated at first, but of course there is a very good reason. It's going to involve a few As, Bs and Ts – in other words we're heading off into the land of tweed jackets and chalk dust again; it won't last long.

When a particle goes from a point A at time zero to a point B at time T, we can calculate what the clock at B will look like by winding the clock at A backwards by an amount determined by the distance of B from A and the time interval, T. In shorthand, we can write that the clock at B is given by $C(A,0)P(A,B,T)$ where $C(A,0)$ represents the original clock at A at time zero and $P(A,B,T)$ embodies the clock-winding and shrinking rule associated with the leap from A to B.[1] We shall refer to $P(A,B,T)$ as the 'propagator' from A to B. Once we know the rule of propagation from A to B, then we are all set and can figure out the probability to find the particle at X. For the example in Figure 4.2, we have lots of starting points so we'll have to propagate from every one of them to X, and add all the resulting clocks up. In our seemingly overkill notation, the resultant clock $C(X,T) = C(X1,0)P(X1,X,T) + C(X2,0)P(X2,X,T) + C(X3,0)P(X3,X,T) + \ldots$ where X1, X2, X3, etc. label all the positions of the particle at time zero (i.e. the positions of the little circles in Figure 4.2). Just to be crystal clear, $C(X3,0)P(X3,X,T)$ simply means 'take a clock from point X3 and propagate it to point X at time T'. Don't be fooled into thinking there is something tricky going on. All we are doing is writing down in a fancy shorthand something we already knew: 'take the clock at X3 and time zero and figure out by how much to turn and shrink it corresponding to the particle making the journey from X3 to X at some time T later and then repeat that for all of the other time-zero clocks and finally add all of the clocks together according to the clock-adding rule'. We're

1. The propagator shrinks the clock as well, in order to make sure that the particle will be found with a probability of 1 somewhere in the Universe at time T.

sure you'll agree that this is a bit of a mouthful, and the little bit of notation makes life easier.

We can certainly think of the propagator as the embodiment of the clock-winding and shrinking rule. We can also think of it as a clock. To clarify that bald statement, imagine if we know for certain that an electron is located at point A at time $T = 0$, and that it is described by a clock of size 1 reading 12 o'clock. We can picture the act of propagation using a second clock whose size is the amount that the original clock needs to be shrunk and whose time encodes the amount of winding we need. If a hop from A to B requires shrinking the initial clock by a factor of 5 and winding back by 2 hours, then the propagator $P(A,B,T)$ could be represented by a clock whose size is $\frac{1}{5} = 0.2$ and which reads 10 o'clock (i.e. it is wound 2 hours back from 12 o'clock). The clock at B is simply obtained by 'multiplying' the original clock at A by the propagator clock.

As an aside for those who know about complex numbers, just as each of the $C(X1, 0)$, $C(X2, 0)$ can be represented by a complex number so can the $P(X1, X, T)$, $P(X2, X, T)$ and they are combined according to the mathematical rules for multiplying two complex numbers together. For those who do not know about complex numbers: it doesn't matter because the description in terms of clocks is equally accurate. All we did was introduce a slightly different way of thinking about the clock-winding rule: we can wind and shrink a clock using another clock.

We are free to design our clock multiplication rule to make this all work: multiply the sizes of the two clocks $(1 \times 0.2 = 0.2)$ and combine the times on the two clocks such that we wind the first clock backwards by 12 o'clock minus 10 o'clock = 2 hours. This does sound a little bit like we are over-elaborating, and it is clearly not necessary when we only have one particle to think about. But physicists are lazy, and they wouldn't go to all this trouble unless it saved time in the long run. This little bit of notation proves to be a very useful way of keeping track of all the winding and shrinking when we come to the more interesting case where there are multiple particles in the problem – the hydrogen atom, for example.

Regardless of the details, there are just two key elements in our method of figuring out the chances to find a lone particle somewhere in the Universe. First, we need to specify the array of initial clocks which codify the information about where the particle is likely to be found at time zero. Second, we need to know the propagator P(A,B,T), which is itself a clock encoding the rule for shrinking and turning as a particle leaps from A to B. Once we know what the propagator looks like for any pair of start and end points then we know everything there is to know, and we can confidently figure out the magnificently dull dynamics of a Universe containing a single particle. But we should not be so disparaging, because this simple state of affairs doesn't get much more complicated when we add particle interactions into the game. So let's do that now.

Figure 10.1 illustrates pictorially all of the key ideas we want to discuss. It is our first encounter with Feynman diagrams, the calculational tool of the professional particle physicist. The task we are charged with is to work out the probability of finding a pair of electrons at the points X and Y at some time T. As our starting point we are told where the electrons are at time zero, i.e. we are told what their initial clock clusters look like. This is important because being able to answer this type of question is tantamount to being able to know 'what happens in a Universe containing two electrons'. That may not sound like much progress, but once we have figured this out the world is our oyster, because we will know how the basic building blocks of Nature interact with each other.

To simplify the picture, we've drawn only one dimension in space, and time advances from left to right. This won't affect our conclusions at all. Let's start out by describing the first of the series of pictures in Figure 10.1. The little dots at T = 0 correspond to the possible locations of the two electrons at time zero. For the purposes of illustration, we've assumed that the upper electron can be in one of three locations, whilst the lower is in one of two locations (in the real world we must deal with electrons that can be located in an infinity of possible locations, but we'd run out of ink if we had to draw that). The upper electron hops to A at some later time whereupon it does

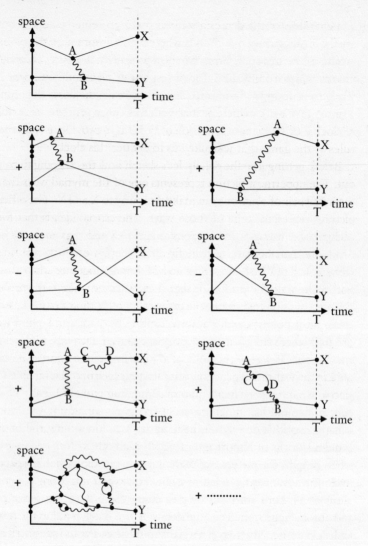

Figure 10.1. Some of the ways that a pair of electrons can scatter off each other. The electrons start out on the left and always end up at the same pair of points, X and Y, at time T. These graphs correspond to some of the different ways that the particles can reach X and Y.

something interesting: it emits a photon (represented by the wavy line). The photon then hops to B where it gets absorbed by the other electron. The upper electron then hops from A to X whilst the lower electron hops from B to Y. That is just one of an infinite number of ways that our original pair of electrons can make their way to points X and Y. We can associate a clock with this entire process – let's call it 'clock 1' or C1 for short. The job of QED is to provide us with the rules of the game that will allow us to deduce this clock.

Before getting into the details, let's sketch how this is going to pan out. The uppermost picture represents one of the myriad ways that the initial pair of electrons can make their way to X and Y. The other pictures represent some of those ways. The crucial idea is that for each possible way that the electrons can get to X and Y we are to identify a quantum clock – C1 is the first in a long list of clocks.[2] When we've got all of the clocks, we are to add them all together and obtain one 'master' clock. The size of that clock (squared) tells us the probability of finding the pair of electrons at X and Y. So once again we are to imagine that the electrons make their way to X and Y not by one particular route, but rather by scattering off each other in every possible way. If we look at the final few pictures in the figure, we can see a variety of more elaborate ways for the electrons to scatter. The electrons not only exchange photons, they can emit and reabsorb a photon themselves, and in the final two figures something very odd is happening. These pictures include the scenario where a photon appears to emit an electron which 'goes in a circle' before ending up where it started out – we shall have more to say about that in a little while. For now, we can simply imagine a series of increasingly complicated diagrams corresponding to cases where the electrons emit and absorb huge numbers of photons before finally ending up at X and Y. We'll need to contemplate the multifarious ways that the electrons can end up at X and Y, but there are two very clear rules: *electrons*

2. We met this idea before, when we tackled the Pauli Exclusion Principle in Chapter 7.

can only hop from place to place and emit or absorb a single photon. That's really all there is to it; electrons can hop or they can branch. Closer inspection should reveal that none of the pictures we have drawn contravenes those two rules because they never involve anything more complicated than a junction involving two electrons and a photon. We must now explain how to go about computing the corresponding clocks, one for each picture in Figure 10.1.

Let's focus on the uppermost picture and explain how to determine what the clock associated with it (clock C1) looks like. Right at the start of the process, there are two electrons sitting there, and they will each have a clock. We should start out by multiplying them together according to the clock multiplication rule to get a new, single clock, which we will denote by the symbol C. Multiplying them makes sense because we should remember that the clocks are actually encoding probabilities, and if we have two independent probabilities then the way to combine them is to multiply them together. For example, the probability that two coins will both come up heads is $\frac{1}{2} \times \frac{1}{2} = \frac{1}{4}$. Likewise, the combined clock, C, tells us the probability to find the two electrons at their initial locations.

The rest is just more clock multiplication. The upper electron hops to A, so there is a clock associated with that; let's call it P(1,A) (i.e. 'particle 1 hops to A'). Meanwhile the lower electron hops to B and we have a clock for that too; call it P(2,B). Likewise there are two more clocks corresponding to the electrons hopping to their final destinations; we shall denote them by P(A,X) and P(B,Y). Finally, we also have a clock associated with the photon, which hops from A to B. Since the photon is not an electron, the rule for photon propagation could be different for the rule for electron propagation so we should use a different symbol for its clock. Let's denote the clock corresponding to the photon hop L(A,B).[3] Now

3. This is a technical point because the clock-winding and -shrinking rule we have used throughout the book to this point does not include the effects of Special Relativity. Including these, as we always must if we are to describe photons, means that the clock-winding rules are different for the electron and photon.

we simply multiply all the clocks together to produce one 'master' clock: $R = C \times P(1,A) \times P(2,B) \times P(A,X) \times P(B,Y) \times L(A,B)$. We are very nearly done now, but there remains some additional clock shrinking to do because the QED rule for what happens when an electron emits or absorbs a photon says that we should introduce a shrinking factor, g. In our diagram, the upper electron emits the photon and the lower one absorbs it – that makes for two factors of g, i.e. g^2. Now we really are done and our final 'clock 1' is obtained by computing $C1 = g^2 \times R$.

The shrinking factor g looks a bit arbitrary, but it has a very important physical interpretation. It is evidently related to the probability that an electron will emit a photon, and this encodes the strength of the electromagnetic force. Somewhere in our calculation we had to introduce a connection with the real world because we are calculating real things and, just as Newton's gravitational constant G carries all the information about the strength of gravity, so g carries all the information about the strength of the electromagnetic force.[4]

If we were actually doing the full calculation, we'd now turn our attention to the second diagram, which represents another way that our original pair of electrons can make their way to the same points, X and Y. The second diagram is very similar to the first in that the electrons start out from the same places, but now the photon is emitted from the upper electron at a different point in space and at a different time and it is absorbed by the lower electron at some other new place and time. Otherwise things run through in precisely the same way and we'll get a second clock, 'clock 2', denoted C2.

Then, on we'd go, repeating the entire process again and again for each and every possible place where the photon can be emitted and each and every possible place where it can be absorbed. We should also account for the fact that the electrons can start out from a variety of different possible starting positions. The key idea is that each and every way of delivering electrons to X and Y needs to be

---

4. g is related to the fine structure constant: $\alpha = \dfrac{g^2}{4\pi}$

considered, and each is associated with its own clock. Once we have collected together all of the clocks, we 'simply' add them all together, to produce one final clock whose size tells us the probability of finding one electron at X and a second at Y. Then we are finished – we will have figured out how two electrons interact with each other because we can do no better than compute probabilities.

What we have just described really is the heart of QED, and the other forces in Nature admit a satisfyingly similar description. We will come on to those shortly, but first we have a little more to discover.

Firstly, a paragraph describing two small, but important, details. Number 1: we have simplified matters by ignoring the fact that electrons have spin and therefore come in two types. Not only that, photons also have spin (they are bosons) and come in three types. This just makes the calculations a little more messy because we need to keep track of which types of photon and electron we are dealing with at every stage of the hopping and branching. Number 2: if you have been reading carefully then you may have spotted the minus signs in front of a couple of the pictures in Figure 10.1. They are there because we are talking about identical electrons hopping their way to X and Y and the two pictures with the minus sign correspond to an interchange of the electrons relative to the other pictures, which is to say that an electron which started out at one of the upper cluster of points ends up at Y whilst the other, lower, electron ends up at X. And as we argued in Chapter 7, these swapped configurations get combined only after an extra 6-hour wind of their clocks – hence the minus sign.

You may also have spotted a possible flaw in our plan – there are an infinite number of diagrams describing how two electrons can make their way to X and Y, and summing an infinite number of clocks might seem onerous to say the least. Fortunately, every appearance of a photon–electron branching introduces another factor of g into the calculation, and this shrinks the size of the resultant clock. This means that the more complicated the diagram, the smaller the clock it will contribute and the less important it will be when we come to add all the clocks up. For QED, g is quite a small number (it's around 0.3), and so the shrinking is pretty severe as the number of branchings increases. Very often, it is enough to consider only diagrams like the first five in the figure, where there are no more than two branchings, and that saves lots of hard work.

This process of calculating the clock (known in the jargon as the 'amplitude') for each Feynman diagram, adding all the clocks together and squaring the final clock to get a probability that the process will happen is the bread and butter of modern particle physics. But there is a fascinating issue hiding away beneath the surface of all that we have been saying – an issue that bothers some physicists a lot and others not at all.

The Quantum Measurement Problem

When we add the clocks corresponding to the different Feynman diagrams together, we are allowing for the orgy of quantum interference to happen. Just as for the case of the double-slit experiment, where we had to consider every possible route that the particle could take on its journey to the screen, we must consider every possible way that a pair of particles can get from their starting positions to their final positions. This allows us to compute the right answer because it allows for interference between the different diagrams. Only at the end of the process, when all of the clocks have been added together and all the interference is accounted for, should we square up the size of the final clock to calculate the probability that the process will happen. Simple. But look at Figure 10.2.

What happens if we attempt to identify what the electrons are doing as they hop to X and Y? The only way we can examine what is going on is to interact with the system according to the rules of the game. In QED, this means that we must stick to the electron–photon branching rule, because there is nothing else. So let's interact with one of the photons that can be emitted from one or other of the electrons, by detecting it using our own personal photon detector; our eye. Notice that we are now asking a different question of the

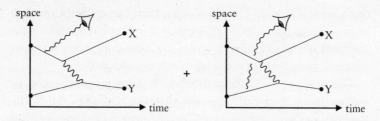

Figure 10.2. A human eye taking a look at what is going on.

theory: 'What is the chance to find an electron at X and another at Y and also a photon in my eye?' We know what to do to get the answer – we should add together all of the clocks associated with the different diagrams that start out with two electrons and end up with an electron at X, another at Y, and also a photon 'in my eye'. More precisely, we should talk about how the photon interacts with my eye. Although that might start out simple enough, it soon gets out of hand. For example, the photon will scatter off an electron sitting in an atom in my eye, and that will trigger a chain of events leading ultimately to my perception of the photon as I become consciously aware of a flash of light in my eye. So to describe fully what is happening involves specifying the positions of every particle in my brain as they respond to the arrival of the photon. We are sailing close to something called the quantum measurement problem.

So far in the book we have described in some detail how to compute probabilities in quantum physics. By that, we mean that quantum theory allows us to calculate the chances of measuring some particular outcome if we conduct an experiment. There is no ambiguity in this process, provided we follow the rules of the game and stick to computing the probabilities of something happening. There is, however, something to feel uneasy about. Imagine an experimenter conducting an experiment for which there are only two outcomes, 'yes' and 'no'. Now imagine actually doing the experiment. The experimenter will record either 'yes' or 'no', and obviously not both at the same time. So far, so good.

Now imagine some future measurement of something else (it

doesn't matter what) made by a second experimenter. Again, we'll assume it is a simple experiment whose outcome is to make a 'click' or 'no click'. The rules of quantum physics dictate that we must compute the probability that the second experiment goes 'click' by summing clocks associated with all of the possibilities that lead to this outcome. Now this may include the circumstance where the first experimenter measures 'yes' *and* the complementary case where they measure 'no'. Only after summing over the two do we get the correct answer for the chances of measuring a 'click' in the second experiment. Is that really right? Do we really have to entertain the notion that, even after the outcome of some measurement, we should maintain the coherence of the world? Or is it the case that once we measure 'yes' or 'no' in the first experiment then the future is dependent only upon that measurement? For example, in our second experiment it would mean that if the first experimenter measures 'yes' then the probability that the second experiment goes 'click' should be computed not from a coherent sum over the 'yes' and 'no' possibilities but instead by considering only the ways in which the world can evolve from 'first experimenter measures yes' to 'second experiment goes click'. This will clearly give a different answer from the case where we are to sum over both the 'yes' and 'no' outcomes and we need to know which is the right thing to do if we are to claim a full understanding.

The way to check which is right is to determine whether there is anything at all special about the measurement process itself. Does it change the world and stop us from adding together quantum amplitudes or rather is measurement part of a vast complex web of possibilities that remain forever in coherent superposition? As human beings we might be tempted to think that measuring something now ('yes' or 'no' say) irrevocably changes the future and if that were true then no future measurement could occur via both the 'yes' and 'no' routes. But it is far from clear that this is the case because it seems that there is always a chance to find the Universe in a future state which can be arrived at via either the 'yes' or 'no' routes. For those states, the laws of quantum physics, taken literally, leave us with no option but to compute the probability of their manifestation by summing over both the 'yes' and 'no' routes. Weird though this may seem, it is no more weird than the summing over histories that we have been performing throughout this book. All that is happening is that we are taking the idea so seriously that we are prepared to do it even at the level of human beings and their actions. From this point of view there is no 'measurement problem'. It is only when we insist that the act of measuring 'yes' or 'no' really changes the nature of things that we run into a problem, because it is then incumbent upon us to explain what it is that triggers the change and breaks the quantum coherence.

The approach to quantum mechanics that we have been discussing, which rejects the idea that Nature goes about choosing a particular version of reality every time someone (or something) 'makes a measurement', forms the basis of what is often referred to as the 'many worlds' interpretation. It is very appealing because it is the logical consequence of taking the laws that govern the behaviour of elementary particles seriously enough to use them to describe all phenomena. But the implications are striking, for we are to imagine that the Universe is really a coherent superposition of all of the possible things that can happen and the world as we perceive it (with its apparently concrete reality) arises only because we are fooled into thinking that coherence is lost every time we 'measure' something. In other words, my conscious perception of the world is fashioned because the alternative (potentially interfering) histories are highly unlikely to lead to the same 'now' and that means quantum interference is negligible.

If measurement is not really destroying quantum coherence then, in a sense, we live out our lives inside one giant Feynman diagram and our predisposition to think that definite things are happening is really a consequence of our crude perceptions of the world. It really is conceivable that, at some time in our future, something can happen to us which requires that, in the past, we did two mutually opposite things. Clearly, the effect is subtle because 'getting the job' and 'not getting the job' makes a big difference to our lives and one cannot easily imagine a scenario where they lead to identical future Universes (remember, we should only add amplitudes that lead to identical outcomes). So in that case, getting and not getting the job do not interfere much with each other and our perception of the world is as if one thing has happened and not the other. However, things become more ambiguous the less dramatic the two alternative scenarios are and, as we have seen, for interactions involving small numbers of particles summing over the different possibilities is absolutely necessary. The large numbers of particles involved in everyday life mean that two substantially different configurations of atoms at some time (e.g. getting the job or not) are simply very unlikely to lead to significantly interfering contributions to some future scenario. In turn, that means we can go ahead and pretend that the world has changed irrevocably as a result of a measurement, even when nothing of the sort has actually happened.

But these musings are not of pressing importance when it comes to the serious business of computing the probability that something will happen when we actually carry out an experiment. For that, we know the rules and we can implement them without any problems. But that happy circumstance may change one day – for now it is the case that questions about how our past might influence the future through quantum interference simply haven't been accessible to experiment. The extent to which meditations on the 'true nature' of the world (or worlds) described by quantum theory can detract from scientific progress is nicely encapsulated in the position taken by the 'shut up and calculate' school of physics, which deftly dismisses any attempt to talk about the reality of things.

Anti-matter

Back in this world, Figure 10.3 shows another way that two electrons can scatter off each other. One of the incoming electrons hops from A to X, whereupon it emits a photon. So far so good but now the electron heads backwards in time to Y where it absorbs another

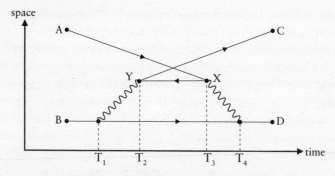

Figure 10.3. Anti-matter . . . or an electron travelling backwards in time.

photon and thence it heads into the future, where it might be eventually detected at C. This diagram does not contravene our rules for hopping and branching, because the electron goes about emitting and absorbing photons as prescribed by the theory. It can happen according to the rules and, as the title of the book suggests, if it can happen, then it does. But such behaviour does appear to violate the rules of common sense, because we are entertaining the idea that electrons travel backwards in time. This would make for nice science fiction, but violating the law of cause and effect is no way to build a universe. It would also seem to place quantum theory in direct conflict with Einstein's Theory of Special Relativity.

Remarkably, this particular kind of time travel for subatomic particles is not forbidden, as Dirac realized in 1928. We can see a hint that all may not be quite as defective as it seems if we reinterpret the goings-on in Figure 10.3 from our 'forwards in time' perspective. We are to track events from left to right in the figure. Let's start at time $T = 0$, where there is a world of just two electrons located at A and B. We continue with a world containing just two electrons until time T_1, whereupon the lower electron emits a photon; between times T_1 and T_2 the world now contains two electrons plus one photon. At time T_2, the photon dies and is replaced by an electron (which will end up at C) and a second particle (which will end up at X). We hesitate to call the second particle an electron because it is 'an electron travelling back in time'. The question is, what does an electron that is travelling back in time look like from the point of view of someone (like you) travelling forwards in time?

To answer this, let's imagine shooting some video footage of an electron as it travels in the vicinity of a magnet, as illustrated in Figure 10.4. Providing that the electron isn't travelling too fast,[5] it will typically travel around in a circle. That electrons can be deflected

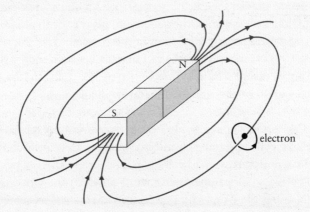

Figure 10.4. An electron, circling near a magnet.

by a magnet is, as we have said before, the basic idea behind the construction of old-fashioned CRT television sets and, more glamorously, particle accelerators, including the Large Hadron Collider. Now imagine that we take the video footage and play it backwards. This is what 'an electron going backwards in time' would look like from our 'forwards in time' perspective. We'd now see the 'backwards in time electron' circle in the opposite direction as the movie advances. From a physicist's perspective, the backwards in time video will look exactly like a forwards in time video shot using a particle which is in every way identical to an electron except that the particle appears to carry positive electric charge. Now we have the answer to our question: electrons travelling backwards in time

5. This is a technical point, to ensure that the electron feels roughly the same sized magnetic force as it moves around.

would appear, to us, as 'electrons of positive charge'. Thus, if electrons do actually travel back in time then we expect to encounter them as 'electrons of positive charge'.

Such particles do exist and they are called 'positrons'. They were introduced by Dirac in early 1931 to solve a problem with his quantum mechanical equation for the electron – namely that the equation appeared to predict the existence of particles with negative energy. Later, Dirac gave a wonderful insight into his way of thinking, and in particular his strong conviction in the correctness of his mathematics: 'I was reconciled to the fact that the negative energy states could not be excluded from the mathematical theory, and so I thought, let us try to find a physical explanation for them.'

Just over a year later, and apparently unaware of Dirac's prediction, Carl Anderson saw some strange tracks in his experimental apparatus while observing cosmic ray particles. His conclusion was that, 'It seems necessary to call upon a positively charged particle having a mass comparable with that of an electron.' Once again, this illustrates the wonderful power of mathematical reasoning. In order to make sense of a piece of mathematics, Dirac introduced the concept of a new particle – the positron – and a few months later it was found, produced in high-energy cosmic ray collisions. The positron is our first encounter with that staple of science fiction, anti-matter.

Armed with this interpretation of time-travelling electrons as positrons, we can finish off the job of explaining Figure 10.3. We are to say that when the photon reaches Y at time T_2 it splits into an electron and a positron. Each head forwards in time until time T_3 when the positron from Y reaches X, whereupon it fuses with the original upper electron to produce a second photon. This photon propagates to time T_4, when it gets absorbed by the lower electron.

This might all sound a little far fetched: anti-particles have emerged from our theory because we are permitting particles to travel backwards in time. Our hopping and branching rules allow particles to hop both forwards and backwards in time, and despite our possible prejudice that this must be disallowed, it turns out that we do not, indeed must not, prevent them from doing so. Quite ironically, it turns out that if we did *not* allow particles to hop back in time then we would have a violation of the law of cause and effect. This is odd, because it seems as if things ought to be the other way around.

That things work out just fine is not an accident and it hints at a deeper mathematical structure. In fact, you may have got the feeling on reading this chapter that the branching and hopping rules all seem rather arbitrary. Could we make up some new branching rules and tweak the hopping rules then explore the consequences? Well, if we did that we would almost certainly build a bad theory – one that would violate the law of cause and effect, for example. Quantum Field Theory (QFT) is the name for the deeper mathematical structure that underpins the hopping and branching rules and it is remarkable for being the *only* way to build a quantum theory of tiny particles that also respects the Theory of Special Relativity. Armed with the apparatus of QFT, the hopping and branching rules are fixed and we lose the freedom to choose. This is a very important result for those in pursuit of fundamental laws because using 'symmetry' to remove choice creates the impression that the Universe simply has to be 'like this' and that feels like progress in understanding. We used the word 'symmetry' here and it is appropriate, because Einstein's theories can be viewed as imposing symmetry restrictions on the structure of space and time. Other 'symmetries' further constrain the hopping and branching rules, and we shall briefly encounter those in the next chapter.

Before leaving QED, we have a final loose end to tie up. If you recall, the opening talk of the Shelter Island meeting concerned the Lamb shift, an anomaly in the hydrogen spectrum that could not be explained by the quantum theory of Heisenberg and Schrödinger. Within a week of the meeting, Hans Bethe produced a first, approximate, calculation of the answer. Figure 10.5 illustrates the QED way to picture a hydrogen atom. The electromagnetic interaction that keeps the proton and the electron bound together can be represented by a series of Feynman diagrams of increasing complexity, just as we saw for the case of two electrons interacting together in

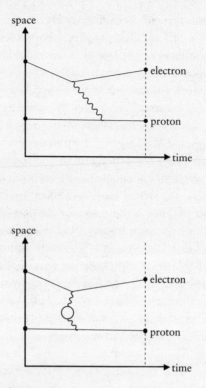

Figure 10.5. The hydrogen atom.

Figure 10.1. We've sketched two of the simplest possible diagrams in Figure 10.5. Pre-QED, the calculations of the electron energy levels included only the top diagram in the figure, which captures the physics of an electron that is trapped within the potential well generated by the proton. But, as we've discovered, there are many other things that can happen during the interaction. The second diagram in Figure 10.5 shows the photon briefly fluctuating into an electron–positron pair, and this process must also be included in a calculation of the possible energy levels of the electron. This, and many other diagrams, enter the calculation as small corrections to the main result.[6]

6. The one first anticipated by Bohr back in 1913.

Bethe correctly included the important effects from 'one-loop' diagrams, like that in the figure, and found that they slightly shift the energy levels and therefore the detail in the observed spectrum of light. His result was in accord with Lamb's measurement. QED, in other words, forces us to imagine a hydrogen atom as a fizzing cacophony of subatomic particles popping in and out of existence. The Lamb shift was humankind's first direct encounter with these ethereal quantum fluctuations.

It did not take long for two other Shelter Island attendees, Richard Feynman and Julian Schwinger, to pick up the baton and, within a couple of years, QED had been developed into the theory we know today – the prototypical quantum field theory and exemplar for the soon-to-be-discovered theories describing the weak and strong interactions. For their efforts, Feynman, Schwinger and the Japanese physicist Sin-Itiro Tomonaga received the 1965 Nobel Prize 'for their fundamental work in quantum electrodynamics, with deep-ploughing consequences for the physics of elementary particles'. It is to those deep-ploughing consequences that we now turn.

11.   *Empty Space Isn't Empty*

Not everything in the world stems from the interactions between electrically charged particles. QED does not explain the 'strong nuclear' processes that bind quarks together inside protons and neutrons or the 'weak nuclear' processes that keep our Sun burning. We can't write a book about the quantum theory of Nature and leave out half of the fundamental forces, so this chapter will make right our omission before delving into empty space itself. As we'll discover, the vacuum is an interesting place, filled with possibilities and obstacles for particles to navigate.

The first thing to emphasize is that the weak and strong nuclear forces are described by exactly the same quantum field theoretic approach that we have described for QED. It is in this sense that the work of Feynman, Schwinger and Tomonaga had deep-ploughing consequences. Taken as a whole, the theory of these three forces is known, rather unassumingly, as the Standard Model of particle physics. As we write, the Standard Model is being tested to breaking point by the largest and most sophisticated machine ever assembled: CERN's Large Hadron Collider (LHC). 'Breaking point' is right because, in the absence of something hitherto undiscovered, the Standard Model stops making meaningful predictions at the energies involved in the collisions of almost light-speed protons at the LHC. In the language of this book, the quantum rules start to generate clock faces with hands longer than 1, which means that certain processes involving the weak nuclear force are predicted to occur with a probability greater than 100%. This is clearly nonsense and it implies that the LHC is destined to discover something new. The challenge is to identify it among the hundreds of millions of proton collisions generated every second a hundred metres below the foothills of the Jura Mountains.

The Standard Model does contain a cure to the malaise of the dysfunctional probabilities and that goes by the name of the 'Higgs mechanism'. If it is correct, then the LHC should observe one more particle of Nature, the Higgs boson, and with it trigger a profound shift in our view of what constitutes empty space. We'll get to the Higgs mechanism later in the chapter, but first we should provide a short introduction to the triumphant yet creaking Standard Model.

The Standard Model of Particle Physics

In Figure 11.1 we've listed all of the known particles. These are the building blocks of our Universe, as far as we know at the time of writing this book, but we expect that there are some more – perhaps we will see a Higgs boson or perhaps a new particle associated with the abundant but enigmatic Dark Matter that seems necessary to explain the Universe at large. Or perhaps the supersymmetric particles anticipated by string theory or maybe the Kaluza-Klein excitations characteristic of extra dimensions in space or techniquarks or leptoquarks or . . . theoretical speculation is rife and it is the duty of those carrying out experiments at the LHC to narrow down the field, rule out the wrong theories and point the way forward.

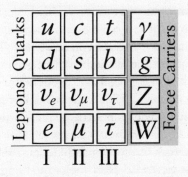

Figure 11.1. The particles of Nature.

Everything you can see and touch; every inanimate machine, every living thing, every rock and every human being on planet Earth, every planet and every star in every one of the 350 billion galaxies in the observable Universe is built out of the particles in the first column of four. You are an arrangement of just three: the up and down quarks and the electron. The quarks make up your atomic nuclei and, as we've seen, the electrons do the chemistry. The remaining particle in the first column, called the electron neutrino, may be less familiar to you but there are around 60 billion of them streaming through every square centimetre of your body every second from the Sun. They mostly sail straight through you and the entire Earth, unimpeded, which is why you've never seen or felt one. But they do, as we will see in a moment, play a crucial role in the processes that power the Sun and, because of that, they make your life possible.

These four particles form a set known as the first generation of matter and, together with the four fundamental forces of Nature, they appear to be all that is needed to build a Universe. For reasons that we do not yet understand, Nature has chosen to provide us with two further generations – clones of the first except that the particles are more massive. They are represented in the second and third columns in Figure 11.1. The top quark in particular is much more massive than the other fundamental particles. It was discovered at the Tevatron accelerator at Fermilab near Chicago in 1995, and its mass has been measured to be over 180 times the mass of a proton. Why the top quark is such a monster, while being point-like in the same way that an electron is point-like, is a mystery. Although these extra generations of matter do not play a direct role in the ordinary affairs of the Universe they do seem to have been crucial players in the moments just after the Big Bang . . . but that is another story.

Also shown in Figure 11.1, in the column on the right, are the force-carrying particles. Gravity is not represented in the table because we do not have a quantum theory of gravity that sits comfortably within the framework of the Standard Model. This isn't to say that there isn't one; string theory is an attempt to bring gravity into the fold but, to date, it has met with limited success. Because gravity is so feeble it plays no significant role in particle physics experiments and for that pragmatic reason we'll say no more about it. We learnt in the last chapter how the photon is responsible for mediating the electromagnetic force between electrically charged particles and that its behaviour was determined by specifying a new branching rule. The W and Z particles do the corresponding job for the weak force while the gluons mediate the strong force. The primary differences between the quantum descriptions of the forces arise because the branching rules are different. It is (almost) that simple and we have drawn some of the new branching rules in Figure 11.2. The similarity with QED makes it easy to appreciate the basics of the weak and strong forces; we just need to know what the branching rules are and then we can draw Feynman diagrams like we did for QED in the last chapter. Fortunately, changing the branching rules makes all the difference to the physical world.

If this were a particle physics textbook, we might proceed to outline the branching rules for each of the processes in Figure 11.2, and many more besides. These rules, known as the Feynman rules, would then allow you, or a computer program, to calculate the probability for some process or other, just as we outlined in the last chapter for QED. The rules capture something essential about the world and it is delightful that they can be summarized in a few simple pictures and rules. But this isn't a particle physics textbook, so we'll instead focus on the top-right diagram, because it is a particularly important branching rule for life on Earth. It shows an up quark branching into a down quark by emitting a W particle and this behaviour is exploited to dramatic effect within the core of the Sun.

The Sun is a gaseous sea of protons, neutrons, electrons and photons with the volume of a million earths, collapsing under its own gravity. The vicious compression heats the solar core to 15 million degrees and at these temperatures the protons begin to fuse together to form helium nuclei. The fusion process releases energy, which increases the pressure on the outer layers of the star, balancing the

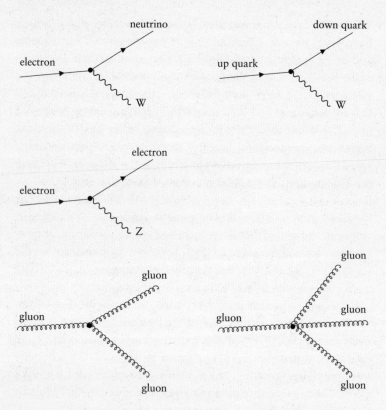

Figure 11.2. Some of the branching rules for the weak and strong forces.

inward pull of gravity. We'll dig deeper into this precarious balancing act in the epilogue, but for now we want to understand what it means to say that 'the protons begin to fuse together'.

This sounds simple enough, but the precise mechanism for fusion in the Sun's core was a source of great scientific debate during the 1920s and 30s. The British scientist Arthur Eddington was the first to propose that the energy source of the Sun is nuclear fusion, but it was quickly pointed out that the temperatures were apparently far too low for the process to occur given the then-known laws of physics. Eddington stuck to his guns, however, issuing the famous retort: 'The helium which we handle must have been put together at some time and some place. We do not argue with the critic who urges that the stars are not hot enough for this process; we tell him to go and find a hotter place.'

The problem is that when two fast-moving protons in the core of the Sun get close, they repel each other as a result of the electro-magnetic force (or, in the language of QED, by photon exchange). To fuse together they need to get so close that they are effectively overlapping and, as Eddington and his colleagues well knew, the solar protons are not moving fast enough (because the Sun is not hot enough) to overcome their mutual electromagnetic repulsion.

The answer to this conundrum is that the W particle steps in to save the day. In a stroke, one of the protons in the collision can con-vert into a neutron by converting one of its up quarks into a down quark, as specified by the branching rule in Figure 11.2. Now the newly formed neutron and remaining proton can get very close, because the neutron carries no electric charge. In the language of quantum field theory, this means there is no photon exchange to push the neutron and proton apart. Freed from the electromagnetic repulsion, the proton and neutron can fuse together (as a result of the strong force) to make a deuteron and this quickly leads to helium formation, releasing life-giving energy for the star. The pro-cess is illustrated in Figure 11.3, which also indicates that the W particle does not stick around for very long; instead it branches into a positron and a neutrino – this is the source of those very same neu-trinos that pass through your body in such vast numbers. Eddington's belligerent defence of fusion as the power source of the Sun was correct, although he could have had no inkling of the solution. The all-important W particle, along with its partner the Z, was eventu-ally discovered at CERN in the 1980s.

To conclude our brief survey of the Standard Model, we turn to the strong force. The branching rules are such that only quarks can branch into gluons. In fact they are much more likely to do that than they are

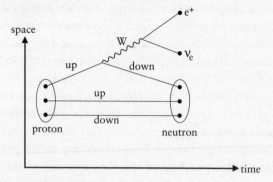

Figure 11.3. Proton conversion into a neutron by weak decay, with the emission of a positron and a neutrino. Without this, the Sun would not burn.

to do anything else. This predisposition to emit gluons is why the strong force is so named and it is the reason why gluon branching is able to defeat the repulsive electromagnetic force that would otherwise cause the positively charged proton to explode. Fortunately, the strong force cannot reach very far. Gluons tend not to travel beyond around 1 femtometre (10^{-15} m) before they branch again. The reason why gluons are so short-ranging in their influence, whilst photons can reach across the Universe, is down to the fact that gluons can also branch into other gluons, as illustrated in the final two pictures in Figure 11.2. This trick of the gluons makes the strong force very different from the electromagnetic force, and effectively confines its actions to the interior of the atomic nucleus. Photons have no such self-branching and that is very fortunate, for if they did you wouldn't be able to see the world in front of your eyes because the photons streaming towards you would scatter off those travelling across your line of sight. It is one of the wonders of life that we can see anything at all, and a vivid reminder that photons very rarely interact with each other.

We have not explained where all of these new rules come from, nor have we explained why the Universe contains the particles that it does. There is a good reason for this: we don't really know the answers to either of these questions. The particles that make up our Universe – the electrons, neutrinos and quarks – are the primary actors in the unfolding cosmic drama, but to date we have no compelling way to explain why the cast should line up as it does.

What is true, however, is that once we have the list of particles then the way they interact with each other, as prescribed by the branching rules, is something we can partially anticipate. The branching rules are not something that physicists have just conjured from nowhere – they are in all cases anticipated on the grounds that the theory describing the particle interactions should be a Quantum Field Theory supplemented with something called gauge symmetry. To discuss the origin of the branching rules would take us too far outside the main line of this book – but we do want to reiterate that the essential rules are very simple: the Universe is built from particles that move around and interact according to a handful of hopping and branching rules. We can take those rules and use them to compute the probability that 'something' *does* happen by adding together a bunch of clocks – there being one clock for each and every way that the 'something' *can* happen.

The Origin of Mass

By introducing the idea that particles can branch as well as hop we have entered into the domain of Quantum Field Theory, and hopping and branching is, to a large extent, all there is to it. We have, however, been rather negligent in our discussion of mass, for the good reason that we have been saving the best until last.

Modern-day particle physics aims to provide an answer to the question 'what is the origin of mass?' and it does so with the help of a beautiful and subtle piece of physics and a new particle – new in the sense that we have not yet really encountered it in this book, and new in the sense that nobody on Earth has ever encountered one 'face to face'. The particle is named the Higgs boson, and the LHC has it firmly in its sights. At the time of writing this book in September 2011, there have been tantalizing glimpses, perhaps, of a Higgs-like object in the LHC data, but there are simply not enough events[1] to decide one way or the other. It may well be that, as you read this book, the situation has changed and the Higgs is a reality. Or it may be that the interesting signals have vanished under further scrutiny. The particularly exciting thing about the question of the origin of mass is that the answer is extremely interesting beyond the obvious desire to know what mass is. Let us now explain that rather cryptic and offensively constructed sentence in more detail.

When we discussed photons and electrons in QED, we introduced the hopping rule for each and said that they are different – we used the symbol $P(A,B)$ for the rule associated with an electron that hops from A to B and the symbol $L(A,B)$ for the corresponding rule for a photon. It is time now to investigate why the rule is different in the two cases. There is a difference because electrons come in two different types (as we know, they 'spin' in one of two different ways), whilst photons come in three different types, but that particular difference will not concern us here. There is another difference, however, because the electron has mass while the photon does not – this is what we want to explore.

Figure 11.4 illustrates one way that we are allowed to think about the propagation of a massive particle. The figure shows a particle hopping from A to B in stages. It goes from A to point 1, from point 1 to point 2 and so on until it finally hops from point 6 to B. What is interesting is that, when written in this way, the rule for each hop is the rule for a particle with *zero* mass, but with one important caveat: every time the particle changes direction we are to apply a new shrinking rule, with the amount of shrinking inversely proportional to the mass of the particle we are describing. This means that, at each kink, the clocks of heavy particles receive less shrinking than the clocks of lighter particles. It is important to emphasize that this

---

1. An 'event' is a single proton–proton collision. Because fundamental physics is a counting game (it works with probabilities) it is necessary to keep colliding protons in order to accumulate a sufficient number of those very rare events in which a Higgs particle is produced. What constitutes a sufficient number depends on how skilful the experimenters are at confidently eliminating fake signals.

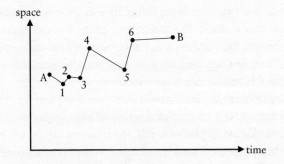

Figure 11.4. A massive particle travelling from A to B.

isn't an ad hoc prescription. Both the zig-zag and the shrink emerge directly from the Feynman rules for the propagation of a massive particle, without any further assumptions.[2] Figure 11.4 shows just one way that our heavy particle can get from A to B, i.e. via six kinks and six shrinkage factors. To get the final clock associated with a massive particle hopping from A to B we must, as always, add together the infinity of clocks associated with all of the possible ways that the particle can zig-zag its way from A to B. The simplest route is the direct one, with no kinks, but routes with huge numbers of kinks need to be considered too.

For particles with zero mass the shrinkage factor associated with each kink is a killer, because it is infinite. In other words, we are to shrink the clock to zero after the first kink. The only route that matters for massless particles is therefore the direct route – there is simply no clock associated with any other route. This is exactly what we would expect: it means that we can use the hopping rule for massless particles when the particle is massless. However, for particles with non-zero mass, kinks are allowed, although if the particle

2. Our ability to think of a massive particle as a massless particle supplemented with a 'kink' rule comes from the fact that $P(A,B) = L(A,B) + L(A,1)L(1,B)S + L(A,1)L(1,2)L(2,B)S^2 + L(A,1)L(1,2)L(2,3)L(3,B)S^3 + \ldots$, where S is the shrinkage factor associated with a kink and it is understood that we should sum over all possible intermediate points 1, 2, 3 etc.

is very light then the shrinking factor imposes a severe penalty on paths with many kinks. The most likely paths are therefore those with very few kinks. Conversely, heavy particles do not get penalized much when they kink, and so they tend to be described by paths with lots of zig-zagging. This seems to suggest that heavy particles really ought to be thought of as massless particles that zig-zag their way from A to B. The amount of zig-zagging is what we identify as 'mass'.

This is all rather nice, for we have a new way to think about massive particles. Figure 11.5 illustrates the propagation from A to B of three different particles of increasing mass. In each case, the rule associated with each 'zig' or 'zag' of the path is the same as that for a massless particle, and for every kink we are to pay a 'the clock must be shrunk' penalty. We should not get overly excited yet because we have not really explained anything fundamental. All we have done is to replace the word 'mass' with the words 'tendency to zig-zag'. We are allowed to do this because they are mathematically equivalent descriptions of the propagation of a massive particle. But even so, it feels like an interesting thing and, as we shall now discover, it may turn out to be rather more than just a mathematical curiosity.

We are now going to move into the realm of speculation –

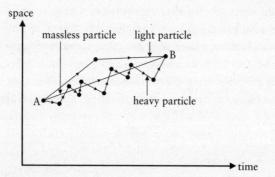

Figure 11.5. Particles of increasing mass propagating from A to B. The more massive a particle is the more it zig-zags.

although by the time you read this book the theory we are about to outline may have been verified. The LHC is currently busy colliding protons together with a combined energy of 7 TeV. 'TeV' stands for Tera electron volts, which corresponds to the amount of energy an electron would have if it were accelerated through a potential difference of 7 million million volts. To get a sense of how much energy this is, it's roughly the energy that subatomic particles would have had about a trillionth of a second after the Big Bang and it is enough energy to conjure out of thin air a mass equal to 7,000 protons (via Einstein's $E = mc^2$). And this is only half the design energy; if needed, the LHC has more gas in the tank.

One of the primary reasons that eighty-five countries around the world have come together to build and operate this vast, audacious experiment is to hunt for the mechanism that is responsible for generating the masses of the fundamental particles. The most widely accepted theory for the origin of mass works by providing an explanation for the zig-zagging: it posits a new fundamental particle that the other particles 'bump into' on their way through the Universe.

That particle is the Higgs boson. According to the Standard Model, without a Higgs the fundamental particles would hop from place to place without any zig-zagging and the Universe would be a very different place. But if we fill empty space with Higgs particles then they can act to deflect particles, making them zig-zag and, as we have just learnt, that leads to the emergence of 'mass'. It is rather like trying to walk through a crowded pub – one gets buffeted from side-to-side and ends up taking a zig-zag path towards the bar.

The Higgs mechanism is named after Edinburgh theorist Peter Higgs and it was introduced into particle physics in 1964. The idea was obviously very ripe because several people came up with the idea at the same time – Higgs of course, and also Robert Brout and François Englert working in Brussels and Gerald Guralnik, Carl Hagan and Tom Kibble in London. Their work was itself built on the earlier efforts of many others, including Heisenberg, Yoichiro Nambu, Jeffrey Goldstone, Philip Anderson and Weinberg. The full realization of the idea, for which Sheldon Glashow, Abdus Salam and Weinberg received the Nobel Prize in 1979, is no less than the Standard Model of particle physics. The idea is simple enough – empty space is not empty, and this leads to zig-zagging and therefore mass. But clearly we have some more explaining to do. How can it be that empty space is jammed full of Higgs particles – wouldn't we notice this in our everyday lives, and how did this strange state of affairs come about in the first place? It certainly sounds like a rather extravagant proposition. We have also not explained how it can be that some particles (like photons) have no mass while others (like W bosons and top quarks) weigh in with masses comparable to that of an atom of silver or gold.

The second question is easier to answer than the first, at least superficially. Particles only ever interact with each other through a branching rule and Higgs particles are no different in that regard. The branching rule for a top quark includes the possibility that it can couple to a Higgs particle, and the corresponding shrinking of the clock (remember all branching rules come with a shrinking factor) is much less than it is in the case of the lighter quarks. That is 'why' a top quark is so much heavier than an up quark. This doesn't explain why the branching rule is what it is, of course. The current answer to that is the disappointing 'because it is'. It's on the same footing as the question 'Why are there three generations of particles?' or 'Why is gravity so weak?' Similarly, photons do not have any branching rule that couples them to Higgs particles and as a result they do not interact with them. This, in turn, means that they do not zig-zag and have no mass. Although we have passed the buck to some extent, this does feel like some kind of an explanation, and it is certainly true that if we can detect Higgs particles at the LHC and check that they couple to the other particles in this manner then we can legitimately claim to have gained a rather thrilling insight into the way Nature works.

The first of our outstanding questions is a little trickier to explain – namely, how can it be that empty space is full of Higgs particles? To get warmed up, we need to be very clear about one thing: quantum physics implies that there is no such thing as empty space. In fact, what we call 'empty space' is really a seething maelstrom of subatomic particles and there is no way to sweep them away and clean it up. Once we realize that, it becomes much less of an intellectual challenge to accept that empty space might be full of Higgs particles. But let's take one step at a time.

You might imagine a tiny region of deep outer space, a lonely corner of the Universe millions of light years from a galaxy. As time passes it is impossible to prevent particles from appearing and then disappearing out of nothing. Why? It is because the process of the creation and annihilation of particle–anti-particle pairs is allowed by the rules. An example can be found in the lower diagram in Figure 10.5: imagine stripping away everything except for the electron loop – the diagram then corresponds to an electron–positron pair spontaneously appearing from nothing and then disappearing back into nothing. Because drawing a loop does not violate any of the rules of QED we must acknowledge that it is a real possibility; remember, everything that can happen does happen. This particular possibility is just one of an infinite number of ways that empty space can fizz and pop, and because we live in a quantum universe the correct thing to do is to add all the possibilities together. The vacuum, in other words, has an incredibly rich structure, made up out of all the possible ways that particles can pop in and out of existence.

That last paragraph introduced the idea that the vacuum is not empty, but we painted a rather democratic picture in which all of the elementary particles play a role. What is it about the Higgs particle that makes it special? If the vacuum were nothing other than a seething broth of matter–antimatter creation and annihilation, then all of the elementary particles would continue to have zero mass – the quantum loops themselves are not capable of delivering it.[3] Instead, we need to populate the vacuum with something different, and this is where the bath of Higgs particles enters. Peter Higgs

3. This is a subtle point and derives from the 'gauge symmetry', which underwrites the hopping and branching rules of the elementary particles.

simply stipulated that empty space is packed with Higgs particles[4] and didn't feel obliged to offer any deep explanation as to why. The Higgs particles in the vacuum provide the zig-zag mechanism and they are working overtime by interacting with each and every massive particle in the Universe, selectively retarding their motion to create mass. The net result of the interactions between ordinary matter and a vacuum full of Higgs particles is that the world goes from being a structureless place to a diverse and wonderful living world of stars, galaxies and people.

The big question of course is where those Higgs particles came from in the first place? The answer isn't really known, but it is thought that they are the remnants of what is known as a phase transition that occurred sometime shortly after the Big Bang. If you are patient and watch the glass in your window as the temperature falls on a winter's evening, you'll see the structured beauty of ice crystals emerge as if by magic from the water vapour in the night air. The transition from water vapour to ice on cold glass is a phase transition – water molecules rearranging themselves into ice crystals; the spontaneous breaking of the symmetry of a formless vapour cloud triggered by a drop in temperature. Ice crystals form because it is energetically more favourable to do so. Just as a ball rolls down the side of a mountain to take up a lower energy in a valley, or electrons rearrange themselves around atomic nuclei to form the bonds that hold molecules together, so the sculpted beauty of a snowflake is a lower energy configuration of water molecules than a formless cloud of vapour.

We think that a similar thing happened early on in the Universe's history. As the hot gas of particles that was the nascent Universe expanded and cooled, so it transpired that a Higgs-free vacuum was energetically disfavoured and a vacuum filled with Higgs particles was the natural state. The process really is similar to the way that water condenses into droplets or ice forms on a cold pane of glass. The spontaneous appearance of water droplets when they con-

4. He was far too modest to call them by that name.

dense on a pane of glass creates the impression that those droplets simply emerged out of 'nothing'. Similarly for the Higgs, in the hot stages just after the Big Bang the vacuum is seething with the fleeting quantum fluctuations (those loops in our Feynman diagrams), as particles and anti-particles pop out of nothing before disappearing again. However, something radical happens as the Universe cools and suddenly, out of nothing, just as the water drops appear on the glass, a 'condensate' of Higgs particles emerges, all held together by their mutual interactions in an ephemeral suspension through which the other particles propagate.

The idea that the vacuum is filled with material suggests that we, and everything else in the Universe, live out our lives inside a giant condensate that emerged as the Universe cooled down, just as the morning dew emerges with the dawn. Lest we think that the vacuum is populated merely as a result of Higgs particle condensation, we should also remark that there is even more to the vacuum than this. As the Universe cooled still further, quarks and gluons also condensed to produce what are, naturally enough, known as quark and gluon condensates. The existence of these is well established by experiments, and they play a very important role in our understanding of the strong nuclear force. In fact, it is this condensation that gives rise to the vast majority of the mass of protons and neutrons. The Higgs vacuum is, however, responsible for generating the observed masses for the elementary particles – the quarks, electrons, muons, taus and W and Z particles. The quark condensate kicks in to explain what happens when a cluster of quarks binds together to make a proton or a neutron. Interestingly, whilst the Higgs mechanism is relatively unimportant when it comes to explaining the mass of protons, neutrons and the heavier atomic nuclei, the converse is true when it comes to explaining the mass of the W and Z particles. For them, quark and gluon condensation would generate a mass of around 1 GeV in the absence of a Higgs particle, but their experimentally measured masses are closer to 100 times this. The LHC was designed to operate in the energy domain of the W and Z, where it can explore the mechanism responsible for

their comparatively large masses. Whether that is the eagerly anticipated Higgs particle, or something hitherto undreamt of, only time
and particle collisions will tell.

To put some rather surprising numbers on all of this, the energy
stored up within 1 cubic metre of empty space as a result of quark
and gluon condensation is a staggering 10^{35} joules, and the energy
due to Higgs condensation is 100 times larger than this. Together,
that's the total amount of energy our Sun produces in 1,000 years.
To be precise, this is 'negative' energy, because the vacuum is lower
in energy than a Universe containing no particles at all. The negative energy arises because of the binding energy associated with the
formation of the condensates, and is not by itself mysterious. It is
no more glamorous than the fact that, in order to boil water (and
reverse the phase transition from vapour to liquid), you have to put
energy in.

What is mysterious, however, is that such a large and negative
energy density in every square metre of empty space should, if taken
at face value, generate a devastating expansion of the Universe such
that no stars or people would ever form. The Universe would literally have blown itself apart moments after the Big Bang. This is what
happens if we take the predictions for vacuum condensation from
particle physics and plug them directly into Einstein's equations
for gravity, applied to the Universe at large. This heinous conundrum goes by the name of the cosmological constant problem and
it remains one of the central problems in fundamental physics. Certainly it suggests that we should be very careful before claiming to
really understand the nature of the vacuum and/or gravity. There is
something absolutely fundamental that we do not yet understand.

With that sentence, we come to the end of our story because we've
reached the edge of our knowledge. The domain of the known is not
the arena of the research scientist. Quantum theory, as we observed
at the beginning of this book, has a reputation for difficulty and
downright contrary weirdness, exerting as it does a rather liberal
grip on the behaviour of the particles of matter. But everything
we've described, with the exception of this final chapter, is known

and well understood. Following evidence rather than common sense, we are led to a theory that is manifestly able to describe a vast range of phenomena, from the sharp rainbows emitted by hot atoms to fusion within stars. Putting the theory to use led to the most important technological breakthrough of the twentieth century – the transistor – a device whose operation would be inexplicable without a quantum view of the world.

But quantum theory is far more than a mere explanatory triumph. In the forced marriage between quantum theory and relativity, anti-matter emerged as a theoretical necessity and was duly discovered. Spin, the fundamental property of subatomic particles that underpins the stability of atoms, was likewise a theoretical prediction required for the consistency of the theory. And now, in the second quantum century, the Large Hadron Collider voyages into the unknown to explore the vacuum itself. This is scientific progress; the gradual and careful construction of a legacy of explanation and prediction that changes the way we live. And this is what sets science apart from everything else. It isn't simply another point of view – it reveals a reality that would be impossible to imagine, even for the possessor of the most tortured and surreal imagination. Science is the investigation of the real, and if the real seems surreal then so be it. There is no better demonstration of the power of the scientific method than quantum theory. Nobody could have come up with it without the most meticulous and detailed experiments, and the theoretical physicists who built it were able to suspend and jettison their deeply held and comforting beliefs in order to explain the evidence before them. Perhaps the conundrum of the vacuum energy signals a new quantum journey, perhaps the LHC will provide new and inexplicable data, and perhaps everything in this book will turn out to be an approximation to a much deeper picture – the exciting journey to understand our Quantum Universe continues.

When we began thinking about writing this book, we spent some time debating how to end it. We wanted to find a demonstration of the intellectual and practical power of quantum theory that would convince even the most sceptical reader that science really does describe, in exquisite detail, the workings of the world. We both agreed that there is such a demonstration, although it does involve some algebra – we have done our best to make it possible to follow the reasoning without scrutinizing the equations, but it does come with that warning. So, our book ends here, unless you want a little bit more: the most spectacular demonstration, we think, of the power of quantum theory. Good luck, and enjoy the ride.

Epilogue: the Death of Stars

When stars die, many end up as super-dense balls of nuclear matter intermingled with a sea of electrons, known as 'white dwarves'. This will be the fate of our Sun when it runs out of nuclear fuel in around 5 billion years time. It will also be the fate of over 95% of the stars in our galaxy. Using nothing more than a pen, paper and a little thought, we can calculate the largest possible mass of these stars. The calculation, first performed by Subrahmanyan Chandrasekhar in 1930, uses quantum theory and relativity to make two very clear predictions. Firstly, that there should even be such a thing as a white dwarf star – a ball of matter held up against the crushing force of its own gravity by the Pauli Exclusion Principle. Secondly, that if we turn our attention from the piece of paper with our theoretical scribbles on it and gaze into the night sky then we should *never* see a white dwarf with a mass greater than 1.4 times the mass of our Sun. These are spectacularly audacious predictions.

Today, astronomers have catalogued around 10,000 white dwarf stars. The majority have masses around 0.6 solar masses, but the largest recorded mass is *just* under 1.4 solar masses. This single number, '1.4', is a triumph of the scientific method. It relies on an understanding of nuclear physics, of quantum physics and of Einstein's Theory of Special Relativity – an interlocking swathe of twentieth-century physics. Calculating it also requires the fundamental constants of Nature we've met in this book. By the end of this chapter, we will learn that the maximum mass is determined by the ratio

$$\left(\frac{hc}{G}\right)^{3/2} \frac{1}{m_p^2}$$

Look carefully at what we just wrote down: it depends on Planck's constant, the speed of light, Newton's gravitational constant and the mass of a proton. How wonderful it is that we should be able to predict the uppermost mass of a dying star using this combination of fundamental constants. The three-way combination of gravity, relativity and the quantum of action appearing in the ratio $(hc/G)^{1/2}$ is called the Planck mass, and when we put the numbers in it works out at approximately 55 micrograms; roughly the mass of a grain of sand. So the Chandrasekhar mass is, rather astonishingly, obtained by contemplating two masses, one the size of a grain of sand and the other the mass of a single proton. From such tiny numbers emerges a new fundamental mass scale in Nature: the mass of a dying star.

We could present a very broad overview of how the Chandrasekhar mass comes about, but instead we'd like to do a little bit more: we'd like to describe the actual calculation because that is what really makes the spine tingle. We'll fall short of actually computing the precise number (1.4 solar masses), but we will get close to it and see how professional physicists go about drawing profound conclusions using a sequence of carefully developed logical steps, invoking well-known physical principles along the way. There will be no leap of faith. Instead, we will keep a cool head and slowly and inexorably be drawn to the most exciting of conclusions.

Our starting point has to be: 'what is a star?' The visible Universe is, to a very good approximation, made up of hydrogen and helium, the two simplest elements formed in the first few minutes after the Big Bang. After around half a billion years of expansion, the Universe was cool enough for slightly denser regions in the gas clouds to start clumping together under their own gravity. These were the seeds of the galaxies, and within them, around smaller clumps, the first stars began to form.

The gas in these first proto-stars became hotter and hotter as they collapsed in on themselves, as anyone who has used a bicycle pump will know, because compressing a gas makes it heat up. When the gas reaches temperatures of around 100,000 degrees, the electrons can no longer be held in orbit around the hydrogen and helium nuclei and the atoms get ripped apart, leaving a hot plasma of bare nuclei and electrons. The hot gas tries to expand outwards and resist further collapse but, for sufficiently massive clumps, gravity wins out. Because protons have positive electric charge they will repel each other but, as the gravitational collapse proceeds and the temperature continues to rise, the protons move faster and faster. Eventually, at a temperature of several million degrees, the protons are moving so fast that they get close enough for the weak nuclear force to take over. When that happens, two protons can react with one another; one of them spontaneously changes into a neutron with the simultaneous emission of a positron and a neutrino (exactly as illustrated in Figure 11.3 on page 202). Freed from the electrical repulsion, the proton and the neutron fuse under the action of the strong nuclear force to make a deuteron. This process releases huge amounts of energy because, just as in the formation of a hydrogen molecule, binding things together releases energy.

The energy release in a single fusion event isn't large by everyday standards. One million proton–proton fusion reactions generate roughly the same amount of energy as the kinetic energy of a mosquito in flight or a 100 watt light-bulb radiates in a nanosecond. But that is huge on atomic scales and, remember, we are talking about the dense heart of a collapsing gas cloud in which there are around 10^{26} protons per cubic centimetre. If all the protons in a cubic centimetre were to fuse into deuterons, 10^{13} joules of energy would be liberated, which is enough to power a small town for one year.

The fusion of two protons into a deuteron is the start of a fusion jamboree. The deuteron itself is eager to fuse with a third proton to make a light version of helium (called helium-3) with the emission of a photon, and those helium nuclei then pair up and fuse into regular helium (called helium-4) with the emission of two protons. At each stage, the fusing together liberates more and more energy. And, just for good measure, the positron, which was emitted right back at the start of the chain, also rapidly fuses with an electron in the surrounding plasma to produce a pair of photons. All of this liberated energy makes for a hot gas of photons, electrons and nuclei that pushes against the in-falling matter and halts any further gravitational collapse. This is a star: nuclear fusion burns up nuclear fuel in the core, and that generates an outward pressure that stabilizes the star against gravitational collapse.

There is, of course, only a finite amount of hydrogen fuel available to burn and, eventually, it will run out. With no more energy released there is no more outward pressure; gravity once again takes control and the star resumes its postponed collapse. If the star is massive enough, the core will heat up to temperatures of around 100 million degrees. At that stage, the helium produced as waste in the hydrogen-burning phase ignites, fusing together to produce carbon and oxygen, and once again the gravitational collapse is temporarily halted.

But what happens if the star is not massive enough to initiate helium fusion? For stars less than about half the mass of our Sun, this is the case, and for them something very dramatic happens. The star heats up as it contracts, but, before the core reaches 100 million degrees, something else halts the collapse. That something is the pressure exerted by electrons due to the fact that they are in the grip of the Pauli Exclusion Principle. As we have learnt, the Pauli principle is crucial to understanding how atoms remain stable, and it underpins the properties of matter. Here is another string to its bow: it explains the existence of compact stars that survive despite the fact that they no longer burn up any nuclear fuel. How does this work?

As the star gets squashed, so the electrons within it get confined to a smaller volume. We can think of an electron in the star in terms of its momentum p and hence its associated de Broglie wavelength, h/p. In particular, the particle can only ever be described by a wave packet that is at least as big as its associated wavelength.[1] This means that, when the star is dense enough, the electrons must be overlapping each other, i.e. we cannot imagine them as being described by

1. Recall from Chapter 5 that particles of definite momentum are in fact described by infinitely long waves and that as we allow for some spread in the momentum so we can start to localize the particle. But this can only go so far and it makes no sense to talk about a particle of a certain wavelength if it is localized to a distance smaller than that wavelength.

isolated wave packets. This in turn means that quantum mechanical effects, and the Pauli principle in particular, are important in describing the electrons. Specifically, they are being squashed together to the point where two electrons are attempting to occupy the same region of space, and we know from the Pauli principle that they resist this. In a dying star, therefore, the electrons avoid each other and this provides a rigidity that resists any further gravitational collapse.

This is the fate of the lightest stars, but what of stars like our Sun? We left them a couple of paragraphs ago, burning helium into carbon and oxygen. What happens when they run out of helium? They too must then start to collapse under their own gravity, which means they will have their electrons squashed together. And, just as for the lighter stars, the Pauli principle can eventually kick in and halt the collapse. But, for the most massive of stars, even the Pauli Exclusion Principle has its limits. As the star collapses and the electrons get squashed closer together, so the core heats up and the electrons move faster. For heavy enough stars, the electrons will eventually be moving so fast that they approach the speed of light, and that is when something new happens. When they close in on light-speed, the pressure the electrons are able to exert to resist gravity is reduced to such an extent that they aren't up to the job. They simply cannot beat gravity any more and halt the collapse. Our task in this chapter is to calculate when this happens, and we've already given away the punchline. For stars with masses greater than 1.4 times the mass of the Sun, the electrons lose and gravity wins.

That completes the overview that will provide the basis for our calculation. We can now go ahead and forget all about nuclear fusion, because stars that are burning are not where our interest lies. Rather, we are keen to understand what happens inside dead stars. We want to see just how the quantum pressure from the squashed electrons balances the force of gravity, and how that pressure becomes diminished if the electrons are moving too fast. The heart of our study is therefore a balancing game: gravity versus quantum pressure. If we can make them balance we have a white dwarf star, but if gravity wins we have catastrophe.

Although not relevant for our calculation, we can't leave things on such a cliff-hanger. As a massive star implodes, two further options remain open to it. If it is not too heavy then it will keep squashing the protons and electrons until they too can fuse together to make neutrons. In particular, one proton and one electron convert spontaneously into a neutron with the emission of a neutrino, again via the weak nuclear force. In this way the star relentlessly converts into a tiny ball of neutrons. In the words of Russian physicist Lev Landau, the star converts into 'one gigantic nucleus'. Landau wrote those words in his 1932 work 'On the Theory of Stars', which appeared in print in the very same month that the neutron was discovered by James Chadwick. It is probably going too far to say that Landau predicted the existence of neutron stars but, with great prescience, he certainly anticipated something like them. Perhaps the credit should go to Walter Baade and Fritz Zwicky, who wrote in the following year: 'With all reserve we advance the view that supernovae represent the transitions from ordinary stars into neutron stars, which in their final stages consist of extremely closely

Figure 12.1. A cartoon from the 19 January 1934 edition of the *Los Angeles Times*.

packed neutrons.' The idea was considered so outlandish that it was parodied in the *Los Angeles Times* (see Figure 12.1), and neutron stars remained a theoretical curiosity until the mid 1960s.

In 1965, Anthony Hewish and Samuel Okoye found 'evidence for an unusual source of high radio brightness temperature in the crab nebula', although they failed to identify it as a neutron star. The positive ID came in 1967 by Iosif Shklovsky and, shortly afterwards, after more detailed measurements, by Jocelyn Bell and Hewish himself. This first example of one of the most exotic objects in the Universe was subsequently named the 'Hewish Okoye Pulsar'. Interestingly, the very same supernova that created the Hewish Okoye Pulsar was also observed by astronomers, a thousand years earlier. The great supernova of 1054, the brightest in recorded history, was observed by Chinese astronomers and, as shown by a famous drawing on an overhanging cliff edge, by the peoples of Chaco Canyon in the south-western United States.

We haven't yet said how those neutrons manage to fend off gravity and prevent further collapse, but you can probably guess how it works. The neutrons (just like electrons) are slaves to the Pauli principle. They too can halt further collapse and so, just like white dwarves, neutron stars represent a possible end-point in the life of stars. Neutron stars are a detour as far as our story goes, but we can't leave them without remarking that these are very special objects in our wonderful Universe: they are stars the size of cities, so dense that a teaspoonful weighs as much as a mountain, held up by nothing more than the natural aversion to one another of spin-half particles.

There is only one option remaining for the most massive stars in the Universe – stars in which even the neutrons are moving close to light-speed. For such giants, disaster awaits, because the neutrons are no longer able to generate sufficient pressure to resist gravity. There is no known physical mechanism to stop a stellar core with a mass of greater than around three times the mass of our Sun falling in on itself, and the result is a black hole: a place where the laws of physics as we know them break down. Presumably Nature's laws don't cease to operate, but a proper understanding of the inner workings of a black hole requires a quantum theory of gravity, and no such theory exists today.

It is time to get back on message and to focus on our twin goals of proving the existence of white dwarf stars and calculating the Chandrasekhar mass. We know how to proceed: we must balance the electron pressure with gravity. This is not going to be a calculation we can do in our heads, so it will pay to make a plan of action. Here's the plan; it's quite lengthy because we want to clear up some background detail first and prepare the ground for the actual calculation.

Step 1: We need to determine what the pressure inside the star is due to those highly compressed electrons. You might be wondering why we are not worrying about the other stuff inside the star – what about the nuclei and the photons? Photons are not subject to the Pauli principle and, given enough time, they'll leave the star in any case. They have no hope of fighting gravity. As for the nuclei, the half-integer spin nuclei are subject to Pauli's rule but (as we shall see) their larger mass means they exert a smaller pressure than do the electrons and we can safely ignore their contribution to the balancing game. That simplifies matters hugely – the electron pressure is all we need, and that is where we should set our sights.

Step 2: After we've figured out the electron pressure, we'll need to do the balancing game. It might not be obvious how we should go about things. It's one thing to say 'gravity pulls in and the electrons push out' but it is quite another thing to put a number on it.

The pressure is going to vary inside the star; it will be larger in the centre and smaller at the surface. The fact that there is a pressure gradient is crucial. Imagine a cube of star matter sitting somewhere inside the star, as illustrated in Figure 12.2. Gravity will act to draw the cube towards the centre of the star and we want to know how the pressure from the electrons goes about countering it. The pressure in the electron gas exerts a force on each of the six faces of the cube, and the force is equal to the pressure at that face multiplied by the area of the face. That statement is precise; until now we have been using the word 'pressure' assuming that we all have sufficient intuitive understanding that a gas at high pressure

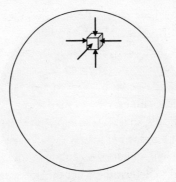

Figure 12.2. A small cube somewhere within the heart of a star. The arrows indicate the pressure exerted on the cube by the electrons within the star.

'pushes more' than a gas at low pressure. Anyone who has had to pump air into a flat car tyre knows that.

Since we are going to need to understand pressure properly, a brief diversion into more familiar territory is in order. Sticking with the tyre example, a physicist would say that a tyre is flat because the air pressure inside is insufficient to support the weight of the car without deforming the tyre: that's why we get to go to all the best parties. We can go ahead and calculate what the correct tyre pressure should be for a car with a mass of 1,500 kg if we want 5 centimetres of tyre to be in contact with the ground, as illustrated in Figure 12.3: it's chalk dust time again.

If the tyre is 20 cm wide and we want a 5 cm length of the tyre to be touching the road, then the area of tyre in contact with the ground will be $20 \times 5 = 100$ square centimetres. We don't know the requisite tyre pressure yet – this is what we want to calculate – so let's represent it by the symbol P. We need to know the downward force on the ground exerted by the air within the tyre. This is equal to the pressure multiplied by the area of tyre in contact with the floor, i.e. $P \times 100$ square centimetres. We should multiply that by four, because our car has four tyres: $P \times 400$ square centimetres. That is the total force exerted on the ground by the air within the tyres. Think of it like this: the air molecules inside the tyre are pound-

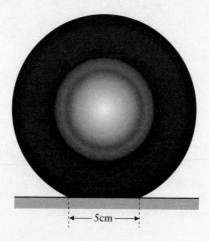

Figure 12.3. A tyre deforming slightly as it supports the weight of a car.

ing the ground (they are, to be pedantic, pounding the rubber in the tyre in contact with the ground, but that isn't important). The ground doesn't usually give way, in which case it pushes back with an equal but opposite force (so we did use Newton's third law after all). The car is being pushed up by the ground and pulled down by gravity and since it doesn't sink into the ground or leap into the air, we know that these two forces must balance each other. We can therefore equate the P × 400 square centimetres of force pushing up with the downward force of gravity. That force is just the weight of the car and we know how to work that out using Newton's second law, $F = ma$, where a is the acceleration due to gravity at the Earth's surface, which is $9.81 \, \text{m/s}^2$. So the weight is $1{,}500 \, \text{kg} \times 9.8 \, \text{m/s}^2 = 14{,}700$ Newtons (1 Newton is equal to $1 \, \text{kg} \, \text{m/s}^2$ and it is roughly the weight of an apple). Equating the two forces implies that

$$\text{P} \times 400 \, \text{cm}^2 = 14{,}700 \, \text{N}$$

This is an easy equation to solve: $\text{P} = (14{,}700/400) \, \text{N/cm}^2 = 36.75 \, \text{N/cm}^2$. A pressure of 36.75 Newtons per square centimetre is probably not a very familiar way of stating a tyre pressure, but we can convert it into the more familiar 'bar'. 1 bar is standard air pressure, and is equal to 101,000 Newtons per square metre. There are 10,000 square centimetres in a square metre, so 101,000 Newtons per square metre is equivalent to 10.1 Newtons per square centimetre. Our desired tyre pressure is therefore $36.75/10.1 = 3.6$ bar (or 52 psi – you can work that one out for yourself). We can also use our equation to deduce that, if the tyre pressure decreases by 50% to 1.8 bar, then we'll double the area of tyre in contact with the ground, which makes for a flatter tyre. After that refresher course on pressure we are ready to return to the little cube of star matter illustrated in Figure 12.2.

If the bottom face of the cube is closer to the centre of the star then the pressure on it should be a little bit bigger than the pressure pressing on the top face. That pressure difference gives rise to a force on the cube that wants to push the cube away from the centre of the star ('up' in the figure) and that is just what we want, because the cube will, at the same time, be pulled towards the centre of the star by gravity ('down' in the figure). If we could work out how to balance those two forces then we'd have developed some understanding of the star. But that is easier said than done because, although step 1 will allow us to work out how much the cube is pushed out by the electron pressure, we still have to figure out by how much gravity pulls in the opposite direction. By the way, we do not need to worry about the pressure pushing against the sides of our cube because the sides are equidistant from the centre of the star, so the pressure on the left side will balance the pressure on the right side and that ensures the cube does not move to the left or right.

To work out the force of gravity on the cube we need to make use of Newton's law of gravity, which tells us that every single piece of matter within the star pulls on our little cube by an amount that decreases in strength the farther the piece is from our cube. So more distant pieces pull less than closer ones. To deal with the fact that the gravitational pull on our cube is different for different pieces of star matter, depending on their distance away, looks like a tricky problem but we can see how to do it, in principle at least – we should chop the star into lots of pieces and then work out the force on the cube for each and every such piece. Fortunately, we do not need to imagine chopping the star up because we can exploit a very beautiful result. Gauss' law (named after the legendary German mathematician Carl Friedrich Gauss) informs us that: (a) we can totally ignore the gravity from all the pieces sitting further out from the centre of the star than our little cube; (b) the net gravitational effect of all of the pieces that sit closer to the centre is exactly as if all of those pieces were squashed together at the exact centre of the star. Using Gauss' law in conjunction with Newton's law of gravity we can say that the cube experiences a force that pulls it towards the centre of the star and that force is equal to

$$G \frac{M_{in} M_{cube}}{r^2}$$

where M_{in} is the mass of the star lying within a sphere whose radius reaches only as far out as the cube, M_{cube} is the mass of the cube and r is the distance of the cube from the star's centre (and G is Newton's constant). For example, if the cube sits on the surface of the star then M_{in} is the total mass of the star. For all other locations, M_{in} is smaller than that.

We're now making progress because to balance the forces on the cube (which we remind you means that the cube doesn't move and that means the star is not going to explode or collapse[2]) we require that

$$(P_{bottom} - P_{top})A = G \frac{M_{in} M_{cube}}{r^2} \qquad (1)$$

where P_{bottom} and P_{top} are the pressures of the electron gas at the upper and lower faces of the cube and A is the area of each side of the cube (remember, the force exerted by a pressure is equal to the pressure multiplied by the area). We have labelled this equation '(1)' because it is very important and we will want to refer back to it.

---

2. We can generalize to the entire star because we are not being specific about where the cube actually is. If we can show that a cube located anywhere in the star does not move then that means all such cubes don't move and the star is stable.

Step 3: Make a cup of tea and feel pleased with ourselves because, after carrying out step 1, we will have figured out the pressures, P_{bottom} and P_{top}, and step 2 has made precise how to balance the forces. The real work is yet to come, though, because we still have to actually carry out step 1 and determine the pressure difference appearing on the left-hand side of equation (1). That is our next task.

Imagine a star packed with electrons and other stuff. How are the electrons scattered about? Let's focus our attention on a 'typical' electron. We know that electrons obey the Pauli Exclusion Principle, which means that no two electrons are likely to be found in the same region of space. What does that mean for the sea of electrons that we've been referring to as the 'electron gas' in our star? Because the electrons are necessarily separated from each other, we can suppose that each electron sits all alone inside a tiny imaginary cube within the star. Actually, that's not quite right because we know that electrons come in two types – 'spin up' and 'spin down' – and the Pauli principle only forbids identical particles from getting too close, which means we can fit two electrons inside a cube. This should be contrasted with the situation that would arise if the electrons did not obey the Pauli principle. In that case the electrons would not be localized two-at-a-time inside 'virtual containers'. Rather they could spread out and enjoy a much greater living space. In fact, if we were to ignore the various ways that the electrons can interact with each other and with the other particles in the star, there would be no limit to their living room.

We know what happens when we confine a quantum particle: it hops about according to Heisenberg's Uncertainty Principle, and the more it is confined the more it hops. That means that, as our would-be white dwarf collapses, so the electrons get increasingly confined and that makes them increasingly agitated. It is the pressure exerted by their agitation that will halt the gravitational collapse.

We can do better than words, because we can use Heisenberg's Uncertainty Principle to determine the typical momentum of an electron. In particular, if we confine the electron to a region of size Δx then it will hop around with a typical momentum $p \sim h/\Delta x$. Actually, in Chapter 4 we argued that this is more like an upper limit on the momentum and that the typical momentum is somewhere between zero and this value; that piece of information is worth remembering for later. Knowing the momentum allows us, immediately, to learn two things. Firstly, if the electrons didn't obey Pauli then they would not be confined to a region of size Δx but rather to some much larger size. That in turn would result in much less jiggling, and less jiggling means less pressure. So it is clear how the Pauli principle is entering the game; it is putting the squeeze on the electrons so that, via Heisenberg, they get a supercharged jiggle. In a moment we'll convert this idea of a supercharged jiggle into a formula for the pressure, but first we should mention the second thing we can learn. Because the momentum $p = mv$, the speed of the jiggle also depends inversely on the mass, so the electrons are jumping around much more vigorously than the heavier nuclei that also make up the star, and that is why the pressure exerted by the nuclei is unimportant. So how do we go from knowing the momentum of an electron to computing the pressure a gas of similar electrons exerts?

What we need to do first is to work out how big the little chunks containing the pairs of electrons must be. Our little chunks have volume $(\Delta x)^3$, and because we have to fit all the electrons inside the star, we can express this in terms of the number of electrons within the star (N) divided by the volume of the star (V). We'll need precisely $N/2$ containers to accommodate all of the electrons because we are allowed two electrons inside each container. This means that each container will occupy a volume of V divided by $N/2$, which is equal to $2(V/N)$. We'll need the quantity N/V (the number of electrons per unit volume inside the star) quite a lot in what follows, so we'll give it its own symbol n. We can now write down what the volume of the containers must be in order to contain all the electrons in the star, i.e. $(\Delta x)^3 = 2/n$. Taking the cube root of the right hand side allows us to conclude that

$$\Delta x = \sqrt[3]{2/n} = (2/n)^{1/3}$$

We can now plug this into our expression from the Uncertainty

Principle to get the typical momentum of the electrons due to their quantum jiggling:

$$p \sim h(n/2)^{1/3} \tag{2}$$

where the $\sim$ sign means 'something like'. Clearly this is a bit vague because the electrons will not all be jiggling in exactly the same way: some will move faster than the typical value and some will move more slowly. The Heisenberg Uncertainty Principle isn't capable of telling us exactly how many electrons move at this speed and how many at that. Rather, it provides a more 'broad brush' statement and says if you squeeze an electron down then it will jiggle with a momentum something like $h/\Delta x$. We are going to take that typical momentum and assume it's the same for all the electrons. In the process, we will lose a little precision in our calculation but gain a great deal of simplicity as a result, and we are certainly thinking about the physics in the right way.[3]

We now know the speed of the electrons and that is enough information to work out how much pressure they exert on the tiny cube. To see that, imagine a fleet of electrons all heading in the same direction at the same speed (v) towards a flat mirror. They hit the mirror and bounce back, again travelling at the same speed but in the opposite direction. Let us compute the force exerted by the electrons on the mirror. After that we can attempt the more realistic calculation, where the electrons are not all travelling in the same direction. This methodology is very common in physics – first think about a simpler version of the problem you want to solve. That way you get to learn about the physics without biting off more than you can chew and gain confidence before tackling the harder problem. Imagine that the electron fleet consists of n particles per cubic metre and that, for the sake of argument, it has a circular cross-section of area 1 square metre – as illustrated in Figure 12.4. In one second nv electrons will hit the mirror (if v is measured in metres

---

3. It is of course possible to compute more precisely how the electrons move around but at the price of introducing more mathematics.

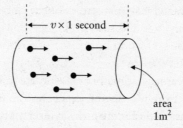

Figure 12.4. A fleet of electrons (the little dots) all heading in the same direction. All of the electrons in a tube this size will smash into the mirror every second.

per second). We know that because all of the electrons stretching from the mirror to a distance $v \times 1$ second away will smash into the mirror every second, i.e. all of the electrons in the tube drawn in the figure. Since a cylinder has a volume equal to its cross-sectional area multiplied by its length, the tube has a volume of v cubic metres and because there are n electrons per cubic metre in the fleet it follows that nv electrons hit the mirror every second.

When each electron bounces off the mirror it gets its momentum reversed, which means that each electron changes its momentum by an amount equal to $2mv$. Now, just as it takes a force to halt a moving bus and send it travelling back in reverse, so it takes a force to reverse the momentum of an electron. This is Isaac Newton once again. In Chapter 1 we wrote his second law as $F = ma$, but this is a special case of a more general statement, which states that the force is equal to the rate at which momentum changes.[4] So the whole fleet of electrons will impart a net force on the mirror $F = 2mv \times (nv)$, because this is the net change in momentum of the electrons every second. Due to the fact that the electron beam has an area of 1 square metre, this is also equal to the pressure exerted by the electron fleet on the mirror.

It is only a short step to go from a fleet of electrons to a gas of electrons. Rather than all the electrons ploughing along in the same

4. Newton's second law can be written as $F = dp/dt$. For constant mass this can be written in the more familiar form: $F = mdv/dt = ma$.

direction, we have to take into account that some travel up, some down, some to the left and so on. The net effect is to reduce the pressure in any one direction by a factor of 6 (think of the six faces on a cube) to $(2mv) \times (nv)/6 = nmv^2/3$. We can replace v in this equation by our Heisenberg-informed estimate of the typical speeds at which the electrons are zipping about (i.e. the previous equation (2)) to get the final result for the pressure exerted by the electrons in a white dwarf star:[5]

$$P = \frac{1}{3} nm \frac{h^2}{m^2} \left(\frac{n}{2} \right)^{2/3} = \frac{1}{3} \left(\frac{1}{2} \right)^{2/3} \frac{h^2}{m} n^{5/3}$$

If you recall, we said that this was only an estimate. The full result, using a lot more mathematics, is

$$P = \frac{1}{40} \left(\frac{3}{\pi} \right)^{2/3} \frac{h^2}{m} n^{5/3} \qquad (3)$$

This is a nice result. It tells us that the pressure at some place in the star varies in proportion to the number of electrons per unit volume at that place raised to the power of $5/3$. You should not be concerned that we did not get the constant of proportionality correct in our approximate treatment – the fact that we got everything else right is what matters. In fact, we did already say that our estimate of the momentum of the electrons is probably a little too big and this explains why our estimate of the pressure is bigger than the true value.

Knowing the pressure in terms of the density of electrons is a good start but it will suit our purposes better to express it in terms of the actual mass density in the star. We can do this under the very safe assumption that the vast majority of the star's mass comes from the nuclei and not the electrons (a single proton has a mass nearly 2,000 times greater than that of an electron). We also know that the number of electrons must be equal to the number of protons in the star because the star is electrically neutral. To get the mass density

5. Here we have combined the exponents according to the general rule $x^a x^b = x^{a+b}$.

we need to know how many protons and neutrons there are per cubic metre within the star and we should not forget the neutrons because they are a by-product of the fusion process. For lighter white dwarfs, the core will be predominantly helium-4, the end product of hydrogen fusion, and this means that there will be equal numbers of protons and neutrons. Now for a little notation. The atomic mass number, A, is conventionally used to count the number of protons + neutrons inside a nucleus and $A = 4$ for helium-4. The number of protons in a nucleus is given the symbol Z and for helium $Z = 2$. We can now write down a relationship between the electron density, n, and the mass density, ρ:

$$n = Z\rho/(m_p A)$$

and we've assumed that the mass of the proton, m_p, is the same as the mass of the neutron, which is plenty good enough for our purposes. The quantity $m_p A$ is the mass of each nucleus; $\rho/m_p A$ is then the number of nuclei per unit volume, and Z times this is the number of protons per unit volume, which must be the same as the number of electrons – and that's what the equation says.

We can use this equation to replace n in equation (3), and because n is proportional to ρ the upshot is that the pressure varies in proportion to the density to the power of $\frac{5}{3}$. The salient physics we have just discovered is that

$$P = \kappa\rho^{5/3} \tag{4}$$

and we should not be worrying too much about the pure numbers that set the overall scale of the pressure, which is why we just bundled them all up in the symbol κ. It's worth noting that κ depends on the ratio of Z and A, and so will be different for different kinds of white dwarf star. Bundling some numbers together into one symbol helps us to 'see' what is important. In this case the symbols could distract us from the important point, which is the relationship between the pressure and the density in the star.

Before we move on, notice that the pressure from quantum jiggling doesn't depend upon the temperature of the star. It only cares about how much we squeeze the star. There will also be an additional contribution to the electron pressure that comes about simply because the electrons are whizzing around 'normally' due to their temperature, and the hotter the star, the more they whizz around. We have not bothered to talk about this source of pressure because time is short and, if we were to go ahead and calculate it, we would find that it is dwarfed by the much larger quantum pressure.

Finally, we are ready to feed our equation for the quantum pressure into the key equation (1), which is worth repeating here:

$$(P_{bottom} - P_{top})A = G\frac{M_{in}M_{cube}}{r^2} \qquad (1)$$

But this is not as easy as it sounds because we need to know the difference in the pressures at the upper and lower faces of the cube. We could re-write equation (1) entirely in terms of the density within the star, which is itself something that varies from place to place inside the star (it must be otherwise there would be no pressure difference across the cube) and then we could try to solve the equation to determine how the density varies with distance from the star's centre. To do this is to solve a differential equation and we want to avoid that level of mathematics. Instead, we are going to be more resourceful and think harder (and calculate less) in order to exploit equation (1) to deduce a relationship between the mass and the radius of a white dwarf star.

Obviously the size of our little cube and its location within the star are completely arbitrary, and none of the conclusions we are going to draw about the star as a whole can depend upon the details of the cube. Let's start by doing something that might seem pointless. We are quite entitled to express the location and size of the cube in terms of the size of the star. If R is the radius of the star, then we can write the distance of the cube from the centre of the star as $r = aR$, where a is simply a dimensionless number between 0 and 1. By dimensionless, we mean that it is a pure number and carries no units. If $a = 1$, the cube is at the surface of the star and if $a = \frac{1}{2}$ it is halfway out from the centre. Similarly, we can write the size of the cube in terms of the radius of the star. If L is the length of a side of the cube, then we can write $L = bR$ where, again, b is a pure number, which will be very small if we want the cube to be small relative to the star. There is absolutely nothing deep about this and, at this stage, it should seem so obvious as to appear pointless. The only noteworthy point is that R is the natural distance to use because there are no other distances relevant to a white dwarf star that could have provided any sensible alternatives.

Likewise, we can continue our strange obsession and express the density of the star at the position of the cube in terms of the average density of the star, i.e. we can write $\rho = f\bar{\rho}$ where f is, once again, a pure number and $\bar{\rho}$ is the average density of the star. As we have already pointed out, the density of the cube depends on its position inside the star – if it is closer to the centre, it will be more dense. Given that the average density $\bar{\rho}$ does not depend on the position of the cube, then f must do so, i.e. f depends on the distance r, which obviously means it depends on the product aR. Now, here is the key piece of information that underpins the rest of our calculation: f is a pure number and R is not a pure number (because it is measuring a distance). This fact implies that f can only depend upon a and not on R at all. This is a very important result, because it is telling us that the density profile of a white dwarf star is 'scale invariant'. This means the density varies with radius in the same way no matter what the radius of the star is. For example, the density at a point $3/4$ of the way out from the centre of the star will be the same fraction of the mean density in every white dwarf star, regardless of the star's size. There are two ways of appreciating this crucial result and we thought we'd present them both. One of us explained it thus: 'That's because any dimensionless function of r (which is what f is) can only be dimensionless if it is a function of a dimensionless variable, and the only dimensionless variable we have is $r/R = a$, because R is the only quantity which carries the dimensions of distance that we have at our disposal.'

The other author feels that the following is clearer: 'f can in general depend in a complicated way on r, the distance of the little cube from the centre of the star. But let's assume for the sake of this paragraph that it is directly proportional to it, i.e. $f \propto r$. In other words, $f = Br$, where B is a constant. Here, the key point is that we want f to be a pure number, whilst r is measured in (say) metres. That means that B must be measured in 1/metres, so that the units of distance cancel each other out. So what should we choose for B? We can't just choose something arbitrary, like '1 inverse metres', because this would be meaningless and has nothing to do with the star. Why not choose 1 inverse light years for example, and get a very different answer? The only distance we have to hand is R, the physical radius of the star, and so we are forced to use this to ensure that f will always be a pure number. This means that f depends only on r/R. You should be able to see that the same conclusion can be drawn if we started out by assuming that $f \propto r^2$ say.' Which is just what he said, only longer.

This means that we can express the mass of our little cube, of size L and volume L^3, sitting at a distance r from the centre of the star, as $M_{cube} = f(a)L^3\bar{\rho}$. We wrote $f(a)$ instead of just f in order to remind us that f really only depends upon our choice of $a = r/R$ and not on the large-scale properties of the star. The same argument can be used to say that we can write $M_{in} = g(a)M$ where $g(a)$ is again only a function of a. For example, the function $g(a)$ evaluated at $a = 1/2$ tells us the fraction of the star's mass lying in a sphere of half the radius of the star itself, and that is the same for all white dwarf stars, regardless of their radius because of the argument in the previous paragraph.[6] You might have noticed that we are steadily working our way through the various symbols which appear in equation (1), replacing them by dimensionless quantities (a, b, f and g) multiplied by quantities that depend only on the mass and radius of the star (the average density of the star is determined in terms of M and R because $\bar{\rho} = M/V$ and $V = 4\pi R^3/3$, the volume of a sphere). To complete the task, we just need to do the same

---

6. For those of a mathematical bent, show that $g(a) = 4\pi R^3 \bar{\rho} \int_0^a x^2 f(x)\,dx$, i.e. that the function $g(a)$ is actually determined once we know the function $f(a)$.

for the pressure difference, which we can (by virtue of equation (4)) write as $P_{bottom} - P_{top} = h(a,b)\kappa\bar{\rho}^{5/3}$ where $h(a,b)$ is a dimensionless quantity. The fact that $h(a,b)$ depends upon both of a and b is because the pressure difference not only depends on where the cube is (represented by a) but also on how big it is (represented by b): bigger cubes will have a larger pressure difference. The key point is that, just like $f(a)$ and $g(a)$, $h(a,b)$ cannot depend upon the radius of the star.

We can make use of the expressions we just derived to rewrite equation (1):

$$(h\kappa\bar{\rho}^{5/3}) \times (b^2 R^2) = G\frac{(gM) \times (fb^3 R^3 \bar{\rho})}{a^2 R^2}$$

That looks like a mess and not much like we are within one page of hitting the jackpot. The key point is to notice that this is expressing a relationship between the mass of the star and its radius – a concrete relation between the two is within touching distance (or desperate grasping distance, depending on how well you handled the mathematics). After substituting in for the average density of the star (i.e. $\bar{\rho} = M/(4\pi R^3/3)$) this messy equation can be rearranged to read

$$RM^{1/3} = \kappa/(\lambda G) \tag{5}$$

where
$$\lambda = \left(\frac{4\pi}{3}\right)^{2/3} \frac{bfg}{ha^2}$$

Now λ only depends upon the dimensionless quantities a, b, f, g and h, which means that it does *not* depend upon the quantities that describe the star as a whole, M and R, and this means that it must take on the same value for all white dwarf stars.

If you are worrying what would happen if we were to change a and/or b (which means changing the locations and/or size of our little cube) then you have missed the power of this argument. Taken at face value, it certainly looks like changing a and b will change λ so that we will get a different answer for $RM^{1/3}$. But that is impossible, because we know that $RM^{1/3}$ is something that depends on the star and not on the specific properties of a little cube that we might or might not care to dream up. This means that any variation in a or b must be compensated for by corresponding changes in f, g and h.

Equation (5) says, quite specifically, that white dwarves can exist. It says that because we've been successfully able to balance the gravity–pressure equation (equation (1)). That is not a trivial thing – because it might have been possible that the equation could not be satisfied for any combination of M and R. Equation (5) also makes the prediction that the quantity $RM^{1/3}$ must be a constant. In other words, if we look up into the sky and measure the radius and the mass of white dwarves, we should find that the radius multiplied by the cube root of the mass will give the same number for every white dwarf. That is a bold prediction.

The argument that we just presented can be improved upon because it is possible to calculate exactly what the value of λ should be, but to do that we would need to solve a second-order differential equation in the density, and that is a mathematical bridge too far for this book. Remember, λ is a pure number: it simply 'is what it is' and we can, with a little higher-level maths, compute it. The fact that we did not actually work it out here should not detract at all from our achievements: we have proven that white dwarf stars can exist and we have managed to make a prediction relating their mass and radius. After calculating λ (which can be done on a home computer), and after substituting in the values for κ and G, the prediction is that

$$RM^{1/3} = (3.5 \times 10^{17}\,\mathrm{kg^{1/3}m}) \times (Z/A)^{5/3}$$

which is equal to $1.1 \times 10^{17}\,\mathrm{kg^{1/3}m}$ for cores of pure helium, carbon or oxygen ($Z/A = 1/2$). For iron cores, $Z/A = 26/56$ and the 1.1 reduces slightly to 1.0. We trawled the academic literature and collected together the data on the masses and radii of sixteen white dwarf stars sprinkled about the Milky Way, our galactic backyard. For each we computed the value of $RM^{1/3}$ and the result is that astronomical observations reveal $RM^{1/3} \approx 0.9 \times 10^{17}\,\mathrm{kg^{1/3}m}$. The agreement between the observations and theory is thrilling – we have succeeded in using the Pauli Exclusion Principle, the Heisenberg Uncertainty Principle and Newton's law of gravity to predict the mass–radius relationship of white dwarf stars.

There is, of course, some uncertainty on these numbers (the theory value of 1.0 or 1.1 and the observational number equal to 0.9). A proper scientific analysis would now start talking about just how likely it is that the theory and experiment are in agreement, but for our purposes that level of analysis is unnecessary because the agreement is already staggeringly good. It is quite fantastic that we have managed to figure all this out to an accuracy of something like 10%, and is compelling evidence that we have a decent understanding of stars and of quantum mechanics.

Professional physicists and astronomers would not leave things here. They would be keen to test the theoretical understanding in as much detail as possible, and to do that means improving on the description we presented in this chapter. In particular, an improved analysis would take into account that the temperature of the star does play some role in its structure. Furthermore, the sea of electrons is swarming around in the presence of positively charged atomic nuclei and, in our calculation, we totally ignored the interactions between the electrons and the nuclei (and between electrons and electrons). We neglected these things because we claimed that they would produce fairly small corrections to our simpler treatment. That claim is supported by more detailed calculations and it is why our simple treatment agrees so well with the data.

We have obviously learnt an awful lot already: we have established that the electron pressure is capable of supporting a white dwarf star and we have managed to predict with some precision how the radius of the star changes if we add or remove mass from the star. Unlike 'ordinary' stars that are eagerly burning fuel, notice that white dwarf stars have the feature that adding mass to a star makes it smaller. This happens because the extra stuff we add goes into increasing the star's gravity, and that makes it contract. Taken at face value the relationship expressed in equation (5) seems to imply that we would need to add an infinite amount of mass before the star shrinks to no size at all. But this isn't what happens. The important thing, as we mentioned at the beginning of the chapter, is that we eventually move into the regime where the electrons are so tightly packed that Einstein's Theory of Special Relativity becomes important because the speed of the electrons starts to approach the speed of light. The impact on our calculation is that we have to stop using Newton's laws of motion, and replace them with Einstein's laws. This, as we shall now see, makes all the difference.

What we're about to find is that as the star gets more massive, the pressure exerted by the electrons will no longer be proportional to the density raised to the power $5/3$; instead, the pressure increases less quickly with density. We will do the calculation in a moment, but straight away we can see that this could have catastrophic consequences for the star. It means that when we add mass, there will be the usual increase in gravity but a smaller increase in pressure. The star's fate hinges on just how much 'less quickly' the pressure varies with density when the electrons are moving fast. Clearly it is time to figure out what the pressure of a 'relativistic' electron gas is.

Fortunately, we do not need to wheel in the heavy machinery of Einstein's theory because the calculation of the pressure in a gas of electrons moving close to light speed follows almost exactly the same reasoning as that we just presented for a gas of 'slow-moving' electrons. The key difference is that we can no longer write that the momentum $p = mv$, because this is not correct any more. What is correct, though, is that the force exerted by the electrons is still equal to the rate of change of their momentum. Previously, we deduced that a fleet of electrons bouncing off a mirror exerts a pressure $P = 2mv \times (nv)$. For the relativistic case, we can write the same expression, but providing that we replace mv by the momentum, p. We are also assuming that the speed of the electrons is close to the speed of light, so we can replace v with c. Finally, we still have to divide by 6 to get the pressure in the star. This means that we can write that the pressure for the relativistic gas as $P = 2p \times nc/6 = pnc/3$.

Just as before, we can now go ahead and use Heisenberg's Uncertainty Principle to say that the typical momentum of the confined electrons is $h(n/2)^{1/3}$ and so

$$P = \frac{1}{3}nch\left(\frac{n}{2}\right)^{1/3} \propto n^{4/3}$$

Again we can compare this to the exact answer, which is

$$P = \frac{1}{16}\left(\frac{3}{\pi}\right)^{1/3} hcn^{4/3}$$

Finally, we can follow the same methodology as before to express the pressure in terms of the mass density within the star and derive the alternative to equation (4):

$$P = \kappa'\rho^{4/3}$$

where $\kappa' \propto hc \times (Z/(Am_p))^{4/3}$. As promised, the pressure increases less quickly as the density increases than it does for the non-relativistic case. Specifically, the density increases with a power of $\frac{4}{3}$ rather than $\frac{5}{3}$. The reason for this slower variation can be traced back to the fact that the electrons cannot travel faster than the speed of light. This means that the 'flux' factor, nv, which we used to compute the pressure saturates at nc and the gas is not capable of delivering the electrons to the mirror (or face of the cube) at a sufficient rate to maintain the $\rho^{5/3}$ behaviour.

We can now explore the implications of this change because we can go through the same argument as in the non-relativistic case to derive the counterpart to equation (5):

$$\kappa'M^{4/3} \propto GM^2$$

This is a very important result because, unlike equation (5), it does not have any dependence upon the radius of the star. The equation is telling us that this kind of star, packed with light-speed electrons, can only have a very specific value of its mass. Substituting in for κ' from the previous paragraph gives us the prediction that

$$M \propto \left(\frac{hc}{G}\right)^{3/2} \left(\frac{Z}{Am_p}\right)^2$$

This is exactly the result we advertised right at the start of this chapter for the maximum mass that a white dwarf star can possibly have. We are very close to reproducing Chandrasekhar's result. All that remains to understand is why this special value is the maximum possible mass.

We have learnt that for white dwarf stars that are not too massive, the radius is not too small and the electrons are not too squashed. They therefore do not quantum jiggle to excess and their speeds are small compared to the speed of light. For these stars, we have seen that they are stable with a mass–radius relationship of the form $RM^{1/3} =$ constant. Now imagine adding more mass to the star. The mass–radius relation informs us that the star shrinks and, as a result, the electrons are more compressed and that means they jiggle faster. Add yet more mass and the star shrinks some more. Adding mass therefore increases the speed of the electrons until, eventually, they are travelling at speeds comparable with the speed of light. At the same time, the pressure will slowly change from $P \propto \bar{\rho}^{5/3}$ to $P \propto \bar{\rho}^{4/3}$ and in the latter case, the star is only stable at one particular value of the mass. If the mass is increased beyond this specific value then the right-hand side of $\kappa' M^{4/3} \propto GM^2$ becomes larger than the left-hand side and the equation is unbalanced. This means that the electron pressure (which resides on the left-hand side of the equation) is insufficient to balance the inward pull of gravity (which resides on the right-hand side) and the star must necessarily collapse.

If we were more careful with our treatment of the electron momentum and had taken the trouble to wheel in the advanced mathematics to compute the missing numbers (again a minor task for a personal computer), we could make a precise prediction for the maximum mass of a white dwarf star. It is

$$M = 0.2 \left(\frac{hc}{G}\right)^{3/2} \left(\frac{Z}{Am_p}\right)^2 = 5.8 \left(\frac{Z}{A}\right)^2 M_\odot$$

where we have re-expressed the bundle of physical constants in terms of the mass of our Sun ($M_\odot$). Notice, by the way, that all the extra hard work that we have not done simply returns the constant of proportionality, which has a value of 0.2. This equation delivers the sought-after Chandrasekhar limit: 1.4 solar masses for $Z/A = 1/2$.

This really is the end of our journey. The calculation in this chapter has been at a higher mathematical level than the rest of the book but it is, in our view, one of the most spectacular demonstrations of the sheer power of modern physics. To be sure, it is not a 'useful' thing, but it is surely one of the great triumphs of the human mind. We used relativity, quantum mechanics and careful mathematical reasoning to calculate correctly the maximum size of a blob of matter that can be supported against gravity by the Exclusion Principle. This means that the science is right; that quantum mechanics, no matter how strange it might seem, is a theory that describes the real world. And that is a good way to end.

Further Reading

We used many books in the preparation of this book, but some deserve special mention and are highly recommended.

For the history of quantum mechanics, the definitive sources are two superb books by Abraham Pais: *Inward Bound* and *Subtle Is the Lord* . . . Both are quite technical but they are unrivalled in historical detail.

Richard Feynman's book *QED: The Strange Theory of Light and Matter* is at a similar level to this book and is more focused, as the title suggests, on the theory of quantum electrodynamics. It is a joy to read, like most of Feynman's writings.

For those in search of more detail, the very best book on the fundamentals of quantum mechanics is, in our view, still Paul Dirac's book *The Principles of Quantum Mechanics*. A high level of mathematical ability is needed to tackle this one.

Online, we should like to recommend two lecture courses that are available on iTunes University: Leonard Susskind's 'Modern Physics: The Theoretical Minimum – Quantum Mechanics' and James Binney's more advanced 'Quantum Mechanics' from the University of Oxford. Both require a reasonable mathematical background.

Index